INSTITUTE ECONOMICS
AND STATISTICS
OXFORD HA 24

HAD

INSTITUTE ... NOMICS
AND STA ... OXFORD
WITHDRAWN

ELEMENTARY
STATISTICS

WITHDRAWN
OXFORD

ELEMENTARY STATISTICS

G. Hadley

University of Hawaii

WITHDRAWN 15 DEC 1969

HOLDEN-DAY, INC.

San Francisco, Cambridge, London, Amsterdam

© Copyright 1969 by Holden-Day, Inc., 500 Sansome Street, San Francisco, California
All rights reserved.
No part of this book may be reproduced in any form, by mimeograph or
any other means, without permission in writing from the publisher.
Library of Congress Catalog Card Number 69–11850
Printed in the United States of America

Preface

This text is intended for a one-semester introductory course in statistics for a broad group of students, including those in the social and biological sciences, economics and education. In many universities it is now the practice to offer a single elementary course for a wide variety of students rather than having separate courses for biologists, psychologists or economists, and the author had in mind these joint courses when writing the text. In order to meet the objectives of such courses, an attempt has been made to provide a very large number of interesting examples from many different fields, including genetics, economics, business, agriculture and everyday life. Similarly, over 750 problems have been included to illustrate the wide range of applicability of the material studied. More material is covered than could normally be put into a one-semester course. This makes it possible for the instructor to have some freedom in the choice of topics to be covered and the depth to which any given topic will be studied. It also makes it possible to use the text in somewhat longer courses, if desired.

The mathematical level of this text is elementary. No calculus is required. However, probability theory is covered in more depth than is true for most texts at this level. Emphasis is placed on the notion of a mathematical model and its relation to the real world. Interest in the newer techniques of statistical decision theory, such as Bayesian statistics, has been growing rapidly. Indeed, interest has grown much more rapidly than the attention the subject has received in elementary texts. An attempt has been made to correct this deficiency here. An entire chapter has been devoted to modern decision theory. The main emphasis is on Bayesian statistics, but other methods which do not make use of prior probabilities are also considered.

v

The classical theory of decision making based on hypothesis testing is also treated in a separate chapter. This makes it possible to cover exclusively the classical approach, or the new decision theory methods, or some combination of both.

The author is indebted to the publisher for various forms of assistance in publishing this work.

Contents

viii

Contents

ELEMENTARY
STATISTICS

CHAPTER 1

Probability

1–1 INTRODUCTION

One of the most significant trends in the social sciences over the last several decades has been the ever increasing effort to make these disciplines more quantitative in nature. The achievements in this regard have been notable and have been accomplished mainly through the application of various mathematical methods to the problems of interest. Two of the most widely useful mathematical tools are probability theory and statistics. In this text we wish to study at an elementary level the interrelated subjects of probability and statistics and illustrate some of the wide variety of applications for the theory.

In broad general terms, the theory of probability is a mathematical subject which is useful in studying real-world situations where uncertainty plays an important role. The meaning of the word statistics has undergone a considerable change in the last 200 or so years, and the meaning used in this text may well be rather different from what the reader has in mind when he uses the term. Originally, statistics referred to the art and science of government. Later it came to refer to the collection of data (initially data used by the government and later to all sorts of data) and more generally to the handling and presentation of empirical data. This interpretation of the word is still used frequently and is probably the one familiar to the reader. The modern meaning of statistics, which we shall use, is that of employing data gathered about some situation involving uncertainty to draw some relevant conclusions about the situation. This definition may appear somewhat vague, but the meaning will become clear as we progress. Statistics is a mathematical subject which is based on probability theory. We shall begin,

therefore, by studying the elements of probability theory. This will be done in Chapters 1 and 2. The remainder of the text will be devoted to a study of a variety of statistical techniques.

1–2 MATHEMATICAL MODELS

Mathematics does not deal directly with the sorts of things met in the real world, namely, physical objects, materials, people, dollars and so forth. It deals only with numbers and symbols and with the relations between such elements. To apply mathematics to a real-world problem, it is necessary to characterize this problem in mathematical terms. This process of representing the real-world situation in mathematical language is referred to as the construction of a *mathematical model* of the situation. After the model is constructed, mathematics is applied to determine a "solution" to the model. The solution to the model is then used as the solution to the real world problem.

A mathematical model consists of nothing but a collection of symbols, a set of statements describing any relevant characteristics of the symbols and a set of mathematical expressions which relate the symbols. In order to construct a model one must keep several things in mind. First of all, the model need not and, indeed, cannot provide a complete description of the system under consideration. It is only necessary to represent those features of the system which are relevant for the problem of interest. For example, the only characteristic of a given item which may be of interest to us is its cost. Its shape, color or materials of construction may be entirely irrelevant and hence need not be represented in the model. One task then, and one that may be difficult, is to determine precisely what characteristics of the real world should be represented.

Another important thing to realize is that it is essentially never true that the nature of the real world can be described with complete accuracy in a model. Certain approximations must always be made. The nature of these approximations can vary widely with the circumstances. Whether or not a given approximation can be considered valid depends on the accuracy needed in the results. One of the most difficult tasks in constructing models is deciding what are realistic and allowable approximations to make. The reader should keep in mind that the process of applying mathematics to real-world situations always requires the introduction of approximations. Thus the probability models and statistical models which we shall be considering throughout the remainder of the text only represent approximations to reality. The theory is so useful, however, because the approximations can often be made sufficiently good to yield solutions to practical problems.

1–3 **PROBABILITY**

We have already indicated that probability theory is a mathematical subject which is useful in studying real-world situations where uncertainty, or equivalently chance, plays an important role. The notions of chance or uncertainty are so basic that it is difficult to describe them in terms of more fundamental ideas. Uncertainty is present in some situation if several alternative outcomes are possible and we are unable to predict in advance which will actually occur. We then say that chance determines the outcome. This sort of situation is to be contrasted with a *deterministic* situation in which the outcome can be forecast exactly from known conditions prevailing at the time the forecast is made. In actuality, chance has some influence on every real-world situation. However, for certain types of situations the influence of chance is so small that the problem can be considered to be deterministic in constructing a model. In this text we shall be interested in constructing models that will be useful in analyzing situations where the effects of uncertainty are sufficiently important that they cannot be ignored.

The word probability is often used in everyday life in a manner which is essentially synonymous with chance and uncertainty. *We shall think of probabilities as numbers which serve to describe in a quantitative way the manner in which chance or uncertainty influences some situation of interest.* Thus we shall use probabilities to describe chance and uncertainty quantitatively in our mathematical models. Let us now explain the sense in which probabilities can describe chance in a quantitative way.

We shall begin with an example. It is common experience that if we toss a coin and catch it in the palm of our hand, we are unable to predict before tossing the coin whether it will land heads or tails up. Sometimes a head will turn up and sometimes a tail. It is common terminology to say that at the time we toss the coin there is uncertainty as to whether it will land heads or tails up, and that chance determines the outcome. This sort of behavior occurs regardless of how careful we may be in attempting to toss the coin in precisely the same way each time. Indeed, it will be observed even if we build a machine to toss the coin.

Why are we unable to predict ahead of time whether the coin will fall heads or tails up? The motion of the coin is described by the laws of mechanics, and thus one should be able to predict the motion of the coin. The difficulty, however, stems from the fact that the final state of the coin, head or tail, is very critically dependent on small changes in the initial velocity of the coin, the initial toss angle, the precise density of the air, the existence of any air currents and so on. It is not possible for us to have a precise knowledge of all these things at the time the coin is tossed, and it is for this reason that we are unable to predict the outcome of any given toss.

If a determination of all the relevant factors could be made before the coin was tossed, then we could compute at the time of tossing whether it would land as a head or tail. To do this would require not only that we predict the launch conditions exactly, but that, in addition, we predict all influences, such as the behavior of the air, at each instant after the coin is tossed. It would not be possible to do this regardless of how much effort was expended. Thus the outcome of tossing a coin cannot be predicted in advance, and we say that chance determines the final state.

In the above example chance refers to the large number of small effects which critically influence the motion of the coin. In general, for real-world situations in which the outcome of something is critically dependent on a variety of small effects that cannot be determined in advance, we expect to find situations in which chance or uncertainty plays a crucial role in determining the outcome.

Although the result of a single toss of a coin is not deterministic, it is a well-known fact obtained from experience that a certain regularity does appear if we examine what happens when the same coin is tossed a large number of times. To illustrate this the author selected a penny from the change in his pocket and tossed it 100 times. The coin was always tossed heads up and an effort was made to toss it in the same way each time. In accordance with the above discussion the coin landed heads up sometimes and tails up other times. It was not possible to predict ahead of time which was going to occur. After each toss a mark was made on a sheet of paper indicating whether a head or a tail was obtained on that toss, thus generating a record of what happened on each toss.

From this information the author next determined the total number of heads n_h obtained after n tosses. This was done for each $n = 1, 2, 3, \ldots ,$ 100, so that the total number of heads obtained after 1, 2, 3, 4, 5, 6, 7, etc., $\ldots ,$ 100 tosses was determined. This yielded 100 numbers n_h. Then n_h/n, the fraction of the n tosses in which the coin landed heads up was determined for $n = 1, 2, \ldots ,$ 100. Thus if 11 heads were obtained in the first 25 tosses, $n_h/n = 11/25 = 0.44$ when $n = 25$. In Figure 1–1 is shown a graph illustrating how the fraction of heads varied with n. Note that the fraction of the total number of tosses on which a head was obtained fluctuates violently at first, and then as n, the number of tosses, increases it becomes more and more stable, that is, the fluctuations die out, and the fraction of the times that a head is obtained appears to be approaching a unique number which is approximately 0.5.

Here we have encountered a general and important characteristic of certain types of phenomena, or experiments as we shall call them, whose outcomes are determined by chance, but which can be repeated over and over again. Even though the outcome on any given performance of the experiment cannot be predicted in advance, it is a characteristic of the real world

that if the experiment is repeated a large number of times, the fraction of
the total number of times that any particular outcome is observed will become more and more nearly a constant as the number of repetitions of the
experiment is increased.

It is now helpful to idealize the situation outlined above to a case where
we imagine that the experiment is repeated unendingly. In the real world,
of course, there is a limit to how often we can repeat an experiment such as
tossing a coin, since if nothing else terminates the experimenting, the coin
ultimately wears out. Let us assume that in the conceptual situation just
suggested, the outcome on one performance of the experiment has no influence on later repetitions. Also assume that conditions do not change from
one repetition to the next. This implies, for example, that in the coin-tossing

FIGURE 1–1

experiment there is never any wear of the coin. Consider one specific outcome, which we shall denote symbolically by A. Then experience in the real
world indicates that if the experiment could be repeated unendingly, the
fraction of the time that outcome A is observed would approach a unique
number p. *The number p is called the probability that on any performance of
the experiment we shall observe the outcome A*, and p tells us what fraction of
the time A would be observed in the limit as the number of repetitions of
the experiment was made arbitrarily large. Also, and more importantly, p
tells us approximately what fraction of the time A will be observed if the
experiment is performed a fairly large number of times.

We shall refer to the fraction of the number of trials in which A is observed as the *relative frequency* of A (one trial referring to one performance
of the experiment). Then the probability of A can be looked upon as the
long-run relative frequency of A, long-run implying that the experiment is
repeated an indefinitely large number of times. For each possible outcome
A of the experiment, we can imagine that there exists a number

interpretable as the long-run relative frequency of A, which we call the probability of A. We shall use this set of probabilities to characterize in a quantitative manner the way in which chance influences the experiment.

It now seems appropriate to ask the following question. Given that we are unable to predict the outcome on any given trial of the experiment, of what value is it to be able to make predictions concerning what will happen if the experiment is performed a large number of times? The answer, of course, is that it is extremely useful to be able to do this, since many real-world situations can be imagined to involve performing the same experiment over and over again. In such situations, an individual frequently wishes to formulate his policies so that things will work out as well as possible in the long run. To obtain such policies, a knowledge of the probabilities for the various outcomes can be extremely useful.

It is quite easy to give practical illustrations of situations of the type just described. Perhaps the best-known are the colorful examples provided by the Las Vegas gambling casinos. Their very existence depends on the long-run stability of the relative frequencies in the various games of chance. Another example is that of insurance companies, whose entire existence, like that of the gambling houses, depends on the ability to predict what will happen in the way of deaths or accidents in very large groups of people, although these predictions cannot be made for any particular individual. The two examples just given are especially striking instances of the relevance of what we have been discussing. However, there are many important situations in the physical and social sciences and in industry which exhibit similar characteristics.

To take a simple, everyday example, the demand for a particular type of bread in a supermarket on a given day cannot be predicted in advance. However, except for special holidays, a specified day of the week, say Tuesday, is like every other Tuesday. If one records the demand for this type of bread on Tuesdays, the relative frequencies for various numbers of loaves being demanded will be stable, and these are very pertinent in deciding how much bread to stock every Tuesday morning. Note that we did not claim that every day was like every other day. The situation on Fridays, for example, might be quite different from Tuesdays.

As another example, it is well known that the sex of a child cannot normally be predicted prior to birth. However, the relative frequency of male and female babies is remarkably stable throughout time. This is illustrated in Table 1–1, where we have shown the total number of live births, the number of live male births and the fraction of live births which were male for the United States in a number of different years. It will be observed that the fraction of male births is remarkably constant from year to year and is about 0.513. This fraction differs slightly among races and also differs slightly from one place to another.

A somewhat different sort of example concerns the behavior of a well-controlled production process involving one or more machines. Experience indicates that, for no apparent reason, a defective item is produced occasionally. The explanation for the defective lies in the chance occurrence of one or more small effects, such as a slight imperfection in the raw material, a fluctuation in the line voltage on a machine or a change in the way the operator handled this unit. Because defectives are caused by small unpredictable effects, it is impossible to determine ahead of time whether any given unit will be defective when it is finished. However, the long-run fraction of defectives in such a situation is stable, and is an important number for the production manager to know.

As a final example, consider the manufacture of color television sets. The manufacturer cannot predict how long any given set will operate before it

TABLE 1–1

LIVE BIRTHS IN THE UNITED STATES IN MILLIONS

Year	Total	Male	Fraction Male
1940	2.360	1.212	0.5131
1945	2.735	1.405	0.5137
1950	3.554	1.824	0.5133
1955	4.047	2.074	0.5125
1958	4.204	2.153	0.5121
1959	4.245	2.174	0.5121
1960	4.258	2.180	0.5120
1961	4.268	2.186	0.5122
1962	4.167	2.132	0.5116
1963	4.098	2.102	0.5129

suffers its first failure. However, the fractions of the sets produced in any month which fail in the first week of operation, the first month and in the first six months are stable from one month to the next because of the huge numbers of sets produced per month. These fractions are clearly of great interest to the manufacturer when writing service guarantees and when planning the operations of company service centers.

In this section we have considered experiments which are capable, in one way or another, of being repeated over and over again, and which have the characteristic that the outcome on any trial is determined by chance. In certain cases, the experiment may always be performed with the same "equipment," for example, we always toss the same penny. In other cases, each trial uses a different piece of equipment, as when failures of television sets or occurrences of automobile accidents are considered. What we have been illustrating then is the fact that individual events, whose outcomes are chance-determined, show in the aggregate, a behavior which comes close

to being deterministic, and which, if we imagine the aggregate to be arbitrarily large, is deterministic. In the hypothetical situation where we imagine that the experiment under consideration is repeated unendingly, we call the limiting relative frequency of some particular outcome A the probability of A, and the probabilities of the various outcomes make it possible to describe chance phenomena mathematically.

In this text we shall usually interpret probabilities intuitively as long-run relative frequencies. There is, however, another interpretation of probability which is widely used in everyday life and which is quite possibly the one the reader would use when speaking of probabilities. For example, when a stockbroker says he believes the probability is 0.60 that the Dow Jones average will be 100 points higher a year from now, or when an individual claims the probability is 0.8 that Hamilton wrote the Federalist Papers, probability is being used by the individual concerned to express what might be called his degree of rational belief. Probability cannot be interpreted as a long-run relative frequency for these situations because they are not of a type which can be repeated. Probabilities which express the degree of rational belief of some individual are often referred to as *personal* or *subjective probabilities*, since it is not necessarily true that two different individuals will assign the same personal probability to a given event. The subjective interpretation of probability has become especially important in connection with the development of modern decision theory as applied to problems in business and economics.

In practice, the distinction between the frequency and subjective interpretations of probability is usually not so sharp as the above discussions might suggest. When actually attempting to determine numerical values for probabilities it is frequently necessary to combine the relative frequency notion with the subjective interpretation, and thus both concepts can be involved in determining a single probability. The mathematical theory of probability is completely independent of the intuitive interpretation of the probabilities. The probabilities are merely numbers which appear in the model. How they were obtained or how they are to be interpreted in the real world is not part of the mathematical theory. In order to apply the theory to the real world, of course, it is most helpful to have a clear intuitive interpretation of the probabilities in mind. Furthermore, the rationale for developing the mathematical theory is based on some sort of intuitive superstructure. The intuitive notion that we shall employ to point the way to an appropriate mathematical theory will be that of interpreting probabilities as long-run relative frequencies. As we have just pointed out, however, once the theory has been developed, it can be applied regardless of how the probabilities are to be interpreted, and the justification for doing this lies in the usefulness of the results obtained.

1–4 **SET THEORY**

The reader has probably at some time studied the elements of set theory. Set theory is very useful in studying probability theory. We shall need to make use of only the most elementary ideas concerning sets, and we shall review these in this section. The notion of a set of elements or a collection of objects is so basic that it is not possible to define it in terms of more fundamental ideas. The elements of a set can be anything at all, and the set is characterized by its elements. Thus we can speak of the set of American Presidents, or the set of positive integers, or the set of stars in the Milky Way. A somewhat odd but legitimate set is one whose elements are the sun, the U.S.S. Enterprise, da Vinci's "Mona Lisa" and an Australian boomerang. A set may contain either a finite or infinite number of elements. It is convenient to introduce symbols to represent sets. Usually, we shall use upper-case letters such as A or E to denote sets. To indicate that some element, say e, belongs to the set A, we use the notation $e \in A$. If a set contains a finite number of elements and we wish to exhibit explicitly what they are, we write down the description of each one, and enclose these in brackets. Thus if A is the set of the first five positive integers we can indicate this by writing $A = \{1, 2, 3, 4, 5\}$. Similarly, if A contains n elements which we represent symbolically as $a_1, a_2, \ldots, a_n$, then $A = \{a_1, a_2, \ldots, a_n\}$.

We shall now introduce several useful definitions.

EQUALITY. *Two sets A and B are said to be equal, written $A = B$, if they contain precisely the same elements, that is, if every element of A is also an element of B, and every element of B is an element of A.*

Thus if $A = \{1, 2, 3, 4\}$ and $B = \{1, 2, 3, 4, 5\}$, A and B are not equal since 5 appears in B but not in A. The sets $A = \{1, 2, 4\}$ and $B = \{2, 4, 1\}$ are equal since they contain the same elements (we do not require that the elements have to appear in the same order when we write down the sets).

SUBSET. *A subset B of a set A is a set all of whose elements are in A. B is called a proper subset of A if there is at least one element in A which is not in B.*

The notation $B \subset A$ is used to indicate that B is a subset of A. A set will, in general, have a number of different subsets. Thus if $A = \{1, 2, 3, 4\}$, then $B = \{1, 2\}$, $C = \{2, 3, 4\}$, $D = \{2\}$ and $F = \{1, 3, 4\}$ are examples of proper subsets of A. By the way a subset is defined, A is always a subset of itself.

INTERSECTION. *The intersection of two sets A and B, written $A \cap B$, is the set of elements common to A and B, that is, $a \in A \cap B$ means $a \in A$ and $a \in B$.*

If $A = \{1, 3, 5, 7\}$ and $B = \{3, 4, 6, 7, 9\}$, then the set of elements in both A and B consists of 3 and 7, so $A \cap B = \{3, 7\}$. The intersection of n sets $A_1, \ldots, A_n$ written $A_1 \cap A_2 \cap \cdots \cap A_n$ is defined to be the set of elements common to $A_1, \ldots, A_n$.

It may turn out that A and B have no elements in common. For example, A might be the set of individuals who fought in the Civil War and B the set of individuals who fought in World War II. There are no individuals who fought in both wars. The reader may now wonder what we call $A \cap B$ if it contains no elements. For convenience, we also refer to this as a set and call it the *empty* or *null set*. The null set, which has no elements, is denoted by $\varnothing$.

UNION. *The union of two sets A and B, written $A \cup B$, is the set of elements which are in A or B or both.*

The union of two sets then contains all elements that are in at least one of the two sets. Thus if $A = \{1, 2, 5, 7\}$ and $B = \{2, 4, 6, 7, 9\}$, then $A \cup B = \{1, 2, 4, 5, 6, 7, 9\}$. If A is the set of all boys in the United States who are in the sixth grade, and B is the set of all girls in the United States who are in the sixth grade, then $A \cup B$ is the set of all boys and girls in the United States who are in the sixth grade. The union of n sets $A_1, \ldots, A_n$, written $A_1 \cup A_2 \cup \cdots \cup A_n$, is the set of elements which are in at least one of the sets $A_1, \ldots, A_n$.

These simple notions will be all we need when using set theory in our future discussions.

1–5 THE EVENT SET

We are now ready to explain how one constructs a mathematical model of some situation where chance plays a role. As we have done previously, we shall find it convenient to refer to the situation as an experiment, even though it may not in any way be considered an experiment in the sense usually thought of in the physical sciences. Experiment here simply refers to some situation we are studying whose outcome is determined by chance. It is conventional to refer to an experiment in the sense we are using the term as a *random experiment*. The word random emphasizes the fact that the outcome of the experiment is determined by chance and is not deterministic. We shall frequently use merely experiment rather than random experiment since we shall be discussing only random experiments.

The first step in constructing a model of a random experiment, which we shall refer to as a *probability model*, is to examine the possible outcomes of the experiment. Imagine that we make a list of all possible outcomes. *Let us list these outcomes in such a way that one and only one of them will actually be observed after the experiment is completed.* We shall call each outcome

defined in this manner a *simple event*. There is not necessarily anything simple about a simple event. The word simple is used because these are the simplest types of events that will be represented in the model. The meaning of this, which probably seems obscure to the reader now, will become clear as we progress. It will be assumed that there are only a finite number of simple events. This is no restriction because everything in the real world is discrete and this automatically guarantees that the number of outcomes is finite. The number may be exceedingly large, but it will always be finite.

In our general discussion we shall use symbols to represent the simple events. Let us list the simple events in any order desired. We shall then denote the first one by e_1, the second by e_2, or in general, the jth one by e_j. Suppose that there are m simple events in total. Consider now the set

$$E = \{e_1, \ldots, e_m\} \tag{1–1}$$

whose elements are the m simple events. The set E will be referred to as the event set; E is simply a listing of the possible outcomes of the random experiment of interest, these outcomes being defined so that one and only one will occur.

It is by no means true that there is only one way to define a set of simple events for an experiment. Generally, there will be many possible ways. We have no basis for making a choice among the possible alternative definitions without knowing what the model is to be used for. Even knowing this we often have the freedom to use several alternative event sets, and in such cases the most convenient one is chosen. Let us now illustrate the determination of event sets by use of some very simple examples.

EXAMPLES. *1.* Consider the random experiment which consists of tossing a coin once and noting whether it lands heads or tails up. How do we describe the outcomes of this experiment? It seems perfectly natural to say that there are two possible outcomes: (1) the coin lands heads up; or (2) the coin lands tails up. One and only one of these will occur and hence these two outcomes can be used as simple events. If we denote the first by e_1 and the second by e_2, the event set is $E = \{e_1, e_2\}$. In this case, rather than using e_1 and e_2 to denote the simple events, we shall often use H for a head and T for a tail, since the latter notation indicates clearly what each simple event is. Then the event set can be written $E = \{H, T\}$. This experiment is so simple that E is the only reasonable event set that can be used.

Our description of the outcome of tossing a coin could have been made much more detailed than that given above. We could have observed how many times the coin turned over when tossed, how high it went and so on. In describing the outcomes of an experiment, it is only necessary to characterize the things which are relevant. We were implicitly assuming in the above description that the only relevant thing to observe was whether the

coin landed heads or tails up. To understand what is relevant and irrelevant, we must thoroughly understand the nature of the real-world situation.

2. Consider next a random experiment which consists of tossing once an ordinary six-sided die. Suppose the only thing that is relevant, so far as the outcome of the experiment is concerned, is the number of dots on the face that turns up. A natural way to characterize the outcomes in this case is to give the number of dots on the face which turns up. There are then six simple events, and we can take e_j to be the event that the face with j dots on it turns up. The event set when the simple events are characterized in this way is $E_1 = \{e_1, e_2, e_3, e_4, e_5, e_6\}$. We use a subscript 1 on E to distinguish it from an alternative event set which we shall now consider.

The above description of the outcomes of tossing a die is not the only one having the characteristic that one and only one outcome will occur. We could use instead a characterization involving only two outcomes, which are: (1) a face with an even number of dots on it turns up; and (2) a face with an odd number of dots turns up. One and only one of these events will occur because no number is both even and odd. If we denote by e^* the occurrence of an even numbered face and by e^+ the occurrence of an odd numbered face, then $E_2 = \{e^*, e^+\}$ can be used as an event set. Yet another subdivision of the outcomes using two simple events would be to take one of the outcomes being a face with three or less dots on it turns up and the other to be a face with more than three dots on it turns up. There are many more ways in which one could subdivide the outcomes into simple events. We might note that the experiment could be described by only two outcomes as follows: (1) the face which turns up has three or less dots on it; and (2) the face which turns up has three or more dots on it. These would not be simple events, since both outcomes will occur if the face with three dots on it turns up. For most purposes, the event set E_1 would be the appropriate one to use for a mathematical model.

3. As a final example, consider the random experiment in which we toss two coins (rather than just a single coin as in the first example). Again we shall only be interested in whether the coins land heads or tails. One way to subdivide the outcomes into simple events is as follows: (1) both coins land heads up; (2) one coin lands heads and the other tails; (3) both coins land tails up. With this subdivision of the outcomes there are three simple events, call them $\bar{e}_1$, $\bar{e}_2$, and $\bar{e}_3$ and the event set is $E_1 = \{\bar{e}_1, \bar{e}_2, \bar{e}_3\}$.

If we think a little more, there comes to mind another eminently reasonable way to describe the outcomes. Let us imagine that we make the coins distinguishable, perhaps by painting a number on them or noting which was tossed first or using coins of different denominations. Then we can imagine that there are four simple events, which are: (1) both coins land heads up; (2) coin 1 lands heads up and 2 tails up; (3) coin 1 lands tails up and 2 heads

up; and (4) both coins land tails up. If we denote these by e_1, e_2, e_3 and e_4 the event set becomes $E_2 = \{e_1, e_2, e_3, e_4\}$. Instead of using e_1, e_2, e_3 and e_4 we shall often use $\langle H, H \rangle$, $\langle H, T \rangle$, $\langle T, H \rangle$ and $\langle T, T \rangle$ to represent these simple events in a way which indicates precisely what the simple event is; $\langle H, T \rangle$, for example, means that coin 1 landed heads and coin 2 tails.

Unlike the previous examples, both event sets E_1 and E_2 obtained above could, for most applications, be used in constructing the mathematical model. It turns out, for reasons that will appear later, that we would normally use E_2 because it is easier to handle. The difference between E_1 and E_2 is one which arises frequently in probability theory. It is essentially that of whether we wish to distinguish between different orders of occurrence or not.

If the reader pauses and thinks about the examples just given, it will be apparent that we have not considered every possible outcome of the experiments discussed, and therefore what we have called event sets are not really event sets according to our previous definition. Consider the first example concerned with the single toss of a coin. It is not necessarily true that the coin will land heads or tails up. Conceivably, it might land standing on its edge, or it might not land at all because a stray shot destroys it in midair, or it may be blown away by a sudden hurricane gust of wind, so that we cannot find it. There are a whole host of alternative possibilities of this sort. Why did we not include these possibilities in setting up the event set? The first thing to note is that we could have included any of them that we wanted to. It would be perfectly valid to include them.

The type of events we are considering here, such as the coin landing on its edge, are what we might call extremely rare events. Probably no one has ever seen a coin land this way on a flat surface. However, there is no physical principle which says that it cannot do so. It cannot be ruled out as physically impossible. Ruling out such events can sometimes be justified by stipulating that, if they occur, the experiment is considered invalid and must be repeated. In other words, we limit the possible outcomes by defining what will be considered a valid experiment.

There are situations, however, where the occurrence of these odd and very rare events would not invalidate the experiment. The justification for omitting them in this case is usually based on the fact that the results which are obtained from the mathematical model will be the same regardless of whether they are included or not. This is normally a valid argument. It is important to note, however, that it is *not* always a valid procedure to exclude these very rare events. For example, suppose that an individual is offered the opportunity to play a game that involves tossing a coin once. If it lands heads up he wins $1000, if it lands tails up he loses $200, and if it lands standing on its edge he faces a firing squad. In this instance, even

though it is extremely unlikely that the coin will land on its edge, the individual would probably not wish to ignore this outcome in constructing a
model. If the consequences of one of these odd and very rare events are
extremely serious, then one is not in general justified in excluding such
outcomes without a rather detailed study. It may quite possibly be the case
that some of them should be included as events. Of course, only such as have
extremely serious consequences need to be included.

What we have really been saying in the last several paragraphs is that the
real world cannot be represented exactly in constructing an event set. It is
not possible to represent or even to conceive of every event which could
occur. An approximation must be made in that only the events which will
occur in the vast majority of cases will be included. Normally, such approximations in no way limit the usefulness of the model so obtained.

1–6 THE PROBABILITY MODEL

Once an event set has been determined, the final step in constructing the
probability model, that is, the mathematical representation of a random
experiment, is to assign a probability p_j to each simple event e_j. It is very
important for the reader to understand that the purpose of the model is *not*
to be of assistance in determining the probabilities of the simple events.
This will no doubt be hard for the reader to grasp at first, since he may
reasonably ask what the model is good for if it cannot be used to determine
the probabilities of the simple events. The value of the model will become
clear as we go along. Let us now note that the task of determining the
probabilities of the simple events is an empirical one and depends on our
understanding of the real world. These probabilities are not things that can
be computed mathematically. Indeed, it is by specifying the number of simple events and the probability of each simple event that the model being
constructed is differentiated from all other probability models.

Let us now introduce a little more notation. It will be convenient to represent random experiments symbolically, and in the future we shall denote
a real-world random experiment by $\Re$, with perhaps a subscript or super-
script on $\Re$. The mathematical model which we construct to represent $\Re$ will
be denoted by $\mathcal{E}$. The model $\mathcal{E}$ consists of nothing but the event set $E =
\{e_1, \ldots, e_m\}$ and the m numbers $p_1, \ldots, p_m$, p_j being the probability of the
simple event e_j.

We shall next examine what restrictions must be placed on the selection of
the numbers p_j. If $\mathcal{E}$ is to be realistic, it must be true that p_j should be
essentially the relative frequency of e_j when $\Re$ is repeated a large number of
times, and indeed, if p_j were determined exactly, it would be the long-run
relative frequency of e_j. If $\Re$ is performed n times and e_j occurs n_j times, then
it must be true that $0 \leq n_j \leq n$, where $\leq$ means less than or equal. Hence

$0 \leq n_j/n \leq 1$, that is, the relative frequency of e_j, being the fraction of the time that e_j occurs, must be a number which is not negative and not greater than 1. Now $0 \leq n_j/n \leq 1$ must hold for every n. However, as n gets larger and larger, n_j/n approaches p_j. Therefore, it follows that

$$0 \leq p_j \leq 1 . \qquad (1\text{--}2)$$

We shall require in our model that the p_j be selected so that (1–2) holds.

Equation (1–2) is not the only condition which must be satisfied if $\mathcal{E}$ is to be realistic. On any given trial of $\mathcal{R}$, one and only one simple event can occur. Hence

$$n_1 + n_2 + \cdots + n_m = n ,$$

where n_1 is the number of times e_1 occurs, n_2 the number of times e_2 occurs and so on. On dividing by n it follows that

$$\frac{n_1}{n} + \frac{n_2}{n} + \cdots + \frac{n_m}{n} = 1 , \qquad (1\text{--}3)$$

so that the sum of the relative frequencies of the simple events must be 1. Now (1–3) must hold for all n, and for very large n, n_j/n is extremely close to p_j. Thus it follows that

$$p_1 + p_2 + \cdots + p_m = 1 , \qquad (1\text{--}4)$$

so that the probabilities of the simple events sum to 1. If $\mathcal{E}$ is to be realistic (1–4) must hold, and hence we shall require that (1–4) does hold in constructing $\mathcal{E}$.

Any set of numbers $p_1, \ldots, p_m$ satisfying (1–2) and (1–4) will be considered to be an allowable assignment of probabilities to the simple events. The precise numbers used must ultimately rest on our understanding of the real world. In selecting the p_j one cannot expect to determine the exact values of the probabilities. The best one can hope to do is determine values which are sufficiently accurate for the purposes at hand. Here again, approximations must be introduced in constructing the model.

One of the main uses for the model $\mathcal{E}$ of $\mathcal{R}$ will be to compute the probabilities of *composite events*. What do we mean by composite events? Let us introduce this concept with an example. Suppose that we toss an ordinary six-sided die once and note which face turns up. For this random experiment we can use the event set E_1 introduced on page 11, having e_j represent the event that the face with j dots on it turns up. Now suppose we wish to determine the probability that the face which turns up will have one dot or two dots on it. We can, of course, refer to this outcome, call it A, as an event. The interesting thing to observe is that A is related to the simple events e_j of E_1. Note that e_1 is the event that the face with one dot turns up and e_2 is the event that the face with two dots turns up. Thus A will occur if e_1 or e_2 occurs and will not occur otherwise.

We shall refer to A as being the *union* of the two simple events e_1 and e_2, and we shall write

$$A = e_1 \cup e_2, \qquad (1\text{-}5)$$

using the set symbol $\cup$ to represent *or*. The use of the union symbol is very appropriate here and suggests some additional valuable terminology. Let $\{e_1\}$ and $\{e_2\}$ be subsets of E_1, each of which contains only a single element. If we now rewrite (1-5) as $A = \{e_1\} \cup \{e_2\}$ and interpret this as being a statement about sets, it says that the set A is the union of the sets $\{e_1\}$ and $\{e_2\}$, so that $A = \{e_1, e_2\}$. What does this mean? It means that the event A occurs when any one of the simple events in the set A occurs. Thus we can think of the description of the event A, in terms of the simple events, by (1-5) and by $A = \{e_1, e_2\}$ as meaning precisely the same thing.

Suppose that instead of desiring to compute the probability that a face with one or with two dots on it turns up, we are interested in the event, call it B, that the face which turns up has an even number of dots on it. Let us note that B is expressible in terms of simple events, just as A was. An even number of dots will appear on the face which turns up if e_2 or e_4 or e_6 occurs, that is, if there are 2, 4 or 6 dots on the face. The event B will not occur otherwise. Thus we say that B is the union of the simple events e_2, e_4 and e_6, and we can write

$$B = e_2 \cup e_4 \cup e_6 \quad \text{or} \quad B = \{e_2, e_4, e_6\}, \qquad (1\text{-}6)$$

so that we can think of B mathematically as a subset of E_1 consisting of the simple events e_2, e_4 and e_6.

Let us now turn to the general case. Consider an arbitrary random experiment $\mathfrak{R}$ for which we are using the event set $E = \{e_1, e_2, \dots, e_m\}$. We can form new events by considering the union of any two or more simple events. We can think of any subset of E which contains two or more simple events as corresponding to an event A in the real world; A is the event that one of the simple events in the subset occurs. Thus we might be interested in the event A that e_2 or e_9 occurs. Then we can characterize A by writing $A = e_2 \cup e_9$ or $A = \{e_2, e_9\}$. *An event which is the union of two or more simple events is called a composite event.* One of the main uses for the probability model is to determine the probability of composite events. Any subset of E which contains two or more simple events can be thought of as defining a composite event, which is the event that one of the simple events in the subset occurs. *More generally, any subset of E will represent an event.* If the subset contains more than one element it is a composite event, and if it contains just one element it is a simple event.

We can now see that the mathematical representation of the notion of a real-world event is simply a subset of the event set. It is possible to give a simple graphical representation of a probability model $\mathcal{E}$ and of any event

A. This is done in Figure 1–2. Each simple event is represented by a dot, and the probability of the simple event is written above the dot (precisely how one arranges the dots in the figure is irrelevant). To represent an event A we merely draw a closed curve which encircles the simple events in A, so that in Figure 1–2, $A = \{e_1, e_6, e_7, e_9, e_{10}\}$.

Our above development now makes it possible for us to explain the logical basis for defining the simple events e_j which make up the event set E. *We must define the simple events in such a way that every other event A whose probability we wish to determine is a subset of E.* To illustrate, consider once again the experiment of tossing a die. Suppose that, among other things, we wish to compute the probability of the event A that the face which turns up has either one dot or six dots on it, and the probability of the

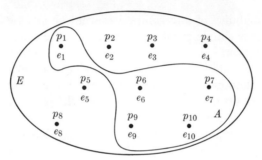

FIGURE 1–2

event B that the face which turns up has either two dots or three dots on it. The event set $E_2 = \{e^*, e^+\}$, where e^* is the simple event that a face with an even number of dots on it appears and e^+ is the simple event that a face with an odd number of dots appears, cannot be used to compute the probability of either A or B, since the only subsets of E_2 are $\{e^*\}$, $\{e^+\}$ and E_2 itself, none of which is either A or B.

To use our mathematical model, we must know how to compute the probability of any composite event A from the probabilities of the simple events. How is this done? To determine the rule which should be used, imagine that the random experiment is performed n times. Consider an event A, say $A = \{e_1, e_2, e_3, e_4\}$. Let n_1, n_2, n_3 and n_4 be the number of times that the simple events e_1, e_2, e_3 and e_4, respectively, occur. Now A occurs if e_1 or e_2 or e_3 or e_4 occurs, and A does not occur otherwise. Thus n_A, the number of times that A occurs, is

$$n_A = n_1 + n_2 + n_3 + n_4,$$

or on dividing by n

$$\frac{n_A}{n} = \frac{n_1}{n} + \frac{n_2}{n} + \frac{n_3}{n} + \frac{n_4}{n}. \qquad (1\text{–}7)$$

As n gets larger and larger, n_1/n, n_2/n, n_3/n and n_4/n approach p_1, p_2, p_3 and p_4, the probabilities of e_1, e_2, e_3 and e_4, respectively; and n_A/n approaches the probability of A, which we shall denote by $p(A)$. From (1–7), we then conclude that the following equation should hold:

$$p(A) = p_1 + p_2 + p_3 + p_4 . \tag{1–8}$$

More generally, if we have any composite event*

$$A = \{e_\alpha, e_\beta, \ldots, e_\zeta\} \tag{1–9}$$

then the same sort of reasoning suggests that

$$p(A) = p_\alpha + p_\beta + \cdots + p_\zeta , \tag{1–10}$$

so that the probability of any composite event should be the sum of the probabilities of the simple events whose union is the composite event. The nature of the real world and the frequency interpretation of probability suggest that (1–10) should hold. Thus if our model is to be realistic, we must compute the probabilities of composite events in accordance with (1–10). We shall then use (1–10) as the rule by which the probability of any composite event is computed in our model.

We have now completed the theoretical discussion concerning the development of a probability model. The method of developing a suitable event set has been discussed. Once a probability is assigned to each simple event the model is complete. To compute the probability of any composite event, (1–10) is used. We have not as yet considered how one estimates the probabilities of the simple events to be used. We shall consider this in the next section.

1–7 **PROBABILITIES OF THE SIMPLE EVENTS**

We have indicated previously that the model builder must supply the probabilities of the simple events, and that their determination is an empirical problem, not a mathematical one. As we go along we shall see that there exist various ways in which the probabilities of the simple events for some model can be determined. The most straightforward way to estimate these probabilities is to perform the associated random experiment a large number of times n. Then for each e_j the relative frequency n_j/n is determined. This should be close to p_j, the probability of e_j, and without any additional

---

* In the future we shall frequently use Greek letters as well as Roman letters. There are two reasons for this. First, it is standard in statistics to use Greek letters for certain quantities, and for this reason it is desirable to do so here. Also, there do not exist enough Roman letters which can be used conveniently, and thus it is often desirable to use Greek letters to avoid ambiguity. A listing of the Greek alphabet is given in the final table at the end of the text.

information one would simply use n_j/n for the probability of e_j in the model. Of course, n_j/n will not be the exact probability, but one should keep in mind that it is never possible to determine the probabilities exactly. The best that can be done is to determine them approximately.

As an illustration of the method just suggested, we might recall that the sex of a baby cannot be predicted before birth, and it can be imagined that the sex is determined by chance. The birth of a child can then be looked upon as a random experiment so far as the sex of the child is concerned, and to build a model of this experiment it is natural to use an event set with two simple events, e_1 being that the child is male and e_2 that it is female. To estimate the probabilities of e_1 and e_2 when the parents are of a specific race, say Caucasian, and living in the United States, one would determine over a long period of time the relative frequencies of male and female babies from parents of the specified type. Data could be generated for millions of births and hence the relative frequencies should provide a very accurate estimate of the probabilities. Studies of this type have been conducted, and the relative frequency of males is about 0.513, so that in the model one would take $p_1 = 0.513$ and $p_2 = 0.487$. These numbers vary slightly depending on the race of the parents.

The example just given makes it appear that it is relatively easy to estimate the probabilities of the simple events. We simply collect a great deal of data and then use the relative frequencies as probabilities. Unfortunately, for most types of problems encountered in business and economics things are not nearly so simple. One of the most important problems is that it is not possible to obtain a large amount of data. There are a variety of reasons for this. For example, it may be too costly to repeat the experiment often enough, or we may not have the freedom to determine when the experiment will be repeated and it occurs only relatively rarely in practice. Another important reason is that data which are very old are of no relevance; in other words, the probabilities of the simple events change with time. For example, in making predictions about future demands from historical data, a company will frequently be unable to use data which are more than two or three years old, since the structure of the market is changing so rapidly that older data are of little relevance.

Other sorts of troubles can frequently arise to complicate the task of estimating the probabilities of the simple events. To illustrate one sort of difficulty which often occurs, consider the supermarket which is concerned with the demand for bread on Tuesdays. The demand can be imagined to be determined by chance, that is, the process of having bread demanded can be looked upon as a random experiment. In constructing a model for this experiment one can use as e_j the event that j loaves of bread are demanded. To estimate p_j the store might keep records of how many loaves of bread are sold on each day and then determine n_j/n, the fraction of the number of

days on which j loaves were sold. It might then proceed to use n_j/n for p_j in the model. However, the number of loaves sold is equal to the number of loaves demanded only if the store does not run out of bread. On any day when the store runs out, the number of loaves demanded will usually be greater than the number of loaves sold. In general, the store has no way of determining how many more loaves were demanded than were sold. What we are pointing out is that sales data are not, as a rule, equivalent to demands. However, if the store only rarely runs out of stock, then there will not be many days when the demand is greater than sales, and then the n_j/n can reasonably be used as estimates of the p_j.

What we have been illustrating in this section is that the task of determining the p_j is not always an easy one, either because extensive data collection procedures must be employed or because it is not possible to obtain very much data at all. In the latter case, we cannot expect to estimate the p_j very accurately. However, even when there is considerable vagueness about the probabilities of the simple events, a probability model still may be of considerable value to a decision maker.

1–8 EQUALLY LIKELY SIMPLE EVENTS

In certain situations we can very conveniently estimate the probabilities of the simple events using an interesting logical procedure referred to as *reasoning by symmetry*. Let us begin with an example. Suppose that we have ten balls, numbered 1 to 10, in a bowl. The random experiment we shall consider consists of picking one ball from the bowl and noting the number which appears on the ball. The natural way to define the simple events is to imagine that there are ten such events, one corresponding to each ball which can be drawn; thus e_j will represent the event that the ball with the number j on it is drawn. What probabilities should be assigned to the e_j? Offhand, we do not know what to say. It depends on how the balls are arranged in the bowl and the manner in which a ball is selected from these.

Suppose next that someone tells us that before the drawing is made the balls are thoroughly mixed, and the person drawing the ball has no opportunity to see the ball being selected. The balls are all of the same size, so it is not possible for him to have any method of differentiating between them. Does this help us to answer the question? Almost everyone would now agree that any one ball is as likely as any other to be chosen, that is, the simple events are equally likely. What equally likely means is that each simple event has the same probability, or in other words, each ball has the same probability of being drawn. What is this probability? Since the probabilities must sum to 1, it follows that if $p_j = p$ (p is the same for each j), then

$$p_1 + p_2 + \cdots + p_{10} = 10p = 1 ,$$

or $p = 0.1$. Consequently, it appears that we should assign a probability of 0.1 to each simple event.

The above illustrates the use of a very interesting logical process for obtaining the probabilities of the simple events, which we shall refer to as reasoning by symmetry. Note that no frequency information was used in obtaining the probabilities. If we actually repeated the experiment a large number of times, would the relative frequency of each ball be essentially 0.1? This depends on how closely the actual situation conforms to the symmetry conditions hypothesized. If the balls are thoroughly mixed, and so forth, then the long-run relative frequency of each simple event should be 0.1. Many experiments have been performed which confirm this.

It is convenient to introduce a special terminology to describe a random experiment with the characteristics just described. If it is indeed true that the probability of selecting any given ball is 0.1, then we say that the experiment is one in which a ball is selected at random. Here random means that each ball has the same probability of being selected. The word random will be used frequently in this sense in the future. Thus we can speak of selecting at random a resistor from a box of resistors or selecting a card at random from a deck of playing cards.

To gain a better understanding of how the symmetry principle can be employed, let us now study several other examples. Suppose that we are shown a coin and asked what the probability is that, if it is tossed, it will land heads up. If we examine the coin and it looks new, showing no signs of wear, we are tempted to say that heads and tails should be equally likely and that the probability of a head (and also of a tail) is 0.5. A coin for which the probability of a head is 0.5 is called a *fair coin*. Does it follow that the coin under consideration will show a long-run relative frequency of heads equal to 0.5 if it is tossed a large number of times? Not at all. It depends on whether the coin is fair or not. Experience in the real world shows that coins frequently behave as if they were fair or very close to being fair. The results obtained by tossing a penny, illustrated in Figure 1–1, indicate that it is not unreasonable to think of the penny as being a fair coin.

However, the author selected another coin, a rather well-worn quarter, from the change in his pocket and tossed it 100 times. The results are shown in Figure 1–3. The relative frequency of heads in 100 tosses was 0.59. Most people would have considerable doubts that this coin is fair. It is not necessarily true that everyone would have such doubts, however. Some people would feel that the difference between the relative frequency of 0.59 and 0.5 in 100 trials is not sufficient to reject the notion that the coin is fair, and such individuals, if they had to build a model for tossing this coin using the information available, would use 0.5 for the probability of a head. The discussion just given, in addition to showing that the results obtained from symmetry arguments may not yield results which accurately portray the real-world situation, also points up the fact that individuals, even in cases

where frequency information is available, may decide not to use directly the relative frequencies as probabilities, but to modify these somewhat to agree more closely with the individual's subjective feelings, based on symmetry or other reasoning. Thus subjective considerations can enter into the determination of a probability even when a fair amount of frequency data are available. Of course, as the amount of frequency data available decreases, the subjective element becomes more important.

Let us next return to the random experiment in which we toss a six-sided die and note which face turns up. Suppose we take e_j to be the event that the face with j dots on it turns up, so that there are six simple events. What probability should be assigned to each simple event? In the absence of any other information, we can reason that by symmetry one face is just as likely to turn up as any other face. Thus all simple events should have the

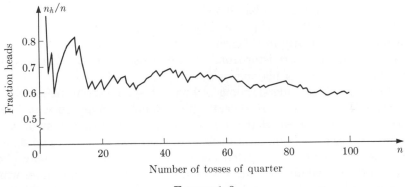

FIGURE 1–3

same probability, that is, $p_j = \frac{1}{6}$ for each j. A die having the characteristic that each face has a probability of $\frac{1}{6}$ of turning up is called a *fair die*. Whether or not the relative frequency of each face will be about $\frac{1}{6}$ for any given die depends on how close it is to being fair. If we enter a game of chance thinking that a die is fair and it is not (a die which is not fair is often referred to as being loaded), we may lose our shirts.

Let us now study an example which shows that we must be extremely careful in applying the symmetry principle if serious errors are to be avoided. Consider the experiment referred to on page 12 where we toss two coins. Suppose that we know each coin is fair. Let us first examine the event set $E_1 = \{\bar{e}_1, \bar{e}_2, \bar{e}_3\}$, where $\bar{e}_1$ is the event that both coins land heads up, $\bar{e}_2$ is the event that one head and one tail are obtained and $\bar{e}_3$ is the event that both coins land tails up. Now one might be convinced that these three outcomes are equally likely and hence each should have the probability $\frac{1}{3}$. But then someone points out that the event set E_2 can be used also, and it appears that the four simple events in E_2 are equally likely. Recall that

$E_2 = \{e_1, e_2, e_3, e_4\}$, where e_1 is the event that both coins land heads up, e_2 is the event that the first yields a head and the second a tail, e_3 is the event that the first yields a tail and the second a head and e_4 is the event that both land tails up. If the simple events in E_2 are equally likely, then each must have the probability $\frac{1}{4}$.

We can now note that it is impossible for both the simple events in E_1 and those in E_2 to be equally likely, since according to the assignment of probabilities in E_1, the probability that both coins land heads up is $\frac{1}{3}$, while the assignment in E_2 gives $\frac{1}{4}$ for this probability. The same discrepancy arises for the event that both coins land tails up. The event that one head and one tail are obtained is a simple event if E_1 is used and has the probability $\frac{1}{3}$ when the simple events in E_1 are taken to be equally likely. However, if E_2 is used, it is not a simple event but is the event $\{e_2, e_3\}$, which has the probability $\frac{1}{4} + \frac{1}{4} = \frac{1}{2}$ when the simple events in E_2 are equally likely. Since both probability assessments cannot be correct, which one, if either, is? Most people would agree, after having it pointed out that one head and one tail can be obtained in two different ways, that the correct assignment is the one which takes the simple events in E_2 to be equally likely. It might not be true, however, that everyone would agree to this. The only way the issue can be resolved is by actually repeating the experiment a large number of times and determining the relative frequencies. These results would show that two heads occur about a quarter of the time, two tails about a quarter of the time and one head and one tail about half of the time. This is in agreement with the assumption that the simple events in E_2 are equally likely and refutes the idea that the simple events of E_1 are equally likely.

In the future we shall occasionally make use of symmetry to determine the probabilities of the simple events. From what we have seen above, the key to the use of this procedure is to find an event set such that if certain symmetry conditions are satisfied, then the simple events are equally likely. Then if there are m simple events, each one is assigned the probability $1/m$. In order for the model so obtained to be a realistic representation of the real world, it is necessary, of course, that the symmetry conditions hypothesized be approximately fulfilled in reality.

EXAMPLE. We shall now use the method of symmetry to obtain a result which will be frequently used in the future. Suppose that we have m balls, numbered 1 to m, in a bowl. Suppose also that balls 1 through k are red and balls $k + 1$ through m are blue. A ball is now selected at random from the bowl. We would like to determine the probability that the ball is red. To construct a model of this experiment it seems perfectly natural to imagine that there are m simple events, e_j being the event that ball j is drawn. By assumption, the ball is drawn at random. This means that each ball has the same chance of being drawn, that is, the probability which should be as-

signed to e_j is $1/m$. Now we are interested in the event R that the ball drawn is red; $R = \{e_1, \ldots, e_k\}$, since by assumption the balls numbered $1, \ldots, k$ are red. Hence by (1–10)

$$p(R) = \underbrace{\frac{1}{m} + \cdots + \frac{1}{m}}_{k \text{ of these}} = \frac{k}{m}, \qquad (1\text{–}11)$$

and the probability of a red ball is simply the fraction of the balls in the bowl which are red.

Let us now imagine that the numbers on the balls referred to above are painted out, so that there are k red and $m - k$ blue balls in the bowl, but no numbers appear on the balls. What now is the probability that a ball drawn is red? Almost everyone would agree that the long-run frequency of drawing a red ball is in no way influenced by whether any numbers appear on the balls or not. In other words, the probability of a red ball is still k/m. This is an important and in some ways slightly subtle point. There were originally m different balls in the bowl. We could have differentiated between these and hence identified each one in any way we desired. For example, each might have been examined very carefully and what we called the first one might have had a particular type of scratch, the second might have had a small bubble in the paint and so forth. By close enough inspection we could always identify each of the balls individually. We can construct an event set with m simple events, each simple event referring to a different ball being drawn, regardless of whether the balls happen to be numbered or not. In the future, when we have m objects, we shall often refer to these as objects $1, \ldots, m$ and construct an event set containing m simple events, e_j being the event that object j is drawn. In doing this, we need not assume that the objects are actually numbered in this way, provided that the event whose probability we wish to compute does not require that the objects be identified in some special way.

After going through the above analysis, we can now see how to introduce a simpler event set for the above random experiment where numbers do not appear at all. We can use the event set $E = \{R, B\}$, in which there are just two simple events. These are: R, a red ball is drawn; and B, a blue ball is drawn. The above reasoning has shown that we should assign a probability of k/m to R and $(m - k)/m$ to B. To see how to obtain these probabilities, however, it was first necessary to use a different event set where reasoning by symmetry could be applied.

If we start out with a bowl containing m balls of three different colors, say red, blue and yellow, and if k_1 are red, k_2 blue and k_3 yellow, then we can use the event set $E = \{R, B, Y\}$, R being the event that a red ball is drawn and so on. By use of the same type of argument employed above we see that the probabilities which should be assigned to R, B and Y are merely the frac-

tions of the total number of balls having the indicated color, that is, k_1/m, k_2/m and k_3/m respectively.

There are many real-world situations which are equivalent to the random experiment just considered, if balls and red and blue are interpreted properly. For example, consider a box of m resistors, k of which are defective. A resistor is selected at random from the box and it is desired to determine the probability that it is defective. If we think of the resistors as balls, defective resistors as red balls and good resistors as blue balls, then this experiment is equivalent to the one just considered, and we see at once that the probability of selecting a defective resistor from the box is k/m, so that if there were 100 resistors, 3 of which were defective, the probability of selecting a defective one would be $3/100 = 0.03$.

1–9.† MODEL BUILDING—AN APPLICATION TO GENETICS

Probability theory has found important applications in the field of genetics, since a child's inheritance of certain traits from his parents is not deterministic, but must be treated as a random experiment. The fact that chance plays a role in inheritance was first pointed out in the 1860's by the Augustinian monk Gregor Mendel as a result of his experiments with pea plants in an Austrian monestary.

The procedure by which a probability model is constructed to represent the actual observed results of Mendel's experiments serves to illustrate the problems in constructing a model. Let us then consider an experiment similar to that which Mendel might have performed. Assume that a particular plant may have either red (R), pink (P) or white (W) flowers. On experimenting with this type of plant it is found that seeds obtained by crossing two red-flowered plants always yield plants with red flowers. This is often represented symbolically by writing $R \times R \to R$. Similarly, it is always true that crossing two plants with white flowers yields a plant with white flowers, that is, $W \times W \to W$, and red-flowered plants crossed with white-flowered plants always yield plants with pink flowers, that is, $R \times W \to P$.

The situation changes, however, when other crosses are considered. When red is crossed with pink $(R \times P)$, the flowers on plants grown from different seeds will not always have the same color (all flowers on a given plant will be of the same color, however). Sometimes the plant will have red and sometimes pink flowers. The color cannot be predicted in advance, but the long-run frequency of red-flowered plants is about $\frac{1}{2}$. Similarly, for white crossed with pink $(W \times P)$, the flowers of the resulting plants will be either white or pink. The color cannot be predicted in advance, but the long-run frequency of plants with white flowers is about $\frac{1}{2}$. The situation is more

† This section may be omitted if desired. However, some later examples make use of the material in this section.

complicated when pink-flowered plants are crossed with pink-flowered ones. In this case the resulting plants may have either red, pink or white flowers, the long-run frequency of plants having red flowers being $\frac{1}{4}$, of those having pink flowers being $\frac{1}{2}$ and of those having white flowers being $\frac{1}{4}$.

We have now presented the results of the experimental studies. To explain such results Mendel constructed a most interesting and remarkable model. He assumed that there were certain discrete entities, today called *genes*, which are responsible for inherited characteristics and which are passed on from plants to their offspring. He supposed, furthermore, that each plant has two genes which together determine the inherited characteristic, the color of the plant's flowers in the case under consideration. One of these genes is obtained from one parent and the other gene from the other parent. In addition, Mendel supposed that there were just two types of genes which determine the color of the flowers. These will be referred to as A and a type genes. Any given plant may then have two A type genes, which will be represented symbolically as AA, two a type genes (aa) or one gene of type A and one gene of type a (Aa). A plant of type AA will have red flowers, one of type Aa will have pink flowers and one of type aa will have white flowers. Plants of type AA or aa are often referred to as being *homozygous*, while those of type Aa are referred to as *heterozygous*.

Finally, Mendel assumed that if a parent is of type Aa (and has pink flowers), then the offspring may get either a gene of type A or of type a from this parent, and the probability of getting either one is $\frac{1}{2}$. Furthermore, there is no relation between the genes an offspring obtains from each parent and, in addition, there is no relation between the gene that a parent gives to one offspring and the gene it gives to any other offspring. The genes have the remarkable characteristic that they can be passed on from one generation to another without being modified in any way. This is sometimes referred to as Mendel's first law.

Let us now proceed to see how the general model just developed can be used to explain the experimental results described at the beginning of this section. If both parents have red flowers and are thus of type AA, the offspring must get an A gene from each parent and hence will be AA and have red flowers, in accordance with the experimental results. Precisely the same conclusion is reached if both parents are aa and have white flowers: the offspring must also have white flowers. If one parent is of type AA and has red flowers and the other is of type aa and has white flowers, the offspring must be of type Aa and will have pink flowers. If one parent is AA and the other Aa, the offspring will always obtain an A type gene from the first parent but may obtain either an A or a type gene from the second. Thus the offspring will either be of type AA or of type Aa, that is, will either have red flowers or pink flowers.

It is imagined that a random experiment is performed which determines whether the offspring will get the A or a gene. We can imagine that there

are two simple events for this experiment which we shall denote by AA and Aa. The probability of each of these is $\frac{1}{2}$, by Mendel's first law, since the probability of AA is the probability that an A gene is obtained from the second parent, which this law claims is $\frac{1}{2}$. Thus we see how the model predicts the correct behavior for crossing plants with red and pink flowers. Precisely the same analysis shows that the model predicts the correct results for plants of type aa crossed with those of type Aa.

Finally, let us consider the most complicated case—that where a plant of type Aa is crossed with a plant of type Aa. Now the offspring can obtain either an A or a type gene from the first parent and either an A or an a type gene from the second. The random experiment leading to an offspring can thus be imagined to consist of four simple events, which are $\langle A, A \rangle$, $\langle A, a \rangle$, $\langle a, A \rangle$, and $\langle a, a \rangle$, where, for example, $\langle A, a \rangle$ means that an A type gene was obtained from the first parent and an a type gene from the second parent.

Now the type of gene received from the first parent can be imagined to be the result of tossing a fair coin, one face being labeled A and the other face a. The face which turns up gives the gene obtained from the first parent. Since it is assumed there is no connection between the gene obtained from the first parent and that from the second, the gene obtained from the second parent can be imagined to be the result of tossing the same coin a second time. The results of these two tosses can be described by the four simple events listed above. We discussed precisely this sort of experiment in the previous section (in our earlier terminology, the simple event would be denoted by $\langle H, H \rangle$, $\langle H, T \rangle$, $\langle T, H \rangle$ and $\langle T, T \rangle$). We noted there that each simple event should have the probability $\frac{1}{4}$. The event that the offspring is of type Aa is the composite event $\{\langle A, a \rangle, \langle a, A \rangle\}$, since both of these simple events yield an Aa type. Thus the probability of an Aa type should be $\frac{1}{2}$, while the probability of an AA or aa type is $\frac{1}{4}$. Once again these conclusions are in agreement with the experimental results.

Mendel found in his experiments that occasionally one of the two types of genes which determine a given characteristic would dominate the other, and as long as one of the two genes in the offspring was of the dominant type, then the offspring would have the characteristic of the dominant gene. For example, the stems of pea plants may be tall or dwarfed. The gene favoring tall plants, call it B, dominates the gene favoring dwarfed plants, call it b. Thus plants of type BB will be tall and those of type bb will be dwarf. However, those of type Bb will also be tall since the tall characteristic dominates the dwarf one. According to the above model, then, when plants of type BB are crossed with those of type Bb, the offspring will always be tall, while if those of type bb are crossed with those of type Bb, about half the plants will be tall and half will be dwarf. When Bb plants are crossed with a Bb type, about $\frac{3}{4}$ of the plants will be tall and $\frac{1}{4}$ will be dwarf, since now $<B, B>$, $<B, b>$ and $<b, B>$ all give tall

plants. The above example dealing with the flower color can also be looked upon as an example of dominance if we only distinguish between white and colored flowers and do not concern ourselves with the precise color of the flower. Any plant possessing at least one A gene will be colored, and thus the A gene can be interpreted as dominating the a type gene.

An interesting illustration of dominance arises in the inheritance of certain types of diseases, mental disorders or crippling defects. It turns out that AA and Aa individuals are healthy and, in fact, that there is no way to differentiate between them. However, aa type individuals will be afflicted. The a type gene is often referred to as *lethal*, since if an offspring receives one of these from each parent he will be affected. An individual of type Aa is called a *carrier*, since although he is healthy he may pass the lethal gene on to his descendents. Let us note that the offspring of two AA parents or one AA and one Aa parent will always be healthy since the offspring will always have at least one A gene. Presumably, an individual of type aa would not have children if the disorder is very serious, either because he would die before reaching maturity or would be recognized to have the disorder. Thus, if this is true, only children each of whose parents is of type Aa may be afflicted, and from our above analysis, the probability that such a child will be afflicted is $\frac{1}{4}$.

Mendel's model is remarkable for its simplicity and its range of applicability. It has been found to apply not only to plants, but to animals and humans as well. It is by no means true that all inherited characteristics are adequately described by Mendel's simple model. Sometimes the inheritance pattern is much more complex, and some inherited characteristics show more or less of a continuous variation rather than the discrete alternatives considered above. Indeed, even today, models have not been worked out for all inherited characteristics, and this is still a field of active research interest. Nonetheless, Mendel's model is extremely useful and has been widely employed.

In later examples, we shall illustrate how the model can be used to answer some realtively complicated questions concerning inheritance. Once the model is presented and its use illustrated, it seems rather natural and simple. In this setting it is hard to appreciate the intellectual achievement necessary to develop it initially. It would not be feasible to try to suggest here a variety of alternatives that Mendel might have considered as models. However, it would be instructive for the reader to investigate this on his own to gain a better appreciation of what real insight Mendel had.

1–10 THE ADDITION LAW

Consider some random experiment $\mathfrak{R}$ for which we are using the model $\mathcal{E}$, having the event set $E = \{e_1, \ldots, e_m\}$, with the probabilities of the simple

events being $p_1, \ldots, p_m$. Let A and B be any two events, that is, any two subsets of E. The events A and B are said to be *mutually exclusive* or simply exclusive if $A \cap B = \varnothing$, that is, if A and B have no simple events in common. In terms of the real world, A and B are mutually exclusive if A and B cannot both occur when $\mathfrak{R}$ is performed. In Figure 1–4 we have represented graphically two events A and B which are mutually exclusive. Let A and B be any two mutually exclusive events and denote the probabilities of A and B computed from $\mathcal{E}$ by summing the probabilities of the simple events in A and B respectively by $p(A)$ and $p(B)$. Consider now the event $A \cup B$, which mathematically is an event because it is that subset of E which contains the simple events in A or in B. The real-world interpretation of $A \cup B$ is the event that either A or B occurs. Let us now determine $p(A \cup B)$, the probability of $A \cup B$. By (1–10), $p(A \cup B)$ is the sum of the probabilities of the simple events in the subset $A \cup B$. In terms of Figure 1–4, $A \cup B$ is the set of simple events which includes those inside the closed curve marked A and those inside the closed curve marked B. Now the sum of the probabilities of the simple events inside the closed curve representing A is $p(A)$, and the sum of those inside the closed curve for B is $p(B)$. The sum of the probabilities of the simple events inside both closed curves is $p(A \cup B)$. Thus we see that

$$p(A \cup B) = p(A) + p(B), \quad A \cap B = \varnothing. \tag{1–12}$$

Equation (1–12) is referred to as the *addition law for probabilities*. It is important to recognize that it holds only when A and B are mutually exclusive events.

To see how (1–12) must be modified when A and B are not mutually exclusive, consider Figure 1–5, which illustrates a case where A and B have some simple events in common. When A and B are not mutually exclusive, then $A \cap B$ contains at least one simple event and hence is an event. It is the event that both A and B occur. For the case illustrated in Figure 1–5, $A \cap B = \{e_6, e_{10}, e_{11}\}$. Now $p(A \cup B)$ is the sum of the probabilities of all the simple events which are in either A or B. In Figure 1–5, $p(A \cup B)$ is

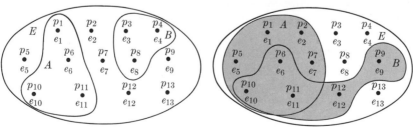

FIGURE 1–4   FIGURE 1–5

the sum of the probabilities of all the simple events in the shaded region. As before, $p(A)$ is the sum of all the probabilities of the simple events in A, and $p(B)$ is the sum of all the probabilities of the simple events in B. Now $p(A) + p(B)$ is not the sum of the probabilities of the simple events in the shaded region of Figure 1–5, because the probabilities of the simple events in $A \cap B$ have been included twice, once in $p(A)$ and once in $p(B)$. To obtain $p(A \cup B)$ it is therefore necessary to subtract from $p(A) + p(B)$ the sum of the probabilities of the simple events in $A \cap B$, that is, $p(A \cap B)$. Thus, when A and B are not mutually exclusive

$$p(A \cup B) = p(A) + p(B) - p(A \cap B). \qquad (1\text{–}13)$$

We might now note that we can make (1–12) a special case of (1–13) if we define $p(\varnothing) = 0$. This is reasonable, since if $A \cap B = \varnothing$, $A \cap B$ cannot occur and thus its probability should be 0. Therefore, we can think of (1–13) as a general result which always holds, and which reduces to (1–12) when A and B are mutually exclusive.

We shall have to wait until the next two sections are covered to give good examples of the usefulness of (1–12) and (1–13). However, we shall now give a very simple example.

EXAMPLE. Consider the random experiment in which a fair die is tossed and the face which turns up noted. Let us use the usual event set, with e_j being the event that the face with j dots on it turns up. Since the die is fair, $p_j = \frac{1}{6}$ for each j. Let $A = \{e_1, e_2\}$ and $B = \{e_4, e_6\}$, so that A is the event that the face which turns up has either one dot or two dots, and B is the event that the face which turns up has either four dots or six dots. By (1–10)

$$p(A) = \frac{1}{6} + \frac{1}{6} = \frac{1}{3}; \quad p(B) = \frac{1}{6} + \frac{1}{6} = \frac{1}{3}.$$

Now $A \cap B = \varnothing$, so that A and B are exclusive events. Thus by (1–12)

$$p(A \cup B) = p(A) + p(B) = \frac{2}{3}.$$

This can easily be checked directly, since $A \cup B = \{e_1, e_2, e_4, e_6\}$ and by (1–10), $p(A \cup B) = \frac{2}{3}$, which agrees with what was obtained from (1–10).

Suppose next that we again take $A = \{e_1, e_2\}$, but that we now take $B = \{e_2, e_6\}$. Once again $p(A) = \frac{1}{3}$ and $p(B) = \frac{1}{3}$. Here, $A \cup B = \{e_1, e_2, e_6\}$, so by (1–10), $p(A \cap B) = \frac{1}{2}$. In this case it is not true that (1–12) holds. Why is this? The reason is that A and B are not exclusive. Instead, $A \cap B = e_2$ and $p(A \cap B) = \frac{1}{6}$. Then according to (1–13)

$$p(A \cap B) = \frac{1}{3} + \frac{1}{3} - \frac{1}{6} = \frac{1}{2},$$

which is indeed what we obtained by summing the probabilities of the simple events in $A \cup B$.

Let A be any event which is a proper subset of E. Next consider the event A^c which contains all simple events in E that are not in A. A^c is referred to as *the complement of* A. Note that $A \cap A^c = \varnothing$. Thus by (1–12)

$$p(A \cup A^c) = p(A) + p(A^c).$$

However, $A \cup A^c = E$ and $p(E) = 1$, since the sum of the probabilities of all the simple events in E is 1. Hence

$$p(A) + p(A^c) = 1 \quad \text{or} \quad p(A) = 1 - p(A^c). \tag{1–14}$$

There are situations where to compute $p(A)$ it is simplest to first compute $p(A^c)$ and to subtract this number from 1. We shall encounter such situations later.

1–11 INDEPENDENCE

Many of the random experiments with which one deals in practice can be imagined to be built up from or, equivalently, decomposable into two or more parts, each of which can itself be considered to be a random experiment. For example, consider the random experiment $\mathcal{R}$ in which a coin is tossed twice. This can, obviously, be thought of as being built up from or decomposable into two random experiments $\mathcal{R}_1$ and $\mathcal{R}_2$, where $\mathcal{R}_1$ is the random experiment in which the coin is tossed the first time and $\mathcal{R}_2$ is the random experiment in which the coin is tossed the second time. In this section and Section 1–13 we wish to study random experiments which are built up from or decomposable into simpler experiments. We shall show that it is often possible to construct a model for the entire experiment from the models for the individual parts, so that no additional investigations need to be made to obtain the probabilities of the simple events for the overall experiment if one has models for the pieces. We shall also show that it is frequently possible to compute the probabilities of certain types of events for the complete experiment without constructing a detailed model. Only the models for the pieces will be required.

In this section we wish to consider cases where $\mathcal{R}$ is built up from two or more *independent* random experiments. Let us now explain what is meant by independent random experiments. Two random experiments $\mathcal{R}_1$ and $\mathcal{R}_2$ are independent if the outcome of $\mathcal{R}_1$ has absolutely no influence on the outcome of $\mathcal{R}_2$ and conversely the outcome of $\mathcal{R}_2$ has absolutely no influence on the outcome of $\mathcal{R}_1$. Similarly, several experiments are independent if the outcome of each of the experiments is not influenced by and does not influence the outcome of any one of the other experiments. Thus if we con-

sider again the random experiment $\Re$, which consists of tossing a coin twice, the two random experiments $\Re_1$ and $\Re_2$ that compose $\Re$ we would feel are independent, since whether the coin lands heads or tails up on the first toss has absolutely no influence on what will happen on the second toss and, of course, the outcome of that toss has no influence on the first.

As another example of an experiment $\Re$ which is built up from two independent random experiments, consider the experiment which consists of selecting at random a resistor from a box of resistors and also selecting at random a capacitor from a box of capacitors. The selection of a resistor from the box of resistors is a random experiment, call it $\Re_1$, and the selection of the capacitor from the box of capacitors is a random experiment, call it $\Re_2$. Now it should certainly be true that the resistor selected will have no influence on the capacitor selected and vice versa. Thus $\Re_1$ and $\Re_2$ are independent random experiments and $\Re$ is built up from two independent random experiments. The notion of independent random experiments is an intuitive one and it is not possible to prove mathematically that two random experiments are independent.

Let us study a random experiment $\Re$ which is built up from two independent random experiments $\Re_1$ and $\Re_2$. Let A be any event of $\Re_1$ (A is also an event of $\Re$, since it is the event that on performing $\Re$, $\Re_1$ yields A). Also let B be any event of $\Re_2$ (B is also an event of $\Re$). Consider now the event $A \cap B$ of $\Re$, which is the event that on performing $\Re$ the experiment $\Re_1$ yields the outcome A and the experiment $\Re_2$ yields the outcome B. Note that $A \cap B$ is not an event of either $\Re_1$ or $\Re_2$ (why?). We shall show however, that it is possible to determine the probability of $A \cap B$ without explicitly constructing a model for $\Re$. The probability of $A \cap B$ can be determined using only the models for $\Re_1$ and $\Re_2$.

To see how to do this we shall use the intuitive meaning of probability as a long-run relative frequency. Suppose that $\Re$ is performed a large number of times n, and let n_{AB} be the number of times that $A \cap B$ occurs. Then $p(A \cap B)$ is the limiting relative frequency n_{AB}/n as n becomes arbitrarily large. Now if n_A is the number of times that A occurs, then

$$\frac{n_{AB}}{n} = \frac{n_A}{n} \frac{n_{AB}}{n_A}. \tag{1-15}$$

In obtaining (1–15) we merely multiplied n_{AB}/n by n_A/n_A, which is 1. Now n_A/n is simply the relative frequency of A in n trials of $\Re$ and hence of $\Re_1$. Since the experiments are independent, the outcome of $\Re_2$ has no influence on the outcome of $\Re_1$ and hence n_A/n must approach $p(A)$, the probability that A occurs on performing $\Re_1$.

Let us next examine the ratio n_{AB}/n_A, which is the fraction of those cases when $\Re_1$ yielded A that $\Re_2$ also yielded B. If the outcome of $\Re_1$ has no influence on the outcome of $\Re_2$, then the long-run frequency of B when $\Re_1$

yields A should not be any different from the long-run frequency of B when $\mathfrak{R}_2$ is performed alone. Thus n_{AB}/n_A should approach $p(B)$, so that (n_A/n) (n_{AB}/n_A) must approach $p(A)p(B)$. However, since n_{AB}/n approaches $p(A \cap B)$ we conclude from (1–15) that

$$p(A \cap B) = p(A)p(B) ,\qquad (1\text{–}16)$$

and $p(A)$ can be determined from the model for $\mathfrak{R}_1$ while $p(B)$ can be determined from the model for $\mathfrak{R}_2$. This is a very important and useful result. We have shown that if $\mathfrak{R}$ consists in performing two independent random experiments $\mathfrak{R}_1$ and $\mathfrak{R}_2$ and if A is an event of $\mathfrak{R}_1$ and B an event of $\mathfrak{R}_2$, then if $p(A)$ and $p(B)$ are the probabilities of A and of B determined from the models of $\mathfrak{R}_1$ and $\mathfrak{R}_2$ respectively, the probability of the event $A \cap B$ of $\mathfrak{R}$ is simply the product of $p(A)$ and $p(B)$. Equation (1–16) is sometimes referred to as the *product rule*.

As a very simple example of the use of (1–16), consider the random experiment $\mathfrak{R}$ in which a fair coin is tossed, and then a fair die is rolled. Let us determine the probability that the coin lands heads up and, in addition, the face of the die with three dots on it turns up. $\mathfrak{R}$ can be considered to be built up from two independent random experiments, $\mathfrak{R}_1$ in which the coin is tossed and $\mathfrak{R}_2$ in which the die is rolled. We wish to determine the probability of the event $A \cap B$, where A is the event that $\mathfrak{R}_1$ yields a head and B is the event that $\mathfrak{R}_2$ yields the face with three dots on it. Since the coin is fair, $p(A)$ computed from a model for $\mathfrak{R}_1$ is $\frac{1}{2}$ and since the die is fair, $p(B)$ computed from a model for $\mathfrak{R}_2$ is $\frac{1}{6}$. Hence by (1–16)

$$p(A \cap B) = \frac{1}{2}\left(\frac{1}{6}\right) = \frac{1}{12} ,$$

so that the long-run relative frequency of having the coin land heads up and also having the three-dot face of the die turn up should be $\frac{1}{12}$.

The rule (1–16) can be conveniently used to help us construct a model $\mathcal{E}$ for a random experiment $\mathfrak{R}$ which is built up from two independent random experiments $\mathfrak{R}_1$ and $\mathfrak{R}_2$. Let us suppose that we have satisfactory models $\mathcal{E}_1$ and $\mathcal{E}_2$ for $\mathfrak{R}_1$ and $\mathfrak{R}_2$, respectively. Assume that the event set for $\mathcal{E}_1$ is $E_1 = \{e_1, \ldots, e_m\}$, with the probabilities of the simple events being $p_1, \ldots, p_m$, respectively. Also let the event set for $\mathfrak{R}_2$ be $E_2 = \{f_1, \ldots, f_s\}$, with the probabilities of the simple events being $q_1, \ldots, q_s$. We use symbols f_j for the simple events and q_j for the probabilities in $\mathcal{E}_2$ because we have already used the familiar symbols e_j and p_j for $\mathcal{E}_1$. A logical way to characterize a simple event for $\mathfrak{R}$ is to give the simple event, say e_i, which occurred on performing $\mathfrak{R}_1$ and also the simple event, say f_j, which occurred on performing $\mathfrak{R}_2$. Thus as a simple event for $\mathfrak{R}$ we can use $\langle e_i, f_j \rangle$, where $\langle e_i, f_j \rangle$ is a symbol which means that e_i occurred on performing $\mathfrak{R}_1$ and f_j occurred on performing $\mathfrak{R}_2$. Thus if $\mathfrak{R}$ refers to tossing a fair coin and then

rolling a fair die, $\langle e_i, f_j \rangle$ might represent the event that the coin lands tails up and the face of the die with six dots on it turns up.

How many simple events are there for $\mathcal{R}$? There are m simple events for $\mathcal{R}_1$ and s for $\mathcal{R}_2$. For each outcome of $\mathcal{R}_1$, s different simple events of $\mathcal{R}_2$ can occur. Since there are m simple events for $\mathcal{R}_1$, there are ms simple events $\langle e_i, f_j \rangle$ for $\mathcal{R}$. Now $\langle e_i, f_j \rangle$ means that the event $e_i \cap f_j$ occurred on performing $\mathcal{R}$. However, by (1–16),

$$p(e_i \cap f_j) = p(e_i)p(f_j) = p_i q_j \,,$$

since $p(e_i)$ is the probability of the simple event e_i in $\mathcal{E}_1$, which is p_i, and $p(f_j)$ is the probability of the simple event f_j in $\mathcal{E}_2$, which is q_j. This analysis shows that if $\mathcal{E}_1$ and $\mathcal{E}_2$ are realistic models of $\mathcal{R}_1$ and $\mathcal{R}_2$, then if $\mathcal{E}$ is to be a realistic model of $\mathcal{R}$, the probability which should be assigned to the simple event $\langle e_i, f_j \rangle$ is $p_i q_j$, the product of the probabilities of the simple events for $\mathcal{R}_1$ and $\mathcal{R}_2$ that make up $\langle e_i, f_j \rangle$.

If $\mathcal{R}$ consists in performing n independent random experiments $\mathcal{R}_1, \ldots, \mathcal{R}_n$, and if A_1 is an event of $\mathcal{R}_1, \ldots,$ and A_n is an event of $\mathcal{R}_n$, then the same reasoning as used above shows that the probability of $A_1 \cap \cdots \cap A_n$ can be computed as follows:

$$p(A_1 \cap \cdots \cap A_n) = p(A_1) \cdots p(A_n) \,, \qquad (1\text{–}17)$$

so that the probability that A_1 and A_2 and . . . and A_n occurs is the probability that A_1 occurs in $\mathcal{R}_1$ times the probability that A_2 occurs in $\mathcal{R}_2$ . . . times the probability that A_n occurs in $\mathcal{R}_n$.

We have been discussing in this section the intuitive notion of independent random experiments. We would now like to introduce the mathematical notion of *independent events* and relate it to the notion of independent experiments. Consider any random experiment $\mathcal{R}$ whatever. $\mathcal{R}$ need not be capable of being thought of as built up from independent random experiments. Let $\mathcal{E}$ be the model we are using for $\mathcal{R}$ and consider any two events A and B, that is, any two subsets of the event set E. Let us determine the probabilities of the events A, B and $A \cap B$ by summing the probabilities of the simple events in these subsets of E. *If it is true that*

$$p(A \cap B) = p(A)p(B) \qquad (1\text{–}18)$$

then the events A and B are said to be independent. The events are not independent if (1–18) does not hold. The notion of independent events is a mathematical notion. Given any model $\mathcal{E}$ and events A and B, one can prove whether they are independent or not by determining whether (1–18) holds.

Let us now note the connection between the intuitive concept of independent experiments and the mathematical notion of independent events.

If $\mathcal{R}$ is made up from two independent random experiments $\mathcal{R}_1$ and $\mathcal{R}_2$ and if A is an event of $\mathcal{R}_1$ and B an event of $\mathcal{R}_2$, then we have seen above that (1–16) should hold. But according to our definition of independent events, it follows that A and B are independent events. Indeed, in almost every case when we deal with independent events A and B, the reason for their independence will be the fact that the random experiment $\mathcal{R}$ is built up from $\mathcal{R}_1$ and $\mathcal{R}_2$, with A being an event of $\mathcal{R}_1$ and B an event of $\mathcal{R}_2$.

1–12 EXAMPLES

1. A visitor to Hong Kong wishes to purchase a new camera and, in addition, to obtain a tailor-made suit. He plans to do his shopping in Kowloon and consults the telephone directory. Listed in the directory are 500 tailors who have their shops in Kowloon, and 100 camera stores which are in Kowloon. The tourist is bewildered by the large number of shops and decides to select at random one of the 100 camera stores and one of the 500 tailors. If the proprietors of 11 of the camera stores and 30 of the tailoring shops are named Lee, what is the probability that the proprietor of both the camera store and tailoring shop selected will be named Lee?

The random experiment $\mathcal{R}$ which involves selecting a camera store and a tailor shop is built up from two independent random experiments $\mathcal{R}_1$ and $\mathcal{R}_2$, $\mathcal{R}_1$ referring to the selection at random of one of the 100 camera stores and $\mathcal{R}_2$ referring to the selection at random of one of the 500 tailoring shops. Let A be the event that the name of the proprietor of the camera shop is named Lee and B the event that the proprietor of the tailor shop is named Lee. Now from the result obtained on page 23, $p(A)$ is the fraction of the camera shop proprietors who are named Lee, so that $p(A) = 11/100$. Similarly, $p(B)$ is the fraction of the tailoring shop proprietors who are named Lee, so that $p(B) = 30/500$. The event whose probability we wish to compute is simply $A \cap B$. Since $\mathcal{R}_1$ and $\mathcal{R}_2$ are independent, it follows from (1–16) that

$$p(A \cap B) = p(A)p(B) = \left(\frac{11}{100}\right)\left(\frac{30}{500}\right) = 0.0066 \,,$$

so that about 6.6 times in 1000 both proprietors would be named Lee.

2. For the situation described in the previous example, let us determine the probability that at least one of the two proprietors is named Lee. In other words, we would like to determine the probability of the event $A \cup B$. Now A and B are not mutually exclusive, since A and B can occur (indeed, we determined the probability of $A \cap B$ in the previous example). To compute

$p(A \cup B)$ we can use (1–13). In the above example we found that $p(A) = \frac{11}{100} = 0.11$, $p(B) = \frac{30}{500} = 0.06$ and $p(A \cap B) = 0.0066$. Thus by (1–13)

$$p(A \cup B) = 0.11 + 0.06 - 0.0066 = 0.1634 .$$

The reader should note that to compute $p(A \cup B)$ it was unnecessary to construct explicitly a model for $\mathfrak{R}$. Only the models for $\mathfrak{R}_1$ and $\mathfrak{R}_2$ were needed. It was (1–13) which made this possible, and this then illustrates why (1–13) is often so useful.

3. One frequently encounters cases where the addition law (1–12) is incorrectly applied. Let us consider an example which the author actually encountered in working with the personnel of a petroleum company. The company was studying two tracts of land which it held for the possibility that they contained oil. One of these, call it 1, was an offshore tract and the other, call it 2, was located thousands of miles away in a foreign country. It had been estimated that the probability that 1 contained significant amounts of crude oil was 0.4 and the probability that 2 did was 0.3. It was then concluded that if drilling were begun on both, the probability that at least one venture would be successful should be $0.3 + 0.4 = 0.7$.

This conclusion is not correct because the executives did not consider the possibility that both ventures might be successful. If A is the event that 1 contains oil and B is the event that 2 does, then $p(A) = 0.4$ and $p(B) = 0.3$. Now A and B are not mutually exclusive. Indeed, we can compute the probability of $A \cap B$. Since the tracts are widely separated, there should be no connection between whether oil appears in 1 or in 2. We can imagine that a random experiment was performed millions of years ago which determined whether oil was placed in 1 and another random experiment which determined whether oil was placed in 2. Now these should have been independent random experiments, and thus

$$p(A \cap B) = p(A)p(B) = 0.3(0.4) = 0.12 .$$

Therefore by (1–13)

$$p(A \cup B) = 0.3 + 0.4 - 0.12 = 0.58 .$$

Hence the probability of $A \cup B$ which the company executives should have used is 0.58, not the larger value of 0.70.

4. Consider the random experiment in which we toss a coin twice. Assume that the probability of a head is p. We are not necessarily assuming that $p = \frac{1}{2}$, that is, we are not necessarily assuming that the coin is fair. Let us now construct a model $\mathcal{E}$ for $\mathfrak{R}$ using the theory developed in the previous section. Here $\mathfrak{R}$ can be thought of as a random experiment in which two independent trials of $\mathfrak{R}^*$ are made, $\mathfrak{R}^*$ being the random experiment in which the coin is tossed once. The model $\mathcal{E}^*$ for $\mathfrak{R}^*$ will be one in which there

are two simple events: H (the coin lands heads) and T (the coin lands tails). The probability of H is p and of T is $1 - p$.

For the model $\mathcal{E}$ of $\mathcal{R}$ we can use as the event set one with the four simple events $\langle H, H \rangle$, $\langle H, T \rangle$, $\langle T, H \rangle$, and $\langle T, T \rangle$, where, for example, $\langle H, T \rangle$ means that the coin landed heads up on the first toss and tails up on the second. The probabilities of each of these simple events can be determined using (1–16) and the model for $\mathcal{R}^*$. Thus the probability of $\langle H, H \rangle$ is $p(H)p(H) = p^2$, since p is the probability of H in $\mathcal{E}^*$. In Table 1–2, we have

TABLE 1–2

Simple event	$\langle H, H \rangle$	$\langle H, T \rangle$	$\langle T, H \rangle$	$\langle T, T \rangle$
Probability	p^2	$p(1-p)$	$p(1-p)$	$(1-p)^2$

given the probability of each simple event. Note that if the coin is fair and $p = \frac{1}{2}$, then each simple event has the probability $\frac{1}{4}$, which is the same result we would obtain using symmetry arguments. The present result is much more general, however, since if the coin is not fair, we cannot determine the probabilities by symmetry. The method introduced in the previous section is a very powerful one for constructing models of complex experiments because it is often the case that such experiments are built up from independent parts.

5. Let us return to the genetic model introduced in Section 1–9. We shall now consider the generalization of this model to include two different inherited characteristics. For example, let A and a genes refer to the color of a plant's flowers. We shall assume that the A gene is dominant and AA and Aa plants have colored flowers, while plants of type aa have white flowers. Also, let B and b genes refer to the length of the plant's stem. It will be assumed that the B gene is dominant and BB and Bb plants are tall, while plants of type bb are dwarf. We shall now construct a probability model to describe the nature of the offspring when a plant of type $AaBb$ is crossed with another plant of type $AaBb$.

To construct the model we shall have to introduce what is called Mendel's second law, which he also deduced from his experimental results. This law says that the random experiment which selects either an A or a type gene from a given parent is independent of the random experiment which selects a B or b gene to pass on to the offspring. Later work has shown that this law is not of universal validity, since many cases have been discovered where the genes are not passed on independently. We shall assume, however, that the second law of Mendel does apply to the case we are considering. There are then four possible sets of two genes which a given parent may pass on to the offspring: AB, Ab, aB and ab. We can think of these as representing the simple events for the random experiment which selects one gene of each type

from the parent to be given to the offspring. This experiment $\mathcal{R}$ is decomposable into two independent random experiments according to Mendel's second law. The first experiment $\mathcal{R}_1$ selects either an A or a gene, and according to Mendel's model the probability is $\frac{1}{2}$ that the gene selected is A. The second experiment $\mathcal{R}_2$ selects either a B or a b gene, and the probability of a B gene is $\frac{1}{2}$. To determine the probability of each simple event in $\mathcal{R}$ we multiply together the probabilities of the simple events from $\mathcal{R}_1$ and $\mathcal{R}_2$, respectively, which make up the simple event of $\mathcal{R}$. Each simple event in $\mathcal{R}_1$ and $\mathcal{R}_2$ has the probability $\frac{1}{2}$. Thus, each of the four simple events of $\mathcal{R}$ has the probability $\frac{1}{2}(\frac{1}{2}) = \frac{1}{4}$. We have thus constructed a model for $\mathcal{R}$.

The status of the offspring is the result of performing $\mathcal{R}$ twice, once for each parent. These trials of $\mathcal{R}$ are independent random experiments according to the Mendel model discussed in Section 1–9. We can then construct a model for the random experiment $\mathcal{R}^*$, which consists of independently performing $\mathcal{R}$ twice; $\mathcal{R}^*$ is the experiment which determines the composition of the genes in the offspring. There are sixteen simple events for $\mathcal{R}^*$: $\langle AABB \rangle$, $\langle AABb \rangle$, $\langle AAbB \rangle$, $\langle AAbb \rangle$, $\langle AaBB \rangle$, $\langle AaBb \rangle$, $\langle AabB \rangle$, $\langle Aabb \rangle$, $\langle aABB \rangle$, $\langle aABb \rangle$, $\langle aAbB \rangle$, $\langle aAbb \rangle$, $\langle aaBB \rangle$, $\langle aaBb \rangle$, $\langle aabB \rangle$ and $\langle aabb \rangle$. The first symbol in each pair gives the genes obtained from the first parent, so that $\langle aABb \rangle$ means that genes a and B were obtained from the first parent and A and b from the second. On multiplying together the probabilities of the simple events from $\mathcal{R}$ which make up a simple event of $\mathcal{R}^*$, we see that each simple event of $\mathcal{R}^*$ should have probability $\frac{1}{16}$.

The nine simple events $\langle AABB \rangle$, $\langle AABb \rangle$, $\langle AAbB \rangle$, $\langle AaBB \rangle$, $\langle AaBb \rangle$, $\langle AabB \rangle$, $\langle aABB \rangle$ and $\langle aABb \rangle$ all correspond to the offspring being a tall plant with colored flowers. The three simple events $\langle AAbb \rangle$, $\langle Aabb \rangle$ and $\langle aAbb \rangle$ correspond to the offspring being a dwarfed plant with colored flowers. The three simple events $\langle aaBB \rangle$, $\langle aaBb \rangle$ and $\langle aabB \rangle$ correspond to the offspring being a tall plant with white flowers. Finally, the single simple event $\langle aabb \rangle$ corresponds to the offspring being a dwarfed plant with white flowers. Thus from our model for $\mathcal{R}^*$, we conclude that the probability is $\frac{9}{16}$ that the offspring will be tall with colored flowers, is $\frac{3}{16}$ that it will be dwarfed with colored flowers, is $\frac{3}{16}$ that it will be tall with white flowers and finally is $\frac{1}{16}$ that it will be dwarfed with white flowers. Experiments show that these predicted relative frequencies are indeed correct. Note that to obtain the model for $\mathcal{R}^*$ we utilized the notion of independent experiments twice.

1–13 TWO-STAGE RANDOM EXPERIMENTS

In Section 1–11 we studied the case where $\mathcal{R}$ was built up from two independent random experiments. Here we wish to study the case where $\mathcal{R}$ again can be imagined to be built up from two random experiments. Now,

however, we shall not assume that the experiments are independent. Instead we shall consider the following sort of situation. $\Re$ will be imagined to be a random experiment in which first a random experiment $\Re_1$ is performed. After $\Re_1$ is completed, another experiment $\Re_2$ is performed. We shall suppose that the model to be used for $\Re_2$ is not independent of the outcome of $\Re_1$. Indeed, in general, a different model for $\Re_2$ will be needed for every possible outcome of $\Re_1$. An experiment $\Re$ of the type just described will be referred to as a two-stage experiment, $\Re_1$ being the first stage and $\Re_2$ the second stage.

As an example of the situation we have in mind, suppose that three bowls numbered 1, 2 and 3 are placed on a table. In bowl 1 are 5 red and 4 blue balls, in bowl 2 are 11 green and 7 yellow balls, and in bowl 3 are 2 green and 12 yellow balls. Let $\Re$ be the experiment in which first a ball is drawn at random from bowl 1. If it is red, a ball is then drawn at random from bowl 2; however, if the ball drawn from bowl 1 is blue, then one draws a ball at random from bowl 3 instead of bowl 2. Thus the bowl from which a ball is drawn at the second stage depends on the outcome of the first stage, and for each of the two possible outcomes of the first stage one needs a different model for the second stage.

We shall now show how to construct a model $\mathcal{E}$ for $\Re$ from the model for $\Re_1$ and the models for $\Re_2$. Let $\mathcal{E}_1$ be the model for $\Re_1$ and suppose that the event set for $\mathcal{E}_1$ is $E_1 = \{e_1, \ldots, e_m\}$, with the probabilities of the simple events being $p_1, \ldots, p_m$. To be specific we shall suppose that the event set for the model of $\Re_2$ will be $E_2 = \{f_1, \ldots, f_s\}$, regardless of which simple event e_i occurs when $\Re_1$ is performed. However, we shall assume that the probabilities of the simple events will depend on the outcome of $\Re_1$. The probabilities of $f_1, \ldots, f_s$ when $\Re_1$ yields the simple event e_i will be denoted by $q_{i1} \ldots, q_{is}$, respectively. These probabilities, in general, will change with i.

As a simple event for $\Re$ we can specify which simple event occurred on performing $\Re_1$ and which simple event occurred on performing $\Re_2$. Thus if $\Re_1$ yielded e_i and $\Re_2$ yielded f_j, the corresponding simple event for $\Re$ will be denoted by $\langle e_i, f_j \rangle$. There will then be ms simple events $\langle e_i, f_j \rangle$ for $\Re$. Let us now determine what probability should be assigned to $\langle e_i, f_j \rangle$. To do this we use the intuitive notion of probability as being a long-run relative frequency. Suppose that $\Re$ is performed n times and suppose that n_{ij} is the number of times $\langle e_i, f_j \rangle$ occurs, that is, the number of times that $\Re_1$ yields e_i and, in addition, $\Re_2$ yields f_j. Then n_{ij}/n approaches the probability of $\langle e_i, f_j \rangle$ as n gets larger and larger. If n_i is the number of times that $\Re_1$ yielded e_i, then we can write

$$\frac{n_{ij}}{n} = \left(\frac{n_i}{n}\right)\left(\frac{n_{ij}}{n_i}\right). \tag{1–19}$$

Now n_i/n is just the relative frequency of e_i in n trials of $\mathfrak{R}$ and hence in n trials of $\mathfrak{R}_1$. However, the long-run relative frequency of e_i on performing $\mathfrak{R}_1$ is p_i, according to the model for $\mathfrak{R}_1$, and thus n_i/n should approach p_i as n is increased unendingly. Next, n_{ij}/n_i is the fraction of those cases where $\mathfrak{R}_1$ yielded e_i that $\mathfrak{R}_2$ yielded f_j, so that n_{ij}/n_i is the relative frequency of f_j in $\mathfrak{R}_2$ in those cases where $\mathfrak{R}_1$ yielded e_i. Now according to the model for $\mathfrak{R}_2$ which is appropriate when $\mathfrak{R}_1$ yields e_i, the probability of f_j is q_{ij} and hence n_{ij}/n_i should approach q_{ij}. Thus from (1–19) we see that n_{ij}/n should approach $p_i q_{ij}$. Hence the probability that we shall assign to $\langle e_i, f_j \rangle$ is $p_i q_{ij}$, and we have developed a model for $\mathfrak{R}$.

In the above we assumed that the event set for $\mathfrak{R}_2$ was independent of the outcome of $\mathfrak{R}_1$ only for simplicity of explanation. If f_k cannot occur when

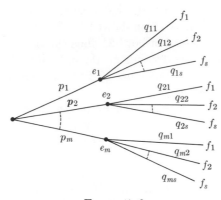

FIGURE 1–6

e_u does, we simply omit the simple event $\langle e_u, f_k \rangle$ in the model for $\mathfrak{R}$ or, alternatively, we assign a probability of zero to this simple event.

A very convenient way to visualize the model for a two-stage experiment $\mathfrak{R}$ is to construct a diagram like that shown in Figure 1–6. This diagram is referred to as a *tree*, and the straight lines are called *branches*. To represent $\mathfrak{R}_1$ we draw m branches, one for each simple event of $\mathfrak{R}_1$. We draw these branches so that they emanate from a single point, called a *node*. Each branch is labeled with the probability that the corresponding simple event will occur. To represent $\mathfrak{R}_2$ for each outcome of $\mathfrak{R}_1$, one draws from the end of each branch representing the outcome of $\mathfrak{R}_1$ a total of s branches, one for each possible outcome f_j of $\mathfrak{R}_2$. These are labeled with the probability that f_j will occur when $\mathfrak{R}_2$ is performed. To obtain the probability $p_i q_{ij}$ of any simple event $\langle e_i, f_j \rangle$, one merely multiplies together the probabilities on the two branches, the first of which represents $\mathfrak{R}_1$ yielding e_i and the second of which represents $\mathfrak{R}_2$ yielding f_j when $\mathfrak{R}_1$ yielded e_i.

Most of the problems involving two-stage experiments which will be of

interest to us will either require the computation of the probability of the event $e_i \cap f_j$, that is, the event that $\mathfrak{R}_1$ yields e_i and $\mathfrak{R}_2$ yields f_j, or the event that $\mathfrak{R}_2$ yields f_j. We can now easily see how to determine these probabilities. The event $e_i \cap f_j$ is nothing but the simple event $\langle e_i, f_j \rangle$ of $\mathfrak{R}$ and has the probability $p_i q_{ij}$. The simple events of $\mathfrak{R}$ for which $\mathfrak{R}_2$ yields f_j are

$$\langle e_1, f_j \rangle, \langle e_2, f_j \rangle, \ldots, \langle e_m, f_j \rangle .$$

Hence by (1–10), the probability that $\mathfrak{R}_2$ yields f_j on performing $\mathfrak{R}$ is

$$p(f_j) = p_1 q_{1j} + p_2 q_{2j} + \cdots + p_m q_{mj} . \tag{1–20}$$

In the next section we shall give examples illustrating how useful these results are.

From what we have noted in the previous paragraph, $p(e_i \cap f_j) = p_i q_{ij}$. Now the probability that $\mathfrak{R}_1$ yields e_i is simply p_i, and this should also be the probability of the event e_i occurring on performing $\mathfrak{R}$, since $\mathfrak{R}_2$ has no influence on the outcome on $\mathfrak{R}_1$. This is indeed true for our model $\mathcal{E}$ since the simple events for which e_i occurs are $\langle e_i, f_1 \rangle, \ldots, \langle e_i, f_s \rangle$. The probability of e_i for $\mathfrak{R}$ is then found by summing the probabilities of these simple events and is

$$p_i q_{i1} + p_i q_{i2} + \cdots + p_i q_{is} = p_i(q_{i1} + q_{i2} + \cdots + q_{is}) = p_i$$

as expected, since

$$q_{i1} + q_{i2} + \cdots + q_{is} = 1 ,$$

inasmuch as the probabilities of the simple events for the model of $\mathfrak{R}_2$ sum to 1. We can thus write $p(e_i \cap f_j) = p_i q_{ij}$ as $p(e_i \cap f_j) = p(e_i)q_{ij}$. Hence

$$q_{ij} = \frac{p(e_i \cap f_j)}{p(e_i)} . \tag{1–21}$$

Recall that q_{ij} is the probability that $\mathfrak{R}_2$ will yield the simple event f_j when $\mathfrak{R}_1$ yields e_i. With respect to the model for $\mathfrak{R}$, q_{ij} is called *the conditional probability that $\mathfrak{R}_2$ will yield f_j given that $\mathfrak{R}_1$ yields e_i*, and an alternative notation for q_{ij} often used is $p(f_j | e_i)$. Thus (1–21) can be written

$$p(f_j | e_i) = \frac{p(e_i \cap f_j)}{p(e_i)} . \tag{1–22}$$

The notion of conditional probability which we have just introduced can be given a more general interpretation as follows. Consider any random experiment $\mathfrak{R}$ (not necessarily a two-stage random experiment) and let A_1 and A_2 be any events of $\mathfrak{R}$, that is, any subsets of the event set being used in the model of $\mathfrak{R}$. Denote by $p(A_1 \cap A_2)$ the probability of $A_1 \cap A_2$ and by $p(A_1)$ the probability of A_1, which we shall assume is positive. *Then the*

number $p(A_1 \cap A_2)/p(A_1)$ is called the conditional probability of A_2 given A_1 and is denoted by $p(A_2|A_1)$. Thus by definition

$$p(A_2|A_1) = \frac{p(A_1 \cap A_2)}{p(A_1)}. \tag{1-23}$$

As we have shown in our above development, conditional probabilities arise naturally in treating two-stage random experiments. A more general intuitive interpretation of conditional probabilities will be given in Section 1–15.

1–14 EXAMPLES

We shall now give some examples to illustrate the usefulness of the notion of a two-stage random experiment introduced in the previous section.

1. A bowl contains five red, four blue and seven yellow balls. Consider the random experiment $\Re$ in which two balls are drawn at random, one after the other. Let us determine the probability that the first ball selected is red and second is yellow. $\Re$ can be visualized as a two-stage random experiment in which the first stage $\Re_1$ consists of selecting at random the first ball from the bowl and the second stage $\Re_2$ involves selecting at random a second ball from the remaining balls. In accordance with the discussion on page 23 we can use as the event set for $\Re_1$ the set $E = \{R, B, Y\}$. Furthermore, after a ball is drawn out of the bowl, there are still red, blue and yellow balls remaining. Thus E can also be used as the event set for $\Re_2$. Now if we use the model developed in the previous section, the event set for $\Re$ will consist of nine simple events, $\langle R, R \rangle$, $\langle R, B \rangle$, $\langle R, Y \rangle$, $\langle B, R \rangle$, $\langle B, B \rangle$, $\langle B, Y \rangle$, $\langle Y, R \rangle$, $\langle Y, B \rangle$ and $\langle Y, Y \rangle$. The meaning of $\langle B, Y \rangle$, for example, is that a blue ball was drawn on performing $\Re_1$ and then a yellow ball on performing $\Re_2$.

We can next note that the event whose probability we wish to compute is the simple event $\langle R, Y \rangle$, since the event of interest is that where a red ball is drawn first and then a yellow one. The probability of $\langle R, Y \rangle$ should be, as we showed in the previous section, the probability of drawing a red ball on performing $\Re_1$ times the probability of drawing a yellow ball on the performing $\Re_2$, given that a red ball was indeed drawn on performing $\Re_1$. If we denote the first of these probabilities by $p(R)$ and the second by $p(Y|R)$, the desired probability is then $p(Y|R)p(R)$. When $\Re_1$ is performed there are sixteen balls in the bowl, five of which are red. Hence, if a ball is drawn at random, $p(R) = \frac{5}{16}$. If a red ball is drawn on performing $\Re_1$, there will be fifteen balls remaining, seven of which are yellow. Hence, if a ball is again drawn at random, the probability of a yellow ball on the second draw should be $\frac{7}{15}$, that is, $p(Y|R) = \frac{7}{15}$. Thus the probability of $\langle R, Y \rangle$ is $(\frac{5}{16})(\frac{7}{15}) = \frac{7}{48} = 0.146$.

2. For the situation considered in Example 1 let us compute the probability that one of the balls drawn will be red and the other yellow. The event of interest now consists of two simple events of $\mathfrak{R}$, $\langle R, Y \rangle$ and $\langle Y, R \rangle$, since we must either get a red ball on the first draw and then a yellow one or first a yellow ball and then a red one. We have already determined the probability of $\langle R, Y \rangle$. Let us next determine the probability of $\langle Y, R \rangle$. Initially there are seven yellow balls so that the probability of drawing a yellow one on performing $\mathfrak{R}_1$ is $p(Y) = \frac{7}{16}$. If a yellow ball is drawn first, there are fifteen balls remaining, five of which are red. Hence the probability of drawing a red ball on performing $\mathfrak{R}_2$ if a yellow one was drawn first is $p(R|Y) = \frac{5}{15}$, so the probability of $\langle Y, R \rangle$ is $(\frac{7}{16})(\frac{5}{15}) = 0.146$. The probability that one ball is yellow and one red is then the sum of the probabilities of $\langle R, Y \rangle$ and $\langle Y, R \rangle$, which is $2(0.146) = 0.292$.

3. For the situation discussed in Example 1, let us determine the probability that two red balls are drawn. Now we wish to compute the probability of the simple event $\langle R,R \rangle$. We have seen previously that the probability of drawing a red ball on performing $\mathfrak{R}_1$ is $\frac{5}{16}$. If a red ball is drawn on performing $\mathfrak{R}_1$, then four of the remaining fifteen balls are red, so that the probability of drawing a red ball on performing $\mathfrak{R}_2$ given that a red one was obtained on performing $\mathfrak{R}_1$ should be $\frac{4}{15}$, and hence the desired probability is $(\frac{5}{16})(\frac{4}{15}) = \frac{1}{12} = 0.0835$.

4. One of two drawers in a desk, call it drawer 1, contains one gold and one silver coin. The other drawer, call it 2, contains one gold and two silver coins. An individual selects one of the two drawers at random, opens the drawer, and selects at random one of the coins there. What is the probability that he selects a gold coin? We can think of this random experiment $\mathfrak{R}$ as a two-stage experiment where the first stage involves the selection at random of a drawer and the second stage involves the selection at random of a coin in the drawer chosen at the first stage. As the simple events for $\mathfrak{R}$ we can use $\langle 1, G \rangle$, $\langle 1, S \rangle$, $\langle 2, G \rangle$, and $\langle 2, S \rangle$, where, for example $\langle 1, S \rangle$ means that drawer 1 was selected at the first stage and then at the second stage a silver coin was chosen from drawer 1. We wish to compute the probability that a gold coin was selected. This is the event $\{\langle 1, G \rangle, \langle 2, G \rangle\}$. Now the probability of selecting a gold coin in drawer 1 is $\frac{1}{2}$, so that the probability of $\langle 1, G \rangle$ is $\frac{1}{4}$. The probability of selecting a gold coin in drawer 2 is $\frac{1}{3}$ so that the probability of $\langle 2, G \rangle$ is $\frac{1}{6}$. Therefore, the probability of selecting a gold coin is

$$\tfrac{1}{4} + \tfrac{1}{6} = 0.250 + 0.167 = 0.417 \, .$$

A graphic representation of $\mathfrak{R}$ using a tree is shown in Figure 1–7. Note that what we were doing here was computing the probability that the second stage yielded a gold coin. Thus, this example represented an application of

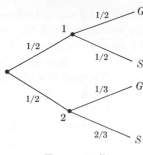

FIGURE 1–7

(1–20). It should be observed that the chances of selecting any one of the five coins are not equal, so that the probability of a gold coin is not $\frac{2}{5} = 0.4$, even though this is close to the correct probability.

1–15 CONDITIONAL PROBABILITY MODELS

Consider some arbitrary random experiment $\Re$ and let $\mathcal{E}$ be a suitable model for $\Re$. Suppose now that $\Re$ is performed, and either while $\Re$ is taking place or after $\Re$ is completed, we are given some information about how $\Re$ is turning out or has turned out, but we are not given complete information about the outcome of $\Re$, that is, we are not told which simple event will or has occurred. Given this additional infomation, our original probability model may be an inappropriate one to use in making any further computations which need to be done. The question then arises as to how we can proceed to develop a new model $\mathcal{E}_c$ which will incorporate the additional information about the outcome of $\Re$. The new model will be called a *conditional probability model*. The word conditional is used because $\mathcal{E}_c$ makes use of information about what has actually happened on performing $\Re$. We shall show that it is normally possible to construct the model $\mathcal{E}_c$ in a straightforward way from $\mathcal{E}$. No new investigations have to be made to determine the probabilities of the simple events in $\mathcal{E}_c$.

The first step in constructing $\mathcal{E}_c$ is to characterize the type of information made available on the outcome of $\Re$. The important thing to note is that the only information which can be made available is that some event A occurred. Now A should be a subset of the event set E in the model $\mathcal{E}$ of $\Re$, if $\mathcal{E}$ is a realistic model of $\Re$. The information that A has occurred then implies that the simple event of $\Re$ that has or will occur is one of the simple events in A. Thus the appropriate event set to use for $\mathcal{E}_c$ is nothing but the set A. Let us next see what probability should be assigned to each simple event e_k which is in A. As usual, the procedure for doing this will be to use the intuitive interpretation of probability. Now the probability to be

assigned to e_k in $\mathcal{E}_c$ should be the long-run fraction of the time that e_k occurs when A does. Note that since $e_k \in A$, e_k cannot occur unless A does.

Imagine that $\mathcal{R}$ is performed a large number of times n and suppose that A and e_k occur n_A and n_k times, respectively. Then n_k/n_A should approach the probability of e_k in $\mathcal{E}_c$, which we shall denote by $p(e_k|A)$. Now

$$\frac{n_k}{n_A} = \frac{n_k}{n} \frac{n}{n_A} = \frac{n_k/n}{n_A/n}, \tag{1–24}$$

and as n gets larger and larger n_k/n approaches p_k, the probability assigned to e_k in $\mathcal{E}$, and n_A/n approaches $p(A)$, the probability of A computed from $\mathcal{E}$. We shall assume $p(A) \neq 0$. Then by (1–24) we see that

$$p(e_k|A) = \frac{p_k}{p(A)}, \quad e_k \in A . \tag{1–25}$$

We have now seen how to construct the model $\mathcal{E}_c$. The event set is A, and the probability of each e_k contained in A is given by (1–25), where p_k and $p(A)$ are determined from the *original model* $\mathcal{E}$.

Let us next note that $e_k = e_k \cap A$ so that (1–25) can be written

$$p(e_k|A) = \frac{p(e_k \cap A)}{p(A)} . \tag{1–26}$$

If we compare this with (1–23), we see that the probability which should be assigned to e_k in $\mathcal{E}_c$ is nothing but what we have previously defined to be the conditional probability of e_k given A. It is instructive to pursue the connection between conditional probabilities and conditional probability models a little bit more. Let B be any event of $\mathcal{R}$ and suppose that we wish to compute the probability of B using the new model $\mathcal{E}_c$. B can occur only if it has at least one simple event in common with A. The set of simple events which B has in common with A is nothing but $A \cap B$. Let us denote by $p(B|A)$ the probability of B computed from $\mathcal{E}_c$. But $p(B|A)$ is the sum of the $p(e_k|A)$ for $e_k \in A \cap B$. However, by (1–26), the sum of the $p(e_k|A)$ for $e_k \in A \cap B$ is the sum of the p_k for $e_k \in A \cap B$ divided by $p(A)$. Now the sum of the p_k for $e_k \in A \cap B$ is $p(A \cap B)$, the probability of $A \cap B$ com- computed from $\mathcal{E}$. Therefore

$$p(B|A) = \frac{p(A \cap B)}{p(A)}, \tag{1–27}$$

so that the probability of B computed from $\mathcal{E}_c$ is the conditional probability of B given A. We can now give a general and very useful intuitive interpretation of the notion of conditional probability introduced in Section 1–12. The conditional probability of B given A can be interpreted as the probability of B which would be computed from a conditional probability model constructed on the knowledge that A has occurred.

Equation (1–27) often provides a useful way to compute $p(A \cap B)$. One writes $p(A \cap B) = p(B|A)p(A)$. In Section 1–12 we showed that if $\mathcal{R}$ is a two-stage random experiment with e_i a simple event of $\mathcal{R}_1$ and f_j a simple event of $\mathcal{R}_2$, then $p(e_i \cap f_j) = p(f_j|e_i)p(e_i)$. From what we have now seen it follows that if A is any event of $\mathcal{R}_1$ and B any event of $\mathcal{R}_2$, then $p(A \cap B) = p(B|A)p(A)$.

1–16 EXAMPLES

We shall next give several examples illustrating the construction and use of conditional probability models.

1. A bowl contains ten balls numbered 1 to 10. Balls 1, 2 and 3 are red and the remainder are blue. Consider the random experiment $\mathcal{R}$ in which a ball is selected at random and the number on the ball is noted. As a model $\mathcal{E}$ for $\mathcal{R}$ we can use one with ten simple events, e_j being the event that ball j is drawn. Since the ball is drawn at random each of these simple events has the probability 0.1. Thus the probability that ball j is drawn is 0.1. Suppose now that a ball is drawn, and we are told that the ball is red. Let us then determine the probability that ball 1 is drawn if it is known that the ball drawn is red. The information given us is that a red ball has been drawn; this is the event $A = \{e_1, e_2, e_3\}$ and $p(A) = 0.3$. The conditional model $\mathcal{E}_c$ will then have A as the event set, and by (1–25), the probabilities of e_1, e_2 and e_3 in $\mathcal{E}_c$ are $\frac{0.1}{0.3} = \frac{1}{3}$. Thus the probability that ball 1 was drawn if we know that the ball selected was red is $\frac{1}{3}$.

2. An individual tosses a fair coin three times and tells us later that precisely one head was obtained in the three tosses. Let us determine the probability that a head was obtained on the first toss. As a model $\mathcal{E}$ for this experiment one can use an event set with the eight simple events: $\langle H, H, H \rangle$, $\langle H, H, T \rangle$, $\langle H, T, H \rangle$, $\langle H, T, T \rangle$, $\langle T, H, H \rangle$, $\langle T, H, T \rangle$, $\langle T, T, H \rangle$, $\langle T, T, T \rangle$, where, for example, $\langle T, H, H \rangle$ means a tail was obtained on the first toss and a head on the others. Either using symmetry or by visualizing $\mathcal{R}$ as being built up from three independent random experiments, we see that each simple event should have the probability $\frac{1}{8}$. We are now told that precisely one head occurred in three tosses. This is the event

$$A = \{\langle H, T, T \rangle, \langle T, H, T \rangle, \langle T, T, H \rangle\},$$

and $p(A) = \frac{3}{8}$.

We wish to compute the probability that a head was obtained on the first, given that one head was obtained in three tosses. The event set for the conditional model $\mathcal{E}_c$ that we shall use to make this computation has A as its event set, and the probability of each simple event of A for $\mathcal{E}_c$ is, by (1–25), $(\frac{1}{8})/(\frac{3}{8}) = \frac{1}{3}$. In particular $\langle H, T, T \rangle$ has the probability $\frac{1}{3}$, and this is what we wished to compute.

3. Imagine that we determine from historical data the fraction of males who die before they are 20, who die between 20 and 40, who die between 40 and 60, who die between 60 and 80 and who die after the age of 80. One can, at the time a new male baby is born, imagine that the time of his death will be the result of a random experiment. Let us construct a model for this experiment by using an event set containing five simple events, where each simple event refers to one of the time intervals referred to above, so that, for example, e_1 is the event that death will occur before the age of 20. Historical data then suggest that $p(e_1) = 0.077$, $p(e_2) = 0.054$; $p(e_3) = 0.191$, $p(e_4) = 0.480$ and $p(e_5) = 0.198$.

Consider a young man who has just turned 20. What is the probability that he will die between 40 and 80? To compute this we must construct a conditional probability model. What information have we been given? We know that the young man did not die before 20, so that the event $A = \{e_2, e_3, e_4, e_5\}$ has occured, that is, the event that he will die after 20. This will then be the event set for $\mathcal{E}_c$. Furthermore, since $p(A) = 0.923$, we see that the probabilities of the simple events in $\mathcal{E}_c$ are

$$p(e_2|A) = \frac{0.054}{0.923} = 0.059; \quad p(e_3|A) = \frac{0.191}{0.923} = 0.206;$$

$$p(e_4|A) = \frac{0.480}{0.923} = 0.520; \quad p(e_5|A) = \frac{0.198}{0.923} = 0.215 .$$

We wish to compute from $\mathcal{E}_c$ the probability of the event $\{e_3, e_4\}$, which is $0.206 + 0.520 = 0.726$.

4. Suppose that an individual grown to maturity has a sister or brother (often abbreviated sib) who is afflicted with a serious hereditary disease, but that the individual himself is not so afflicted. Let us determine the probability that the individual is a carrier of the lethal gene. We shall assume that neither the mother or father of the individual was afflicted. However, as we noted in Section 1–9, both the mother and father must be carriers, that is, must be of Aa types. We showed in Section 1–9 that, at the time of his birth, the probabilities that the individual is of type AA, Aa or aa are respectively $\frac{1}{4}$, $\frac{1}{2}$ and $\frac{1}{4}$. We are now given the additional information that the individual has grown to maturity and is not afflicted. Thus the event $\{AA, Aa\}$ has occurred, the probability of this event being $\frac{3}{4}$. The conditional probability model for the individual's genotype will then have $\{AA, Aa\}$ as the event set. The probability of AA is, by (1–25), $\frac{1}{4}/\frac{3}{4} = \frac{1}{3}$ and the probability of Aa is then $\frac{2}{3}$. Hence the probability that he is a carrier is $\frac{2}{3}$.

5. We shall now give an example from genetics which illustrates how the use of independence, two-stage experiments and conditional probability models can all enter into the solution of a single problem. Suppose that it is known that the aunt of an individual (his mother's sister in this case) is afflicted

with a serious hereditary disease. However, neither the individual himself nor his mother is so afflicted. Furthermore, his father is not afflicted and, so far as is known, none of his father's relatives is afflicted. Let us determine the probability that the individual under consideration is a carrier of the lethal gene.

Since the individual's mother has an afflicted sib, we know from the previous example that the appropriate model for the mother is that given in Table 1–3. How shall we build a corresponding model for the father? Since

TABLE 1–3

MODEL FOR MOTHER

Genotype	AA	Aa
Probability	$\frac{1}{3}$	$\frac{2}{3}$

there is no known history of the disease in the father's family, the best model we can construct is merely to assume that the father was selected at random from the entire healthy population. Denote by f the fraction of the healthy population which are carriers. Then the model to be used for the father will be that given in Table 1–4.

TABLE 1–4

MODEL FOR FATHER

Genotype	AA	Aa
Probability	$1 - f$	f

Now presumably there is no connection between the genotypes of mother and father. Consequently, we can imagine that the experiment which determines the combined genotypes of the mother and father is built up from two independent random experiments, the models for which are given in Tables 1–2 and 1–3. The theory of Section 1–1 then yields the model shown in Table 1–5 for the father–mother pair. The first pair of genes refers to the father and the second to the mother, so that $AAAa$ means that the father is AA and the mother Aa. The probability of each simple event is found by the product rule, multiplying together the appropriate probabilities from Tables 1–3 and 1–4. We shall refer to the experiment whose model is given in Table 1–5 as $\Re_1$.

TABLE 1–5

MODEL FOR FATHER–MOTHER COMBINATION

Event	$AAAA$	$AAAa$	$AaAA$	$AaAa$
Probability	$\frac{1}{3}(1 - f)$	$\frac{2}{3}(1 - f)$	$\frac{1}{3}f$	$\frac{2}{3}f$

We are interested in determining the probability that the individual under consideration is a carrier. This can be thought of as the result of a two-stage experiment. First the experiment $\mathfrak{R}_1$ determines the genotypes of the father and mother. The model for $\mathfrak{R}_1$ is given in Table 1–5. Then there is the experiment $\mathfrak{R}_2$ which determines the genotype of the individual under consideration. The model to be used for $\mathfrak{R}_2$ will depend on the outcome of $\mathfrak{R}_1$, that is, on the genotypes of the mother and father. Four different models are needed for $\mathfrak{R}_2$, one for each possible outcome of $\mathfrak{R}_1$. These models are given in Table 1–6. The model of $\mathfrak{R}_2$ corresponding to any par-

TABLE 1–6

MODELS FOR $\mathfrak{R}_2$

$\mathfrak{R}_1$ \ $\mathfrak{R}_2$	AA	Aa	aa
$AAAA$	1	0	0
$AAAa$	$\frac{1}{2}$	$\frac{1}{2}$	0
$AaAA$	$\frac{1}{2}$	$\frac{1}{2}$	0
$AaAa$	$\frac{1}{4}$	$\frac{1}{2}$	$\frac{1}{4}$

ticular outcome of $\mathfrak{R}_1$ is given in the line corresponding to that outcome of $\mathfrak{R}_1$. The models are obtained immediately from the theory of Section 1–9. The model for $\mathfrak{R}_2$ has three simple events, in general, which are that the individual is AA, Aa or aa. However, for some parental combinations not all of these can actually occur. This is indicated in Table 1–5 by assigning a probability of zero to the corresponding event.

There are a total of twelve simple events for the two-stage experiment

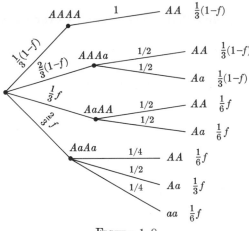

FIGURE 1–8

$\mathfrak{R}^*$ if those with a zero probability are included, or eight simple events if only those with a positive probability are included. The tree for $\mathfrak{R}^*$ is shown in Figure 1–8. Only branches with positive probabilities are shown. The probabilities of each of the eight simple events are shown at the ends of the corresponding branches. These probabilities are obtained by multiplying together the numbers on the two branches which lead from the initial node to the desired number.

To determine the probabilities at birth that the individual under consideration will be AA, Aa or aa, we add together the probabilities of all the simple events of $\mathfrak{R}^*$ having the desired characteristic. Thus the probability that the individual will be AA is

$$\frac{1}{3}(1-f) + \frac{1}{3}(1-f) + \frac{1}{6}f + \frac{1}{6}f = \frac{1}{3}(2-f)$$

and the probability that he will be Aa is

$$\frac{1}{3}(1-f) + \frac{1}{6}f + \frac{1}{3}f = \frac{1}{6}(2+f).$$

Finally, the probability he will be aa is $\frac{1}{6}f$.

Now the probability we wish to compute is not $\frac{1}{6}(2+f)$, because this is the probability that the individual will be Aa at birth. We have been given the information that the individual has reached maturity and is not afflicted. Thus he must be either AA or Aa, and we desire the conditional probability of Aa given that he is healthy. The probability of the event $\{AA, Aa\}$ is

$$\frac{1}{3}(2-f) + \frac{1}{6}(2+f) = 1 - \frac{1}{6}f.$$

Therefore the probability that the individual is a carrier is

$$\frac{\frac{1}{3} + \frac{1}{6}f}{1 - \frac{1}{6}f},$$

since this is the probability of Aa for a model in which $\{AA, Aa\}$ is known to have occurred. Frequently, f is very small, and when this is true, the probability is about $\frac{1}{3}$ that the individual is a carrier.

1–17 COMBINATORIAL ANALYSIS

Before going on it will be desirable to examine some simple counting problems, which as a class are referred to as *combinatorial analysis*. Imagine

that n balls numbered 1 to n are placed in a box. Now a ball is drawn from the box and the number on it is written down. The ball is then placed aside (it is not put back in the box). Next a second ball is drawn, the number noted, and the ball placed aside. This is continued until a total of r balls have been drawn from the box. The process just described will generate a sequence of r numbers which we shall symbolize by $a_1, \ldots, a_r$; a_1 gives the number on the first ball drawn and a_r the number on the final one drawn. For example, if there are six balls and four are drawn, the sequence might be 3, 1, 2, 6. Let us now proceed to compute how many different sequences of numbers can be generated in this way.

The problem is easily solved. The first ball can have any one of n different numbers on it, so that a_1 can have n different values. Once the first ball is drawn, there are $n - 1$ balls left. Hence once a_1 is determined a_2 can have any of $n - 1$ values. Since this is true for each value of a_1, there are $n(n - 1)$ possible sequences for the first two elements. Given a_1 and a_2, there are $n - 2$ balls remaining in the box, and hence a_3 can have $n - 2$ different values. This is true for every pair of values of a_1 and a_2. Hence there are $n(n - 1)(n - 2)$ possible sequences for the first three elements. When we are ready to draw out the rth ball, $r - 1$ have been drawn out previously and $n - r + 1$ remain. Given $a_1, \ldots, a_{r-1}$, there are then $n - r + 1$ possible values that a_r can have, or there is a total of

$$n(n - 1)(n - 2) \cdots (n - r + 1) \qquad (1\text{–}28)$$

different sequences $a_1, \ldots, a_r$ which can be generated.

Suppose that all the balls are drawn out, so that $r = n$. Then (1–28) reduces to $n(n - 1) \cdots (2)(1)$. We use a special symbol $n!$ to denote this particular product of n numbers. Thus by definition

$$n! = n(n - 1)(n - 2) \cdots (2)(1) . \qquad (1\text{–}29)$$

The symbol $n!$ is read n factorial and is called a factorial symbol. As an example of (1–29),

$$5! = 5(4)(3)(2)(1) = 120 .$$

Note that $n!$ is defined for every positive integer n. It is also convenient to introduce the definition

$$0! = 1 . \qquad (1\text{–}30)$$

Now observe that (1–29) can be written

$$n! = n(n - 1) \cdots (n - r + 1)(n - r) \cdots (2)(1)$$
$$= n(n - 1) \cdots (n - r + 1)(n - r)! \qquad (1\text{–}31)$$

when $1 \le r \le n$. On comparison of (1–28) with (1–31), we see that (1–28) can be written $n!/(n - r)!$ The number of sequences of r numbered balls

that can be selected from n balls is called the *number of permutations of n balls taken r at a time* and is sometimes denoted symbolically by P_r^n or $(n)_r$. Any particular sequence of balls is referred to as a permutation. From what we have just shown

$$P_r^n = (n)_r = \frac{n!}{(n-r)!}. \tag{1–32}$$

EXAMPLES. *1.* Suppose that there are three balls in the box referred to above and we draw out two of them. According to (1–28), there are $3(2) = 6$ sequences of numbers which could be obtained. These are $(1, 2)$, $(2, 1)$, $(1, 3)$, $(3, 1)$, $(2, 3)$ and $(3, 2)$. If all three balls are drawn out, there are $3(2)(1) = 6$ sequences again, which are $(1, 2, 3)$, $(1, 3, 2)$, $(2, 1, 3)$, $(2, 3, 1)$, $(3, 1, 2)$ and $(3, 2, 1)$.

2. Suppose that there are four balls in the box and two of them are drawn out. According to (1–28), there are $4(3) = 12$ possible sequences of numbers which could be obtained. These are $(1, 2)$, $(2, 1)$, $(1, 3)$, $(3, 1)$, $(2, 3)$, $(3, 2)$, $(1, 4)$, $(4, 1)$, $(2, 4)$, $(4, 2)$, $(3, 4)$ and $(4, 3)$.

3. A great many different types of problems are mathematically equivalent to the type of problem we have been studying in this section. For example, the number of different ways one could assign r executives to n regional offices, $r \leq n$, and one executive to an office is P_r^n, since the first executive can be assigned to any one of n offices, and given this assignment, there are $n - 1$ offices left to which the next executive can be assigned, and so on. The offices correspond to the numbered balls, and the executives serve to define the order of drawing the balls.

In determining the number of permutations of n balls taken r at a time, ordering is important. Thus in the first example above where two balls were drawn, $(2, 3)$ and $(3, 2)$ were different permutations. Let us now study a slightly different problem. Suppose that once again we draw r balls from a box containing n balls numbered 1 to n. Now, however, we shall concern ourselves only with the numbers on the balls drawn and not with the order in which the numbers were obtained. We then wish to determine how many different sets of numbers are possible when we do not care about the order in which the numbers were obtained. When order is irrelevant, $(2, 3)$ and $(3, 2)$ yield the same pair of numbers and would not be considered to be different sets. When two balls are drawn from a box of three balls there are just three pairs of numbers which can be obtained. These are 1 and 2, 2 and 3, and 1 and 3. The number of sets of numbers which can be obtained on selecting r balls from n is called the *number of combinations of n balls taken r at a time*. To determine the number of combinations all we need to do is determine the number of permutations and then combine all permutations

which involve the same set of numbers. There are precisely $r!$ permutations of any given set of r numbers so that there are $r!$ permutations for each combination. Inasmuch as there are a total of P_r^n permutations, we see that if C_r^n is the number of combinations then $r!\,C_r^n = P_r^n$ or

$$C_r^n = \frac{P_r^n}{r!} = \frac{n!}{r!(n-r)!}, \qquad (1\text{--}33)$$

and we have determined the number of different sets of numbers which can appear on the r balls drawn.

EXAMPLES. *1.* From a box containing five balls numbered 1 to 5, three balls are drawn. How many different combinations of three numbers may be observed on the balls drawn? According to (1–33), this number is

$$C_3^5 = \frac{5!}{3!2!} = \frac{5(4)(3)(2)(1)}{3(2)(1)(2)(1)} = 10\,.$$

Let us list these ten combinations. We shall use a format such as 123. No order is implied here; we are simply listing the numbers observed on the balls. The combinations are then: 123, 124, 125, 134, 135, 145, 234, 235, 245 and 345. There are 60 permutations in this case, since for each combination such as 145, there are six permutations or orders in which the numbers could be obtained. For 145 these are $(1, 4, 5)$, $(1, 5, 4)$, $(4, 1, 5)$, $(4, 5, 1)$, $(5, 1, 4)$ and $(5, 4, 1)$.

2. The reader should note that the results we have obtained are independent of what numbers are painted on the balls, so long as a different number appears on each ball. Suppose that we had four balls numbered 2, 5, 7 and 9. If two of them were selected, the number of combinations of numbers which could be obtained is $4!/2!2! = 6$. These are 25, 27, 29, 57, 59 and 79.

3. Suppose that we have ten numbered balls in a box and we draw out six of them. How many combinations of numbers can be observed? Here

$$C_6^{10} = \frac{10!}{6!4!} = \frac{10(9)(8)(7)6!}{4(3)(2)6!} = 210\,,$$

while the number of permutations is $6! = 720$ times 210 or 151,200. If we had 100 balls and selected six of them, the number of combinations would be

$$C_6^{100} = \frac{100!}{6!94!} = 5(33)(49)(97)(16)(95)\,,$$

which is a number larger than 50 million.

Let us now study a problem which looks somewhat different from the one just studied but which will turn out to be mathematically equivalent

to it. Consider the random experiment $\mathfrak{R}$ which consists in tossing the same coin n times and observing whether the coin lands heads or tails up on each toss. Let us concentrate our attention on outcomes which yield precisely r heads. How many different sequences of heads and tails are there which yield a total of r heads? If we tossed a coin three times the following three sequences, $\langle H, H, T \rangle$, $\langle H, T, H \rangle$ and $\langle T, H, H \rangle$, would yield two heads, so that the answer is three when $n = 3$ and $r = 2$. We would like to determine an expression for the number of sequences for any n and r, $r \leq n$. To see how to do this note that $\mathfrak{R}$ consists of n independent random experiments $\mathfrak{R}_j$, where $\mathfrak{R}_j$ is the experiment in which the jth toss of the coin is made. Consider any outcome of $\mathfrak{R}$ which yields precisely r heads. If we give the numbers of the experiments $\mathfrak{R}_j$ which yielded a head, this completely describes the sequence of heads and tails in $\mathfrak{R}$, since all other experiments $\mathfrak{R}_j$ must have yielded a tail. Each set S of r numbers selected from the numbers 1 to n will then serve to define a sequence in which r heads are obtained if we imagine that each experiment $\mathfrak{R}_j$, with $j \in S$ yields a head and no other $\mathfrak{R}_j$ does. Hence if we imagine the experiments to be numbered balls, we see that the number of sequences having r heads is the number of combinations of n things taken r at a time and is given by (1–33).

EXAMPLE. A coin is tossed five times. How many different sequences will yield three heads? This was solved in Example 1 on page 53 and the answer is 10. From the combinations listed there we can immediately write down the corresponding sequences. They are: $\langle H, H, H, T, T \rangle$, $\langle H, H, T, H, T \rangle$, $\langle H, H, T, T, H \rangle$, $\langle H, T, H, H, T \rangle$, $\langle H, T, H, T, H \rangle$, $\langle H, T, T, H, H \rangle$, $\langle T, H, H, H, T \rangle$, $\langle T, H, H, T, H \rangle$, $\langle T, H, T, H, H \rangle$ and $\langle T, T, H, H, H \rangle$. We have given these in the same order that the combinations were listed on page 53 and the reader should check the correspondence.

1–18 SAMPLING

Consider the random experiment $\mathfrak{R}$ in which n balls are selected from a box containing N balls numbered 1 to N (the reason why we are now using n and N rather than r and n as in the previous section will appear below). Let us now construct a model $\mathcal{E}$ of $\mathfrak{R}$ from which we can compute the probability of various events of interest. A convenient way to characterize a simple event is to give the numbers on the n balls drawn from the box. We could do this either by giving the numbers in the sequence they were obtained, in which case there are P_n^N simple events, or we can merely note the numbers without being concerned about the order in which they appear, and in this case there are C_n^N simple events. In problems that will be of interest to us, the order of the numbers will be irrelevant and thus it is convenient to use the event set containing C_n^N simple events, each simple event representing one possible set of n numbers which could be obtained on drawing the n balls.

Let us now decide what probability should be assigned to each simple event. This clearly depends on how the n balls are selected. If the balls are well mixed before the drawing and the person drawing them has no way of differentiating between them, then by symmetry it should be true that the simple events are equally likely and each has the probability $1/C_n^N$. If it is indeed true that each simple event has the probability $1/C_n^N$, then we say that the n balls were drawn *at random* from the box. We shall assume that the n balls are drawn at random from the box. Then as a model $\mathcal{E}$ for $\mathcal{R}$ we can use one with C_n^N simple events. A simple event will be denoted by $\alpha_1\alpha_2 \cdots \alpha_n$; $\alpha_1, \ldots, \alpha_n$ are the numbers on the balls drawn (no order implied). The probability of each simple event is $1/C_n^N$.

To proceed, we shall now imagine that balls $1, \ldots, M$ are red and balls $M + 1$ to N are blue. Let us compute the probability $p(r)$ that when n balls are selected at random from the box precisely r of them turn out to be red. This probability is k/C_n^N, where k is the number of simple events having the characteristic that $\alpha_1\alpha_2 \ldots \alpha_n$ contains precisely r numbers from the set $1, \ldots, M$, since each such simple event has the characteristic that precisely r balls are red. Consider then the problem of determining k. There are C_r^M different sets of numbers possible for r balls drawn from the red ones. There are also C_{n-r}^{N-M} sets of numbers possible on $n - r$ balls drawn from the blue balls. There are thus C_r^M simple events $\alpha_1\alpha_2 \ldots \alpha_n$ which have a given set of $n-r$ numbers selected from $M + 1, \ldots, N$. But there are C_{n-r}^{N-M} different sets of $n - r$ numbers which can be selected from $M + 1, \ldots, N$ for each set of r selected from $1, \ldots, M$. Hence $k = C_r^M C_{n-r}^{N-M}$ and

$$p(r) = \frac{C_r^M C_{n-r}^{N-M}}{C_r^N} = \frac{\dfrac{M!}{r!(M-r)!} \dfrac{(N-M)!}{(n-r)!(N-M-n+r)!}}{\dfrac{N!}{n!(N-n)!}}. \quad (1\text{--}34)$$

Hence we have obtained a very complicated-looking expression from which we may compute the probability that precisely r of the n balls drawn are red.

Equation (1–34), in spite of its complexity, is an extremely important one and arises frequently in practice. We shall now try to explain why it is so important. First, however, it will be convenient to introduce some additional terminology which will be used frequently in the future. We shall call the set of n balls drawn at random from the box a *sample*. Frequently, it is also convenient to refer to the N balls from which the sample was drawn as the *population* from which the sample was drawn.

Clearly, no practical problems are directly concerned with drawing colored balls from a box. However, many practical problems are mathematically equivalent to this, and there are a number of such problems which are not only mathematically equivalent but also physically quite

similar to drawing balls from a box. The type of situation encountered in practice can be described in a general way as follows. We are concerned with a population containing N members. What the members of the population happen to be can vary widely with circumstances. In our above development they were balls. Each member of the population can be classified as having one of two possible characteristics, call them 1 and 2, of interest (each ball was either red or blue). We may be interested in this population because we are trying to decide which one of two or more actions should be selected in some decision situation, and the appropriate action to select depends on the fraction of the population which has characteristic 1. For example, a businessman may be considering marketing a new product, and the appropriate action to take depends on the fraction of potential customers who will buy the product if it is marketed.

One way to determine exactly the fraction of the population which has characteristic 1 would be to examine each member and ascertain which characteristic it possesses. In practice, it may be impossible or inordinately expensive to do this. Instead, a sample is selected at random from the population, and based on the number of members in the sample having characteristic 1, a decision is made. It is desired to have the sample as representative of the population as possible. This is the reason that an effort is made to select a sample from the population at random thus giving each member an equal chance of being selected. If the sample is not selected at random the sample may be *biased* and not representative of the population from which it is drawn. In practice, very serious errors have been made in marketing surveys or in opinion polls because of a bias introduced into the sampling procedure. We shall not try to study these matters in detail now, since currently we are only concerned with the problem of how to compute the probability that a sample selected at random from a population will have a given characteristic. It is important to note, however, that even when the sample is selected at random, the fraction of the members of the sample which have characteristic 1 can, due to chance, be different and even very different from the fraction of the members of the population which have characteristic 1.

1–19 EXAMPLES

We shall now study some very simple examples illustrating the use of (1–34). Equation (1–34) is complicated to work with because of the difficulty in evaluating all the factorials. Later we shall see how to avoid these problems and then we shall be able to give some more complicated examples.

1. Two senators are chosen at random from the 100 senators in the United States Senate. What is the probability that neither is from Pennsylvania? The situation can be imagined as one in which a sample of two balls is

chosen at random from a box containing 100 balls, 2 of which are red. We want the probability that neither is red, that is, $p(0)$ in (1–34). Thus

$$p(0) = \frac{\dfrac{2!}{0!2!}\dfrac{98!}{2!96!}}{\dfrac{100!}{2!98!}} = \frac{98(97)}{100(99)} = 0.960 \ .$$

The probability that both senators selected are from Pennsylvania is

$$p(2) = \frac{\dfrac{2!}{2!0!}\dfrac{98!}{0!98!}}{\dfrac{100!}{2!98!}} = \frac{1}{50(99)} = 0.000202 \ .$$

2. A large company operating in India uses sampling methods to check yearly on the incidence of TB in its employees. Four girls are selected at random from 100 secretaries employed in Calcutta and are given an x-ray test. If five of the secretaries actually have TB, what is the probability that two or fewer girls in the sample will have the disease? Let A_0, A_1 and A_2 be the events that precisely 0, 1 and 2 girls in the sample have TB. The probability that two or fewer have TB is then the probability of the event $A_0 \cup A_1 \cup A_2$. Now A_0 and $A_1 \cup A_2$ are mutually exclusive events, so by (1–12)

$$p(A_0 \cup A_1 \cup A_2) = p(A_0) + p(A_1 \cup A_2) \ .$$

However, A_1 and A_2 are mutually exclusive, so that

$$p(A_1 \cup A_2) = p(A_1) + p(A_2) \ .$$

Hence

$$p(A_0 \cup A_1 \cup A_2) = p(A_0) + p(A_1) + p(A_2) \ .$$

By (1–34)

$$p(A_0) = p(0) = \frac{\dfrac{5!}{0!5!}\dfrac{95!}{4!91!}}{\dfrac{100!}{4!96!}} = \frac{95}{100}\left(\frac{94}{99}\right)\left(\frac{93}{98}\right)\left(\frac{92}{97}\right) = 0.809$$

$$p(A_1) = p(1) = \frac{\dfrac{5!}{1!4!}\dfrac{95!}{3!92!}}{\dfrac{100!}{4!96!}} = 4\left(\frac{5}{100}\right)\left(\frac{95}{99}\right)\left(\frac{94}{98}\right)\left(\frac{93}{97}\right) = 0.176$$

$$p(A_2) = p(2) = \frac{\dfrac{5!}{2!3!}\dfrac{95!}{2!93!}}{\dfrac{100!}{4!96!}} = 6\left(\frac{5}{100}\right)\left(\frac{4}{99}\right)\left(\frac{95}{98}\right)\left(\frac{94}{97}\right) = 0.011 \ .$$

The reasoning used in computing $p(A_1)$, for example, is that since there are a total of five girls who have TB, $M = 5$ and $N - M = 95$. Four girls are selected, so $n = 4$, and we are interested in the probability that $r = 1$, that is, precisely one of the four girls selected has TB. Given the above, the probability that two or less of the girls selected are affected is

$$p(A_0 \cup A_1 \cup A_2) = 0.810 + 0.176 + 0.011 = 0.997 ,$$

so that it would be quite unusual if this event did not occur. Indeed, since $p(A_0) = 0.810$, the probability is high that none of the girls selected will be affected.

REFERENCES

1. Hadley, G., *Introduction to Probability and Statistical Decision Theory*. Holden-Day, San Francisco, 1967.
Gives a more complete development of probability theory at a slightly more advanced level than that given here.

2. Hodges, J. L. Jr., and E. L. Lehmann, *Basic Concepts of Probability and Statistics*. Holden-Day, San Francisco, 1964.
A good and also elementary discussion of probability theory is provided.

3. Mosteller, F., R. E. K. Rourke and G. B. Thomas Jr., *Probability with Statistical Applications*. Addison-Wesley, Reading, Mass., 1961.

PROBLEMS

Section 1-3

1. Toss a penny 100 times. On each toss record whether it landed heads or tails up. Determine the total number of tails n_t, obtained after n tosses, for each n, and also determine the fraction of the tosses which yielded a tail. Construct a figure like Figure 1–1 which shows the fraction of tails in n tosses for each n.

2. Take five pennies and mark on them the numbers 1 through 5, one number per penny. Place the pennies in a container and shake it so that they are thoroughly mixed. Now select one penny from the container without looking. Record the number on the penny selected. Repeat this experiment 100 times. Determine the relative frequency of each penny in the 100 trials.

3. In some parts of the country weather reports include statements such as "There is a 3 in 10 chance that it will rain this afternoon." What does this statement mean?

Section 1-4

In problems 1 through 5 list all subsets of the given set.

1. $A = \{1, 3, 5, 8\}$; **2.** $A = \{a_1, a_2, a_3\}$;

3. $X = \{4, 5, 8, 12\}$; **4.** $E = \{e_1, e_2, e_3, e_4\}$;

5. $R = \{\alpha, \beta, 1, 2, 3\}$.

In Problems 6 through 9 determine the intersection and union of the given sets.

6. $A = \{1, 3, 5, 7\}$ **7.** $A = \{1, 3, 5, 7, 9\}$
      $B = \{2, 4, 5, 8, 9\}$; $B = \{1, 5, 9\}$.

8. A is the set of students at a given university who have taken the introductory physics course, and B is the set of students at the same university who have taken introductory calculus.

9. A is the set of all persons who filed a California State income tax form in 1965, and B is the set of all persons who filed either a New York State or Massachusetts State income tax form in 1965.

10. Show that the following always hold if $\varnothing$ is considered to be a subset of every set.

    (a) $A \cap B \subset A$;   (b) $A \cap B \subset B$;   (c) $(A \cap B) \subset (A \cup B)$.

Illustrate each with an example.

11. Show that the following always hold:

    (a) $A \subset A \cup B$;   (b) $B \subset A \cup B$.

Illustrate each with an example.

12. We define the set of all elements in A which are not in B to be the difference between A and B, written $A - B$. If $A = \{1, 3, 4, 5, 7\}$ and $B = \{3, 5, 7, 11, 13\}$, determine

    (a) $A - B$;   (b) $B - A$;   (c) $(A - B) \cup (B - A)$.

Section 1–5

1. Consider a random experiment in which we place three balls in a bowl, mix the balls thoroughly and then, without looking, select one of the balls. The balls are identical, except that each ball has a different number painted on it (the numbers are 1, 2 and 3). List all the ways you can think of for subdividing the outcomes of this experiment in such a way that one and only one outcome will occur. Each such subdivision can be used to construct an event set. What are these event sets? Give a characterization of the simple events in each set. Is there one set that seems most natural and generally useful?

2. Consider a random experiment in which we place three balls, numbered 1, 2, and 3 in an urn, mix them well and then, without looking, select two of the balls. List all the ways you can think of for subdividing the outcomes of this experiment in such a way that one and only one will occur. Each such subdivision can be used to define an event set. Determine each such event set and characterize the simple events in each case.

3. Consider a random experiment in which three pennies are tossed. List all the ways you can think of for subdividing the outcomes of this experiment in such a way that one and only one will occur. Each can be used to define an event set. Are there one or two event sets which seem especially natural to use?

4. Consider an experiment in which two six-sided dice are tossed and it is noted how many dots are on each face which turns up. Let us use as a characterization of the simple events the number of dots on each of the faces turned up, for example, one die turns up two dots and the other turns up five dots. How many simple events are there? What is the event set? Suppose now that instead of using the simple events just referred to, we distinguish between the dice, perhaps by using one red and one green die. Now let the simple events be characterized by the number of dots turned up on the red die and the number of dots turned up on the green die. How many simple events are there in this case? What is the event set? What is the difference between the two event sets here obtained?

Section 1–6

1. Consider the random experiment that involves the tossing of a die and noting what face turns up. Let a simple event be characterized by the number of dots on the face which turns up. Express each of the following events as subsets of E, and explain what the subset means in each case.

   (a) A face with four or less dots on it turns up.
   (b) A face with an odd number of dots on it turns up.
   (c) A face with six or less dots on it turns up.
   (d) The face which turns up has two or less dots on it or five or more dots on it.

2. Consider an experiment in which we toss three pennies and note how many heads and tails there are. Let us use an event set E in which the simple events are characterized by the number of heads and tails observed. Express each of the following as subsets of E, and explain what the subset means in each case.

   (a) At least two heads appear.   (b) Precisely one tail appears.
   (c) No tail appears.   (d) One or more tails appear.

3. Suppose that for the situation described in Problem 2, we now paint a number on each coin, so that they are numbered 1, 2 and 3. Let us use as a simple event the specification of whether coin 1 lands heads or tails up and similarly for the other two coins. Answer Problem 2 using this event set.

4. Consider Problem 4 for Section 1–5. If we use the first event set referred to there, express each of the following events as subsets of that event set.

   (a) The total number of points (dots) on the two faces which turn up is 4.
   (b) The total number of points on the two faces is less than or equal to 4.
   (c) The total number of points is either 4 or 10.

5. Answer Problem 4 using the second event set referred to in Problem 4 for Section 1–5.

6. Answer Problem 4 for the following events.

   (a) At least one of the faces has four dots on it.
   (b) One face has four dots on it, and the other has either one dot or five dots.

7. Answer Problem 5 for the events listed in Problem 6.

8. Indicate which of the following represent an allowable assignment of probabilities to the simple events in $E = \{e_1, e_2, e_3\}$. For those which are not allowable, explain why.

(a) $p_1 = 0.5$; $p_2 = 0.3$; $p_3 = 0.2$.
(b) $p_1 = 0.4$; $p_2 = 0.6$; $p_3 = 0$.
(c) $p_1 = -0.4$; $p_2 = 0.8$; $p_3 = 0.6$.
(d) $p_1 = 0.6$; $p_2 = 0.3$; $p_3 = 0.2$.
(e) $p_1 = 0$; $p_2 = 1$; $p_3 = 0$.

9. Explain whether or not each of the following claims could be correct.

(a) A supplier claims that the long-run fraction of the resistors he produces which are defective is 0.001. In one lot of 10,000 resistors obtained from this supplier 30 defectives were discovered.

(b) A plant engineer claims the probability that a machine will not fail in a one-month period is 0.20, the probability that it will fail exactly once is 0.50, the probability that it will fail twice is 0.30 and the probability that it will fail more than twice is 0.30.

(c) A market analyst claims that the probability that sales of less than 4 million pounds in the next year is 0.3, of sales between 4 and 6 million pounds is 0.4 and sales of more than 6 million pounds is 0.2.

10. Consider a random experiment having the event set $E = \{e_1, e_2, e_3, e_4, e_5\}$, with the probabilities of the simple events being $p_1 = 0.1$, $p_2 = 0.2$, $p_3 = 0.1$, $p_4 = 0.3$ and $p_5 = 0.3$. Compute the probabilities of the events.

(a) $A = \{e_1, e_2\}$;   (b) $B = \{e_1, e_3, e_4\}$;
(c) $C = \{e_1, e_4, e_5\}$;   (d) $B \cap C$;
(e) $A \cup B$;   (f) $A \cap B \cap C$;
(g) $A \cup (B \cap C)$.

11. Consider a random experiment having the event set $E = \{e_1, e_2, e_3, e_4, e_5, e_6\}$, with the probabilities of the simple events being $p_1 = 0.05$, $p_2 = 0.25$, $p_3 = 0.10$, $p_4 = 0.20$, $p_5 = 0.30$ and $p_6 = 0.10$. Compute the probabilities of the events:

(a) $A = \{e_1, e_2, e_3\}$;   (b) $B = \{e_2, e_3, e_4, e_5\}$;   (c) $C = \{e_1, e_6\}$;
(d) $A \cap B$;   (e) $B \cup C$;   (f) $(A \cap C) \cup B$;
(g) $(A \cap C) \cup (A \cap B)$;   (h) $(A \cup B) \cap C$.

Section 1–8

1. Consider the experiment in which three pennies are tossed. To characterize a simple event suppose that we specify the number of heads and tails obtained. List the simple events and denote this event set by E_1. Now imagine that we characterize a simple event by indicating whether the first coin landed heads or tails up, and similarly for the other two coins. List the simple events and denote this event set by E_2. Show that the simple events in E_1 and E_2 cannot both be equally likely. In which of the sets should the simple events be equally likely? Why? What symmetry conditions must be satisfied? What is the probability of each simple event?

2. Use the results of Problem 1 to compute the probabilities of the following events:

(a) Two coins land heads and the other one tails up.
(b) Two or more coins land heads up.
(c) No more than two coins land heads up.
(d) Precisely one coin lands tails up, or precisely one coin lands heads up.

3. Consider Problem 2 of Section 1–5. Suggest an event set in which the simple events might be considered equally likely. List the simple events and determine the probability of each. What symmetry conditions must be satisfied if the simple events are to be equally likely? Can you think of two different event sets in which the simple events are equally likely if the above-mentioned symmetry conditions are met?

4. Use the results of Problem 3 to compute the probability that
  (a) One of the balls is numbered 1.
  (b) The ball with the number 3 on it is not drawn.
  (c) Either the ball numbered 1 or the ball numbered 2 is drawn or both are drawn.

5. Consider Problem 4 for Section 1–5. Show that it is impossible for the simple events in both the event sets developed there to be equally likely. If certain symmetry conditions are satisfied, it seems reasonable to assume that the simple events in one of these event sets are equally likely. Which one is it and what are the symmetry conditions needed? Determine the probability of each simple event for this set in which the simple events are equally likely.

6. Use the results of Problem 5 to compute the probability that
  (a) At least one of the dice turns up a face with only one dot.
  (b) Neither die turns up a face with six dots.
  (c) One die turns up a face with two dots, and the other turns up a face with three dots.

7. Use the results of Problem 5 to compute the probability that
  (a) The sum of the number of points (dots) on the two faces is 8.
  (b) The sum of the number of points on the two faces is 11.
  (c) The sum of the number of points on the two faces is less than or equal to 6.

8. Use the results of Problem 5 to compute the probability that
  (a) The sum of the number of points (dots) on the two faces is 12 or is less than or equal to 3.
  (b) The sum of the number of points on the two faces is less than or equal to 8, and one of the faces has 4 dots on it.

9. Consider the random experiment in which we select, without looking, two balls from an urn containing two red balls, a white ball and a green ball. First make a list of the simple events when a simple event is characterized by the color of each of the balls selected. Call this event set E_1. Next imagine that we paint numbers on the balls, 1 and 2 on the red balls, 3 on the white one and 4 on the green one, and we characterize a simple event by giving the numbers on the balls drawn. List the simple events in this case and call the event set so obtained E_2. Finally, consider the case of a simple event characterized by specifying the number on the first ball and the second ball drawn. List the simple events for this case and call the event set E_3. Show that if the simple events in E_3 are equally likely then so are those in E_2. Show that it is inconsistent to assume both that the simple events in E_2 and E_3 are equally likely and that the simple events in E_1 are equally likely. One normally feels that the simple events in E_2 and E_3 are equally likely if certain symmetry conditions are satisfied. Why is this? What are the symmetry conditions? Determine the

probability of each simple event in E_2 and in E_3 when the symmetry conditions are satisfied.

10. Use the results of Problem 9 to determine the probability that
(a) One of the balls drawn is red.
(b) One of the balls is red and the other green.
(c) No red ball is drawn.

11. Use the results of Problem 9 to compute the probability that
(a) At least one red ball is drawn.
(b) The red ball numbered 2 is drawn.
(c) One of the balls is red or one is white.

12. A resistor is selected at random from a lot of 1000 resistors, 15 of which are defective. Set up a model for this experiment and determine the probability that the resistor is defective.

13. An executive is selected at random from a pool of 25 executives to fill a particular position. A total of 18 executives have the competence to do a good job while the remainder do not. What is the probability that the executive selected will do a good job? Describe the model used in making the computation.

Section 1–9

1. What sort of information would be needed to show that an Aa parent did not give an a type gene to one offspring and then an A type gene to the next offspring, or vice versa? To be specific, we might assume for a physical model that the pair of genes A and a exist in the plant as little bars, as shown in Figure 1–9, much like a magnet with a north and south pole, the A gene being at one end and the a gene at the other. When a gene is about to be passed on to an offspring this bar is broken into two pieces as shown in Figure 1–9 and the piece to be passed on is selected at random. However, once this is done, the remaining piece is passed on to the next offspring rather than breaking a new bar. After every two offspring a new bar is broken. This model is not correct, but it might conceivably have seemed reasonable in Mendel's time. Part of it can be refuted from frequency information. How?

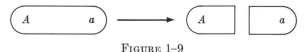

FIGURE 1–9

2. What information from Mendel's experiments, if any, contradicts the fact that just one of the two parents has an influence on the inherited characteristics of the offspring? For example, why might not the offspring receive two genes from a single parent, these being selected randomly from the type that the parent possesses with the parent who contributes the genes also being selected at random? Try to determine the implications of such a model and relate these to the experimental results. Do not be concerned if you find it difficult to determine the probabilities in every case.

3. What experimental results rule out the possibility that each plant has only a

single gene? What would be wrong in assuming that there are three types of genes, say R, P and W, with each plant having only a single gene, and the parent which contributes the gene to the offspring being selected at random? What are the implications of such a model which are not in accord with experimental results?

4. Suppose that each plant had four genes, with two being contributed from each parent, the two being selected independently and at random from the four genes possessed by the parent. Suppose that there are just two types of genes, A and a. Assume also that $AAAA$ types are red, genotypes with three, two or one A type genes are pink (three different shades of pink) and those of type $aaaa$ are white. Is this model consistent with the experimental results? Do not attempt to determine the probabilities in every case.

5. Suppose that an a gene is lethal, but it is possible for aa individuals to have children, as would be true with diabetics, for example. What fraction of the children would be afflicted if an aa individual married an AA individual? What if he married an Aa individual?

Section 1–10

For Problems 1 through 5 take the event set to be $E = \{e_1, e_2, e_3, e_4, e_5\}$, with the probabilities of the simple events being $p_1 = 0.15$, $p_2 = 0.10$, $p_3 = 0.20$, $p_4 = 0.05$ and $p_5 = 0.50$.

1. Are the following events exclusive? Prove whatever statement you make.

(a) $A = \{e_1, e_3\}, B = \{e_2, e_4\}$;

(b) $A = \{e_1, e_3\}; B = \{e_2, e_3, e_4\}$;

(c) $C = A \cup B$ and $D = \{e_2, e_4\}$, where $A = \{e_1, e_3\}$ and $B = \{e_1, e_2, e_3\}$.

2. Let $A = \{e_1, e_3\}$ and $B = \{e_2, e_5\}$. Compute $p(A \cup B)$ using (1–12) as well as directly by summing the probabilities of the simple events in $A \cup B$. Represent this situation graphically.

3. Let $A = \{e_1, e_3\}$ and $B = \{e_2, e_3, e_5\}$. Compute $p(A \cup B)$ using (1–13) as well as directly by summing the probabilities of the simple events in $A \cup B$. Represent this situation graphically.

4. Let $A = \{e_1, e_2, e_3\}$ and $B = \{e_2, e_3, e_4\}$. Compute $p(A \cup B)$ using (1–13) as well as directly by summing the probabilities of the simple events in $A \cup B$. Represent this situation graphically.

5. Let $A = \{e_1, e_2, e_3, e_4\}$ and $B = \{e_1, e_2\}$. Compute $p(A \cup B)$ using (1–13) as well as directly by summing the probabilities of the simple events in $A \cup B$. Represent this situation graphically.

6. Explain whether or not each of the following claims could be correct.

(a) A businessman claims the probability that he will get contract A is 0.15 and that he will get contract B is 0.20. Furthermore, he claims that the probability of getting A or B is 0.50.

(b) A market analyst claims that the probability of selling ten million pounds of plastic A or five million pounds of plastic B is 0.60. He also claims that the probability of selling ten million pounds of A and five million pounds of B is 0.45.

7. Prove that if $B \subset A$, then $p(A) \geq p(B)$.

8. Prove that if $B \subset A$, then $p(A \cup B) = p(A)$.

9. Consider three events A, B and C. Suppose that $A \cap B = \varnothing$, $A \cap C = \varnothing$ and $B \cap C = \varnothing$. Show that it is not possible for two or more of these events to occur simultaneously. Also show that

$$p(A \cup B \cup C) = p(A) + p(B) + p(C) .$$

10. Consider Problem 9 for Section 1–8. Compute the probability of the event A that one of the balls drawn is red and the probability of the event B that one of the balls drawn is green. Then compute the probability of $A \cup B$ using the methods of this section.

11. Consider the event set $E = \{e_1, e_2, e_3, e_4\}$ with the probabilities of the simple events given by $p_1 = 0.20$, $p_2 = 0.30$, $p_3 = 0.40$ and $p_4 = 0.10$. Compute $p(A^c)$ when

(a) $A = \{e_1, e_2\}$;   (b) $A = \{e_1, e_3, e_4\}$.

Sections 1–11 and 1–12

1. One box contains three red, two green and five yellow balls, while another box contains five red, three green and four yellow balls. Let $\Re$ be the random experiment in which one ball is selected at random from each box. What is the probability that both balls are red?

2. For the situation described in Problem 1, what is the probability that one ball is red and one ball is green? Hint: The red ball could be drawn from either one of the boxes.

3. For the situation described in Problem 1, what is the probability that neither ball is red?

4. For the situation described in Problem 1, what is the probability that at least one red ball is drawn? Hint: The easy way to solve this is to use A^c.

5. Two mothers who have recently had babies meet on the street and begin to discuss their children. What is the probability that both had girls if historical data show that about 0.49 of all births are girls?

6. An executive believes that the probability that he will get contract A is 0.50 and the probability that he will get contract B is 0.30. If the contracts are awarded independently, what is the probability that he will get both contract A and contract B?

7. One roulette wheel has five equally-spaced numbers on it (the numbers 1 through 5). Another roulette wheel has ten equally-spaced numbers on it (the numbers 1 through 10). Each roulette wheel is spun, and the number at which it stops is observed. A casino operates a game using these two roulette wheels which is of this form: If both wheels stop on the same number, the player wins \$50. If they stop on different numbers, he pays the casino \$10.50. What is the probability that the player will win \$50 if he plays the game once? What is the probability that the player does not lose money if he plays the game twice?

8. A box contains 1000 vacuum tubes, 150 of which are defective. Another box contains 100 resistors, 3 of which are defective. A third box contains 500 capacitors, 30 of which are defective. Suppose that an engineer selects at random a vacuum

tube, a resistor and a capacitor. What is the probability that none of the elements is defective?

9. For the situation described in Problem 8, what is the probability that at least one of the elements is defective?

10. For the situation described in Problem 8, what is the probability that the vacuum tube or the resistor is defective?

11. For the situation described in Problem 8, what is the probability that the resistor and the capacitor are defective?

12. Many electronic systems now used in missiles and space-exploration work contain thousands of elements such as transistors and resistors. Each of these elements must perform satisfactorily if the whole system is to do so. The reliability of a system (or element) is defined to be the probability that it will perform in a satisfactory way during a complete mission, the mission being specified. Suppose that a given system contains 1000 elements. A mission can then be characterized, so far as success or failure of the system goes, as a random experiment consisting of 1000 random experiments, the ith of which has two outcomes, that is, the ith element fails or it does not fail. Let us imagine, as is often done in practice, that these experiments are independent. If each element has the same reliability p, what value must p have if the reliability of the system is to be 0.95? The system will be assumed to fail if any one of the elements fails.

13. A fair coin is tossed three times. Construct a model of this experiment using the method introduced in this section.

14. A coin, which has a probability p of yielding a head when tossed, is tossed three times. Construct a model of this experiment. If $p = 0.3$, what is the probability that two or less heads will be obtained?

15. Two fair dice are tossed, and it is observed which face turns up on each. Set up a mathematical model for this random experiment and determine what seems to be a proper estimate of the probability of each simple event.

16. Two loaded dice are tossed, and it is observed which face turns up on each. Set up a mathematical model for this random experiment if the probabilities that the various faces which turn up for each of the individual dice are those given in Table 1–7.

TABLE 1–7

Dots on face turned up	1	2	3	4	5	6
Probability for die 1	0.10	0.20	0.25	0.05	0.30	0.10
Probability for die 2	0.20	0.10	0.20	0.10	0.20	0.20

17. One bowl contains two red, three green and four yellow balls. A second bowl contains three red, one green and two yellow balls. A third bowl contains one yellow, five red and six green balls. Consider a random experiment ℜ in which one ball is selected at random from each of the bowls and the color is noted. Let a simple event be one which gives the color of the ball obtained from each of the three bowls. List all the simple events and determine the probability of each.

18. For the situation described in Problem 17, what is the probability that two of the balls are green and the other is yellow?

19. For the situation described in Problem 17, what is the probability that one ball is red, one is green and one is yellow?

20. For the situation described in Problem 17, what is the probability that at least one ball is red?

21. For the situation described in Problem 17, what is the probability that precisely two different colors appear on the three balls drawn?

22. For the situation described in Problem 17, what is the probability that no yellow ball is drawn?

23. Consider Example 5 for Section 1–12. Construct models to describe the outcome of:

    (a) Crossing an $AABB$ plant with an $AaBb$ plant;
    (b) Crossing an $AAbb$ plant with an $AaBb$ plant;
    (c) Crossing an $AABb$ plant with an $AaBb$ plant;
    (d) Crossing an $AaBB$ plant with an $AABb$ plant.

24. Consider Example 5 for Section 1–12 and let us now add a third characteristic to the two considered there. Let the third characteristic be the smoothness of the peas. The C gene favors smooth peas while the c gene favors rough-surfaced peas. The C gene dominates the c variety. Determine what kinds of plants can arise from crossing an $AaBbCc$ plant with an $AaBbCc$ plant. Construct a probability model to represent this and determine the probability of each type of offspring.

25. Re-solve Problem 24 to characterize the outcome of:

    (a) Crossing an $AaBBCc$ plant with an $AaBbCc$ plant;
    (b) Crossing an $AABbCC$ plant with an $AaBBCc$ plant.

26. Show that if A and B are mutually exclusive they cannot be independent when $p(A)$ and $p(B)$ are both different from zero.

27. Consider Problem 2 for Section 1–9. For the case where an AA is crossed with an Aa plant, determine the probabilities of the various outcomes if the model suggested there applies.

28. Consider Problem 4 for Section 1–9. For each possible crossing in the model suggested there, determine the probabilities of the various outcomes.

Sections 1–13 and 1–14

1. A bowl contains five balls numbered 1 through 5. Consider the random experiment $\Re$ in which two balls are withdrawn from the bowl, one after the other. What is the probability that ball 2 is drawn first and then ball 3? What is the probability that balls 2 and 3 are drawn?

2. A bowl contains five red, seven green and three yellow balls. Two balls are selected from the bowl, one after the other. Each is drawn at random. What is the probability that both are red? What is the probability that the first is red and the second is green? What is the probability that a red and a green ball are drawn?

3. In a box are fifty light bulbs, ten of which are defective. One bulb is selected at random from the box, then a second one. What is the probability that both bulbs are defective? What is the probability that at least one bulb is defective?

4. A bowl contains six red, four green and three yellow balls. Three balls are selected at random from the bowl. What is the probability that one is red, one green and one yellow?

5. Two boxes sit on a table. In the first box are four red and three green balls. In the second box are two red and six green balls. Consider the random experiment $\mathfrak{R}$ in which first a box is selected at random and then one of the balls in the box is selected at random. Construct a model for $\mathfrak{R}$ and draw a tree to represent the situation. What is the probability that a red ball is drawn?

6. A desk contains three drawers. In the first are one gold and one silver coin. In the second are one gold and two silver coins and in the third are two gold and one silver coin. Consider the random experiment $\mathfrak{R}$ in which first a drawer is selected at random and then one of the coins in the drawer. Construct a model for $\mathfrak{R}$ and draw a tree to represent the situation. What is the probability that the coin selected is gold?

7. A bowl contains six red and three blue balls. Someone draws out at random one of the balls but does not tell us what color it was. We now select at random one of the remaining eight balls. What is the probability that it is red? How do you explain the result obtained?

8. A production run of a seldom used, very expensive part is normally for only five units of the part. If the setup is made correctly, the probability that any given unit produced will be defective is 0.05, but if it is not made correctly, the probability that any given unit will be defective is 0.15. Setups are made correctly 90 percent of the time. In either case, the production process behaves as if each unit produced is the result of an independent trial of a random experiment. What is your estimate of the probability that no defectives will be produced on any given production run?

9. Seven cards having printed on them the letters A, G, H, J, M, N, O, one letter per card, are shuffled thoroughly and then the first four cards are turned face up and placed side by side in the order in which they were drawn. What is the probability that they will spell out the name JOHN?

10. An individual wishes to call a friend from a public phone booth. The friend has an unlisted number. The individual can remember all but the last digit of his friend's number, but unfortunately cannot recall the final digit. He has two dimes in his pocket. He decides to select the last digit at random, and if it is wrong he will select another digit at random from the unused digits. What is the probability that he will dial the correct number before running out of dimes?

11. An antiaircraft gun crew is about to fire at a target aircraft. The probability of hitting the aircraft on the first shot is 0.25. If the aircraft is not hit on the first shot, the gun can be reloaded and fired again. The probability of a hit on the second shot is 0.15 (the probability is lower since the aircraft is now in a less favorable position). If the second shot is also a miss, the gun can be reloaded once again and a third shot fired. The probability of a hit on the third shot is 0.07. There is no possi-

bility of firing a fourth shot because the plane will be out of range. What is the probability that the target will be hit?

12. A petroleum company is bidding on two parcels of land, call them A and B. Its executives feel that the probabilities of winning the bids on A and B are 0.4 and 0.6 respectively. It is also believed that the probability that A has oil is 0.2 and that B has oil is 0.1. What is the probability that the company will obtain more crude oil as a result of these bids?

13. A company is planning to market a new product. If its competitor does not market a similar product, then the probability that the product will do well is 0.8. If the competitor also markets a similar product, the probability that the product under consideration will do well is 0.4. It is believed that the probability that the competitor will market a similar product is 0.6. What is the probability that the product will do well if it is marketed?

14. A man and woman who are healthy and have no family history of a hereditary disease marry and have a child. If the fraction of the healthy adult population which are carriers of the lethal gene is f, determine the probability that the child will be a carrier. Also determine the probability that the child will be afflicted. Hint: Assume that the probability that the mother is a carrier is f and that the probability that the father is a carrier is also f. Why is this reasonable?

15. A casino operates the following sort of game. First a roulette wheel colored red and blue is spun. The probability of stopping on red is 0.7. If the roulette wheel stops on red, then another roulette wheel colored white, black and green is spun. The probability of stopping on white, black and green are respectively 0.2, 0.3 and 0.5. If the original roulette wheel stops on blue, instead of red, then still a different roulette wheel colored brown, black and yellow is spun. The probabilities of this wheel stopping on brown, black, and yellow are respectively 0.2, 0.4 and 0.4. The individual loses if the second roulette wheel stops at black, and he wins if it stops on yellow or brown. What are the probabilities that the individual will lose or win?

16. One student is selected at random from a class of 100 boys, 70 of which are white and 30 of which are negro. Also a student is selected at random from a group of 200 girls, 125 of whom are white and 75 of which are negro. Of the white boys, 30 have fathers who attended college, and of the negro boys, 10 have fathers who attended college. Of the white girls, 60 have fathers who attended college, and of the negro girls, 25 have fathers who attended college. What is the probability that at least one of the two students selected has a father who attended college?

Sections 1–15 and 1–16

1. A fair die is tossed and we are told that a face with an even number of dots turned up. Develop a conditional probability model which could be used for making any additional computations concerning this random experiment. What is the probability that the face with two dots on it turned up?

2. A bowl contains eight red and six blue balls. Four of the red balls and three of the blue balls have a gold dot painted on them. A ball is selected at random. We are not told its color, but we are told that it has a gold dot on it. Set up a conditional probability model which could be used for making additional computations con-

cerning this random experiment. What is the probability that the ball drawn was red?

3. Reconsider Problem 2 under the assumption that two balls are drawn and we are told that both have a gold dot on them. What is the probability that both are red? What is the probability that one is red and one is blue?

4. A lot of vacuum tubes contains 1000 tubes, 10 of which have a defective grid and no other defects, and 20 of which have both a defective grid and a defective heating element. A tube is drawn at random from the lot and we are told that it has a defective grid. What is the probability that it also has a defective heating element? What model did you use in computing this probability?

5. If the data given in Example 3 for Section 1–16 apply, what is the probability that an individual who has lived to the age of 40 will die after the age of 60? What is this probability at the time of the individual's birth?

6. Consider a random experiment $\Re$ for which we are using the mathematical model ε having the event set $E = \{e_1, \ldots, e_{10}\}$, with $p_1 = 0.05$, $p_2 = 0.10$, $p_3 = 0.20$, $p_4 = 0.07$, $p_5 = 0.03$, $p_6 = 0.15$, $p_7 = 0.12$, $p_8 = 0.08$, $p_9 = 0.14$ and $p_{10} = 0.06$. Imagine that $\Re$ is performed, and we are told that the event $A = \{e_1, e_6, e_8, e_{10}\}$ has occurred. Construct a conditional probability model that would be appropriate for making additional computations about $\Re$.

7. A fair coin is tossed four times, and three heads are obtained. What is the probability that a head and a tail were obtained on the first two tosses (in either order)?

8. A fair coin is tossed five times and three heads are obtained. What is the probability that a head was obtained on each of the first two tosses?

9. A production process which turns out transistors has a long-run fraction defective of 0.005. A testing device is used to check each transistor produced. It has been found that the device always indicates that a defective is indeed defective, but for about 1 in every 100 transistors produced it indicates that a good transistor is defective. If the device indicates that a given transistor is defective, what is the probability that it is actually defective? Describe carefully the model you are using.

10. A fair die is rolled twice, and the sum of the number of points obtained on the two tosses is ten. What is the probability that the face with five dots on it turned up on the first toss?

11. If $p(A) = 0$, why is it not meaningful to introduce conditional probabilities $p(B|A)$? In other words, show that there is no definition of $p(B|A)$ in terms of the probabilities in ε for this case which would seem appropriate. Hint: If $p(A) = 0$, can we construct a conditional model ε_c from ε, or if A occurs when $p(A) = 0$ in ε, does this mean that ε is completely irrelevant and we must start over to obtain a model ε_c?

12. Show that when A and B are independent events, then $p(B|A) = p(B)$. What is the intuitive interpretation of this result?

13. If A and B are mutually exclusive events for some random experiment and $p(A \cup B) \neq 0$, then show that

$$p(A|A \cup B) = \frac{p(A)}{p(A) + p(B)} .$$

14. Use $p(A \cap B) = p(B|A)p(A)$ to show that

$$p(A_1 \cap A_2 \cap \cdots \cap A_n)$$
$$= p(A_n|A_1 \cap \cdots \cap A_{n-1})p(A_{n-1}|A_1 \cap \cdots \cap A_{n-2}) \cdots p(A_2|A_1)p(A_1) .$$

Explain how this can be used in practice. Hint: First let $B = A_1 \cap \cdots \cap A_{n-1}$.

15. Show that when $p(A)$ and $p(B)$ are different from 0, then

$$p(B|A)p(A) = p(A|B)p(B) .$$

16. A company gives an intelligence and aptitude test to new junior executives hired from outside the company. Past history has shown that the probability that an individual will do well on these tests is 0.4, and the probability that he will be a successful executive is 0.3. Furthermore, the fraction of those who do well on the tests which turn out to be successful executives is 0.6. What is the probability that an individual who will be a successful executive will also do well on the test? Hint: Use the result of Problem 15.

17. Past experience with a production line for turning out color television sets has shown that the fraction of sets which have a defective speaker as well as some other defect is 0.05 and the fraction of those which have a defective speaker is 0.08. A set is found to have a defective speaker. What is the probability that it also has some other defect?

18. Consider Example 5 for Section 1–16. Show that if f is negligible, it is possible to construct a very simple model which yields the probability 1/3 that the individual is a carrier.

19. Consider Example 5 of Section 1–16. Suppose that in addition to the individual's mother having an afflicted sister, his father also had an afflicted sister. Construct a new model from which the probability that the individual is a carrier can be determined.

20. For Example 5 of Section 1–16, suppose that instead of the individual's mother having an afflicted sister, his mother's mother had an afflicted sister. Assume there is no history of the disease in the father's or in the mother's father's family. Construct a model to represent the situation, and determine the probability that the individual is a carrier.

21. Suppose that, as in ancient Egypt, a brother marries his sister. What is the probability that both are carriers of a lethal gene, thus making it dangerous for them to have children? Assume that the mother and father of the brother and sister under consideration are unrelated and that there is no history of the disease in either the father's or mother's families. Let f be the fraction of the healthy population who are carriers. Answer: $(\frac{1}{2}f - \frac{1}{4}f^2)/(1 - \frac{7}{16}f^2)$. How does this compare with the case where two unrelated individuals who have no family history of the disease marry?

22. Re-solve Problem 21 if a half-brother and half-sister marry. Answer: $f/4$ approximately, if f is small.

Section 1–17

1. A box contains five balls numbered 1 through 5. List the possible sequences, that is, permutations obtainable when three balls are drawn from the box. Show that this number is $(5)_3$.

2. Re-solve Problem 1 when the balls are numbered 2, 5, 7, 8, 9. Thus show that in this case the number of permutations does not depend on the particular set of five numbers printed on the balls, so long as they are all different.

3. List the possible permutations of the numbers 1, 2, 3, 4.

4. For the situation described in Problem 1, determine how many different combinations of numbers there are; list each one. Show that this number is C_3^5.

5. For the situation described in Problem 2, determine how many different combinations of numbers there are; list each one. Show that this number is C_3^5.

6. In how many ways may five boys and four girls be seated in a row of nine chairs? What is the answer if boys and girls must alternate?

7. A firm has the possibility of obtaining seven contracts. How many different combinations of three of these could be obtained?

8. A coin is tossed six times. List each possible sequence of heads and tails for the case where two heads are obtained. Show that the number of sequences is C_2^6.

9. Roughly estimate the number of combinations of 100 numbered balls taken 50 at a time.

10. A company wishes to fill two positions from a pool of ten executives. In how many different ways could this be done?

11. A company is planning to build two plants of the same size in two of six different cities. In how many different ways could this be done?

12. A company is planning to build two plants, one large and one small, in two of six different cities. In how many ways could this be done?

Sections 1–18 and 1–19

1. A box contains five balls numbered 1 to 5, the first three of which are red and the remaining two are blue. Three balls are drawn from the box. List all the sets of numbers which correspond to having two red balls drawn. Show that the number of such sets is $C_2^3 C_1^2$.

2. A box contains ten balls numbered 1 to 10. Five of the balls are selected at random from the box. What is the probability that the number 1 appears on one of the balls drawn?

3. Consider the situation outlined in Problem 2. What is the probability that one ball drawn is numbered 1 and that another ball drawn is numbered 2?

4. Consider the situation outlined in Problem 2. What is the probability that the balls numbered 1, 2 and 3 are drawn?

5. Consider the situation outlined in Problem 2. What is the probability that the balls numbered 1, 2, 3 and 4 are drawn?

6. Consider the situation outlined in Problem 2. What is the probability that the balls numbered 1, 2, 3, 4 and 5 are drawn?

7. One box contains ten balls numbered 1 to 10, and another box contains eight balls numbered 1 to 8. Consider the random experiment in which we select at random five balls from the first box and select at random three balls from the second box. What is the probability that two balls drawn are numbered 1?

8. For the situation described in Problem 7, what is the probability that at least one of the balls drawn is numbered 1?

9. For the situation described in Problem 7, what is the probability that two balls numbered 1 and two balls numbered 2 were drawn?

10. For the situation described in Problem 7, what is the probability that at least one ball is numbered 1 and at least one ball is numbered 2?

11. For the situation described in Problem 7, what is the probability that the balls numbered 1 and 2 appear precisely once (and are not repeated)?

12. For the situation described in Problem 7, what is the probability that a ball numbered 1 does not appear?

13. Suppose that there are n balls in a box, m of which are red. One ball is now selected at random. Show that the probability that the ball is red as computed from (1–34) is m/n, as would be expected in this case.

14. Two students are selected at random from a group of ten students, six of whom are boys with the remainder being girls. What is the probability that one boy and one girl are selected?

15. Answer Problem 14 in the case where there are 100 students, 60 of whom are boys. Can you explain intuitively the reason for the difference in the answers?

16. A bowl contains six red and five blue balls. Four balls are selected at random. What is the probability that precisely two of them are red?

17. For the situation outlined in Problem 16, what is the probability that at least one red ball is obtained?

18. For the situation described in Problem 16, what is the probability that there are precisely two red balls or three blue balls?

19. A box contains 1000 transistors, 25 of which are defective. A sample of four transistors is selected at random from the lot. What is the probability that none of those selected is defective?

20. An oil company has rights on ten different offshore tracts, but it has the resources to develop only four of them. They all appear equally good and thus the company decides to select at random the four to be developed. If five of the tracts contain oil, what is the probability that all four selected will be oil bearing? What is the probability that two or more will be oil bearing?

21. The marketing manager of a large food-products manufacturer plans to select at random five stores from a group of fifteen in a given city to test market a new product. Three of these stores happen to be owned by a supermarket chain. What is the probability that all three of these stores will be selected?

22. A consumer testing organization obtains four color television sets from a lot of 100 which a supplier has. Ten sets in the lot are defective. The model will be rated as good if no defectives are found in the models tested, of doubtful quality if one defective is found and of poor quality if two or more sets are defective. What are your estimates of the probabilities that the model will be characterized as good, doubtful and poor?

23. A customs inspector in Bombay, India, selects at random three of the 100 passengers from a jet plane which has just arrived from Nairobi, Kenya, for a very detailed baggage examination. If two of the 100 passengers are smuggling gold into India, what is the probability that at least one will be caught, if the probability that a smuggler will be caught when his baggage is examined in detail is 0.8?

CHAPTER 2

Random Variables

2-1 THE NOTION OF A RANDOM VARIABLE

Suppose that we toss a fair coin ten times and note the number of heads obtained. If this experiment is repeated many times, the number of heads obtained will not always be the same. Sometimes there will appear four heads, other times six, and so on. We can think of the number of heads which turn up on tossing the coin ten times as a variable. The number of heads cannot be predicted before the experiment is performed, since the number obtained depends on the outcome of the experiment. Similarly, if an individual plays a game of chance the amount he wins cannot be predicted in advance but is determined by the outcome of the game of chance. It can thus be considered to be a variable associated with the game of chance. Equally well, the profit that a firm will receive from some venture whose outcome is determined by chance will, in general, depend on what the outcome is and can thus be considered to be a variable. Any variable of the type we have been considering is often referred to as a *random variable*, the word random being used to emphasize the fact that its value cannot be selected in advance but is determined by the outcome of a random experiment. The value that a random variable takes on when the relevant experiment $\mathfrak{R}$ is performed is determined once the outcome of $\mathfrak{R}$ is specified, that is, the simple event which has occurred is specified. Normally, however, the value of the random variable will not be the same for each simple event, but will change from one simple event to another.

In general, we shall define a random variable as follows:

RANDOM VARIABLE: *Any numerical quantity whose value is determined by the outcome of a random experiment is called a random variable.*

The notion of a random variable is a very important one in probability and statistics, and indeed most of the remainder of the text will be concerned in one way or another with random variables. We have given some examples of random variables above; we shall encounter many more as we progress.

To characterize a random variable, we must give the value it takes on for each simple event of ℜ. Once this is done we have specified completely the random variable. Note that in order for the notion of a random variable to have any meaning, we must have in mind a random experiment with which it is associated. We shall denote random variables by symbols in much the same manner as was done in algebra. Thus we might use x or y to denote some random variable of interest. Whenever we associate with each simple event e_j for some random experiment ℜ, a number ξ_j, then by this process, we have automatically defined a random variable, call it x, which can take on one of the values $\xi_1, \ldots, \xi_m$ and such that $x = \xi_j$ if e_j occurs on performing ℜ. Let us now give several examples illustrating how a random variable can be associated with some random experiment.

EXAMPLES. *1*. Imagine that we have 100 cards and on each we write the name and age of a United States senator, a different name on each card. Consider the random experiment in which we place the cards in a bowl and select one at random. To construct a model of this experiment we can use as a simple event the name of the senator on the card drawn. We have, in writing on each card the age of the senator as well as his name, associated with each simple event a number which is the age of the senator. In this way we have defined a random variable x which is the age of the senator whose name is drawn.

2. A coin is tossed, and it is noted whether it lands heads or tails up. Suppose that as the event set for this experiment we use $E = \{H, T\}$. Consider now the random variable x which has the value 1 if H occurs and has the value 0 if T occurs. What is the meaning of this random variable? It is nothing but the number of heads obtained on tossing the coin. The number of heads is 1 if the coin lands heads and is 0 if it lands tails.

3. A casino operates a game of chance which involves spinning a roulette wheel and noting the color on which it stops. The roulette wheel is painted red (R), green (G), yellow (Y), black (B), white (W) and brown (N). As a simple event we can use the color at which the wheel stops. The areas devoted to the various colors are such that the probabilities of the simple events are those shown in Table 2–1. The amount that the player wins for each color at which the wheel stops is also shown. A negative number means that the player loses this amount and a zero means that he neither wins nor loses. The amount that the player wins on playing this game once can be looked upon as a random variable. Before the game is played, the player

TABLE 2–1

Simple event	R	G	Y	B	W	N
Probability	0.20	0.10	0.15	0.20	0.10	0.25
Winnings	2	5	0	−3	−10	2.8

does not know what his winnings will be, although he does know that the amount won will be one of the values 2, 5, 0, −3, −10, 2.8. After the game is played, however, then the value of the random variable, that is, the amount won, is determined.

2–2 EXPECTED VALUE OF A RANDOM VARIABLE

Generally speaking, we shall be especially interested in what will be called the *expected value* of a random variable. Let us begin by explaining the intuitive meaning of expected values. Let $\Re$ be a random experiment for which we are using the model $\mathcal{E}$ with event set $E = \{e_1, \ldots, e_m\}$ and with the probabilities of the simple events being $p_1, \ldots, p_m$. Assume that x is a random variable associated with $\Re$ which takes on the value ξ_j if e_j occurs. Let us now perform $\Re$ a total of n times, and let n_j be the number of times that e_j occurs. Each time $\Re$ is performed, some value of x is determined. Assume now that we add up the n values of x so obtained and then divide by n. What does this give us? It gives what we call the arithmetic average of the values taken on by the random variable in n trials of the experiment. The reader will be well aware that arithmetic averages are often used in the real world. As an illustration, suppose that we play a game of chance five times and our winnings are 2, −5, −1, −5 and 3. Our average winnings per play are then

$$\frac{1}{5}(2 - 5 - 1 - 5 + 3) = -\frac{6}{5} = -1.2,$$

that is, we lost an average of 1.2 per game.

Now what is the average of the values of x for the general situation we have been discussing? Since e_j occurred precisely n_j times, $x = \xi_j$ precisely n_j times. Thus the average of the values of x in n trials, which we shall denote by $\bar{\xi}$, is

$$\bar{\xi} = \frac{1}{n}[\xi_1 n_1 + \xi_2 n_2 + \cdots + \xi_m n_m] \qquad (2\text{–}1)$$

or

$$\bar{\xi} = \xi_1 \frac{n_1}{n} + \xi_2 \frac{n_2}{n} + \cdots + \xi_m \frac{n_m}{n}. \qquad (2\text{–}2)$$

To determine $\bar{\xi}$ we multiply ξ_j by the relative frequency of e_j and add the results.

Next imagine that instead of performing the above experiment a finite number of times, it is repeated unendingly. As n becomes larger and larger, the relative frequency n_j/n must approach p_j, the probability of e_j. Since (2–2) holds for any n, we see that $\bar{\xi}$ must then approach a unique number we shall denote by μ_x, which is

$$\mu_x = \xi_1 p_1 + \xi_2 p_2 + \cdots + \xi_m p_m . \qquad (2\text{–}3)$$

The number μ_x is called the *expected value* of the random variable x; μ_x can be computed directly from the model $\mathcal{E}$ for $\mathcal{R}$, and μ_x can be interpreted intuitively as the long-run average value of x. *To compute the expected value of x, we multiply together the value x takes on if e_j occurs and the probability of e_j and add up the numbers so obtained for all possible values of j.* We have in our above discussion suggested that the number μ_x given by (2–3) should be of interest in practical applications. For our mathematical development we simply use (2–3) to define the notion of the expected value of a random variable. Thus we introduce the following definition.

EXPECTED VALUE OF A RANDOM VARIABLE. *Let $\mathcal{E}$ be a probability model for some experiment $\mathcal{R}$. Denote by $E = \{e_1, \ldots , e_m\}$ the event set for $\mathcal{E}$ and assume p_j is the probability of e_j. Suppose now that a random variable x is defined with respect to $\mathcal{E}$ by specifying that when e_j occurs $x = \xi_j$. The expected value of x, denoted by μ_x, is then defined to be the number computed from (2–3).*

We shall see as we go along that the expected value of a random variable is often a very useful number to know.

EXAMPLE. The use of expected values was first employed in studying games of chance. Consider the game described in Example 3 of the previous section. Let us compute the player's expected winnings μ_x per play of the game. On using (2–3) and the data from Table 2–1 we see that

$$\mu_x = 2(0.20) + 5(0.10) + 0(0.15) + (-3)(0.20) + (-10)(0.10) + 2.8(0.25)$$
$$= 0.4 + 0.5 - 0.6 - 1.0 + 0.7 = 0 .$$

The expected winnings per play are 0. A game having 0 as the expected winnings per play is called a fair game. A gambling house could not operate a game for which the expected winnings were positive because gamblers would quickly discover this and in time bankrupt the casino. Normally, the casino operates games in which the expected winnings of the players are slightly negative, that is, they lose money in the long run. In this way the casino gains enough to cover expenses and make a profit.

2–3 DISTRIBUTION OF A RANDOM VARIABLE

In Section 2–1 we explained what was meant by a random variable. A random variable x is completely specified by giving for each j the value ξ_j

that x takes on if the simple event e_j occurs. We have not required that the ξ_j all be different numbers, that is, that x take on a different value for each simple event. For example, it might be true that $x = 16$ if e_1 occurs and also $x = 16$ if e_{17} occurs. Equally well, in Table 2–1 we could have had the winnings be 2 if the roulette wheel stopped at red and also if it stopped at green. Assume that the *different* values which x can take on are $x_1, x_2, \ldots,$ x_k. Let us next subdivide E into k subsets $S_1, \ldots, S_k$ which have the property that for each simple event $e_j \in S_i$, $x = x_i$, that is, $\xi_j = x_i$. Now suppose that we have numbered the x_i so that $x_1 < x_2 < x_3 < \cdots < x_k$. Then the set

$$X = \{x_1, \ldots, x_k\} \qquad (2\text{–}4)$$

is one whose elements are the different values the random variable x may take on.

The subsets S_i which we introduced above are events; S_i is the event that $x = x_i$. The probability of S_i is the sum of the probabilities of the simple events in S_i. We shall denote this probability by $p(x_i)$; thus $p(x_i)$ is the probability that the random variable x will take on the value x_i. Associated with each possible value of x_i of x is a number $p(x_i)$ which is the probability of x.

At this point it is convenient to introduce a new concept referred to as a *function*. A function is simply another name for a rule which associates with each element in a given set A one and only one element of another set B. For example, suppose that A is the set of students in a given class and B is the set of family names of the students. Consider the rule which associates with each student his family name. This is an example of a function. Another function would be one which associated with each student his weight. Still another function would be one which associated with each student his age. We have already used implicitly the notion of a function on several occasions in our earlier development. Thus the rule which assigns to each simple event e_j in E the probability p_j of e_j defines a function with A being the event set E and B being a set of real numbers. Similarly, to define a random variable x we give for each j the value ξ_j it takes on if e_j occurs. This association with each $e_j \in E$ a number ξ_j defines a function with $A = E$ and B being a set of real numbers.

The set A in the definition of a function is called the *domain* of the function. If a is any element of A, the element $b \in B$ which is associated with a by the function is called the *image* of a. To indicate that b is the image of a, a notation such as $b = f(a)$, or $b = g(a)$ is used. For the function which assigns to each student in some class his family name, the domain is the set of students in the class, and the image of any student is his family name. For the function which defines a random variable x, the domain is E, and the image of e_j is the value ξ_j which x takes on if e_j occurs. The notion of a func-

tion is basically a very simple one and is used a great deal in everyday life. We shall be using it frequently in the future.

We shall now use the notion of a function to define what is called the probability function for a random variable x. The probability function for x is merely the function which associates with each element $x_i \in X$, X being the set of different values which x can take on, the probability $p(x_i)$ that $x = x_i$. More precisely, we make the following definition.

PROBABILITY FUNCTION FOR A RANDOM VARIABLE. *Consider some random variable x and let $X = \{x_1, \ldots, x_k\}$ be the set of different values that x may take on. Now consider the function which associates with each element $x_i \in X$ the probability $p(x_i)$ that $x = x_i$. This function is called the probability function for the random variable x.*

The domain of the probability function for x is the set X of different values which x can take on. Any element of X is a real number. The image of any element $x_i \in X$ is $p(x_i)$, the probability that $x = x_i$; $p(x_i)$ is also a real number and is sometimes referred to as the *value of the function* when $x = x_i$. It is convenient to have a symbolism to denote the entire probability function in addition to the symbolism $p(x_i)$ used to denote its value when $x = x_i$. The reader will note that in the past we have always used separate symbols for a random variable and the values which it can take on. It is not standard to do this in algebra, but it is very convenient to do so in probability theory and statistics, because it helps to clarify considerably the discussion. We shall always in the future use different symbols for random variables and the values they take on.

This convention also makes it possible to introduce a simple, clear notation for the probability function. The notation we shall use is $p(x)$. All we have done is modified slightly the symbolism for the value of the function at $x = x_i$ by replacing x_i by x. Therefore, $p(x)$ represents a function and $p(x_i)$ is a number which is the value of the function at $x = x_i$. The reader familiar with functions from his other work in mathematics will recognize that $p(x)$ is the notation normally used in elementary mathematics to represent a function, and this is the reason we have used it here. Let us now give two examples illustrating the determination of the probability function for a random variable.

EXAMPLES. *1.* Consider a game of chance which involves tossing a fair die. The amount that the player wins depends on what face turns up. Let a simple event give the number of dots on the face which turns up so that e_j is the event that the face with j dots on it turns up. The amount won by the player for each simple event is shown in Table 2–2. The amount won can be looked upon as a random variable x; x can take on four different values, which are $x_1 = -6$, $x_2 = 0$, $x_3 = 1$ and $x_4 = 2$, so $X = \{-6, 0,$

TABLE 2–2

Simple event	e_1	e_2	e_3	e_4	e_5	e_6
Amount won	2	−6	1	2	1	0

1, 2}. Consider now the event S_i that $x = x_i$. We see that $S_1 = \{e_2\}$, $S_2 = \{e_6\}$, $S_3 = \{e_3, e_5\}$ and $S_4 = \{e_1, e_4\}$, since, for example, the event $x = x_1$ occurs only if e_2 occurs, and the event $x = x_4$ occurs if e_1 or e_4 occurs. Because the die is fair, each simple event has the same probability $\frac{1}{6}$. Then, on summing the probabilities of the simple events in S_i to obtain $p(x_i)$, the probability that $x = x_i$, we see that $p(x_1) = p(-6) = \frac{1}{6}$, $p(x_2) = p(0) = \frac{1}{6}$, $p(x_3) = p(1) = \frac{1}{3}$ and $p(x_4) = p(2) = \frac{1}{3}$. A convenient way to represent the probability function for x in this case is to give a table listing each element in the domain and its image. This is done in Table 2–3.

TABLE 2–3

Value of x	−6	0	1	2
Probability	$\frac{1}{6}$	$\frac{1}{6}$	$\frac{1}{3}$	$\frac{1}{3}$

2. Consider the random experiment $\mathcal{R}$ in which a fair coin is tossed three times. Let us introduce a random variable x for $\mathcal{R}$ which is the number of heads obtained in the three tosses. To define a simple event let us indicate whether a head or a tail was obtained on each toss. Then there are eight simple events. These are listed in Table 2–4 along with the corresponding value of x.

TABLE 2–4

Simple event	x	Simple event	x
$\langle H, H, H \rangle$	3	$\langle H, T, T \rangle$	1
$\langle H, H, T \rangle$	2	$\langle T, H, T \rangle$	1
$\langle H, T, H \rangle$	2	$\langle T, T, H \rangle$	1
$\langle T, H, H \rangle$	2	$\langle T, T, T \rangle$	0

The random variable x then takes on one of the four values 0, 1, 2, 3, so $X = \{0, 1, 2, 3\}$. Since the coin is fair, the simple events should be equally likely, and each should have the probability $\frac{1}{8}$. The event that $x = 0$ is $\langle H, H, H \rangle$, so $p(0) = \frac{1}{8}$. The event $x = 1$ is $\{\langle H, H, T \rangle, \langle H, T, H \rangle, \langle T, H, H \rangle\}$, so $p(1) = \frac{3}{8}$. The event $x = 2$ is $\{\langle H, T, T \rangle, \langle T, H, T \rangle, \langle T, T, H \rangle\}$, so $p(2) = \frac{3}{8}$. Finally, the event $x = 3$ is $\langle T, T, T \rangle$, so $p(3) = \frac{1}{8}$. The probability function for x is then summarized in Table 2–5. If $\mathcal{R}$ is repeated a number of times the same value of x will not be observed in each instance. Indeed, from Table 2–5, about $\frac{1}{8}$ of the time there will be no heads, about $\frac{3}{8}$ of the time precisely one head, about $\frac{3}{8}$ of the time precisely two heads, and

about $\frac{1}{8}$ of the time precisely three heads. Once one has the probability function for x then it is possible to determine immediately what the long-run relative frequency of any particular value of x should be, if the model is accurate.

TABLE 2–5

Value of x	0	1	2	3
Probability	$\frac{1}{8}$	$\frac{3}{8}$	$\frac{3}{8}$	$\frac{1}{8}$

Let us now note that the events corresponding to the different value $x_1, \ldots, x_k$ which the random variable x can take on have the characteristic that when $\Re$ is performed one and only one of these events will occur. In other words x cannot take on two or more different values on the same trial of $\Re$ but it will always take on some value. Thus it would be possible to use as an event set for $\Re$ one in which the simple events correspond to the values taken on by the random variable x. Very frequently one can conveniently use numbers to characterize the simple events in some model. These numbers can be looked upon as the values taken on by some random variable. The probability function for the random variable then becomes the function which assigns the probabilities to the simple events. For example, we have often considered the random experiment which involves tossing a die and noting the face which turns up. We normally use the event set in which the simple event e_j corresponds to the face with j dots on it turning up. Thus each simple event can be represented by a number, this number being the number of dots on a particular face of the die.

When the simple events for a random experiment are characterized by numbers we shall, in general, think of these numbers as being the possible values of some random variable. In such cases it is convenient to introduce some simplifications in terminology. Instead of using the symbols e_j for the simple events and x_i for the values of the random variable, we shall use the values x_i of the random variable to denote the simple events also. Then $X = \{x_1, \ldots, x_k\}$ can be used as the event set, and the probability of the simple event x_i, that is, the event that $x = x_i$, is $p(x_i)$.

A graphical representation of the probability function $p(x)$ for some random variable x is often very useful. A convenient way to give such a graphical representation is to use what is called a *bar diagram*, an example of which is given in Figure 2–1. A point 0 is selected on a horizontal line, called the horizontal axis, and then a unit of length is chosen. The points corresponding to the numbers x_i are next located on this line, positive numbers lying to the right of 0 and negative numbers to the left. Now at each number x_i, a vertical bar is erected whose length is $p(x_i)$. In order to have a scale for measuring the lengths of the bars, a vertical line is passed through 0 and a scale is placed on this line. At the right-hand end of the horizontal axis is

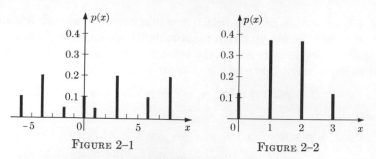

FIGURE 2–1   FIGURE 2–2

placed the symbol for the random variable or sometimes its name. At the top of the vertical axis we place the symbol for the function being studied. In Figure 2–1 we have represented a possible probability function for a random variable x which can take on the values $-6, -4, -2, 0, 1, 3, 6$ and 8. It is important to note that for any probability function

$$p(x_1) + p(x_2) + \cdots + p(x_k) = 1, \tag{2–5}$$

since if we think of the x_i as being simple events, then by (1–4) the probabilities of the simple events must sum to 1. Of course (2–5) holds even if we do not use the x_i as simple events. What (2–5) says is that the lengths of the bars in Figure 2–1 must sum to 1. The bar diagram for the probability function of Table 2–5 is shown in Figure 2–2, while that of Table 2–3 is shown in Figure 2–3.

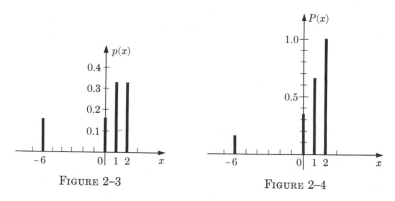

FIGURE 2–3   FIGURE 2–4

The probability function completely describes the manner in which chance influences the value the random variable will take on, and thus is an important function to know. There is another function obtainable from the probability function which we shall use frequently. It is called the *cumulative function* or *cumulative distribution* for the random variable x. The probability function associates with each $x_i \in X$ the probability that $x = x_i$. The cumulative function associates with each $x_i \in X$ the probability of the

event $x \leq x_i$, that is, of the event that x takes on the value x_i or a smaller value. We shall denote by $P(x_i)$ the probability that $x \leq x_i$. Now $x \leq x_i$ if $x = x_1$ or $x = x_2, \ldots$, or $x = x_i$. In other words, if X is used for the event set, the simple events for which $x \leq x_i$ are $x_1, x_2, \ldots, x_i$. Thus we see that

$$P(x_i) = p(x_1) + p(x_2) + \cdots + p(x_i), \quad i = 1, \ldots, k. \qquad (2\text{--}6)$$

In particular

$$P(x_1) = p(x_1); \quad P(x_2) = p(x_1) + p(x_2); \quad \ldots; \quad P(x_k) = 1. \qquad (2\text{--}7)$$

We shall use the symbolism $P(x)$ to denote the cumulative function, just as $p(x)$ is used to represent the probability function.

EXAMPLE. The cumulative function for the random variable whose probability function is given by Table 2–3 is easily determined. It is

$$P(-6) = p(-6) = \frac{1}{6}; \quad P(0) = p(-6) + p(0) = \frac{1}{3};$$
$$P(1) = p(-6) + p(0) + p(1) = \frac{2}{3};$$
$$P(2) = p(-6) + p(0) + p(1) + p(2) = 1.$$

It is often helpful to represent the cumulative function graphically using a bar diagram just as was done for the probability function. The bar diagram for the cumulative function just obtained is shown in Figure 2–4.

We have shown how the cumulative function can be determined from the probability function. It is also possible to go in the reverse direction and obtain the probability function from the cumulative function, since

$$p(x_i) = P(x_i) - P(x_{i-1}), \quad i = 2, \ldots, k \qquad (2\text{--}8)$$

and

$$p(x_1) = P(x_1). \qquad (2\text{--}9)$$

From (2–8) it follows that $P(x_i) \geq P(x_{i-1})$ inasmuch as $p(x_i) \geq 0$. Thus as one moves from left to right on the bar diagram for $P(x)$, the length of the bars can never decrease.

Either one of the functions $p(x)$ or $P(x)$ completely describes the manner in which chance influences the value that x takes on. The manner in which x behaves can be thought of intuitively as being controlled by a certain type of chance mechanism. We shall refer to the particular chance mechanism that controls x as *the distribution of the random variable* x. Either function $p(x)$ or $P(x)$ provides a quantitative representation of the distribution of x. Thus we can use the terminology that x is a random variable with the distribution $p(x)$ or x is a random variable with the distribution $P(x)$.

Recall that the expected value of x is

$$\mu_x = \xi_1 p_1 + \xi_2 p_2 + \cdots + \xi_m p_m, \qquad (2\text{--}10)$$

where ξ_j is the value of x if the simple event e_j occurs. Consider now the sets S_i, S_i being the event that $x = x_i$. For each $e_j \in S_i$, $\xi_j = x_i$; furthermore, the sum of the probabilities of the simple events is $p(x_i)$. Thus if in (2–10) we collect together terms j with $\xi_j = x_i$, the sum of these is $x_i p(x_i)$, so

$$\mu_x = x_1 p(x_1) + x_2 p(x_2) + \cdots + x_k p(x_k) . \qquad (2\text{–}11)$$

This is, of course, what we would conclude that μ_x should be if we use X as the event set; however, as we have just shown, it holds regardless of how the simple events are defined. To compute μ_x it is only necessary to know the probability function for x. There is no need to make use of the simple events or even to know how the simple events are defined. To compute μ_x, each possible value x_i of x is multiplied by $p(x_i)$ and the resulting numbers are summed. Thus the expected value of the random variable x of Table 2–3 is

$$\mu_x = (-6)\left(\frac{1}{6}\right) + 0\left(\frac{1}{6}\right) + 1\left(\frac{1}{3}\right) + 2\left(\frac{1}{3}\right) = 0 .$$

The expected value μ_x of x is also often referred to as the *mean* of x. Also, μ_x is referred to as the mean of the distribution of x. In the future we shall frequently refer to the mean of a probability function $p(x)$. The mean of $p(x)$ is the expected value of x. The reason that it is convenient to associate the number μ_x with the function $p(x)$ as well as with the random variable x is that many random variables may have the same probability function, and we shall be interested in studying the characteristics of certain probability functions without being concerned about how the associated random variable is to be interpreted.

2–4 THE BINOMIAL DISTRIBUTION

In practice one frequently encounters situations where a given random experiment $\Re^*$ is repeated a number of times, in such a way that the outcome of the experiment on any given trial neither influences nor is influenced by the outcomes on any other trial. In these circumstances we refer to the individual performances of $\Re^*$ as *independent trials*. Frequently there are only two outcomes of $\Re^*$ which are of interest, so that as a model for $\Re^*$ we can use one with an event set $E = \{e_1, e_2\}$ consisting of only two simple events. Let p be the probability of e_1 and $q = 1 - p$ be the probability of e_2. Consider now the random experiment $\Re$ which consists in performing n independent trials of $\Re^*$. Let us construct a model $\mathcal{E}$ for $\Re$.

As a simple event for $\Re$ we can merely give the simple event that occurred on each trial of $\Re^*$. Thus each simple event for $\Re$ will have the form $\langle \alpha_1, \ldots, \alpha_n \rangle$, where each α_j is either e_1 or e_2. Since α_1 can have two values, and α_2 can have two values for each value of α_1, and so on, we see that there are 2^n simple events in the event set for $\mathcal{E}$. Now $\Re$ can be imagined to be

built up from n independent random experiments, each of which corresponds to a trial of $\mathfrak{R}^*$. Thus according to what we obtained in Section 1–11, we should assign the probability $p(\alpha_1)p(\alpha_2) \cdots p(\alpha_n)$ to the simple event $\langle \alpha_1, \alpha_2, \ldots, \alpha_n \rangle$. Furthermore, $p(\alpha_j) = p$ if $\alpha_j = e_1$, and $p(\alpha_j) = q$ if $\alpha_j = e_2$. Therefore, every simple event $\langle \alpha_1, \alpha_2, \ldots, \alpha_n \rangle$ in which e_1 appears precisely r times has the probability $p^r q^{n-r}$, since e_2 then appears $n - r$ times. We have now constructed a model for $\mathfrak{R}$.

Frequently the characteristic of the outcome of $\mathfrak{R}$ that is of interest is the number of times that e_1 occurred. Now the number of times that e_1 occurs on performing $\mathfrak{R}$ is a random variable, call it x. We can easily determine the probability function for x. Consider the event that e_1 occurs precisely r times in n trials of $\mathfrak{R}^*$, that is, the event $x = r$. As we have noted above, each simple event for $\mathfrak{R}$ in which e_1 appears r times has the probability $p^r q^{n-r}$. How many simple events are there for which $x = r$? As we showed in Section 1–17 this is merely C_r^n, the number of combinations of n numbers taken r at a time. Thus the probability that $x = r$ is

$$p(r) = C_r^n p^r q^{n-r} = \frac{n!}{r!(n - r)!} p^r q^{n-r}. \tag{2-12}$$

The random variable x can take on $n + 1$ different values $0, 1, \ldots, n$. The probability function for x is a very important one and is encountered frequently. The interpretation of x can vary considerably with the circumstances. The probability function $p(x)$ for x is thus given a special name and is called the *binomial probability function* or *binomial distribution*, and x is said to have a binomial distribution. It will be convenient to use a special notation $b(x; n, p)$ to denote the binomial probability function rather than $p(x)$. Thus, by definition

$$b(r; n, p) = \frac{n!}{r!(n - r)!} p^r q^{n-r}. \tag{2-13}$$

The probability that $x = r$ clearly depends on how many trials of $\mathfrak{R}^*$ were made, that is, on n and on the probability p that e_1 will be observed on a given trial. The symbols n and p have been included in $b(x; n, p)$ to remind us constantly that the binomial probability function depends on these two numbers or *parameters*, as we shall call them.

In Figures 2–5, 2–6 and 2–7 we have shown the bar diagrams for $b(x; 10, p)$ where $p = 0.1, 0.45$ and 0.83 respectively. It is helpful to keep in mind the manner in which the bar diagram changes with p for a given n. When p is close to 0, so that e_1 does not occur very frequently, one expects small values of x to have the highest probability of occurrence and large values to have very low probabilities. This is indeed the situation in Figure 2–5. Precisely the reverse is true when p is close to 1. This is illustrated in Figure 2–7. In Figure 2–5, no bars are shown for $x = 5$ through 10; $b(x; 10, 0.1)$ is actually positive for these values, but is so small that the bars

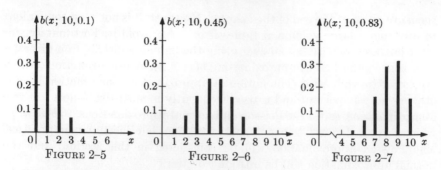

FIGURE 2–5   FIGURE 2–6   FIGURE 2–7

could not be shown on the figure. The same holds for $x = 0$, 9 and 10 in Figure 2–6 and $x = 0$ through 4 in Figure 2–7.

The distribution shown in Figure 2–6 appears to be much more symmetric than those shown in Figures 2–5 and 2–7. Intuitively, Figure 2–6 looks symmetric because if we draw a vertical line at $x = 4.5$, the distribution on one side of this line is essentially the mirror image of that on the other side of the line. The precise definition of symmetry can be given as follows. *A distribution $p(x)$ is said to be symmetric with ξ as a point of symmetry if whenever $\xi + u$ is a possible value of the random variable x, so is $\xi - u$ and $p(\xi + u) = p(\xi - u)$.* The reader will understand intuitively what we mean by symmetric even if the precise definition seems a little complicated. A distribution which is not symmetric is said to be *skewed*. Thus the distributions shown in Figures 2–5 and 2–7 are called skewed. The closer p comes to 0.5 the closer $b(x; n, p)$ comes to being symmetric; the distribution is symmetric when $p = 0.5$.

For any probability function $p(x)$, the value of x (which is not necessarily unique) where $p(x)$ takes on its largest value is called the *mode* of the distribution. Thus $x = 9$ is the mode of $b(x; 10, 0.83)$ shown in Figure 2–7 and $x = 1$ is the mode of $b(x; 10, 0.1)$ shown in Figure 2–5. In Figure 2–6 it appears that $x = 4$ and $x = 5$ are modes. However, the probability is slightly greater for 4 than for 5 and thus $x = 4$ is the mode of $b(x; 10, 0.45)$. As one would expect intuitively, the mode of $b(x; n, p)$ tends to increase with p for a given n.

It is very clumsy to make by hand numerical computations which involve the binomial distribution, since it is a laborious task to evaluate $C_r^n p^r q^{n-r}$. It is even somewhat complicated to do this on a computer. This would suggest that tables of the binomial distribution could be useful to eliminate the need for such computations. However, it is also difficult to tabulate the binomial distribution in a convenient form. The reason for this is that $b(x; n, p)$ involves two parameters n and p, and for each n and p there are $n + 1$ values of x. Thus covering the entire range of n and p values which might be of interest leads to tables which fill a very large volume. Two well-known tables of this type have been computed and references to

them are given at the end of the chapter. However, it is not very convenient to use such a large volume of tables if one can avoid it. Fortunately, for most purposes one can avoid ever using the binomial distribution directly. This is accomplished by approximating $b(x; n, p)$ by other functions which are easier to work with. The notion of approximating one function by another is very important and is used frequently in statistics. There are two approximations, called the Poisson and normal approximations, to $b(x; n, p)$ which serve rather well to handle most situations one encounters in practice. The Poisson approximation will be considered in this section, and the normal approximation will be introduced later.

Before going on to consider the Poisson approximation, we might mention that some very brief tables of $b(x; n, p)$ have been included at the end of the text (Tables B and C) to illustrate what tables of $b(x; n, p)$ are like. Normally, tables of $b(x; n, p)$ only show values of p in the interval $0 \leq p \leq 0.5$ (a finite number of such values) and no values of $p > 0.5$. Nonetheless, these tables can be used to evaluate $b(x; n, p)$ for $x = r$ and $p > 0.5$ because of the following. If e_1 occurs r times in n trials of $\mathfrak{R}^*$, then e_2 must occur $n - r$ times, so that the probability that $x = r$ is also the probability that e_2 occurred $n - r$ times. However, the probability that e_2 occurs $n - r$ times is the binomial probability $b(n - r; q, n)$. Thus

$$b(r; n, p) = b(n - r; n, q) . \qquad (2\text{--}14)$$

Suppose that we wish to evaluate $b(4; 9, 0.7)$ so that $r = 4, p = 0.7$ and $n = 9$. Then $n - r = 5$ and $q = 0.3$ and by (2–14)

$$b(4; 9, 0.7) = b(5; 9, 0.3),$$

and the latter value can be found in the tables. Although we have not included one, tables of the cumulative binomial distribution, which we shall denote by $B(x; n, p)$, are frequently useful to avoid the necessity for summing entries in the tables for $b(x; n, p)$.

Let us now consider the Poisson approximation to the binomial distribution. Assume that β is any positive number. Then for each $r = 0, 1, 2, \ldots$ it is possible to compute a number $(\beta^r e^{-\beta})/r!$, where $e = 2.7182818 \cdots$ is the base for the natural logarithms. This association of a number with each nonnegative integer r defines a function which we shall represent symbolically by $p(x; \beta)$. The value of the function for $x = r$ is

$$p(r; \beta) = \frac{\beta^r}{r!} e^{-\beta} . \qquad (2\text{--}15)$$

The function $p(x; \beta)$ is called the *Poisson function*. The Poisson function can be looked upon as the probability function for a random variable x which can take on any non-negative integral value, and therefore we shall frequently refer to the Poisson function as the *Poisson distribution*. The

reader should not be concerned about trying to remember (2–15), since when using the Poisson distribution we shall always use a table of $p(x; \beta)$ rather than (2–15).

Let us now explain how and under what conditions $b(r; n, p)$ can be determined approximately using the Poisson distribution. The way in which $b(r; n, p)$ is approximated using $p(x; \beta)$ is very simple. We compute np and then determine $p(x; \beta)$ for $x = r$ and $\beta = np$. Thus $p(r; np)$ is used as the approximation to $b(r; n, p)$, so that

$$b(r; n, p) \doteq p(r; np) , \qquad (2\text{–}16)$$

where $\doteq$ means approximately equal to. The conditions under which $p(r; np)$ can be used for $b(r; n, p)$ depend on how accurately we desire to compute $b(r; n, p)$. Generally speaking, for most statistical computations, an error of five percent or so for probabilities of the order of 0.10 or greater is quite acceptable, with even a larger error being tolerable for smaller probabilities. For such accuracy it is usually quite acceptable to use the Poisson approximation for $p < 0.1$ and $np \geq 5$. The important thing to keep in mind is that the Poisson approximation is used only when p is small, say $p < 0.1$. The value that np should have is much more indefinite. The Poisson approximation may be quite adequate even for np as small as 0.5, as we shall see in an example below. We shall see later that np is the mean of the distribution $b(x; n, p)$. It is also true that β is the mean of the Poisson distribution. Thus one approximates the binomial distribution by a Poisson distribution with the same mean.

It is not much easier to compute $p(r; \beta)$ than it is $b(r; n, p)$. What then has been gained by approximating $b(r; n, p)$ with $p(r; \beta)$? The thing gained is that the Poisson distribution involves only a single parameter β, and thus it is possible to give rather complete tables of the Poisson distribution in a small volume and very useful tables in just a few pages. A brief table of $p(x; \beta)$ is given in Table D at the end of the text. It is much too brief to be generally useful, but it will be adequate for our purposes. In Table 2–6 we have given $b(r; 10, p)$ and the Poisson approximation $p(r; 10p)$ for the case where $p = 0.1$ and 0.05 for each $r = 0, 1, \ldots$. The values of $b(r; 10, p)$ were determined from Table B at the end of the text and $p(r; 10p)$ from Table D. For example, to determine $b(3; 10, 0.1)$, we first look in Table D for $n = 10$. Then we find the $r = 3$ row and move over in it to the column for $p = 0.1$. There we read 0.0574, which is the value listed in Table 2–6. The Poisson approximation to $b(3; 10, 0.1)$ is $p(3; 1)$ since $np = 10(0.1) = 1$. To determine $p(3; 1)$ in Table D we first look for $\beta = 1$. Then we move down in this column to the $r = 3$ row and read the number 0.0613, which is what is given in Table 2–6. It will be observed that the Poisson approximation does very well both for $p = 0.1$ and $p = 0.05$, even though np is much less than 5. It should also be noted that the accuracy of the approximation is

TABLE 2–6

COMPARISON OF $b(r; 10, p)$ and $p(r; 10p)$

	$p = 0.10$			$p = 0.05$	
r	$b(r; 10, 1)$	$p(r; 1)$	r	$b(r; 10, 0.05)$	$p(r; 0.5)$
0	0.3487	0.3679	0	0.5987	0.6065
1	0.3874	0.3679	1	0.3151	0.3033
2	0.1937	0.1839	2	0.0746	0.0758
3	0.0574	0.0613	3	0.0105	0.0126
4	0.0112	0.0153	4	0.0010	0.0016
5	0.0015	0.0031	5	0.0001	0.0002
6	0.0001	0.0005	6	0.0000	0.0000

not the same for all values of r. It is much better for some values of r than for others.

The binomial distribution occurs very frequently in practice. Let us now provide some simple examples illustrating this.

2–5 EXAMPLES

1. In Table 2–7 we have given the explicit representation of $b(x; n, p)$ for the case where $n = 5$.

TABLE 2–7

EXPLICIT REPRESENTATION OF $b(x; 5, p)$

r	0	1	2	3	4	5
$b(r; n, p)$	q^5	$5pq^4$	$10p^2q^3$	$10p^3q^2$	$5p^4q$	p^5

2. The fraction of color television sets produced by a given manufacturer which fail in the first month of operation is 0.2. A dealer receives ten of these sets. What is the probability that precisely four of them will fail in the first month of operation? It seems reasonable here to assume that the operation of each of the sets during the first month can be looked upon as a random experiment in which the probability that the set will fail is 0.2. The random experiments representing the operation of different sets should be independent (if nothing took place at the factory which caused an entire lot to be defective). Based on this model, the probability of having exactly four out of ten fail is $b(4; 10, 0.2)$, which from Table B is 0.0881.

3. A chemical company has initiated research work on ten new products. There is no connection between the research projects, and past experience suggests that the probability that any one will yield a marketable product is 0.1. What is the probability that none of these projects will yield a mar-

ketable product? In the absence of any additional information, the best model which can be constructed here is one where the ten projects are imagined to be ten independent trials of a random experiment in which the probability of success on any single trial is 0.1. We then wish to compute the probability of zero successes, which is $b(0; 10, 0.1) = 0.3487$.

4. The Air Force has fifty Minuteman missiles in readiness at a given launch site with their electronic systems always operating. These missiles are all checked periodically by electronic checkout equipment, perhaps once per day, to determine if any failures have occurred in the electronic systems. Suppose that the probability that any given missile fails in the period between tests is 0.01. What is the probability that no missiles fail in the period between tests? If we make the reasonable assumption that a failure in one missile in no way influences the others, then the probability is $b(0; 50, 0.01)$. This number cannot be determined from Table B, since this table does not go up to $n = 50$. However, $p = 0.01$ here, which is very small, and thus one might hope to use the Poisson approximation. Now $np = 50(0.01) = 0.5$, so the Poisson approximation to $b(0; 50, 0.01)$ is $p(0, 0.5)$. From Table D one finds that $p(0; 0.5) = 0.6065$. The Poisson approximation works very well here, since from extensive tables of the binomial distribution one finds that $b(0; 50, 0.01) = 0.6050$.

Let us also determine the probability that two or more missiles fail in the time period. Denote this event by A. Then A^c, the complement of A, is the event that one or less fail. Recall that

$$p(A) = 1 - p(A^c),$$

so

$$p(A) = 1 - b(0; 50, 0.01) - b(1; 50, 0.01),$$

since A^c is the event that $x \leq 1$, x being the number of missiles that failed. We have already determined $b(0; 50, 0.01)$. The Poisson approximation to $b(1; 50, 0.01)$ is $p(1; 0.5) = 0.3033$. Thus the probability that two or more missiles fail should be

$$p(A) = 1 - 0.6065 - 0.3033 = 0.0902.$$

The correct result obtained from detailed binomial tables is 0.0894. Thus once again the Poisson approximation does very well.

2–6 THE HYPERGEOMETRIC DISTRIBUTION

In Section 1–18 we studied random experiments which involved selecting at random a sample of size n from a population containing N elements. Depending on the nature of the situation the elements might be resistors, people or balls. We now wish to study experiments of this sort in a little

more detail. Let us suppose that a fraction p of the elements in the population have some characteristic of interest, while the remainder do not. To be specific in the theoretical discussions we shall suppose, as in Section 1–18, that the elements are balls and that a fraction p of them are red, while the remainder are blue. Consider the experiment $\mathcal{R}$ in which n of the balls are selected at random. We can now introduce the random variable x which is the number of red balls obtained in the sample. The probability that $x = r$ will depend on p, n and N. We shall denote the probability function for x by $h(x; n, p, N)$. This probability function is given the odd name of *hypergeometric probability function* or *hypergeometric distribution*. Since the number of red balls is $m = pN$ and the number of blue balls is qN, $q = 1 - p$, we see from (1–34) that the probability that $x = r$ is

$$h(r; n, p, N) = \frac{C_r^{pN} C_{n-r}^{qN}}{C_n^N} . \qquad (2\text{--}17)$$

The random variable x can take on the values $0, 1, \ldots, n$ if $n \leq pN$ and $0, 1, \ldots, pN$ if $n > pN$.

The hypergeometric distribution appears frequently in practice, just as does the binomial distribution. However, it is even more difficult to make computations using the hypergeometric distribution than it is the binomial. It is also much more cumbersome to tabulate the hypergeometric distribution, since it involves three parameters rather than two. Thus when using the hypergeometric distribution one almost always proceeds by approximating it with some other distribution. The Poisson distribution, interestingly enough, can be used to approximate the hypergeometric distribution if p, the fraction of red balls, is small (note that p is not a probability here). Let us now explain how this is done. In practice the sample size n is almost always small compared to the population size N. We shall assume here that this is true. Then if p is small, say less than 0.1, and np is moderately large,

$$h(r; n, p, N) \doteq p(r; np) , \qquad (2\text{--}18)$$

so that the Poisson probability $p(r; np)$ is used to approximate $h(r; n, p, N)$. Often the Poisson approximation will be adequate even if np is as small as 1, and certainly should be adequate when $np > 5$. By use of the Poisson approximation one can compute approximately hypergeometric probabilities that would be extremely cumbersome to compute using (2–17). We shall now illustrate this with an example. Before doing so, however, we might mention that there is another approximation, called the normal approximation, which is useful for larger values of p. It will be introduced later.

EXAMPLE. The government has purchased a lot of 10,000 special batteries for use in mobile radio units. In this lot are 100 defective batteries. The re-

mainder will function properly. To check on the quality of the lot a government inspector selects at random 100 batteries from the lot and tests each battery in the sample. What is the probability that one or fewer batteries in the sample are defective? Here $N = 10,000$, $n = 100$ and $p = 100/10,000 = 0.01$. If x is the number of defective batteries in the sample, we wish to determine the probability that $x \leq 1$. By (2–17) and (2–6), this probability is

$$h(0; 100, 0.01, 10,000) + h(1; 100, 0.01, 10,000) . \qquad (2\text{--}19)$$

For the case under consideration p is small and $np = 1$ is moderately large, while n is small with respect to N. Thus we should be able to compute (2–19) approximately using the Poisson approximation to each term, that is, using

$$p(0; 1) + p(1; 1) = 0.3679 + 0.3679 = 0.7358 .$$

The numerical values were obtained from Table D. Thus the probability that one or less batteries in the sample selected is defective is about 0.74.

2–7 THE SUMMATION NOTATION

In the past we frequently found it necessary to add together a number of probabilities. To avoid continually writing out long sums, it is convenient to introduce a special notation for summations, called the summation notation. To characterize completely the sum of n elements $a_1, \ldots, a_n$, which we have been writing $a_1 + a_2 + \cdots + a_n$, all we need to do is specify the general form of the elements to be added, here a_j, indicate what values j takes on and introduce a symbol to represent the operation of summation. A special symbol Σ (the capital Greek letter sigma) is used in mathematics to represent the operation of summation. It is called the *summation sign*. We shall represent the sum under consideration as

$$\sum_{j=1}^{n} a_j .$$

Immediately after the summation sign we place the symbol a_j for the general form of the elements being summed. We call j the *summation index*, and when j takes on all integral values between two specified integers, we give the smallest value of j (1 in our sample) below the summation sign and the largest value (n in our example) on top of the summation sign. These largest and smallest values of j are referred to as the *limit of the summation*. Thus, by definition,

$$\sum_{j=1}^{n} a_j = a_1 + a_2 + \cdots + a_n . \qquad (2\text{--}20)$$

We read the left-hand side of (2–20) as "the sum from j equal one to j equal n of a_j." More generally, if m and n are any two integers such that $n > m$, then

$$\sum_{j=m}^{n} a_j = a_m + a_{m+1} + \cdots + a_n. \qquad (2\text{--}21)$$

The symbol used for the summation index is irrelevant; it need not be j. Thus

$$\sum_{j=1}^{n} a_j, \quad \sum_{i=1}^{n} a_i, \quad \sum_{k=1}^{n} a_k$$

all mean the same thing, that is, $a_1 + \cdots + a_n$.

EXAMPLES. *1.* $\displaystyle\sum_{j=1}^{n} j = 1 + 2 + \cdots + n$.

2. $\displaystyle\sum_{j=1}^{m} j^2 = 1^2 + 2^2 + \cdots + m^2$.

3. $\displaystyle\sum_{j=1}^{n} (1 + j) = (1 + 1) + (1 + 2) + \cdots + (1 + n)$.

4. $\displaystyle\sum_{j=1}^{n} \lambda = \lambda + \lambda + \cdots + \lambda = n\lambda$. $\qquad (2\text{--}22)$

Here $a_j = \lambda$. Note carefully that the sum is $n\lambda$ and not λ.

5. $\displaystyle\sum_{i=1}^{k} p(y_i) = p(y_1) + p(y_2) + \cdots + p(y_k)$.

6. $\displaystyle\sum_{j=1}^{n} a_{ij} = a_{i1} + \cdots + a_{in}; \quad \sum_{i=1}^{k} a_{ij} = a_{1j} + \cdots + a_{kj}$.

Sometimes two subscripts are needed, and summation can be carried out with respect to either one.

7. Consider the following array of mn numbers:

$$\begin{matrix} a_{11} & a_{12} & \cdots & a_{1n} \\ a_{21} & a_{22} & \cdots & a_{2n} \\ \cdot & & & \cdot \\ \cdot & & & \cdot \\ \cdot & & & \cdot \\ a_{m1} & a_{m2} & \cdots & a_{mn}. \end{matrix}$$

The sum of the elements in the ith row is

$$r_i = \sum_{j=1}^{n} a_{ij}$$

and the sum of all mn numbers is

$$\sum_{i=1}^{m} r_i = \sum_{i=1}^{m} \left[\sum_{j=1}^{n} a_{ij} \right],$$

since to sum all elements we can add the numbers in each row and then add these sums. The sum of the elements in column j is

$$c_j = \sum_{i=1}^{m} a_{ij}$$

and the sum of all the numbers in the array is then

$$\sum_{j=1}^{n} c_j = \sum_{j=1}^{n} \left[\sum_{i=1}^{m} a_{ij} \right].$$

Thus we have shown that

$$\sum_{i=1}^{n} \left[\sum_{j=1}^{n} a_{ij} \right] = \sum_{j=1}^{n} \left[\sum_{i=1}^{m} a_{ij} \right].$$

Hence the summation signs can be interchanged. This summation is often written without the brackets as

$$\sum_{i=1}^{m} \sum_{j=1}^{n} a_{ij} \quad \text{or} \quad \sum_{j=1}^{n} \sum_{i=1}^{m} a_{ij}.$$

There are two properties of summations which are used continually and we shall now derive these. Observe that

$$\sum_{j=1}^{n} (a_j + b_j) = (a_1 + b_1) + \cdots + (a_n + b_n)$$

$$= (a_1 + \cdots + a_n) + (b_1 + \cdots + b_n)$$

$$= \sum_{j=1}^{n} a_j + \sum_{j=1}^{n} b_j,$$

that is,

$$\sum_{j=1}^{n} (a_j + b_j) = \sum_{j=1}^{n} a_j + \sum_{j=1}^{n} b_j. \qquad (2\text{–}23)$$

Thus when each of the terms in the summation can be written as the sum of two terms, the resulting sum can be split into two summations, as indicated in (2–23). The same reasoning shows, of course, that if each term in the summation can be written as the sum of r terms, then the summation can be split up into r summations.

To obtain the other property note that

$$\sum_{j=1}^{n} \lambda a_j = \lambda a_1 + \cdots + \lambda a_n = \lambda(a_1 + \cdots + a_n) = \lambda \sum_{j=1}^{n} a_j ,$$

so

$$\sum_{j=1}^{n} \lambda a_j = \lambda \sum_{j=1}^{n} a_j . \qquad (2\text{--}24)$$

In words, all that (2–24) says is that a common factor may be taken outside the summation sign.

2-8 EXPECTED VALUES AGAIN

In working with a random experiment, we shall frequently be dealing not with a single random variable, but with several random variables simultaneously. For example, if we consider the random experiment $\Re$ in which a coin is tossed twice, there are three random variables associated with the occurrence of heads which may be of interest. The first is the total number of heads obtained on performing $\Re$, the second is the number of heads on the first toss, and the third is the number of heads obtained on the second toss.

Consider now an arbitrary random experiment $\Re$ for which we are using the model $\mathcal{E}$ having the event set $E = \{e_1, \ldots, e_m\}$, with the probability of e_j being p_j. Suppose that there are two random variables x and y associated with $\Re$ which are of interest, such that if e_j occurs then $x = \xi_j$ and $y = \zeta_j$. The values of both random variables are determined, of course, when the simple event that occurred on performing $\Re$ is specified. Now for each j, suppose that we compute the number

$$\theta_j = a\xi_j + b\zeta_j + c, \quad j = 1, \ldots, m , \qquad (2\text{--}25)$$

where a, b and c are some specified numbers. In this way we associate with each e_j a new number θ_j. But this rule of association automatically defines a new random variable z, such that $z = \theta_j$ when e_j occurs. An alternative way to write (2–25) is $z = ax + by + c$. What do we mean by writing $z = ax + by + c$? We mean that whatever simple event of $\Re$ occurs, the value of z is a times the value of x plus b times the value of y plus c.

In practice one frequently finds it convenient to define new random variables in terms of other random variables by expressions such as $z = ax + by + c$. For example, consider some project whose outcome is determined by chance, and imagine that both the revenues to be received x and the cost incurred y are random variables. The profit z is also a random variable and $z = x - y$, since the profit is equal to the revenues received minus the costs incurred. Here $z = x - y$ is a special case of $z = ax + by + c$ with $a = 1$, $b = -1$, and $c = 0$. As yet another example, consider the experiment involving two tosses of a coin and let x be the number of heads

on the first toss and y the number of heads obtained on the second toss. Then $z = x + y$ is the total number of heads obtained on performing $\Re$.

We shall now show that if $z = ax + by + c$ there is a very simple relation between the expected value of z and the expected values of x and y. Denote the expected values of x, y and z by μ_x, μ_y and μ_z, respectively. Then using the definition (2–3) of the expected value of any random variable, we see that

$$\mu_x = \sum_{j=1}^{m} \xi_j p_j; \quad \mu_y = \sum_{j=1}^{m} \zeta_j p_j; \quad \mu_z = \sum_{j=1}^{m} \theta_j p_j . \qquad (2\text{–}26)$$

Here we have conveniently used the summation notation introduced in the previous section. If we now use in the expression for μ_z the value of θ_j given by (2–25) we have

$$\mu_z = \sum_{j=1}^{m} (a\xi_j + b\zeta_j + c) p_j = \sum_{j=1}^{m} (a\xi_j p_j + b\zeta_j p_j + c p_j) . \qquad (2\text{–}27)$$

However, on using (2–23), we obtain

$$\mu_z = \sum_{j=1}^{m} a\xi_j p_j + \sum_{j=1}^{m} b\zeta_j p_j + \sum_{j=1}^{m} c p_j . \qquad (2\text{–}28)$$

Equation (2–24) is used next to yield

$$\mu_z = a \sum_{j=1}^{m} \xi_j p_j + b \sum_{j=1}^{m} \zeta_j p_j + c \sum_{j=1}^{m} p_j . \qquad (2\text{–}29)$$

However, the definitions of μ_x and μ_y can now be utilized to yield

$$\mu_z = a\mu_x + b\mu_y + c \sum_{j=1}^{m} p_j . \qquad (2\text{–}30)$$

Recall that the probabilities of the simple events sum to 1, so that $\Sigma_{j=1}^{m} p_j = 1$ and

$$\mu_z = a\mu_x + b\mu_y + c . \qquad (2\text{–}31)$$

Equation (2–31) is important and will be used often in the future. We have shown that if $z = ax + by + c$, then μ_z is related to μ_x and μ_y by (2–31). Thus if z is interpreted as the profit from the project referred to above, (2–31) says that the expected profit is equal to the expected revenues minus the expected costs. If z is the number of heads obtained on two tosses of a coin (2–31), indicates that the expected number of heads is the expected number of heads obtained on the first toss plus the expected number of heads obtained on the second toss.

The results just obtained can easily be generalized to cases where a number of random variables is involved. When dealing with a sizable number of random variables it is desirable to use a notation different from x, y and z to represent the random variables, since one soon uses up every possible letter. Instead, to represent n different random variables we shall use a notation

such as $x\{1\}, \ldots, x\{n\}$. We cannot use $x_1, \ldots, x_n$ because we have used this notation previously to denote the different values which a given random variable x takes on.

Suppose then that we have n random variables $x\{1\}, \ldots, x\{n\}$ associated with a random experiment $\Re$, and that a new random variable z is defined by the equation

$$z = a_1 x\{1\} + \cdots + a_n x\{n\} + c .\tag{2-32}$$

Then precisely the same reasoning used above shows that

$$\mu_z = a_1\mu_1 + \cdots + a_n\mu_n + c ,\tag{2-33}$$

where μ_j is the expected value of $x\{j\}$.

Equation (2–33) can now be used to derive in a very simple way the mean of the binomial distribution. By (2–11) the mean of $b(x; n, p)$ is

$$\mu_x = \sum_{r=0}^{n} r b(r; n, p) .\tag{2-34}$$

One can compute the mean directly using (2–34), but it is easier to use (2–33). Let us show how this can be done. Recall that if x has the distribution $b(x; n, p)$, then x can be thought of as the number of times e_1 occurs in n independent trials $\Re^*$ when as the model for $\Re^*$ we are using $\mathcal{E}^*$, with the event set $E^* = \{e_1, e_2\}$ and with the probability of e_1 being p. Now let $y\{j\}$ be the random variable which gives the number of times e_1 occurs on the jth trial of $\Re$. Thus $y\{j\} = 1$ if e_1 occurs on the jth trial of $\Re$ and $y\{j\} = 0$ if e_2 occurs. The random variable x can be expressed in terms of the $y\{j\}$ as

$$x = y\{1\} + y\{2\} + \cdots + y\{n\} ,\tag{2-35}$$

and thus by (2–33)

$$\mu_x = \mu_1 + \mu_2 + \cdots + \mu_n ,\tag{2-36}$$

where μ_j is the expected value of $y\{j\}$.

It is very easy to compute μ_j. To see this note that $y\{j\}$ only takes on the values 1 and 0. It takes on the value 1 if e_1 occurs on the jth trial. The probability that e_1 occurs is p. It takes on the value of 0 if e_2 occurs; the probability of e_2 is q. Thus the probability function for $y\{j\}$ is given by $p(1) = p$ and $p(0) = q$. Hence by (2–11)

$$\mu_j = 1p + 0q = p ,\tag{2-37}$$

so $\mu_j = p$ for each $j = 1, \ldots, n$. Thus by (2–36)

$$\mu_x = p + p + \cdots + p = np .\tag{2-38}$$

We have thus proved that the mean of $b(x; n, p)$ is np. It also happens to be true that the mean of the Poisson distribution $p(x; \beta)$ is β. We shall not

attempt to prove this. However, given that this is true, we see that to approximate $b(x; n, p)$ with the Poisson distribution we use a Poisson distribution with the same mean as the binomial, because as we noted earlier $b(r; n, p)$ is approximated by $p(r; np)$.

Recall that μ_x can be interpreted intuitively as the long-run average value of x when $\Re$ is repeated a large number of times. Thus if we repeat a large number of times the experiment $\Re$ which involves making n independent trials of $\Re^*$, and each time we note the number of times e_1 occurred, that is, the value of x, then the average of these values should get closer and closer to np as the number of times $\Re$ is repeated is increased unendingly. It is very important to note that $\mu_x = np$ is not the value of x we expect to be observed on any given trial of $\Re$. Indeed, μ_x is not necessarily even one of the values which x can take on. If we consider $b(x; 9, 0.5)$, for example, $\mu_x = 9(0.5) = 4.5$. This is not one of the possible values of x, that is, e_1 will never occur 4.5 times on performing $\Re^*$ a total of 9 times. Even when μ_x is an allowable value of x, the probability that $x = \mu_x$ may be extremely small. Consider $b(x; 30, 0.5)$, where $\mu_x = 15$. The probability that $x = 15$ is 0.1445, from Table C, so only about 15 times in 100 will x take on its expected value when $\Re$ is performed.

2–9 FUNCTIONS OF A RANDOM VARIABLE

We noted in the previous section that frequently we must be concerned not with just a single random variable but with several different random variables. Often it turns out that all of the random variables of interest are functions of one particular random variable. We wish to study such situations in this section.

What do we mean by saying that a random variable y is a function of another random variable x? To explain this, suppose that we associate with each possible value x_i which the random variable x can take on a unique number y_i. This association defines a function whose domain is the set $X = \{x_1, \ldots, x_k\}$ of different values that x can take on. If we denote the image of x_i by $g(x_i)$, then $y_i = g(x_i)$, since y_i is the image of x_i. Let Y be the set of different values of the y_i. We shall then show that Y can be considered as the set of values which some random variable y can take on. To do this, we need only show that associated with each simple event e_j there is a unique number in Y. If e_j occurs then a value ξ_j of x will be determined. Furthermore, ξ_j is one of the x_i, so $\xi_j \in X$. Let ζ_j be the image of ξ_j so that $\zeta_j = g(\xi_j)$ and $\zeta_j \in Y$. Consequently, in this way we have associated with each e_j a number $\zeta_j \in Y$, and this automatically defines a random variable y which can take on the values in Y.

The value of y is determined once the value of x is known, and to indicate this we shall write $y = g(x)$, and say that the random variable y is a func-

tion of the random variable x. Situations where one variable is a function of another arise naturally in practice. For example, the profit a firm receives from some product will be a function of the number of units sold. If the number of units sold is imagined to be a random variable x, then the profit is also a random variable y which is a function of x. As another example, the income tax of an individual is a function of his taxable income. If the individual under consideration must treat his taxable income for the coming year as a random variable (perhaps he is a gambler), then his next year's income tax will also be a random variable which is a function of another random variable, his taxable income.

If $y = g(x)$, then μ_y, the expected value of y, is, by definition

$$\mu_y = \sum_{j=1}^{m} \zeta_j p_j = \sum_{j=1}^{m} g(\xi_j) p_j , \qquad (2\text{--}39)$$

where p_j is the probability of the simple event e_j, and where ζ_j and ξ_j are defined as above. Let S_i be the event that $x = x_i$. Then for each $e_j \in S_i$, $x = x_i = \xi_j$ and $g(\xi_j) = g(x_i)$. The probability of S_i is what we have previously called $p(x_i)$. Thus if we collect together in (2–39) all $g(\xi_j) p_j$ for which $\xi_j = x_i$ and sum these we obtain $g(x_i) p(x_i)$. Hence we see that

$$\mu_y = \sum_{i=1}^{k} g(x_i) p(x_i) . \qquad (2\text{--}40)$$

Equation (2–40) tells us how to compute the expected value of a random variable y, which is a function of x, using the probability function for x. We compute the image $g(x_i)$ of each $x_i \in X$, multiply this by the probability that $x = x_i$ and sum the results. This is a result that will be used frequently in the future. Let us now provide some examples.

EXAMPLES.   *1.* If x is a random variable, then some examples of functions which serve to define a new random variable y are

$$y = 2x; \quad y = x^2; \quad y = 3x^2 - 2x + 1; \quad y = \frac{4 - x}{x^2 + 2} .$$

2. Suppose that x is a random variable which takes on the values indicated in Table 2–8. Let $y = x^2$. Then the value of y taken on for each simple event is that given in Table 2–8.

TABLE 2–8

Simple event	e_1	e_2	e_3	e_4	e_5
x	-2	0	4	-2	1
$y = x^2$	4	0	16	4	1

3. Let x be a random variable with probability function $p(x)$. Suppose that a new random variable y is defined by $y = ax + b$, so that y is a function of x. Then by (2–40), the expected value of y is

$$\mu_y = \sum_{i=1}^{k} (ax_i + b)p(x_i) = a\sum_{i=1}^{k} x_i p(x_i) + b\sum_{i=1}^{k} p(x_i)$$

$$= a\mu_x + b .   \tag{2–41}$$

Here we have obtained in a slightly different way a special case of the result obtained in the previous section.

4. Let x be a random variable with expected value μ_x. Consider now the random variable $y = x - \mu_x$ so that y gives the deviation of x from its expected value. This is just a special case of the situation studied in Example 4 where $a = 1$ and $b = -\mu_x$. Hence from (2–41), $\mu_y = \mu_x - \mu_x = 0$. In other words, when $y = x - \mu_x$, then $\mu_y = 0$. This is what we would expect intuitively, since if μ_x is the long-run average value of x, the long-run average deviation of x from its average value should be 0.

5. If $y = x^2$, and x has the probability function $p(x)$, then the expected value of y is, by (2–40),

$$\mu_y = \sum_{i=1}^{k} x_i^2 p(x_i) .$$

Suppose that the probability function for x is that given in Table 2–9. Then μ_y is given by

$$\mu_y = 9(0.2) + 1(0.1) + 0(.03) + 4(0.3) + 25(0.1)$$
$$= 1.8 + 0.10 + 0 + 1.2 + 2.5 = 5.6 .$$

The reader should check that $\mu_x = 0.4$. Thus $\mu_y \neq \mu_x^2$ even though $y = x^2$. In other words the expected value of the square of a random variable is not in general equal to the square of the expected value of the random variable.

TABLE 2–9

x	-3	-1	0	2	5
$p(x)$	0.2	0.1	0.3	0.3	0.1

Just as we can introduce one random variable which is a function of another, we can make use of the concept of a random variable z which is a function of two random variables x and y. To indicate that z is a function of x and y we can write $z = g(x, y)$. By this we mean that the value of z is determined by the values of x and y. Examples of function of this sort are $z = xy$, $z = x^2 + y^2$, $z = 3x - 2y$ and $z = x^2 - xy$.

2–10 THE VARIANCE OF A RANDOM VARIABLE

Let x be a random variable having the expected value μ_x. If the random experiment $\mathcal{R}$ with which x is associated is repeated a number of times, a set of values of x will be generated, one for each trial. Suppose now that the value of x obtained on each trial is represented graphically as shown in Figure 2–8.

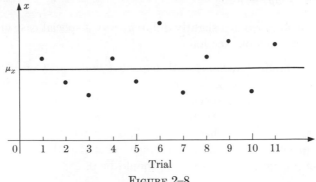

FIGURE 2–8

The value of x observed will not, in general, be the same on each trial. Suppose we draw a horizontal line which crosses the vertical axis at μ_x, as shown in Figure 2–8. Some of the observed values of x will then be above this line and some will be below it, since the long-run average of the values of x is μ_x. It is very useful to have a numerical measure of the fluctuations of the values of x about μ_x that will be observed if $\mathcal{R}$ is repeated many times. What one is looking for normally is not the maximum deviation which x could take from μ_x, but instead a measure of the typical deviations.

In the previous section we introduced a random variable $y = x - \mu_x$ which is the deviation of x from μ_x. We showed that $\mu_y = 0$, the reason for this being that some values of y are positive while others are negative. When examining the spread of the values of x about μ_x, one does not care whether $x > \mu_x$ or $x < \mu_x$. Now the random variable $z = y^2 = (x - \mu_x)^2$ has the characteristic that $z > 0$ whenever $x \neq \mu_x$. Thus z does provide a measure of how much x deviates from μ_x which is independent of whether $x > \mu_x$ or $x < \mu_x$. The expected value of z turns out to be a useful measure of the typical fluctuations of the values of x about μ_x that will be exhibited when $\mathcal{R}$ is repeated a large number of times.

The expected value of the random variable $z = (x - \mu_x)^2$ is called the *variance of x* and is usually denoted by σ_x^2; σ_x^2 is also referred to as the variance of the distribution of x. Note that z is a function of x. Hence if we know the probability function $p(x)$ for x, then by (2–40), σ_x^2 is given by

$$\sigma_x^2 = \sum_{i=1}^{k} (x_i - \mu_x)^2 p(x_i) . \qquad (2\text{–}42)$$

If σ_x^2 provides a numerical measure of the fluctuations that x will exhibit about μ_x, the reader will no doubt feel that we should be able to characterize precisely the way in which σ_x^2 does this. From the definition of z and the definition of expected values, we can say that σ_x^2 is the long-run average of the squares of the deviations of x from μ_x. This is the intuitive interpretation of σ_x^2. However, it does not give us a very clear idea of precisely how σ_x^2 might be usefully employed to measure the fluctuations. It is not easy to provide a simple characterization of what any particular numerical value of σ_x^2 implies. The usefulness of the variance concept will become clearer as we go along and for this reason we shall not attempt at this point to discuss in detail its precise interpretation.

Most random variables of interest in the real world have some physical dimensions associated with them, such as dollars, pounds or feet. Note that μ_x has the same physical dimensions as x. However, σ_x^2 does not have the same physical dimensions as x since $z = (x - \mu_x)^2$, so z and therefore σ_x^2 have physical dimensions which are the square of those of x. Thus if x has dimensions of dollars, σ_x^2 has the dimensions of dollars squared. It is desirable to have the numerical measure of the fluctuations of x about μ_x have the same physical dimensions as x. This can be accomplished by using instead of σ_x^2 the square root of this quantity, which is denoted by σ_x; $\sigma_x = \sqrt{\sigma_x^2}$ is called the *standard deviation of the random variable* x. The standard deviation σ_x has the same physical dimensions as x, and for this reason it is often more convenient to use σ_x as a measure of the fluctuations rather than σ_x^2. Note however, that in order to obtain the standard deviation σ_x it is first necessary to compute the variance σ_x^2. We can now introduce the notions of variance and standard deviation into our probability model with the following definitions.

VARIANCE OF A RANDOM VARIABLE. *The variance of a random variable x with distribution $p(x)$ is the unique nonnegative number σ_x^2 computed according to (2–42).*

STANDARD DEVIATION OF A RANDOM VARIABLE. *The standard deviation of the random variable x with distribution $p(x)$ is the nonnegative number $\sigma_x = \sqrt{\sigma_x^2}$, where σ_x^2 is the variance of x.*

EXAMPLE. Consider two random variables x and y which have the probability functions shown in Figures 2–9 and 2–10 respectively. Both random variables have precisely the same expected value of 0.75, since from the figures we see that

$$\mu_x = 0.25(-2) + 0.50(1) + 0.25(3) = 0.75$$
$$\mu_y = 0.25(-5) + 0.50(1) + 0.25(6) = 0.75 .$$

From (2–42), we see that the variances of x and y are

$$\sigma_x^2 = 0.25(-2 - 0.75)^2 + 0.50(1 - 0.75)^2 + 0.25(3 - 0.75)^2$$
$$= 1.89 + 0.03125 + 1.27 = 3.19$$

and

$$\sigma_y^2 = 0.25(-5 - 0.75)^2 + 0.50(1 - 0.75)^2 + 0.25(6 - 0.75)^2$$
$$= 8.29 + 0.03125 + 6.88 = 15.20 .$$

Hence $\sigma_x = \sqrt{3.19} = 1.78$ and $\sigma_y = \sqrt{15.2} = 3.88$. Thus $\sigma_y > \sigma_x$. The reason for this can be seen from Figures 2–9 and 2–10. The bar diagram for y is more spread out than that for x, so that the typical deviations of y from μ_y will be greater than those of x from μ_x. This also points out that the standard

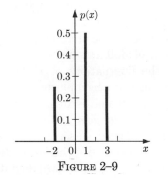

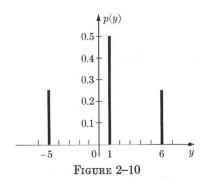

FIGURE 2–9   FIGURE 2–10

deviation tells us something about the bar diagram for a random variable, in the sense that if we have two random variables with the same mean, the bar diagram for the one with the larger standard deviation will be more spread out than the one with the smaller standard deviation. The standard deviation does not, however, give us any detailed information about the nature of the bar diagram.

It is often a little inconvenient to use (2–42) to compute σ_x^2 because one must first compute $x_i - \mu_x$ for each i. We shall now show that (2–42) can be converted to an alternative form which is easier to use for computational purposes. First observe that

$$(x_i - \mu_x)^2 = (x_i - \mu_x)(x_i - \mu_x) = x_i^2 - 2\mu_x x_i + \mu_x^2 .$$

Thus (2–42) can be written

$$\sigma_x^2 = \sum_{i=1}^{k} (x_i^2 - 2\mu_x x_i + \mu_x^2)p(x_i) , \tag{2–43}$$

or by (2–23) and (2–24)

$$\sigma_x^2 = \sum_{i=1}^{k} x_i^2 p(x_i) - 2\mu_x \sum_{i=1}^{k} x_i p(x_i) + \mu_x^2 \sum_{i=1}^{k} p(x_i) . \qquad (2\text{--}44)$$

However

$$\sum_{i=1}^{k} x_i p(x_i) = \mu_x; \qquad \sum_{i=1}^{k} p(x_i) = 1,$$

so

$$\sigma_x^2 = \sum_{i=1}^{k} x_i^2 p(x_i) - 2\mu_x^2 + \mu_x^2$$

or

$$\sigma_x^2 = \sum_{i=1}^{k} x_i^2 p(x_i) - \mu_x^2 . \qquad (2\text{--}45)$$

Equation (2–45) provides a convenient way to compute σ_x^2. Note that (2–45) says that σ_x^2 is equal to the expected value of x^2 minus the expected value of x squared.

2–11 VARIANCE OF SUMS OF RANDOM VARIABLES

Consider a random experiment which has associated with it two random variables x and y. Suppose now that a new random variable z is defined by writing $z = ax + by + c$. From (2–31) we know that $\mu_z = a\mu_x + b\mu_y + c$. Let us now determine the variance of z, σ_z^2. By definition σ_z^2 is the expected value of $(z - \mu_z)^2$. However

$$\begin{aligned}
(z - \mu_z)^2 &= (ax + by + c - a\mu_x - b\mu_y - c)^2 \\
&= [a(x - \mu_x) + b(y - \mu_y)]^2 \\
&= a^2(x - \mu_x)^2 + 2ab(x - \mu_x)(y - \mu_y) + b^2(y - \mu_y)^2 \\
&= a^2 w + 2ab v + b^2 u , \qquad (2\text{--}46)
\end{aligned}$$

where $w = (x - \mu_x)^2$, $v = (x - \mu_x)(y - \mu_y)$ and $u = (y - \mu_y)^2$. We now have the random variable $(z - \mu_z)^2$ written in the form (2–32). Hence by (2–33)

$$\sigma_z^2 = a^2 \mu_w + 2ab \mu_v + b^2 \mu_u . \qquad (2\text{--}47)$$

However, the expected value of $w = (x - \mu_x)^2$ is σ_x^2 and the expected value of $u = (y - \mu_y)^2$ is σ_y^2. We shall denote the expected value of $v = (x - \mu_x)(y - \mu_y)$ by σ_{xy} rather than μ_u; σ_{xy} is called the *covariance* of the random variables x and y. Thus

$$\sigma_z^2 = a^2 \sigma_x^2 + 2ab \sigma_{xy} + b^2 \sigma_y^2 , \qquad (2\text{--}48)$$

and we have expressed the variance of z in terms of the variance of x and the variance of y and the covariance of x and y. Note that while μ_z can be expressed in terms of μ_x and μ_y, σ_z^2 cannot be expressed in terms of σ_x^2 and σ_y^2 only; σ_{xy} appears also.

There is an important case where $\sigma_{xy} = 0$. When $\sigma_{xy} = 0$ then σ_z^2 can be expressed in terms of σ_x^2 and σ_y^2. Suppose that $\mathcal{R}$ is built up from two independent random experiments $\mathcal{R}_1$ and $\mathcal{R}_2$, and x is associated with $\mathcal{R}_1$ while y is associated with $\mathcal{R}_2$. Random variables which are associated with different independent random experiments are called *independent random variables*. We are thus now considering a case where x and y are independent random variables. Let us now compute σ_{xy} when x and y are independent random variables. Let us use as a model for $\mathcal{R}_1$ one in which the event set is X, the different values which x can assume, and as a model for $\mathcal{R}_2$ one in which the event set is Y, the different values which y can assume. Then from Section 1–11 we can use as a model for $\mathcal{R}$ one in which the simple events are of the form $\langle x_i, y_j \rangle$, and this is the event that $x = x_i$ and $y = y_j$. Its probability is $p_1(x_i)p_2(y_j)$, when $p_1(x)$ is the probability function for x and $p_2(y)$ is the probability function for y. If x can take on k different values and y can take on s different values, there will be ks simple events $\langle x_i, y_j \rangle$. Then $\sigma_{xy} = \mu_v$ is simply the sum of the products of the value of v for each simple event $\langle x_i, y_j \rangle$ times the probability of this simple event, that is,

$$
\begin{aligned}
\sigma_{xy} = & (x_1 - \mu_x)(y_1 - \mu_y)p_1(x_1)p_2(y_1) + \cdots + (x_k - \mu_x)(y_1 - \mu_y)p_1(x_k)p_2(y_1) \\
& + (x_1 - \mu_x)(y_2 - \mu_y)p_1(x_1)p_2(y_2) \\
& + \cdots + (x_k - \mu_x)(y_2 - \mu_y)p_1(x_k)p_2(y_2) \\
& + \cdots + (x_1 - \mu_x)(y_s - \mu_y)p_1(x_1)p_2(y_s) \\
& + \cdots + (x_k - \mu_k)(y_s - \mu_y)p_1(x_k)p_2(y_s) \\
= & \left[\sum_{i=1}^{k} (x_i - \mu_x)p_1(x_i) \right] [(y_1 - \mu_y)p_2(y_1) + \cdots + (y_s - \mu_y)p_2(y_s)] .
\end{aligned}
$$

$$(2\text{–}49)$$

However $\sum_{i=1}^{k}(x_i - \mu_x)p_1(x_i)$ is the expected value of $y = x - \mu_x$, which we have noted previously is 0. Hence $\sigma_{xy} = 0$. We have thus proved that $\sigma_{xy} = 0$ when x and y are independent random variables, and so

$$\sigma_z^2 = a^2\sigma_x^2 + b^2\sigma_y^2, \quad x \text{ and } y \text{ independent} . \tag{2–50}$$

This is a result which we shall use many times.

The above analysis is easily generalized to the case where

$$z = a_1 x\{1\} + \cdots + a_n x\{n\} + c . \tag{2–51}$$

Then

$$
\sigma_z^2 = a_1^2\sigma_1^2 + \cdots + a_n^2\sigma_n^2 + 2a_1a_2\sigma_{12} + \cdots + 2a_1a_n\sigma_{1n}
$$
$$
+ \cdots + 2a_{n-1}a_n\sigma_{n-1,n} , \tag{2–52}
$$

where σ_j^2 is the variance of $x\{j\}$ and σ_{ij} is the covariance of $x\{i\}$ and $x\{j\}$, that is, the expected value of $(x\{i\} - \mu_i)(x\{j\} - \mu_j)$. Equation (2–52) reduces to a simpler form if the random variables $x\{j\}$ are independent. The definition of n independent random variables can be made using a generalization of the definition for two variables as follows.

INDEPENDENT RANDOM VARIABLES. *Let $\Re$ be a random experiment which has associated with it n random variables $x\{1\}, \ldots, x\{n\}$. Suppose that $\Re$ is built up from n independent random experiments $\Re_1, \ldots, \Re_n$ and, in addition, each random variable $x\{j\}$ is associated with a different one of these independent random experiments, so that $x\{1\}$ is associated with $\Re_1, \ldots,$ and $x\{n\}$ is associated with $\Re_n$. Then the random variables $x\{1\}, \ldots, x\{n\}$ are said to be independent random variables.*

When the random variables $x\{1\}, \ldots, x\{n\}$ in (2–51) are independent, then precisely the same proof used above shows that $\sigma_{ij} = 0$, that is, the covariance of $x\{i\}$ and $x\{j\}$ is 0, $i \neq j$. In this case, (2–52) reduces to the much simpler form

$$\sigma_z^2 = a_1^2 \sigma_1^2 + \cdots + a_n^2 \sigma_n^2, \quad x\{1\}, \ldots, x\{n\} \text{ independent} . \quad (2\text{–}53)$$

As an example of the usefulness of (2–53), we shall employ it to determine the variance of the binomial distribution. Recall that if x has a binomial distribution $b(x; n, p)$, then x can be written as the sum of n random variables $y\{j\}$ as in (2–35), where $y\{j\}$ is the number of occurrences of e_1 on trial j. We can now note that since $\Re$ consists of n independent trials of $\Re^*$, and each $y\{j\}$ refers to a different trial of $\Re^*$, the $y\{j\}$ are independent random variables. Hence by (2–53)

$$\sigma_x^2 = \sigma_1^2 + \cdots + \sigma_n^2 , \quad (2\text{–}54)$$

where σ_j^2 is the variance of $y\{j\}$; here $a_j^2 = 1$ since $a_j = 1$. Now it is easy to determine σ_j^2. Recall that we have shown previously that $\mu_j = p$. Thus by (2–45), since $y\{j\} = 1$ with probability p and $y\{j\} = 0$ with probability q,

$$\sigma_j^2 = 1^2 p + 0^2 q - p^2 = p(1 - p) = pq . \quad (2\text{–}55)$$

Each $y\{j\}$ has the same variance pq. Thus from (2–54)

$$\sigma_x^2 = pq + \cdots + pq = npq , \quad (2\text{–}56)$$

and we see that the variance of $b(x; n, p)$ is npq. The standard deviation of the binomial distribution is then

$$\sigma_x = \sqrt{npq} . \quad (2\text{–}57)$$

We have now shown that the mean of $b(x; n, p)$ is $\mu_x = np$ and the standard deviation is $\sigma_x = \sqrt{npq}$. We shall use these results frequently in our future work.

2–12 **RANDOM VARIABLES WITH MANY VALUES**

Often one encounters random variables which can take on a large number of different values. It is usually exceedingly clumsy to make probability computations involving such random variables if one proceeds directly using the techniques we have discussed previously. For example, to compute the value of the cumulative function for a given value of the random variable, it might be necessary to add up 100,000 different values of the probability function. This would be totally out of the question to carry out by hand. In this section and the next we wish to study two ways in which these complications can be avoided.

Before proceeding let us give some examples of random variables which can take on a large number of different values. Consider a geologist who is examining some structure for the possibility that it contains oil. For a structure of the type he is studying, the volume of oil in place might vary from 0 to 10 million barrels. We can think of the quantity of oil in place as being a random variable x. The quantity would never be measured in units smaller than a barrel. Nonetheless, x can take on about 10 million different values. As another example, the return on a project might vary from a loss of one million to a profit of 10 million dollars. If we think of the return as being a random variable x, and if x is measured to the nearest cent, then x can take on more than one billion values. To take a less extreme example, consider a process which turns out ball bearings. The diameter of any particular ball bearing can be thought of as a random variable. The diameter can only be measured to the nearest hundred thousandth of an inch, but to this accuracy the variations in diameter make it possible for x to take on a thousand or so different values. As a final example, suppose that we were concerned with the random variable x giving the number of gallons of gasoline sold at a particular station during a month. This might vary from 50,000 to 80,000 gallons, so that if x is measured to the nearest gallon, x could take on about 30,000 different values.

There are basically two alternative approaches that can be used to simplify the treatment of random variables which can take on a large number of values. These two approaches work in opposite directions. One decreases the number of alternatives. The other increases the number of alternatives unendingly. Both procedures are useful, and the one which is most appropriate depends on the situation.

Consider first the procedure which reduces the number of alternatives. Frequently when we have a random variable which takes on a very large number of values, a little thought will indicate that the random variable provides us with too much detail about the outcome of the experiment. We are not really interested in every possible value x can take on. All we are really interested in is the probability that x will be in a relatively small

number of intervals, that is, we can imagine that the total range of variation of x is subdivided into a number of intervals by the values $\gamma_0, \gamma_1, \ldots, \gamma_m$, and all we are really interested in is the probability that $\gamma_{j-1} \leq x < \gamma_j$ for $j = 1, \ldots, m$. Let e_j be the event that x lies in the interval $\gamma_{j-1} \leq x < \gamma_j$. Then one and only one e_j will occur, and we can construct a model for the experiment using the e_j as simple events. Given that there are a relatively small number of e_j the problem has then been reduced to one that can be handled more conveniently. Thus, for example, instead of being concerned about how many barrels of oil the structure might contain, to the nearest barrel, the geologist would probably find it quite satisfactory to be concerned instead with a relatively small number of ranges of values. He might, therefore, want to estimate the probabilities that there are less than 100,000 barrels, between 100,000 and 200,000, between 200,000 and 500,000, between 500,000 and one million and of million-barrel increments after this.

There is another way in which one can view the process of reducing the number of alternatives. This is simply to change the dimensions in which x is measured to a larger unit. Thus in the example dealing with the quantity of gasoline sold by a given station per month it might be quite satisfactory to measure sales only to the nearest thousand gallons, so that x would then be the number of thousands of gallons sold. In this case x would take on the values 50, 51, $\ldots$, 80, so that instead of taking on about 30,000 different values x would now take on only 31 values.

Let us now illustrate with two examples how one can proceed to construct models where ranges of a random variable are used rather than values of the variable. Suppose that we have a population consisting of 100 Air Force officers and we consider the random experiment in which one man is selected at random. Assume that we are interested in the height of the individual selected. Now we can think of the height of the individual selected as being a random variable x. No two men will have precisely the same height and hence for this experiment x can take on 100 different values. Let us suppose that we are really only interested in the height to the nearest inch. Imagine that the height of each man has been measured to the nearest 0.1 inches and the results given in Table 2–10. We have given in column 2 the number of men whose height lies in each of the intervals given in column 1. In column 3 is given the fraction of the men lying in the given interval. In constructing a model of this experiment the numbers in the third column would then be used as the probabilities that the height of the man selected would be in each of the intervals given in the first column, the intervals of the first column representing the simple events. The nature of this model can be conveniently represented geometrically using a diagram similar to that used in representing the probability function for a random variable. One starts off with a horizontal line on which the values of x can be indicated. However, instead of erecting a vertical bar at each value of x, one erects a rectangle

TABLE 2–10

HEIGHT DATA

Height	Number of Men	Fraction of Men
5'2"–5'3"	2	0.02
5'3"–5'4"	3	0.03
5'4"–5'5"	3	0.03
5'5"–5'6"	8	0.08
5'6"–5'7"	10	0.10
5'7"–5'8"	14	0.14
5'8"–5'9"	18	0.18
5'9"–5'10"	21	0.21
5'10"–5'11"	12	0.12
5'11"–6'0"	5	0.05
6'0"–6'1"	3	0.03
6'1"–6'2"	1	0.01

over each subinterval, the height of this rectangle being the probability that x will be in the subinterval. The resulting diagram for the data given in Table 2–10 is shown in Figure 2–11. We shall call a diagram of this type a *relative frequency diagram*.

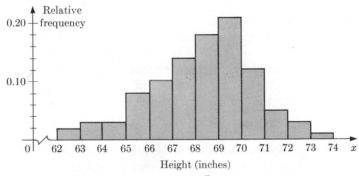

FIGURE 2–11

The intervals into which the range of variation of x is subdivided are often referred to as *class intervals*, and the γ_j are referred to as the *boundary values* for the class intervals. The interval determined by γ_{j-1} and γ_j is called the jth class interval and γ_{j-1} and γ_j are called the boundary values of the jth class interval; $\gamma_j - \gamma_{j-1}$ is called the length of the jth class interval. Sometimes it is helpful to select the γ_j so that they are not one of the possible values of x. It is, however, not always convenient to do this. It was not done in the above example. When this is not done it may be true that $x = \gamma_j$, and one must then decide whether this value is to be included in the jth or

$(j + 1)$th class interval. We adopted the convention above that the jth class intervals includes γ_{j-1} but not γ_j.

As another example let us study a production process which turns out ball bearings. Suppose that we are interested in the diameters of the ball bearings produced. If the diameters of a number of these ball bearings are measured, it will be found that the diameter varies somewhat from one to

TABLE 2–11

DATA FOR BALL BEARINGS

Using class intervals of length 0.001 inches

Class interval	Number of ball bearings	Fraction of ball bearings
$(-5, -4)$	90	0.0090
$(-4, -3)$	420	0.0420
$(-3, -2)$	895	0.0895
$(-2, -1)$	1455	0.1455
$(-1, 0)$	1950	0.1950
$(0, 1)$	2008	0.2008
$(1, 2)$	1622	0.1622
$(2, 3)$	800	0.0800
$(3, 4)$	551	0.0551
$(4, 5)$	209	0.0209

another. It can be imagined that the production of each unit represents a trial of a random experiment. Associated with this experiment is a random variable x, which is the diameter of the ball bearing produced. Now x can take on a very large number of different values. Suppose, however, that we are only interested in the probabilities that x lies in different intervals of length 0.001 inches.

To construct a model for the random experiment representing the production of a ball bearing, we shall use as simple events ten different class intervals of length 0.001 inches. The only way to determine the probabilities of the simple events is to measure a large number of ball bearings, and use the relative frequency of each simple event as the probability of that event. The nominal diameter (diameter aimed for) of the ball bearings under consideration is 0.2500 inches. The results of measuring the diameters of 10,000 ball bearings are given in Table 2–11. In column 1, instead of indicating the interval whose class boundaries are 0.2530 and 0.2540, we simply use (3, 4). Similarly, $(-5, -4)$ refers to the interval whose class boundaries are 0.2450 to 0.2460. In column 2 is given the number of ball bearings whose diameter lay in the indicated class interval, and in column 3 the fraction of the 10,000 bearings whose diameter was in the particular

class interval. The numbers in column 3 would then be used as the probabilities for the class intervals (simple events). The graphical representation of the resulting probability model by use of a relative frequency diagram is shown in Figure 2–12.

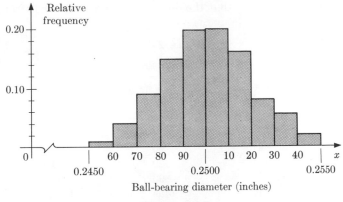

FIGURE 2–12

We have now explained how one can proceed to eliminate the problems caused by a random variable which takes on a large number of values through the procedure of reducing the number of alternatives through the use of class intervals.

2–13 CONTINUOUS RANDOM VARIABLES

Let us now turn to the second procedure for handling random variables which can take on a large number of values. We shall use the ball bearing data summarized in Table 2–11 to motivate the development. In Figure 2–12 the height of each rectangle is the fraction of the ball bearings whose diameter lay in the class interval over which the rectangle is constructed. Thus the height of the rectangle is what we would use as the probability of the event that the diameter x of any ball bearing will be in the corresponding class interval. Now it turns out that it is often convenient to use diagrams in which probabilities are represented by areas. In other words, it is of interest to convert Figure 2–12 to a form where the area of each rectangle represents the probability to be associated with the class interval and not its height. The area of a rectangle is equal to its base times its altitude. If this is to be the probability of the class interval, then the height of the rectangle must be the probability divided by the length of the class interval.

For the case under consideration the length of the class interval is 0.001. Thus if every number in the third column of Table 2–11 is divided by 0.001 or equivalently multiplied by 1000, one obtains the heights of the rectangles

having the characteristic that the area of each is the probability that x will be in the corresponding class interval. The resulting diagram is shown in Figure 2–13. Diagrams in which probabilities are measured by areas are called *histograms*. Figure 2–13 then represents a histogram for the ball bearing data. It might be worth pointing out that some authors use the word histogram both in referring to what we have called a histogram and what we called a relative frequency diagram.

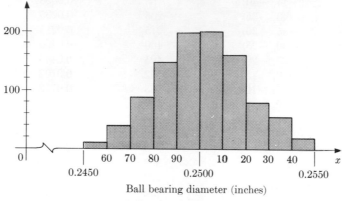

Ball bearing diameter (inches)

FIGURE 2–13

Suppose now that it is decided to construct a new ball bearing model in which the lengths of the class intervals are half those for the model just developed. In other words, each class interval in Table 2–11 is subdivided into two class intervals, each of which has length 0.0005. After the ball bearing data were reclassified on the basis of the new class intervals, Table 2–12 resulted. In the table (15, 20), for example, refers to the class interval whose class boundaries are 0.2515 and 0.2520. Note that if the number of ball bearings in the class intervals (10, 15) and (15, 20) are summed in Table 2–12, one obtains the number of ball bearings in the class interval (1, 2) in Table 2–11. This is true because each class interval of Table 2–11 was split in half to obtain the class intervals for Table 2–12. The same result is then obtained if the numbers of ball bearings are summed for any two class intervals in Table 2–12 representing the two parts into which a class interval of Table 2–11 was divided. Based on the information in Table 2–12 it is possible to construct the relative frequency diagram shown in Figure 2–14, where the probabilities are measured by the heights of the rectangles as in Figure 2–12. It is also possible to construct a histogram in which the probabilities are given by the areas of the rectangles. This is shown in Figure 2–15.

It is now of interest to compare Figures 2–12 and 2–14 and then 2–13 and 2–15. When Figures 2–12 and 2–14 are examined it will be noted that the

TABLE 2–12

DATA FOR BALL BEARINGS

Using class intervals of length 0.0005 inches

Class interval	Number of ball bearings	Fraction of ball bearings
(−50, −45)	32	0.0032
(−45, −40)	58	0.0058
(−40, −35)	205	0.0205
(−35, −30)	215	0.0215
(−30, −25)	400	0.0400
(−25, −20)	495	0.0495
(−20, −15)	702	0.0702
(−15, −10)	753	0.0753
(−10, −05)	900	0.0900
(−05, 0)	1050	0.1050
(0, 05)	1100	0.1100
(05, 10)	908	0.0908
(10, 15)	850	0.0850
(15, 20)	772	0.0772
(20, 25)	432	0.0432
(25, 30)	368	0.0368
(30, 35)	300	0.0300
(35, 40)	251	0.0251
(40, 45)	130	0.0130
(45, 50)	79	0.0079

heights of all the rectangles in Figure 2–14 are approximately one-half the height of the corresponding ones in Figure 2–12. This is illustrated more clearly in Figure 2–16, where we have shown on the same figure the outlines of the shaded regions in Figures 2–12 and 2–14. The situation is quite different for the histograms of Figures 2–13 and 2–15. In Figure 2–17 we have shown on the same figure the outlines of the shaded regions from Figures 2–13 and 2–15. It will be observed that these two curves are almost identical. In other words, there was almost no change in the outlines of the histograms on doubling the number of class intervals. If the number of class intervals were doubled once again by subdividing each subinterval into two of length 0.00025, then the outline of the diagram in which the heights of the rectangles measure probabilities would be roughly half the height of the dotted curve shown in Figure 2–16, while the outline of the histogram would be essentially the same as that shown in Figure 2–17.

Let us now explain the reasons why the outline of the histogram remains unchanged as the number of class intervals is increased while the outline of the relative frequency diagram, as shown in Figure 2–16, gets lower and

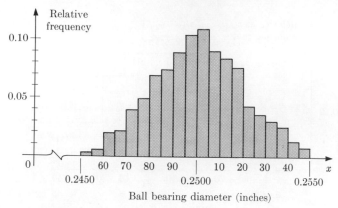

Ball bearing diameter (inches)

FIGURE 2–14

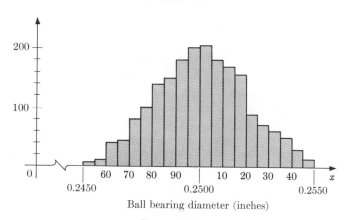

Ball bearing diameter (inches)

FIGURE 2–15

lower. When a class interval is subdivided into two class intervals, each half the length of the original one, then if the length of the original class interval is small compared with the range of variation of x, we expect that roughly half the number of values of x which were in the original class interval will lie in each of the two new class intervals. Thus if n_j values of x were in the original class interval and n'_j and n'' are the number of values of x in each of the two new class intervals formed from it, then $n_j = n'_j + n''_j$ and we expect both n'_j and n''_j to be approximately $n_j/2$. Thus on a relative frequency diagram where the heights of the rectangles give the relative frequency of x lying in the corresponding class intervals, we see that these heights should decrease as the number of class intervals is increased.

Consider next what happens to a histogram when each class interval is subdivided into two class intervals of equal length. Recall that on a histogram probabilities are measured by areas, and the height of any rectangle

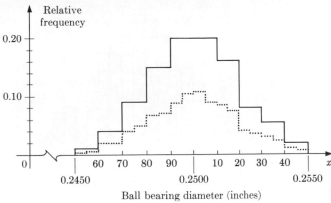

FIGURE 2–16

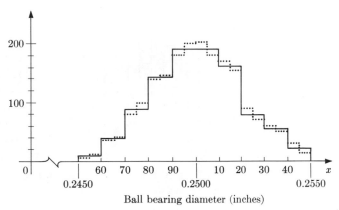

FIGURE 2–17

is the relative frequency divided by the length of the class interval. Now we have already noted that n'_j and n''_j should both be roughly $n_j/2$. However, the lengths of the new class intervals are half the length of the original ones. Thus the heights of the rectangles for each of the new class intervals will be roughly the same as the height of the rectangle for the original class interval. This explains why the outline of the histogram does not change much when the number of class intervals is increased.

Figure 2–16 points out the important fact that as the number of class intervals is increased, the probability that x lies in any one of the class intervals decreases. If we now go all the way and consider all the different values which x can take on, it will turn out that each of these values has a very small probability of occurring. What we are pointing out is that when a random variable can take on a very large number of different values, it is typically true that the probability of x being equal to any one of these

values is very small. In other words, if we constructed a bar diagram for $p(x)$, the height of each of the bars would be very small and a greatly enlarged scale would be required for the vertical axis in order to see them at all.

Suppose, however, that instead of using a bar diagram to represent $p(x)$ we use a histogram. The histogram will be constructed as follows. Let $x_1, \ldots, x_k$ be the values x can take on. Imagine that these values of x are now represented by points on a horizontal axis just as if we were going to construct a bar diagram. Next let γ_j be the point halfway between x_j and x_{j+1}. Also take γ_0 to be the same distance to the left of x_1 that γ_1 is to the right of x_1 and γ_k the same distance to the right of x_k as γ_{k-1} is to the left of x_k. Consider now the k class intervals determined by $\gamma_0, \gamma_1, \ldots, \gamma_k$. The jth class interval has γ_{j-1} and γ_j as the class boundaries and has length $\gamma_j - \gamma_{j-1}$. The only value of x which lies in the jth class interval is x_j. Let us now erect a rectangle over the jth class interval whose area is $p(x_j)$. The height of this rectangle is $p(x_1)/(\gamma_j - \gamma_{j-1})$. This is done for each j. The resulting diagram, which might look like that shown in Figure 2–18, will be called the histogram for $p(x)$. For clarity only a small number of values of x are shown in Figure 2–18. There will be one rectangle corresponding to each value of x. The probability of some event such as $x_2 \leq x \leq x_6$ is then the sum of the areas of the rectangles corresponding to x_2, x_3, x_4, x_5 and x_6. This is the area of the shaded region shown in Figure 2–18.

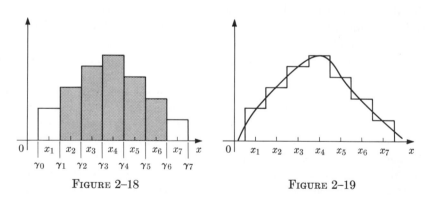

FIGURE 2–18   FIGURE 2–19

The outline of the histogram of Figure 2–18 is shown in Figure 2–19. We can now note that there would be little loss in accuracy if when computing areas we used instead of the outline of the actual histogram a smooth curve such as that shown in Figure 2–19. We can observe, in addition, that once we agree that the histogram can be well approximated by a smooth curve, there is really no need to be concerned in detail with precisely what values x can take on between a and b if x takes on a great number of different values. No serious errors will be introduced in constructing a model if we imagine

that x can take on every value between a and b. Here, then, we have at last obtained the second procedure for treating random variables which can take on a large number of values in some interval. We shall use as a model for this situation one in which it is imagined that x can take on all values in the interval. To make probability computations, we no longer use a probability function for x but instead introduce another function, call if $f(x)$, which is represented graphically by a smooth curve, and which intuitively can be thought of as approximating the outline of the histogram for $p(x)$. To compute the probability of any event such as $\alpha \leq x \leq \beta$, we determine the area under the curve representing $f(x)$ between α and β, as shown in Figure 2–20.

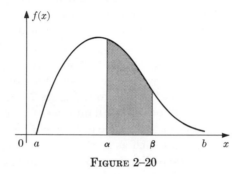

FIGURE 2–20

The thing that makes this method so useful is that these areas can be readily summarized in a table, and hence the need to add up the values of a probability function over many values of x is eliminated.

A random variable which can take on every value in some interval is called a *continuous random variable*. We have just seen that often one can conveniently treat random variables which can take on a large number of values as if they were continuous random variables. No random variables in the real world are continuous random variables. The notion of a continuous random variable is a mathematical idealization. Nonetheless, it is one that has great intuitive appeal as well as being extremely useful. In several of the examples studied previously, it may seem to the reader that the random variables introduced are indeed continuous random variables and can take on any value in some interval. Thus, for example, it seems plausible that the diameter of a ball bearing or the height of a man are examples in nature of continuous random variables. This is not true, however, because in both cases the objects are built up from discrete building blocks, atoms or molecules, so that they are not infinitely divisible and the random variables can take on only a finite number of values (the number of different values is enormous, of course). For all practical purposes, however, treating them as continuous random variable yields an extremely good approximation.

To construct a model of a random experiment ℜ involving a random variable x which it is desired to treat as continuous, taking on all values between a and b, we use as the event set the set X of all numbers lying between a and b, that is, the set of all values which x can take on. Then a function $f(x)$ is introduced whose domain is X, having the characteristics that $f(x) \geq 0$ for each $x \in X$, and the probability of any event $\alpha \leq x \leq \beta$ is the area under the curve representing $f(x)$ from α to β, as shown in Figure 2–20. The function $f(x)$ is called the *probability density function* for the random variable x.

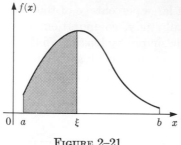
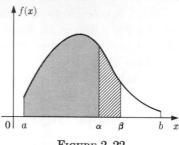

FIGURE 2–21   FIGURE 2–22

Random variables which can take on only a finite number of values are called *discrete random variables*. When dealing with discrete random variables it is convenient to use a probability function $p(x)$, where $p(x_i)$ is the probability that $x = x_i$ and the probability that $x_i \leq x \leq x_k$ is $\Sigma_{j=i}^{k} p(x_j)$. For continuous random variables, a probability function is not used. Instead a probability density function $f(x)$ is used, and the probability that $\alpha \leq x \leq \beta$ is the area under $f(x)$, such as the shaded region shown in Figure 2–20. Intuitively, the graph of $f(x)$ can be thought of as approximating the outline of the histogram for x if we were completely accurate and treated x as being discrete.

Let us now explain how probability computations are normally made using a model for a continuous random variable. Denote by $F(\xi)$ the probability that $x \leq \xi$. Then if the curve representing the probability density function $f(x)$ is that shown in Figure 2–21, $F(\xi)$ is the area of the shaded region. For each $\xi \in X$ it is possible to compute a number $F(\xi)$ which is the probability that $x \leq \xi$. This association defines a new function with domain X, which we shall denote by $F(x)$. $F(x)$ is called the cumulative function for x and has the same interpretation as $P(x)$, the cumulative function for a discrete random variable. Imagine now that $F(x)$ has been computed for a large number of different values of x, and these results have been given to us in the form of a table. This table is all we need to make probability computations with the model. Suppose that we wish to compute the probability of the event $\alpha \leq x \leq \beta$, which is the area of the crosshatched re-

gion in Figure 2–22. Now $F(\beta)$, the probability that $x \leq \beta$, is the area to the left of β, that is, the area of the shaded region plus the area of the cross-hatched region. Furthermore $F(\alpha)$, the probability that $x \leq \alpha$, is the area of the shaded region. Therefore, the area of the crosshatched region is $F(\beta) - F(\alpha)$, and this is the probability that $\alpha \leq x \leq \beta$. Thus it is only necessary to look up two numbers in the table of $F(x)$ to determine the probability that $\alpha \leq x \leq \beta$.

Now interestingly enough, it turns out that, in practice, only a relatively small number of different types of density functions $f(x)$ are used in building models. Thus by tabulating a relatively small number of cumulative functions $F(x)$ it is possible to take care of a large percentage of the cases one normally deals with. In all cases we shall study, computations involving continuous random variables will be made simply by looking up the appropriate values of the cumulative function (or some equivalent function) in one of the tables given at the end of the text.

2–14 THE NORMAL DISTRIBUTION

There is one density function which arises with much greater frequency in constructing models of continuous random variables than any other. This is referred to as the *normal density function* or *normal distribution*. A continuous random variable with mean μ and standard deviation σ is said to be *normally* distributed if its density function is

$$n(x; \mu, \sigma) = \frac{1}{\sigma\sqrt{2\pi}} e^{-(x-\mu)^2/2\sigma^2}, \tag{2–58}$$

and $n(x; \mu, \sigma)$ is called the normal density function or normal distribution with mean μ and standard deviation σ. We shall use the special symbolism $n(x; \mu, \sigma)$ for the normal density function rather than $f(x)$. In (2–58), e is the same number ($e = 2.718$ approximately) which appeared in the Poisson distribution, and π is the number ($\pi =$ approximately 3.14159) encountered in geometry.

When a model is built in which it is assumed that x is normally distributed, one more idealization is introduced in addition to the notion of a continuous random variable. The additional idealization is the assumption that x can take on any value, however large or small. In other words, the event set X is taken to be the set of all real numbers. No actual random variable can be arbitrarily large or small, but this additional approximation does not introduce any significant errors if the probability that x lies outside the interval of actual interest is very small. It is for mathematical reasons that it is convenient to treat the event set as the set of all real numbers.

In Figure 2–23 we have shown what the curves representing $n(x; \mu, \sigma)$ look like for a given value of μ and two different values of σ. The value of μ

determines where the peak of the curve occurs, and σ determines how spread out the curve is. We shall never need to use (2–58) directly in working with the normal distribution. As shown in the previous section, all we shall need is a table of the cumulative function, which we shall write $N(x; \mu, \sigma)$. Now the normal distribution involves two parameters, and thus it would appear that the cumulative function would need to be tabulated for a variety of values of μ and σ. Interestingly enough, this does not need to be done. Let us now see why.

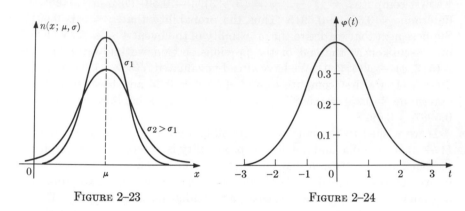

FIGURE 2–23   FIGURE 2–24

Consider a random variable t which is normally distributed with mean $\mu = 0$ and standard deviation $\sigma = 1$; t will be said to have a *standardized normal distribution*, and instead of using $n(t; 0, 1)$ to denote the density function we shall use $\varphi(t)$. From (2–58), with $\mu = 0$ and $\sigma = 1$ we see that

$$\varphi(t) = \frac{1}{\sqrt{2\pi}} \, e^{-t^2/2} . \qquad (2\text{–}59)$$

The graphical representation of $\varphi(t)$ is shown in Figure 2–24. The cumulative distribution for t will be denoted by $\Phi(t)$, and $\Phi(\tau)$ is the probability that $t \leq \tau$. Now if a table of $\Phi(t)$ is available, it is possible to evaluate $N(\xi; \mu, \sigma)$ for any ξ, μ and σ. In fact, if we compute the number $t_1 = (\xi - \mu)/\sigma$, then

$$N(\xi; \mu, \sigma) = \Phi(t_1) . \qquad (2\text{–}60)$$

We shall not try to prove this since the proof requires the use of calculus. Equation (2–60) is, however, the key to using the normal distribution. We can write (2–60) in an alternative way as

$$N(\xi; \mu, \sigma) = \Phi\!\left(\frac{\xi - \mu}{\sigma}\right) . \qquad (2\text{–}61)$$

A table of $\Phi(t)$ is to be found in Table E at the end of the text. What (2–61) says is that if the random x has a normal distribution with mean μ_x and

standard deviation σ_x, then the random variable $t = (x - \mu_x)/\sigma_x$ has a standardized normal distribution. We know from our previous work that $\mu_t = 0$ and $\sigma_t = 1$. The additional fact presented here is that if x is normally distributed, then so is t.

Let us now explain some of the details concerning the use of Table E. Suppose that x has a normal distribution with mean 3 and standard deviation 4. Imagine that we wish to determine the probability that $x \leq 5$. Hence we wish to evaluate $N(\xi; \mu, \sigma)$ with $\xi = 5$, $\mu = 3$ and $\sigma = 4$. To do so we first compute $t_1 = (\xi - \mu)/\sigma = (5 - 3)/4 = 0.50$. Then in Table E, we determine $\Phi(0.50) = 0.6915$. Thus the probability that $x \leq 5$ is 0.6915. Suppose next that we desire the probability of the event $4 \leq x \leq 5$. From our discussion at the end of the previous section we know that this is $N(5; 3, 4) - N(4; 3, 4)$. We have already evaluated $N(5; 3, 4)$. To evaluate $N(4; 3, 4)$ we first compute $t_1 = (4 - 3)/4 = 0.25$ and then in Table E determine $\Phi(0.25) = 0.5987$. Then the desired probability is $0.6915 - 0.5987 = 0.0928$.

If we wished to compute the probability that $x \leq 1$, we find that $t_1 = (1 - 3)/4 = -0.5$, and the desired probability is $\Phi(-0.5)$. However, if we look in Table E it is seen that no negative values of t are given. How, then, do we determine $\Phi(-0.5)$? To do this we must make use of two properties of $\varphi(t)$. One of these is that the entire area under the curve for $\varphi(t)$ is 1, that is, $\Phi(t)$ approaches 1 as t becomes large. Table E clearly shows that this is the case. The fact that the area under the curve for $\varphi(t)$ is 1 follows from the fact that the probabilities of the simple events must sum to 1. Consider next Figure 2–25. Let t_1 be any positive number; then $-t_1$ is a negative number.

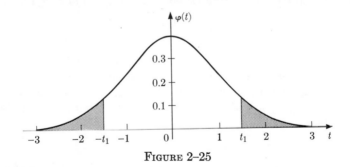

FIGURE 2–25

It happens to be a property of $\varphi(t)$ that the area of the shaded region lying to the right of t_1 is precisely equal to the area of the shaded region to the left of $-t_1$. The shaded area to the left of $-t_1$ is $\Phi(-t_1)$. The area to the left of t_1 is $\Phi(t_1)$ and hence the area to the right of t_1 is $1 - \Phi(t_1)$. Since the two shaded regions have the same area

$$\Phi(-t_1) = 1 - \Phi(t_1). \tag{2-62}$$

Here we have the means to evaluate $\Phi(t)$ for $t < 0$ using a table which only gives $\Phi(t)$ for positive t. Thus to compute $\Phi(-0.5)$ we see from (2–62) that $\Phi(-0.5) = 1 - \Phi(0.5)$. Now $\Phi(0.5) = 0.6915$ so $\Phi(-0.5) = 1 - 0.6915 = 0.3085$.

On comparing (2–58) with (2–59) we see that

$$n(\xi; \mu, \sigma) = \frac{1}{\sigma} \varphi \left(\frac{\xi - \mu}{\sigma} \right).$$

Thus, given a table of $\varphi(t)$, such as Table F at the end of the text, it is possible to evaluate $n(x; \mu, \sigma)$ for any given value of x. We have now explained everything the reader needs to know to make computations using the normal distribution. Before illustrating in more detail the usefulness of the normal distribution, let us first explain why the normal distribution is of such great practical importance.

2–15 THE CENTRAL LIMIT THEOREM

One of the most remarkable results in probability theory is what is known as the central limit theorem. This theorem makes a statement about what happens in situations of the following sort. Suppose that we consider a random experiment $\mathcal{R}$ which has associated with it some random variable y having expected value μ_y and variance σ_y^2. The random variable may be continuous or discrete. It does not matter which. Consider next the random experiment $\mathcal{R}_n$ which consists of making n independent trials of $\mathcal{R}$. Now on each trial of $\mathcal{R}$ some value of y will be observed. Denote by $y\{j\}$ the random variable representing y for the jth trial. Then each $y\{j\}$ has the same distribution and hence has expected value μ_y and variance σ_y^2. Furthermore, since the $y\{j\}$ refer to different independent random experiments they are independent random variables. Consider next the random variable

$$z = y\{1\} + \cdots + y\{n\} \tag{2–63}$$

associated with $\mathcal{R}_n$. By (2–33) and the fact that $\mu_j = \mu_y$, we see that

$$\mu_z = \mu_y + \cdots + \mu_y = n\mu_y , \tag{2–64}$$

and by (2–53), since $\sigma_j^2 = \sigma_y^2$,

$$\sigma_z^2 = \sigma_y^2 + \cdots + \sigma_y^2 = n\sigma_y^2 \tag{2–65}$$

or

$$\sigma_z = \sqrt{n} \, \sigma_y . \tag{2–66}$$

The central limit theorem then states that *as the number of trials of $\mathcal{R}$ is increased unendingly, that is, as n is increased unendingly, the distribution of z*

approaches a normal distribution with mean $n\mu_y$ and standard deviation $\sqrt{n}\,\sigma_y$.
This is true regardless of what the distribution of y happens to be. A more
precise way to state the central limit theorem is to say that the distribution
of the random variable t

$$t = \frac{z - n\mu_y}{\sigma_y\sqrt{n}} \tag{2–67}$$

approaches $\varphi(t)$, the standardized normal distribution, as n increases unendingly.

The proof of the central limit theorem requires quite advanced mathematics and we shall not consider it. Furthermore, the mere fact that the
distribution of t approaches $\varphi(t)$ for very large t is not necessarily of any
practical interest. The question of practical interest is how rapidly the distribution of t approaches $\varphi(t)$. If n had to be one million or greater the
central limit theorem would have little practical relevance. The surprising
and very important thing, which does not follow directly from the central
limit theorem, is that the distribution of t or of z is often very well approximated by a normal distribution for small values of n, often as small as
$n = 5$ or 6. This is, as we shall see, of great practical value.

The central limit theorem can be used to show that the binomial distribution $b(x; n, p)$ should be well approximated by the normal distribution
for large n. If x has the binomial distribution $b(x; n, p)$, then as we have
noted on page 98, $x = \Sigma_{j=1}^{n} y\{j\}$, where the $y\{j\}$ are independent random variables each having the same distribution, with $\mu_j = p$ and $\sigma_j^2 = pq$.
Recall that $y\{j\}$ is the number of times e_1 occurs on trial j. Thus we have
written x in the form (2–63) and the conditions of the central limit theorem
are satisfied. Hence for large n, $b(x; n, p)$ will be well approximated by the
normal distribution, or the distribution of

$$t = \frac{x - np}{\sqrt{npq}} \tag{2–68}$$

will be essentially $\varphi(t)$.

Very frequently, sums of independent random variables will approach the
normal distribution even if they are not identically distributed. This can
occur even if they are not independent. Given that sums of random variables are often essentially normally distributed under rather general conditions, we can see why many random variables occurring in practice are
essentially normally distributed. For example, the variations in the diameters of the ball bearings produced by a process can be thought of as due to
the sum of a number of small chance effects, such as variations in steel
hardness and machine settings. For this reason it would not be surprising
to find that the distribution for the random variable representing the diameter of a ball bearing was essentially normal. This would certainly appear

to be true for the data shown in Figure 2–15. We shall investigate this in more detail later.

2–16 NORMAL APPROXIMATION TO BINOMIAL AND HYPER-GEOMETRIC DISTRIBUTIONS

In the previous section we concluded on the basis of the central limit theorem that for large n the binomial distribution $b(x; n, p)$ is well approximated by the normal distribution $n(x; np, \sqrt{npq})$, and the distribution of $t = (x - np)/\sqrt{npq}$ is essentially $\varphi(t)$. What do we mean by saying that $b(x; n, p)$ is well approximated by $n(x; np, \sqrt{npq})$ and that the distribution of t is well approximated by $\varphi(t)$? What is meant here is what we discussed in Section 2–13. In other words, if we construct a histogram for $b(x; n, p)$, where probabilities are measured by areas, and consider the outline of this histogram as in Figure 2–19, the curve representing $n(x; np, \sqrt{npq})$ will essentially be superimposed on the outline of the histogram. Similarly, if a histogram for t is constructed, the outline of this histogram will be essentially the same as the curve representing $\varphi(t)$.

Let us illustrate this behavior by constructing a histogram for t when $p = 0.5$ and for two values of n, $n = 10$ and $n = 30$. If $n = 10$ and $p = 0.5$, $np = 5$ and $\sqrt{npq} = \sqrt{2.5} = 1.58$ and

$$ t = \frac{x - 5}{1.58} = 0.633x - 3.17 \,. $$

Since x can take on the values $0, 1, \ldots , 10$, t can take on the eleven values

$$ -3.17, \; -3.17 + 0.633 = -2.54, \ldots , 2.54, 3.17 \,. $$

The spacing between successive values of t is 0.633. The probability that $t = -3.17$ is $b(0; 10, 0.5)$, the probability that $x = 0$. Similarly, the probability that $t = -2.54$ is the probability that $x = 1$, and so on. In constructing the histogram for t, the altitude of the rectangle will be the probability of t divided by 0.633, since the base of each rectangle will have length 0.633. The outline of the resulting histogram is shown in Figure 2–26. The outline of the histogram for t when $p = 0.5$ and $n = 30$ is shown in Figure 2–27. It will be observed that the outline of the histograms does indeed appear to be approaching the curve representing $\varphi(t)$ shown in Figure 2–24. In Figure 2–28 we have shown on the same diagram the outline of the histogram of Figure 2–26 and the curve for $\varphi(t)$.

Let us now explain how the normal distribution can be used conveniently to approximate the binomial distribution. In Figure 2–29 we have shown the histogram for $b(x; 10, 0.5)$. The interesting thing about this histogram is that it is also a relative frequency diagram since the height of the rec-

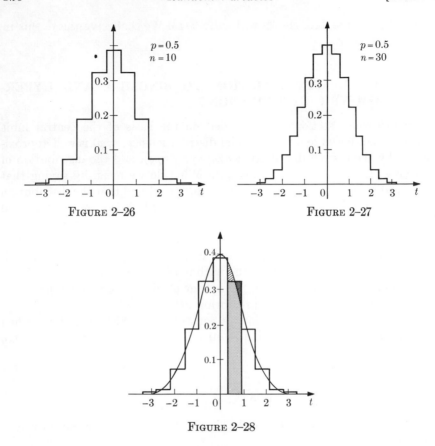

FIGURE 2–26   FIGURE 2–27

FIGURE 2–28

tangle corresponding to $x = r$ is $b(r; 10, 0.5)$. The reason for this is that the spacing between the values of x is 1, so that the length of the base of each rectangle is 1 and hence there is no difference between the relative frequency diagram and histogram in this case. When thought of as a histogram, the probability that x has a given value, say 4, is then the area of the shaded rectangle in Figure 2–29. The base of this rectangle begins at $3\frac{1}{2}$ and extends

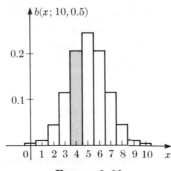

FIGURE 2–29

to $4\frac{1}{2}$. To approximate $b(4; 10, 0.5)$ we then merely compute the area under the normal curve corresponding to $n(x; 5, 1.58)$ from $x = 3\frac{1}{2}$ to $x = 4\frac{1}{2}$. The reason for this is that we observed from the central limit theorem that $b(x; n, p)$ is approximated by a normal distribution with the same mean and standard deviation, that is, $n(x; np, \sqrt{npq})$. Instead of using the outline of the histogram for $b(x; n, p)$ to compute areas, we use the normal curve. Thus, for example, $b(5; 10, 0.5)$ could be determined approximately using

$$b(5; 10, 0.5) \doteq \Phi\left(\frac{5.5 - 5}{1.58}\right) - \Phi\left(\frac{4.5 - 5}{1.58}\right)$$

$$= \Phi(0.32) - \Phi(-0.32)$$
$$= \Phi(0.32) - [1 - \Phi(0.32)]$$
$$= 2\Phi(0.32) - 1$$
$$= 2(0.6255) - 1 = 0.2510 .$$

The correct answer from Table C is 0.2461. Thus the normal approximation does quite well in this case.

In general, to approximate $b(r; n, p)$ by the normal distribution we note that the rectangle corresponding to $x = r$ extends from $r - \frac{1}{2}$ to $r + \frac{1}{2}$. Hence

$$b(r; n, p) \doteq \Phi\left(\frac{r + \frac{1}{2} - np}{\sqrt{npq}}\right) - \Phi\left(\frac{r - \frac{1}{2} - np}{\sqrt{npq}}\right). \qquad (2\text{–}69)$$

To compute $B(r; n, p)$, the probability that $x \leq r$, we sum the areas of the rectangles for $x = 0,1,2,\ldots, r$, that is, we determine the area of the histogram lying to the left of $r + \frac{1}{2}$. The normal approximation to this is the area under the normal curve to the left of $r + \frac{1}{2}$. Hence

$$B(r; n, p) \doteq \Phi\left(\frac{r + \frac{1}{2} - np}{\sqrt{npq}}\right). \qquad (2\text{–}70)$$

Finally, the probability that $r_1 \leq x \leq r_2$ is the area of the histogram from $r_1 - \frac{1}{2}$ to $r_2 + \frac{1}{2}$. Thus

$$\sum_{r=r_1}^{r_2} b(r; n, p) \doteq \Phi\left(\frac{r_2 + \frac{1}{2} - np}{\sqrt{npq}}\right) - \Phi\left(\frac{r_1 - \frac{1}{2} - np}{\sqrt{npq}}\right). \qquad (2\text{–}71)$$

Let us next consider the conditions under which the binomial distribution can be adequately approximated by the normal distribution. The central limit theorem gives no information about this and to answer the question it is necessary to make an empirical study. The answer, of course, depends on the accuracy desired. If, as is usually true, an error of five percent or so is not serious, with even a larger error being tolerable if the probabilities are very small, the normal approximation to $b(x; n, p)$ is adequate if the mean np is greater than 25. This is true regardless of what value p has. However,

for values of p close to 0.5, the approximation may be quite adequate for np as small as 10 or even less. We noted above that for $np = 5$ and $p = 0.5$ we obtained a very good approximation. The closer p is to 0.5, the smaller np can be while still obtaining a good approximation.

Not only is the normal distribution very useful for approximating the binomial distribution, it is also of great value for approximating the hypergeometric distribution $h(x; n, p, N)$. The fact that the normal distribution can be used to approximate $h(x; n, p, N)$ does not follow directly from the central limit theorem because x cannot be written as the sum of independent random variables. However, it does follow on using a somewhat more extensive analysis which we shall not give. The mean and variance of the hypergeometric distribution are

$$\mu_x = np; \quad \sigma_x^2 = npq \left(\frac{N - n}{N - 1} \right). \tag{2-72}$$

We shall not attempt to verify these. Their derivation is not quite so simple as for the binomial distribution. To approximate $h(x; n, p, N)$ one simply uses a normal distribution of the same mean and variance, applying precisely the same type of formulas as for the binomial distribution. For example,

$$h(r; n, p, N) \doteq \Phi \left(\frac{r + \frac{1}{2} - \mu_x}{\sigma_x} \right) - \Phi \left(\frac{r - \frac{1}{2} - \mu_x}{\sigma_x} \right), \tag{2-73}$$

where μ_x and σ_x are given by (2–72).

The normal approximation to $h(x; n, p, N)$ can be used if n is small with respect to N, say $n/N = 0.2$ or less, and $np > 25$. If p is close to 0.5, then the normal approximation can be used even if np is as small as 5. When making computations involving the hypergeometric distribution, one has little choice but to use either the Poisson approximation (for small p) or the normal approximation otherwise. Fortunately, one or the other of these is almost always quite adequate.

The above discussion centered on the problem of computing approximately $b(r; n, p)$, $h(r; n, p, N)$ or the cumulative functions by using the normal approximations. In this approach the basic random variable x was assumed to have a binomial or hypergeometric distribution, but it was desired to use the normal approximation to compute the probabilities of interest. With this approach x is always treated as a discrete random variable which can take on the values $0, 1, \ldots, n$. Now there is another way in which the normal distribution can be applied when dealing with x. This alternative procedure is actually to treat x as being normally distributed, that is, as a continuous random variable. In this case the discreteness of x is ignored in the model. In other words, even though a very accurate model would treat x as discrete, having either a binomial or hypergeometric distribution,

it is instead sometimes convenient to treat x as continuous, having a normal distribution. This, then, forms an alternative way of making a normal approximation. We shall use this latter procedure, as well as the former, quite frequently in the future. In summary, the two methods by which a normal approximation can be used are: (1) Treat x as discrete having a binomial or hypergeometric distribution, but compute the probabilities in the manner discussed above, using the normal approximation. (2) Treat x as continuous, having a normal distribution. When the latter procedure is used, the probability that $x \leq r$ would be

$$\Phi\left(\frac{r - np}{\sqrt{npq}}\right) \quad \text{not} \quad \Phi\left(\frac{r + \frac{1}{2} + np}{\sqrt{npq}}\right),$$

and r would not need to be an integer here. Generally speaking, when n is fairly large these two alternative approaches will yield totally negligible differences in any results which are obtained.

2-17 EXAMPLES

1. Let us estimate $b(6; 10, 0.5)$ using the normal approximation and illustrate the situation geometrically. We have shown previously that $\mu_x = 5$ and $\sigma_x = 1.58$. Thus (2–69) is used with $r = 6$:

$$\frac{r + \frac{1}{2} - \mu_x}{\sigma_x} = \frac{6.5 - 5}{1.58} = 0.95; \quad \frac{r - \frac{1}{2} - \mu_x}{\sigma_x} = \frac{5.5 - 5}{1.58} = 0.32 \,.$$

Therefore from (2–69)

$$b(6; 10, 0.5) = \Phi(0.95) - \Phi(0.32)$$
$$= 0.8289 - 0.6255 = 0.2034 \,.$$

From Table C the correct value of $b(6; 10, 0.5)$ is 0.2051. Thus the normal approximation does well here just as it did for the case of $x = 5$ studied earlier.

Now $\Phi(0.95) - \Phi(0.32)$ is the area under the $\varphi(t)$ curve from $t = 0.32$ to $t = 0.95$ and is the area of the lightly shaded region plus the crosshatched region shown in Figure 2–28. On the other hand $b(6; 10, 0.5)$ is the area of the rectangle on the histogram for $t = (x - 5)/1.58$ which extends from $t = 0.32$ to $t = 0.95$. The area of this rectangle is the area of the lightly- and darkly-shaded regions in Figure 2–28. The crosshatched area is very close to being equal to the area of the darkly-shaded region and this is why the normal approximation yields an answer which is quite close to being correct.

2. We shall now estimate $B(6; 10, 0.5)$ using the normal approximation. According to (2–70)

$$B(6; 10, 0.5) \doteq \Phi\left(\frac{6 + 0.5 - 5}{1.58}\right) = \Phi(0.95) = 0.8289 .$$

If we sum the values of $b(r; 10, 0.5)$ in Table C from $r = 0$ to 6 we obtain 0.8282 as the exact value of $B(6; 10, 0.5)$. Thus the normal approximation again gives a value which is very close to being correct.

3. Let us next estimate $B(1; 10, 0.1)$ using the normal approximation. We use (2–70) with $r = 1$. Now $np = 1$ and $\sqrt{npq} = \sqrt{0.9} = 0.948$. Therefore

$$B(1; 10, 0.1) = \Phi\left(\frac{1 + 0.5 - 1}{0.948}\right) = \Phi(0.53) = 0.7019 .$$

If the values of $b(r; 10, 0.1)$ for $r = 0$ and 1 given in Table B are summed, the correct value $B(1; 10, 0.1) = 0.7361$ is obtained. The normal approximation does not do quite so well here. The reason is that p is far from 0.5, and the histogram for t is skewed and does not resemble $\varphi(t)$ very closely. The outline of the histogram for t is shown in Figure 2–30. The Poisson ap-

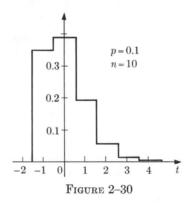

FIGURE 2–30

proximation does better than the normal here, since $p(0; 1) + p(1; 1) = 0.7358$. To use the normal approximation for p as small as 0.1, n should be larger than 10.

4. In a given city 10,000 white children are born in a twelve-month period. If the long-run fraction of white male babies is 0.514, what is the probability that the actual fraction of the 10,000 babies who were male will differ by at least one percentage point from the long-run fraction, that is, will be greater than or equal to 0.524 or less than or equal to 0.504? Let x be the random variable representing the number of the 10,000 babies that were male. Then we wish to determine the probability that $x \geq 0.524(10,000) = 5240$ or $x \leq 0.504(10,000) = 5040$. As a model for the situation we shall imagine that each birth represents an independent trial of a random experiment in which the probability of a male birth is 0.514.

With this model in mind, x has a binomial distribution with $\mu_x = 0.514(10,000) = 5140$ and

$$\sigma_x^2 = 0.486(5140) = 2500 \quad \text{or} \quad \sigma_x = 50 \ .$$

Now no tables of the binomial distribution go up to $n = 10,000$. However, the normal approximation to the binomial distribution should be very good here since μ_x is quite large. If A_1 is the event that $x \leq 5040$ and A_2 is the event that $x \geq 5240$, we wish to compute $p(A_1 \cup A_2)$. Since A_1 and A_2 are mutually exclusive it follows from (1–12) that

$$p(A_1 \cup A_2) = p(A_1) + p(A_2)$$
$$= B(5040; 10,000, 0.514) + 1 - B(5239; 10,000, 0.514) \ .$$

In obtaining the above we used the fact that the probability $x \geq 5240$ is one minus the probability that $x \leq 5239$.

Now, on using the normal approximation, we see

$$B(5040; 10,000, 0.514) \doteq \Phi\left(\frac{5040.5 - 5140}{50}\right) = \Phi(-1.99)$$
$$= 1 - \Phi(1.99) = 1 - 0.9767 - 0.0233$$

and

$$B(5239; 10,000, 0.514) \doteq \Phi\left(\frac{5239.5 - 5140}{50}\right) = \Phi(1.99) = 0.9767 \ .$$

Hence

$$p(A_1 \cup A_2) = 2(0.0233) = 0.0466 \ .$$

Consequently, only about five times in 100 would the actual fraction w of male births not satisfy the inequality $0.504 < w < 0.524$, that is, would differ by at least one percentage point from the long-run average.

5. Consider once again a process which turns out ball bearings. The diameter x of any bearing coming off the production line can be looked upon as a random variable. We shall suppose that previous studies have shown that x can be assumed to be normally distributed with an expected value of 0.4020 inches and a standard deviation of 0.0020 inches. Assume that in order to be considered acceptable the diameter of a ball bearing must lie in the interval

$$0.3975 \leq x \leq 0.4070 \ . \tag{2–74}$$

We can now compute the probability that any given ball bearing produced will be defective. Let A_1 be the event that $x < 0.3975$ and A_2 the event $x > 0.4070$. The ball bearing will be defective if either one of the events occurs, that is, if the event $A = A_1 \cup A_2$ occurs. Now $A_1 \cap A_2 =$

$\varnothing$. Thus if p is the probability of a defective, it follows from (1–12) that $p = p(A_1) + p(A_2)$. However,*

$$p(A_1) = N(0.3975; 0.4020, 0.0020) = \Phi\left(\frac{0.3975 - 0.4020}{0.0020}\right)$$

$$= \Phi(-2.25) = 1 - \Phi(2.25) = 1 - 0.9878 = 0.0122 .$$

Consider next $p(A_2)$. This is the area under the normal curve $n(x; 0.4020, 0.0020)$ to the right of $x = 0.4070$. The area to the right of 0.4070 is simply 1 minus the area to the left of 0.4070, so that

$$p(A_2) = 1 - N(0.4070; 0.4020, 0.0020) = 1 - \Phi\left(\frac{0.4070 - 0.4020}{0.0020}\right)$$

$$= 1 - \Phi(2.5) = 1 - 0.9938 = 0.0062 .$$

Hence $p = 0.0122 + 0.0062 = 0.0184$. The probability that any given ball bearing will be defective is, by the intuitive interpretation of probability, the long-run fraction of the ball bearings produced that will be defective. We have just shown that about 1.8 percent of the output will be defective. In this example we have shown that it is sometimes possible to compute the probability of a defective from a knowledge of some more basic random variable, in this case, the diameter of the ball bearing.

6. Let us study in a little more detail the process for making ball bearings which was considered in Example 5. By adjusting the settings on one or more machines, it is possible to adjust μ, the expected diameter of a ball bearing. Such adjustments may or may not simultaneously influence the standard deviation σ. Frequently they do not. Let us then examine a situation where, by setting the machines, we are free to select the expected diameter of a ball bearing, and where σ remains constant at the value 0.0020 independently of the value selected for μ. We shall continue to suppose that if a ball bearing is to be acceptable, its diameter must satisfy (2–74). Now, the fraction of defectives will depend on the value of μ. From the results of Example 5 we see that the long-run fraction defective p for the process is related to μ by

$$p = 1 + \Phi\left(\frac{0.3975 - \mu}{0.0020}\right) - \Phi\left(\frac{0.4070 - \mu}{0.0020}\right) = p(A_1) + p(A_2) . \quad (2\text{–}75)$$

For each μ there is determined a p. Thus p is a function of μ. We have illustrated how p varies with μ in Figure 2–31. The minimum value of the fraction defective occurs, as one would expect intuitively, when μ is halfway between 0.3975 and 0.4070, that is, for $\mu = 0.40225$, and the minimum

* The careful reader will note that we computed the probability of $x < 0.3975$ as if it were the event $x \leq 0.3975$. Why is the probability that $x < \xi$ the same as $x \leq \xi$ when x is a continuous random variable?

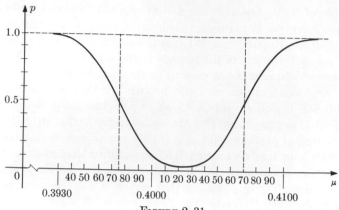

FIGURE 2–31

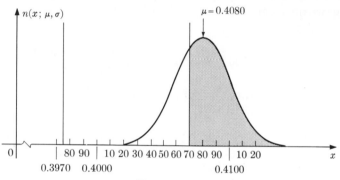

FIGURE 2–32

fraction defective is $p = 0.0176$. When μ is equal to 0.3975 or 0.4070, the fraction defective is 0.5. To check one value on the graph, let us compute p when $\mu = 0.4080$. The situation is illustrated in Figure 2–32, where we have shown the curve representing $n(x; 0.4080, 0.0020)$; p is the area of the shaded set of points which is the area under the normal curve lying to the right of 0.4070 and to the left of 0.3975. To compute p we use (2–75) with $\mu = 0.4080$. In this case

$$\Phi\left(\frac{0.3975 - 0.4080}{0.0020}\right) = \Phi(-5.25) = 1 - \Phi(5.25) \doteq 0 \,.$$

In Table E, $\Phi(t)$ is not tabulated for $t = 5.25$ because $\Phi(5.25)$ is 1 to four decimal places. In other words, the area to the left of 0.3975 is essentially 0. Next

$$1 - \Phi\left(\frac{0.4070 - 0.4080}{0.0020}\right) = 1 - \Phi(-0.5) = \Phi(0.5) = 0.6915 \,.$$

Thus by (2–75), $p = 0.6915$.

7. A sample of 5000 persons is selected at random from one million registered voters in a given city. Precisely half of the voters in the city are Democrats and half are Republicans. What is the probability that 49 percent or less or 51 percent or more of the persons in the sample will be Democrats? Let x represent the number of persons in the sample that are Democrats. We are then asked to compute the probability that $x \leq 5000(0.49)$ or $x \geq 5000(0.51)$. In other words, we wish to compute the probability of the event $x \leq 2450$ or $x \geq 2550$. The situation here is mathematically equivalent to selecting at random 5000 balls from a box of one million balls when half of the balls in the box are red. The probability that $x = r$ is then the hypergeometric probability $h(r; n, p, N)$ when $n = 5000$, $p = 0.5$ and $N =$ one million. The probability that $x \leq r$ is the cumulative hypergeometric probability $H(r; n, p, N)$.

The only convenient way to solve this problem is through the use of the normal approximation. The normal distribution used is one with μ and σ^2 given by (2–72). Thus

$$\mu = 5000(0.5) = 2500; \quad \sigma^2 = 1250 \left(\frac{1,000,000 - 5000}{1,000,000 - 1} \right) \doteq 1250 \,,$$

so $\sigma = 35.3$. Using the same sort of reasoning as in Example 5, we then see that the probability desired is

$$p = \Phi \left(\frac{2540 + 0.5 - 2500}{35.3} \right) + 1 - \Phi \left(\frac{2549 + 0.5 - 2500}{35.3} \right)$$
$$= \Phi(-1.41) + 1 - \Phi(1.41) = 2[1 - \Phi(1.41)]$$
$$= 2[1 - 0.9207] = 2(0.0793) = 0.1586 \,.$$

In the above we treated x as discrete, but used the normal approximation in estimating the desired probability. If instead we treat x as a continuous random variable which is normally distributed, that is, we use the alternative method of making the normal approximation, the probability becomes

$$p = \Phi \left(\frac{2450 - 2500}{35.3} \right) + 1 - \Phi \left(\frac{2550 - 2500}{35.3} \right)$$
$$= 2[1 - \Phi(1.42)] = 0.1556 \,,$$

which is negligibly different from the result obtained above. Thus the probability that the fraction of the Democrats in the sample differs by 0.01 or more on either side from the fraction of the Democrats in the population is quite small.

2–18 RANDOM NUMBERS

Consider a random experiment $\mathfrak{R}$ which has associated with it some discrete or continuous random variable x. Let the distribution of x be characterized

by the probability function $p(x)$ if x is discrete, or by the density function $f(x)$ if x is continuous. When $\Re$ is performed we observe some value of x, say ξ. We shall now introduce a new terminology which is very useful. We shall say that the performance of $\Re$ *generates a random number from the distribution* $p(x)$ *or* $f(x)$. The random number is ξ, the value x takes on. If n independent trials of $\Re$ are made, the n values $\xi_1, \ldots, \xi_n$ of x observed on the n trials are said to be n random numbers from the distribution $p(x)$ or $f(x)$.

We are frequently interested in obtaining a set of random numbers from a given distribution for a variety of reasons which will be explained below. We would like to be able to generate a sequence of random numbers from any given distribution without performing a number of independent trials of a complicated random experiment to do so. We shall see in this section how this can be done. It is important to note before going on, however, that the notion of a random number has no meaning without reference to some distribution from which it is imagined to be generated.

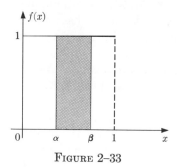

FIGURE 2–33

Let us first consider the problem of generating random numbers representing the values of a continuous random variable x, which can take on values in the interval $0 \leq x \leq 1$ and whose probability density function is $f(x) = 1$. Thus $f(x)$ is a constant and x is said to have a *uniform distribution*. The density function for x then looks like that shown in Figure 2–33. The probability that x lies between α and β is the area of the shaded rectangle, which is $\beta - \alpha$. What this says is that the probability that x lies in any interval of length h is h, and this is independent of where the initial point for the interval is, so long as the interval lies in the event set. A random variable with a uniform distribution is, then, the continuous analog of the notion of equally likely for the discrete case. The probability that x lies in the event set $0 \leq x \leq 1$ is 1, as desired. In order to have this probability be 1, it is necessary that $f(x) = 1$ if $f(x)$ is to be constant over the event set. This is the reason we originally wrote $f(x) = 1$. The random numbers from this distribution will be called *random numbers drawn from a uniform distribution on the unit interval*. It is not possible to generate such random numbers in a rigorous sense because no physical process can represent a continuous ran-

dom variable exactly. However, one can come arbitrarily close to doing this.

Suppose then that we are content to generate random numbers of this sort which have only ten digits after the decimal point. Not every number in the interval $0 \leq x \leq 1$ can be written as a ten-decimal number. There are, however, ten billion of them, and for most purposes, this can be considered to be a quite adequate approximation for all numbers in the interval. One can proceed to generate ten-digit numbers which are essentially random numbers from a uniform distribution on the unit interval as follows.

Imagine that we have ten balls, numbered 0, 1, 2, . . . , 9, in a bowl. We now mix the balls thoroughly and draw one. We write down the number on the ball drawn. The ball is then replaced in the box, the balls are mixed thoroughly again and another one is drawn. We write down the number on this ball after the number on the first ball and repeat the procedure. After ten repetitions we have generated ten digits, which when written down one after the other might be 8003157821. If we place a decimal point in front of the number to yield 0.8003157821, this is a number of the type desired. Each ten-digit number so generated should have the same probability of occurrence. If we continually repeat this experiment, every ten digits generated serve to define a random number from the uniform distribution on the unit interval. In this way a sequence of random numbers from the uniform distribution on the unit interval can be generated.

It would be very clumsy to have to go through a procedure of the type just described every time one wanted random numbers of this form. Fortunately, this is not necessary. One reason is that tables of random numbers from the uniform distribution on the unit interval have been published. A well-known table is that prepared by the RAND Corporation, to which a reference is given at the end of the chapter. Another reason that it is unnecessary to go through the above procedure is that digital computers can be programmed to generate these random numbers. In actuality, the computer cannot perform any random experiments. Only deterministic operations can be carried out. For this reason the random numbers generated are not random in the strict sense of the word because they are generated by a deterministic process. They are referred to as pseudo-random numbers. The numbers so generated are close enough to being random, however, that they can be used as such for almost all purposes. Thus we can, if we wish, have a computer generate for us a sequence of random numbers from the uniform distribution over the unit interval.

Let us next turn to the problem of generating random numbers from an arbitrary distribution. Interestingly enough, we shall show that if it is possible to generate random numbers from the uniform distribution over the unit interval, we can generate from them random numbers from any distribution. In other words, if we have a sequence of random numbers from the uniform distribution over the unit interval, we can convert them into a

sequence of random numbers from any distribution. To see how to do this, suppose first that we are interested in generating random numbers from the distribution $p(x)$ for a discrete random variable x. Let $P(x)$ be the cumulative function for x, and suppose that the bar diagram for $P(x)$ looks like that shown in Figure 2–34. We shall imagine x can take on the values $x_1, \ldots, x_k$. Let us now generate a random number ζ from the uniform distribution over the unit interval. In Figure 2–34 let us draw a horizontal line through the point representing ζ on the vertical axis and note the first bar this line intersects. This is the bar for $x = x_5$ in Figure 2–34. We then claim that x_5 is a random number generated from $p(x)$, and if we repeat this procedure, first generating a random number from the uniform distribution and then determining in the manner just described a value of x, this generates a sequence of random numbers from $p(x)$.

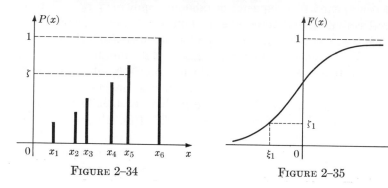

FIGURE 2–34   FIGURE 2–35

To see that the process just described does generate a sequence of random numbers from $p(x)$, let us determine the probability that the value x_i of x will be selected. Now x_i will be selected when the value ζ satisfies $p(x_{i-1}) < \zeta \leq P(x_i)$, because then the bar corresponding to x_i will be the first one intersected. Note that $P(x_i) - P(x_{i-1}) = p(x_i)$, so that the length of the interval containing values of ζ which will yield x_i is $p(x_i)$. However, the probability that the number ζ generated from the uniform distribution will lie in any interval is simply the length of the interval, that is, $p(x_i)$ for the interval under consideration. But this is precisely what we want. The probability of selecting x_i is $p(x_i)$, and the value selected on any given trial is independent of what happened on previous trials. Thus the values of x so generated will be random numbers from the distribution $p(x)$.

Consider now the case where we wish to generate random numbers from the density function $f(x)$, with domain X corresponding to a continuous random variable x. Let $F(x)$ be the cumulative function for x, so that $F(\xi)$ is the probability that $x \leq \xi$. If we perform an experiment $\mathfrak{R}$ having associated with it the random variable x a number of times, we shall generate a sequence of values of x, which we can denote by $\xi_1, \xi_2, \xi_3, \ldots$. These are a

set of random numbers from the distribution $f(x)$, and they have the characteristic that if $\mathcal{R}$ is repeated a very large number of times, the fraction of these numbers which are less than or equal to ξ will be $F(\xi)$. We are then looking for a process which corresponds to independent trials of $\mathcal{R}$ which will generate a set of numbers with the characteristics just referred to. Suppose that the graph of $F(x)$ looks like that shown in Figure 2–35. Let us now select a particular value of x, say ξ, and let $\zeta = F(\xi)$. Every value ξ_1 of x which is less than ξ has the characteristic that if $\zeta_1 = F(\xi_1)$, then $\zeta_1 < \zeta$. Imagine now that we generate a random number ζ_1 from the uniform distribution over the unit interval. The probability that $\zeta_1 \leq \zeta$. Consider the unique number ξ_1 having the property that $\zeta_1 = F(\xi_1)$. Now $\xi_1 \leq \xi$ if and only if $\zeta_1 \leq \zeta$. Thus the probability that $\xi_1 \leq \xi$ is $\zeta = F(\xi)$. This is true regardless of what value of ξ is selected.

Thus if we generate a sequence of random numbers $\zeta_1, \zeta_2, \ldots$, from the uniform distribution and determine the numbers $\xi_1, \xi_2, \ldots$, such that $\zeta_j = F(\xi_j)$, then the ξ_j form a sequence of random numbers from the distribution $f(x)$. This follows, since the long-run fraction of the numbers which are less than or equal to ξ is $F(\xi)$ for any ξ, which is precisely what we want, and the numbers do also have the property that they are generated by independent trials of an experiment. We can easily find graphically the number ξ_1 corresponding to ζ_1 by drawing a horizontal line through ζ_1 on the vertical axis, as shown in Figure 2–35, and determining where it crosses the curve for $F(x)$; the value of x so determined is ξ_1. If random numbers from $f(x)$ were generated using a computer, the computer would proceed in much the same way that we have just described. It would first generate a random number ζ_1 from the uniform distribution over the unit interval, and then it would determine the number ξ_1 such that $\zeta_1 = F(\xi_1)$.

Frequently situations are encountered where it is desired to generate a set of random numbers. Usually, the problem corresponds to one where it is desired to select at random n balls from a set of N balls. Many opinion polls operate in this way. For example, suppose that we wished to select at random two students from a class of 50. This could easily be done using a table of random numbers from the uniform distribution. We first take an alphabetic list of the students and assign the first the number 00, the second the number 01, . . . , and the last the number 49. Next we take the table of random numbers and select any one of the numbers. The last two digits of this number, which might be 14, are used to select one of the students, that is, the student numbered 14 is selected. If the last two digits yield a number which is greater than 49 we move on to the next random number in the table and repeat the process. This is continued until the last two digits yield a number less than 50. Once the first student is selected one merely continues in the table using the next number in sequence until another number is obtained whose last two digits yield a number less than 50 which is different

from the number used to select the first student. This illustrates how random-number tables can be used to select random samples. In the next section we wish to study yet another important use for random numbers.

2-19 SIMULATION

Many problems encountered in practice are very complicated and it may be extremely difficult or impossible to solve these problems mathematically. The advent of large-scale digital computers has made possible the use of an alternative method, referred to as *simulation,* to determine a practical numerical solution to many very complicated problems in which chance plays a role. To see how the simulation procedure works, we shall begin by considering a very simple example.

Suppose that x is a continuous random variable with a known density $f(x)$. Imagine now that we define a new random variable y as a function of x by writing $y = \psi(x)$, and we wish to determine the density function for y. It can be extremely complicated to do this mathematically. Let us then consider an alternative procedure by which a good approximation to the density function for y can be obtained. We have shown in the previous section how a computer can be used to generate random numbers from the distribution $f(x)$. Suppose, then, we have the computer generate a large quantity of random numbers, perhaps 100,000 or so, from $f(x)$. For each such number ξ_j, the computer determines $\zeta_j = \psi(\xi_j)$, and in this way a large number of values of y are generated, that is, a large quantity of random numbers from the density function for y.

Suppose that next the computer aggregates these values of y according to class intervals and in this way constructs a histogram for y. The outline of this histogram should be very close to the curve representing the density function for y, and this curve can be used to give a table of the density function for y. The computer could be programmed to print out such a table. It could also be programmed to display the outline of the histogram on the curve-plotting device. In this way the density function for y could be obtained.

Simulation, as the name suggests, is a procedure whereby real-world experiments are simulated by generating random numbers and making whatever other computations are needed. The real-world experiment can be simulated over and over again by generating more random numbers. To use the simulation technique it is often necessary to perform a great deal of arithmetic computation, and this is why computers are often needed.

Simulation techniques are widely used in industry today. For example, a firm may wish to study the implications of using a new set of policies for controlling the inventories in its warehouses. This can be done without actually implementing these policies in the real world by the use of simula-

tion. The entire warehouse system can be "operated" on a computer using the new policies. The computer will generate random numbers representing the times of arrivals of orders and order sizes. Then the new set of rules will be used to set up shipping schedules, place orders for additional inventory and so forth. More random numbers are generated to determine when stock ordered arrives in the warehouse and when orders are received by customers. In this way insight can be gained into how the new rules will work without taking the dangerous step of actually implementing them without such an understanding. A number of years of warehouse operation can easily be simulated on the computer in a relatively few minutes.

The following represent other situations where simulation can be used advantageously. A company may wish to study alternative rules for truck dispatching. A change in policy regarding the handling of a company's cash balance is about to be made and it is desired to study the implications of the new policy. It is desired to evaluate a new proposal for routing orders through a job shop. The great advantage of simulation in situations like this is that it is possible to evaluate what will be likely to happen under a variety of circumstances without ever actually implementing the new rules in practice. We shall use small scale simulations in several examples in the following chapters.

REFERENCES

1. Burington, R. S. and D. C. May, *Handbook of Probability and Statistics with Tables*. Handbook Publishers, Sandusky, Ohio, 1953.

A handbook which contains a somewhat more extensive collection of tables than is given at the end of this text.

2. General Electric Co., *Tables of the Individual and Cumulative Terms of the Poisson Distribution*. Van Nostrand, Princeton, N.J., 1962.

3. Hertz, D. B., "Risk Analysis in Capital Investment," *Harvard Business Review*, **42**, 95–106 (January 1964).

Illustrates how simulation can be used in certain types of decision problems.

4. Lieberman, G. J. and D. B. Owen, *Tables of the Hypergeometric Probability Distribution*. Stanford University Press, Stanford, Calif., 1961.

Gives hypergeometric probabilities up to N = 50.

5. RAND Corporation, *A Million Random Digits with* 100,000 *Normal Deviates*, Free Press, Glencoe, Ill., 1955.

6. *Table of the Cumulative Binomial Probabilities*. Ordnance Corps Pamphlet ORDP 20–1, 1952. (577 pages.)

7. *Tables of the Cumulative Binomial Probability Distribution*. Harvard University Press, Cambridge, Mass., 1955.

PROBLEMS

Section 2–1

1. Consider a random experiment in which a fair coin is tossed twice. Construct a model for this experiment. Let x be the number of heads obtained on the two tosses. Show that x is a random variable and determine the value x takes on for each simple event.

2. A box contains ten balls numbered 1 to 10. Consider the random experiment in which a ball is selected at random from the box. Construct a model of this experiment. Let x be the number on the ball drawn. Show that x is a random variable and determine the value x takes on for each simple event.

3. A box contains ten balls numbered 11 to 20. Consider the random experiment in which two balls are selected at random from the box. Construct a model of this experiment. Let x be the sum of the numbers on the two balls drawn. Show that x is a random variable and determine the value x takes on for each simple event.

4. Two fair dice are tossed. Construct a model of this experiment. Let x be the sum of the number of dots on the two faces which turn up. Show that x is a random variable and determine the value x takes on for each simple event.

5. A fair coin is tossed three times. Construct a model for this experiment. Let x be the number of heads obtained on the three tosses. Show that x is a random variable and determine the value x takes on for each simple event.

6. Ten balls numbered 1 to 10 are in a bowl. Two balls are selected at random from the bowl. Construct a model for this experiment. Let x be the larger of the numbers on the two balls drawn. Show that x is a random variable and determine the value x takes on for each simple event.

7. The rainfall in a given location during some specified future month can be looked upon as a random variable. Why?

8. The number of people who die from cancer in Chicago in the coming month can be looked upon as a random variable. Why?

9. The number of letters which a postman will have to deliver next week can be looked upon as a random variable. Why?

10. Consider an experiment in which a plant of type AA is crossed with a plant of type Aa. Suppose that these genes determine the color of the offspring's flowers.
  (a) Is the number of A genes received by the offspring from the AA parent a random variable? Explain.
  (b) Is the number of A genes received by the offspring from the Aa parent a random variable? Explain.
  (c) Is the color of the offspring's flowers a random variable? Explain.

Section 2–2

1. A fair coin is tossed once. Let the random variable x be the number of heads obtained. Compute the expected value of x. What is the intuitive meaning of μ_x?

2. Suppose that as a model for some experiment we are using one with $E = \{e_1, e_2, e_3\}$ and $p_1 = 0.25$, $p_2 = 0.60$ and $p_3 = 0.15$. Let x be a random variable defined for this experiment such that if $x = \xi_j$ when e_j occurs then $\xi_1 = -17$, $\xi_2 = 2$ and $\xi_3 = 1$. What is the expected value of x?

3. A jewelry store is considering stocking seven of an expensive pin for the Christmas season. The demand may turn out to be for 0, 1, ..., 8 pins. The owner estimates the probabilities of each of these events to be 0.05, 0.10, 0.10, 0.10, 0.10, 0.20, 0.20, 0.05 and 0.10 and the corresponding profits are $-\$200$, $-\$150$, $-\$100$, $\$0$, $\$10$, $\$200$, $\$250$, $\$300$, and $\$250$. Compute the expected profit if seven pins are stocked.

4. Compute the expected value of the random variable introduced in Problem 1 for Section 2–1. What is the intuitive meaning of this expected value?

5. Compute the expected value of the random variable introduced in Problem 2 for Section 2–1. What is the intuitive meaning of this expected value?

6. Compute the expected value of the random variable introduced in Problem 3 for Section 2–1. What is the intuitive meaning of this expected value?

7. Compute the expected value of the random variable introduced in Problem 4 for Section 2–1. What is the intuitive meaning of this expected value?

8. Compute the expected value of the random variable introduced in Problem 5 for Section 2–1. What is the intuitive meaning of this expected value?

9. Compute the expected value of the random variable introduced in Problem 6 for Section 2–1. What is the intuitive meaning of this expected value?

Section 2–3

1. A model for some experiment consists of the event set $E = \{e_1, e_2, e_3\}$ with $p_1 = 0.25$, $p_2 = 0.60$ and $p_3 = 0.15$. Consider the random variable x for which $\xi_1 = -17$, $\xi_2 = 2$ and $\xi_3 = -1$. Determine the probability function for this random variable and construct a bar diagram to illustrate it graphically. Determine the cumulative function for x and illustrate it graphically.

2. A model for some experiment consists of the event set $E = \{e_1, e_2, e_3\}$ with $p_1 = 0.20$, $p_2 = 0.50$ and $p_3 = 0.30$. Consider the random variable x for which $\xi_1 = 2$, $\xi_2 = -4$ and $\xi_3 = 2$. Determine the probability function for this random variable and construct a bar diagram to illustrate it geometrically. Determine the cumulative function for x and illustrate it graphically. Compute the expected value of x using (2–3) and (2–11) and verify that the same result is obtained in both cases.

3. A model for some experiment consists of the event set $E = \{e_1, e_2, e_3, e_4, e_5\}$ with $p_1 = 0.05$, $p_2 = 0.15$, $p_3 = 0.10$, $p_4 = 0.50$ and $p_5 = 0.20$. Consider the random variable x for which $\xi_1 = 9$, $\xi_2 = 1$, $\xi_3 = 3$, $\xi_4 = 5$ and $\xi_5 = 9$. Determine the probability function for x and construct a bar diagram to illustrate it graphically. Determine the cumulative function for x and illustrate it graphically. Compute the expected value of x from (2–3) and (2–11), and verify that the same result is obtained in both cases.

4. A model for some experiment consists of the event set

$$E = \{e_1, e_2, e_3, e_4, e_5, e_6, e_7, e_8, e_9, e_{10}\} \,,$$

with $p_1 = 0.10$, $p_2 = 0.15$, $p_3 = 0.05$, $p_4 = 0.07$, $p_5 = 0.13$, $p_6 = 0.20$, $p_7 = 0.03$, $p_8 = 0.06$, $p_9 = 0.10$ and $p_{10} = 0.11$. Consider the random variable x for which $\xi_1 = 0$, $\xi_2 = -3$, $\xi_3 = 3$, $\xi_4 = 5$, $\xi_5 = 3$, $\xi_6 = 3$, $\xi_7 = 3$, $\xi_8 = 5$, $\xi_9 = 7$ and $\xi_{10} = 0$. Determine the probability function for x and construct a bar diagram to illustrate it graphically. Determine the cumulative function for x and illustrate it graphically. Compute the expected value of x from (2–3) and (2–11), and verify that the same result is obtained in both cases.

5. Determine the probability function and cumulative function for the random variable x introduced in Problem 1 for Section 2–1; construct bar diagrams to represent graphically each of these functions.

6. Determine the probability function and cumulative function for the random variable x introduced in Problem 2 for Section 2–1; construct bar diagrams to represent graphically each of these functions.

7. Determine the probability function and cumulative function for the random variable x introduced in Problem 3 for Section 2–1; construct bar diagrams to represent graphically each of these functions.

8. Determine the probability function and cumulative function for the random variable x introduced in Problem 4 for Section 2–1; construct bar diagrams to represent graphically each of these functions.

9. Determine the probability function and cumulative function for the random variable x introduced in Problem 5 for Section 2–1; construct bar diagrams to represent graphically each of these functions.

10. Determine the probability function and cumulative function for the random variable x introduced in Problem 6 for Section 2–1; construct bar diagrams to represent graphically each of these functions.

Sections 2–4 and 2–5

1. A fair coin is tossed five times. Compute the probability that precisely three heads are obtained.

2. A fair coin is tossed ten times. Compute the probability that precisely five heads are obtained. What is the probability that five or less heads are obtained?

3. Ten percent of the autos assembled on a given line are defective. Each auto produced can be looked upon as an independent trial of a random experiment. Ten cars are shipped to a dealer. What is the probability that none is defective? What is the probability that two or less are defective?

4. A petroleum company averages 1.5 strikes (that is, oil is found) per 10 wildcat wells drilled. Seven wildcat wells are to be drilled in widely separated locations. What is the probability that two or more will strike oil?

5. A clothing store has found that of those customers who come in to look at suits only one in five actually makes a purchase. What is the probability that out of the

next ten customers who look at suits precisely four will buy one? What is the probability that three or less will buy a suit?

6. Twenty percent of the light bulbs produced by a given manufacturer fail before the end of a two-week period when left burning continuously. Seven bulbs are installed in the hall lights of an apartment building where they burn continuously. What is the probability that exactly five of the bulbs will be burned out at the end of two weeks? What assumption was made in making the computation?

7. Past records show that half of the job applicants pass a general aptitude test that a given company requires them to take. What is the probability that at least three of the next five applicants will pass the test? What assumption was made in making the computations?

8. Use Table B at the end of the text to obtain a bar diagram for $b(x; 7, 0.3)$.

9. Use Table B at the end of the text to obtain a bar diagram for $b(x; 7, 0.1)$.

10. Use Table B at the end of the text to obtain a bar diagram for $b(x; 9, 0.2)$.

11. Use Table B at the end of the text to obtain a bar diagram for the cumulative binomial distribution $B(x; 10, 0.2)$.

12. Eight passengers board a train consisting of three cars. Each passenger selects at random which car he will sit in. What is the probability that there will be three people in the first car?

13. A man buys two boxes of matches and places them in his pocket. Every time he has to light a match, he selects at random one of the two boxes. After some time the man takes one of the boxes from his pocket and on opening it finds it empty. Observe that this statement implies that at the point in time when the individual used the last match in this box, he did not throw it away, but instead put it back in his pocket. What is the probability that there are at the moment k matches left in the other box if each box originally contained n matches and we assume that this is the first time that he has encountered an empty box? This is a famous problem which is referred to as Banach's match-box problem, after the Polish mathematician of the same name.

14. Suppose that the probability that a batter gets a hit whenever he comes up to bat is $\frac{1}{5}$. Some people feel this implies that he is sure to get at least one hit in five times at bat. What is the probability of at least one hit in five times at bat? What is the expected number of hits in five times at bat? Why does the expected value not imply that he gets one hit in five?

15. Consider an electronic system which consists of n parts. Suppose that each part has the same probability p of failing when the system is used for a given mission. Assume that whether or not a given part fails is independent of what happens to other parts. What is the probability that the system does not fail, that is, that no part fails? The following argument is often given to obtain this probability. Since the probability that any one part fails is p, the probability that at least one part fails is np, and hence the probability that the system does not fail is $1 - np$. Show that the reasoning and resulting probability are incorrect. Even though the probability of successful operation is not $1 - np$, there are certain conditions under

which the number $1 - np$ is very close to the probability of successful operation. Explain what these conditions are, and why things work out this way.

16. Two ground-to-air missiles are fired simultaneously at a target aircraft. If the probability that a given missile hits the target is $\frac{1}{3}$, what is the probability that at least one of the missiles hits the target? What is the probability of a hit if three missiles are fired?

17. Make a table comparing $b(r; 9, 0.1)$ and $p(r; 0.9)$ for $r = 0, \ldots, 9$.

18. Make a table comparing $b(r; 10, 0.2)$ and $p(r; 2)$ for $r = 0, \ldots, 10$.

19. Make a table comparing $b(r; 8, 0.1)$ and $p(r; 0.8)$ for $0, \ldots, 8$.

20. Use a Poisson approximation to estimate $b(10; 100, 0.05)$ and $b(5; 1000, 0.0001)$.

21. Use a Poisson approximation to estimate $b(15; 500, 0.02)$ and $b(8; 200, 0.05)$.

22. Verify that the sum over r of $p(r; 0.2)$ is unity to the accuracy given in Table D at the end of the text.

23. Construct a bar diagram for $p(x; 0.4)$.

24. Construct a bar diagram for $p(x; 4)$.

25. Estimate the following using the Poisson approximation.
(a) $b(4; 1000, 0.003)$;   (b) $b(5; 10,000, 0.0007)$.

26. Ten seeds obtained from crossing an AA type plant with an Aa type are planted. The A and a genes under consideration determine the color of the plants' flowers. If all ten seeds grow into plants, what is the probability that precisely four plants will have red flowers, the remaining ones being pink? What is the probability that four or fewer plants will have red flowers?

27. Ten seeds obtained from crossing an Aa type plant with an Aa type are planted. The A and a genes under consideration determine the color of the plant's flowers. If all ten seeds grow into plants, what is the probability that precisely seven of the plants will have colored flowers (red or pink)? What is the probability that at least seven plants will have colored flowers?

Section 2–6

1. In South Africa, a medical team selects at random 100 natives from a region of Zululand which has a population of 10,000. If seven percent of the natives have TB, what is the probability that 15 or fewer in the sample selected have this disease. Use the Poisson approximation in carrying out the computations.

2. For the situation described in Problem 1, what is the probability that the fraction of the natives in the sample who have TB differs by more than 0.02 from the fraction of the natives in the population who have the disease?

3. Sometimes an attempt is made to estimate the fraction of a wild animal population which is diseased by trapping a number of the animals and checking them. In Kenya, 100 impala were trapped and checked for a certain type of parasite infection. What is the probability that the fraction of infected animals in the sample differs by more than 0.03 from the fraction of the entire impala population which

are infected if the actual fraction infected is 0.06? Use the Poisson approximation. What assumptions are needed concerning the trapping procedures?

4. A box contains 1000 transistors, 25 of which are defective. A sample of four transistors is selected at random from the lot. Compute the exact probability that none of those selected is defective, and also compute the Poisson approximation to this.

5. A box contains 1000 transistors, 50 of which are defective. A sample of 100 transistors is selected at random from the lot. Use the Poisson approximation to estimate the probability that nine of the transistors in the sample are defective.

6. A box contains 1000 transistors, 10 of which are defective. A sample of 100 transistors is selected at random from the lot. Use the Poisson approximation to estimate the probability that two or less defectives will be found in the sample.

7. When N is large with respect to n,

$$h(r; n, p, N) \doteq b(r; n, p) .$$

Can you explain why this is true? Hint: Suppose that instead of selecting n balls from N, one ball is selected and its color noted. Then before the next one is selected the first one is replaced and the balls are well mixed. Then a second ball is selected. Show that the probability of obtaining r red balls in n draws when the ball drawn at the previous step is replaced before the next ball is drawn is $b(r; n, p)$. How does this procedure differ from that where n balls are selected at random from the N balls? What happens as N is allowed to become large when n and p are held constant? Use the notion of conditional probabilities to help in explaining the desired result. What this problem says is that when N is very large, the selection of n balls corresponds to n independent trials of an experiment having probability p of selecting a red ball.

8. A box contains two red and six green balls. The balls are well mixed and a blindfolded individual chooses two of them. Both are red. What is the probability that the individual will repeat exactly three times this performance of selecting precisely two red balls if the experiment is repeated a total of five times?

Section 2–7

In Problems 1 through 8 express the sums using summation notation.

1. $b_1 + b_2 + b_3 + b_4 + b_5$.

2. $1 + 2 + \cdots + 10$.

3. $a_1 b_1^2 + a_2 b_2^2 + \cdots + a_n b_n^2$.

4. $a_1 b_2^2 + a_2 b_3^2 + \cdots + a_n b_{n+1}^2$.

5. $(a_1^2 + b_1) + (a_2^2 + b_2) + \cdots + (a_m^2 + b_m)$.

6. $(a_6 b_9 - c_7) + (a_7 b_{10} - c_8) + \cdots + (a_{25} b_{28} - c_{26})$.

7. $a_1 b_2 + a_2 b_4 + a_3 b_6 + \cdots + a_{100} b_{200}$.

8. $a_1 b^3 + a_2 b^5 + a_3 b^7 + \cdots + a_{30} b^{61}$.

In Problems 9 through 13 write out explicitly the indicated summations.

9. $\displaystyle\sum_{i=1}^{5} i a_i$.

10. $\displaystyle\sum_{j=3}^{7} (j+1)^2 a^j$.

11. $\displaystyle\sum_{j=0}^{8} (6a_j - j^2 b)$.

12. $\displaystyle\sum_{j=0}^{5} a_j b^{3j}$.

13. $\displaystyle\sum_{j=1}^{7} (-1)^j a_{2j}$.

14. Show that

$$\sum_{j=1}^{n} a_{j+3} = \sum_{k=4}^{n+3} a_k .$$

15. Show that

$$\sum_{j=3}^{n} a^j = a^3 \sum_{j=0}^{n-3} a^j .$$

16. Write out

$$\sum_{i=1}^{3} \sum_{j=1}^{6} a_{ij}$$

17. Write out

$$\sum_{i=1}^{3} \sum_{j=1}^{6} i j^2 .$$

18. Show that the following hold.

(a) $\displaystyle\sum_{i=1}^{m} \sum_{j=1}^{n} (a_{ij} + b_{ij}) = \sum_{i=1}^{m} \sum_{j=1}^{n} a_{ij} + \sum_{i=1}^{m} \sum_{j=1}^{n} b_{ij}$;

(b) $\displaystyle\sum_{i=1}^{m} \sum_{j=1}^{n} \lambda a_{ij} = \lambda \sum_{i=1}^{m} \sum_{j=1}^{n} a_{ij}$.

Section 2–8

1. A model for some experiment consists of the event set $E = \{e_1, e_2, e_3, e_4\}$, with $p_1 = 0.15$, $p_2 = 0.30$, $p_3 = 0.25$ and $p. = 0.30$. Let x be a random variable which takes on the values $\xi_1 = -3$, $\xi_2 = 1$, $\xi_3 = 0$ and $\xi_1 = -2$, and y a random variable which takes on the value $\varsigma_1 = 2$, $\varsigma_2 = 0$, $\varsigma_3 = 4$ and $\varsigma_4 = -1$. Suppose that we define a new random variable z by writing $z = 4x + y - 2$. Determine the value z takes on for each simple event. Compute the expected value of z using (2–31). Also, determine the probability function for z and the expected value of z using this function.

2. A model for some experiment consists of the event set $E = \{e_1, e_2, e_3, e_4, e_5, e_6, e_7\}$, with $p_1 = 0.08$, $p_2 = 0.12$, $p_3 = 0.14$, $p_4 = 0.06$, $p_5 = 0.20$, $p_6 = 0.30$ and $p_7 = 0.10$. Let x be a random variable which takes on the values $\xi_1 = 0$, $\xi_2 = 4$, $\xi_3 = 7$, $\xi_4 = 1$, $\xi_5 = 5$, $\xi_6 = 9$ and $\xi_7 = 0$, and y a random variable which takes on the values $\varsigma_1 = 5$, $\varsigma_2 = 0$, $\varsigma_3 = -2$, $\varsigma_4 = -5$, $\varsigma_5 = 1$, $\varsigma_6 = 3$ and $\varsigma_7 = -5$. Suppose

that we define a new random variable z by writing $z = 3x - 2y + 11$. Determine the value z takes on for each simple event. Compute the expected value of z using (2–31). Also, determine the probability function for z and the expected value of z using this function.

3. Consider the random experiment in which two fair dice are tossed. What is the expected value of the random variable representing the sum of the number of dots on the two faces turned up? Solve the problem for the case where n fair dice are tossed and the random variable is the sum of the number of dots on the n faces turned up.

4. A petroleum company plans to bid on five tracts of land. The prices which it plans to bid on each of these are, respectively, 2, 1.5, 0.7, 1.8 and 0.82 million dollars. The company has estimated that the probability of winning the bid on the first tract when bidding 2 million dollars is 0.4. The corresponding probabilities of winning the other bids are 0.7, 0.5, 0.3 and 0.8. The company has to pay out money only on those bids won. What is the expected amount that the company will have to pay out on these bids? Hint: Let $y\{j\} = 1$ if the bid on tract j is won and $y\{j\} = 0$ if it is not won.

5. A company is involved in two projects. The expected return from the first is 2.5 million dollars and the expected return from the second is 4.1 million dollars. What is the expected return from both projects combined? What is the justification for your answer?

6. For a given project a company estimates that the expected revenues received will be 18 million dollars and the expected total costs incurred will be 16.2 million dollars. What is the expected profit from this project? What is the justification for your answer?

7. A company plans to send in a bid of $0.85 per pound for one million pounds of government liquid oxygen business. It has two competitors A and B. It feels that the probability that A will bid less than $0.85 is 0.3 and the probability that B will bid less than $0.85 is 0.7. Lowest bidder gets all the business. Each company submits its bid independently of the others. What is the expected dollar volume of business that the company under consideration will get under these circumstances? What does the expected value mean intuitively in this case?

8. Consider the situation outlined in Problem 7. Suppose now that it is not true that A and B can handle all the business. A can supply 500,000 pounds and B can supply 800,000 pounds. Under these circumstances the lowest bidder gets all the business he can handle and the next lowest bidder gets all the remaining business. Assume that it is believed that A will bid only one of the values 0.82, 0.84 and 0.88 with the probabilities 0.05, 0.25 and 0.70, respectively. Assume also that B will bid only one of the four values 0.80, 0.83, 0.86 and 0.88 with the probabilities 0.30, 0.40, 0.20 and 0.10, respectively. What is the expected dollar volume of business that the company under consideration will obtain in this case?

9. Determine μ_x when x has the distribution:

  (a) $b(x; 4, 0.3)$;   (b) $b(x; 100, 0.7)$;   (c) $b(x; 1000, 0.003)$

10. A biased coin which has a probability 0.65 of landing heads up is tossed 15 times. What is the expected number of heads that will be obtained? What is the expected number of tails?

11. A given company has found in the past that its profit h in millions of dollars can be expressed in terms of sales s in millions of units by the equation

$$h = 1.2s - 750,000.$$

It is believed that sales in the coming year will be one of the four following values: 700,000, 900,000, 1,000,000 or 1,300,000, with the probabilities of these being 0.20, 0.40, 0.30 and 0.10 respectively. What is the company's expected profit for the coming year? If sales only take on the values indicated, what are the possible values for the profit in the coming year?

Section 2–9

1. Consider a random variable x which can take on values in the set $X = \{-2, 0, 1, 4, 8, 10\}$. For each of the following functions determine the set of values which the random variable y can take on.

(a) $y = 2x^2 + 3x$; (b) $y = \dfrac{3x - 2}{x + 1}$; (c) $y = x + \dfrac{1}{x + 1}$.

2. A model for some random experiment consists of an event set $E = \{e_1, e_2, e_3, e_4, e_5\}$, with $p_1 = 0.25$, $p_2 = 0.10$, $p_3 = 0.30$, $p_4 = 0.05$ and $p_5 = 0.30$. Let x be a random variable which takes on the values $\xi_1 = 6$, $\xi_2 = -1$, $\xi_3 = 1$, $\xi_4 = 2$ and $\xi_5 = 1$. Suppose that a new random variable y is defined by writing $y = 3x^2 - 2x$. Compute μ_y by first determining the probability function for x and using (2–40). Next, determine the value ζ_i of y for each e_i and then the probability function for y. Finally, determine μ_y using the probability function for y.

3. A model for some random experiment consists of an event set $E = \{e_1, e_2, e_3, e_4, e_5, e_6\}$, with $p_1 = 0.05$, $p_2 = 0.10$, $p_3 = 0.20$, $p_4 = 0.30$, $p_5 = 0.25$ and $p_6 = 0.10$. Let x be a random variable which takes on the values $\xi_1 = 0$, $\xi_2 = 2$, $\xi_3 = 3$, $\xi_4 = 4$, $\xi_5 = 5$ and $\xi_6 = 7$. Suppose we now define a new random variable y by writing $y = (x - 2)/(x + 3)$. Compute μ_y by first determining the probability function for x and using (2–40). Next, determine the value ζ_i of y for each e_i and then the probability function for y. Finally, determine μ_y using the probability function for y.

4. Consider the random variable x which can take on the values in the set $X = \{0, 1, 2, 3, 4, 5\}$ and whose probability function is $p(0) = 0.10$, $p(1) = 0.20$, $p(2) = 0.30$, $p(3) = 0.20$, $p(4) = 0.10$ and $p(5) = 0.10$. Compute μ_x and consider the random variable $y = x - \mu_x$. Determine the probability function for y, and illustrate graphically both the probability function for y and the probability function for x. What is the relation between the two probability functions?

5. Consider the random variable x which can take on the values in the set $X = \{0, 1, 2, 3, 4, 5\}$ and whose probability function is $p(0) = 0.40$, $p(1) = 0.30$, $p(2) = 0.20$, $p(3) = 0.08$, $p(4) = 0.01$ and $p(5) = 0.01$. Compute μ_x. Consider the random variable $y = 2x + 3$. Compute μ_y from (2–40). Also, determine the probability function for y, and compute μ_y directly using it. Illustrate graphically both the probability function for y and the probability function for x. What is the relation between the two probability functions?

6. Let x be a random variable defined over E with expected value μ_x. Also let $y = g(x)$. Show by an example different from that given in the text that it is not true in general that $\mu_y = g(\mu_x)$. Give an intuitive explanation for this.

7. An individual plays a game of chance which involves the tossing of a fair die. Suppose that he wins a dollar if a face with an even number of dots turns up and loses a dollar if a face with an odd number of dots turns up. Let x be the random variable representing the number of dots showing and y the random variable representing the amount won or lost. Show that y is a function of x. Determine the expected value of y.

8. An individual plays a game of chance which involves the tossing of a fair die. He wins the square of the number of the dots turned up if this number is odd, and wins eight dollars less the square of the number of dots turned up if this number is even (when the sum is negative, he loses the corresponding amount). If x is the random variable representing the number of dots turned up and y the amount won, show that y is a function of x. What is the expected value of y?

9. The demand for a particular high-priced cake stocked by a given bakery on Saturdays only can be looked upon as a random variable which will take on one of the values 0, 1, 3, 4 or 5, with probabilities $p(0) = 0.10$, $p(1) = 0.10$, $p(2) = 0.20$, $p(3) = 0.30$, $p(4) = 0.20$ and $p(5) = 0.10$. Each cake costs the bakery \$2.50 and sells for \$5.75. Any cakes left over at the end of the day are a total loss. Determine the bakery's expected profit from this cake on a Saturday when it stocks four cakes.

10. Re-solve Problem 9 if any cakes left over on Saturday can, on Monday, be placed on a stale-goods counter and sold for \$2.00 each. All cakes on this counter will be sold.

11. Re-solve Problem 11 if when the demand is greater than the number of cakes sold, each disappointed customer buys less than usual of other things resulting in a \$1.00 loss in profits to the bakery on these items.

12. A company has found in the past that its profit h is related to sales s by the equation

$$h = 160\sqrt{s} - 500{,}000 .$$

It is believed that sales in the coming year will only take on one of the values 30, 60, 90 or 100 million units, with the probabilities being 0.2, 0.3, 0.4 and 0.1 respectively. What is the company's expected profit for the coming year? What are the possible values that the profit can actually take on under these circumstances?

Section 2–10

1. Consider a random variable x which can take on the values in $X = \{-3, 0, 1, 2, 6\}$, with the probability function $p(-3) = 0.10$, $p(0) = 0.20$, $p(1) = 0.40$, $p(2) = 0.20$ and $p(6) = 0.10$. Compute μ_x and σ_x. Illustrate graphically the probability function for x. On the horizontal axis of this bar diagram, indicate the points μ_x, $\mu_x - \sigma_x$, $\mu_x + \sigma_x$, $\mu_x - 2\sigma_x$, $\mu_x + 2\sigma_x$, $\mu_x - 3\sigma_x$ and $\mu_x + 3\sigma_x$.

2. It is possible to construct as follows a random experiment $\Re$, having associated with it a random variable x whose probability function is that shown in Figure 2–9. Mark the number -2 on one penny, the number 1 on two pennies, and the number 3

on one penny. Place the four pennies in a container, mix them well, and without looking, select one. Repeat ℛ a total of 50 times and on each trial observe the value of x obtained. Then construct a diagram like that shown in Figure 2–8.

3. Repeat Problem 2 using a random variable having the probability function shown in Figure 2–10.

4. Show that if the random variable x takes on the value ξ_j when e_j occurs then

$$\sigma_x^2 = \sum_{j=1}^{m} (\xi_j - \mu_x)^2 p_j .$$

5. Compute the variance and standard deviation of the random variable introduced in Problem 1 for Section 2–1.

6. Compute the variance and standard deviation of the random variable introduced in Problem 2 for Section 2–1.

7. Compute the variance and standard deviation of the random variable introduced in Problem 3 for Section 2–1.

8. Compute the variance and standard deviation of the random variable introduced in Problem 4 for Section 2–1.

9. Compute the variance and standard deviation of the random variable introduced in Problem 5 for Section 2–1.

10. Compute the variance and standard deviation of the random variable introduced in Problem 6 for Section 2–1.

11. Consider the random variable x with probability function $p(x)$. Denote the variance of x by σ_x^2. Consider now the random variable $y = ax + b$. Prove that the variance of y is given by $\sigma_y^2 = a^2\sigma_x^2$. Verify this directly in a specific case by considering the random variable x defined in Problem 1 and writing $y = 2x + 3$. Make the verification by determining the probability function for y and computing σ_y^2 directly.

12. Consider a random variable x with probability function $p(x)$, expected value μ_x and standard deviation σ_x. Consider the random variable $y = (x - \mu_x)/\sigma_x$. Show that $\mu_y = 0$ and $\sigma_y = 1$.

13. Apply the results of Problem 12 to the random variable x defined in Problem 1 to obtain a random variable y with $\mu_y = 0$ and $\sigma_y = 1$. Determine the probability function for y and illustrate it graphically.

14. Under what conditions can $\sigma_x = 0$?

15. Let x be a random variable with expected value μ_x and variance σ_x^2. Also let $y = |x - \mu_x|$ so that y measures the magnitude of the deviation of x from its expected value. If $p(y \geq \epsilon)$ is the probability that $y \geq \epsilon$, prove that

$$p(y \geq \epsilon) \leq \frac{\sigma_x^2}{\epsilon^2} .$$

This is referred to as the *Chebyshev inequality*. Hint: Denote by $\lambda_1, \ldots, \lambda_u$ the values of x for which $y < \epsilon$ and by $\gamma_1, \ldots, \gamma_v$ the values of x for which $y \geq \epsilon$. Then

$$\sigma_z^2 = \sum_{i=1}^{u} (\lambda_i - \mu_x)^2 p(\lambda_i) + \sum_{j=1}^{v} (\gamma_j - \mu_x)^2 p(\gamma_j)$$

$$\geq \sum_{j=1}^{v} (\gamma_j - \mu_x)^2 p(\gamma_j) \geq \epsilon^2 \sum_{j=1}^{v} p(\gamma_j) \,.$$

But

$$p(y \geq \epsilon) = \sum_{j=1}^{v} p(\gamma_j) \,.$$

The Chebyshev inequality is often very conservative in the sense that $p(y \geq \epsilon)$ is frequently very much smaller than σ_z^2/ϵ^2.

Section 2–11

1. A model for some random experiment consists of an event set $E = \{e_1, e_2, e_3\}$ with $p_1 = 0.20$, $p_2 = 0.50$ and $p_3 = 0.30$. Let x be a random variable which takes on the values $\xi_1 = -3$, $\xi_2 = 0$ and $\xi_3 = 5$, and y be a random variable which takes on the values $\zeta_1 = 2$, $\zeta_2 = 3$ and $\zeta_3 = -1$. Suppose now that a new random variable z is defined by writing $z = 3x - 2y$. Determine σ_z^2 using (2–48). Check the result by determining the probability function for z and computing σ_z^2 directly. What is σ_{xy} in this case?

2. A model for some random experiment consists of an event set $E = \{e_1, e_2, e_3, e_4, e_5\}$ with $p_1 = 0.10$, $p_2 = 0.15$, $p_3 = 0.25$, $p_4 = 0.35$ and $p_5 = 0.15$. Let x be a random variable which takes on the values $\xi_1 = 3$, $\xi_2 = 1$, $\xi_3 = 0$, $\xi_4 = 3$ and $\xi_5 = 4$, and y be a random variable which takes on the values $\zeta_1 = 4$, $\zeta_2 = -1$, $\zeta_3 = 4$, $\zeta_4 = 2$ and $\zeta_5 = 3$. Suppose now that a new random variable z is defined by writing $z = 2x - 5y$. Determine σ_z^2 using (2–48). Check the result by determining the probability function for z and computing σ_z^2 directly. What is σ_{xy} in this case?

3. Let x and y be independent random variables with expected values μ_x and μ_y and variances σ_x^2 and σ_y^2. Show that if $z = x - y$, then $\mu_z = \mu_x - \mu_y$ and $\sigma_z^2 = \sigma_x^2 + \sigma_y^2$.

4. Consider a random experiment in which a fair coin is tossed twice. Let x be the total number of heads obtained and y the number of heads obtained on the first toss. Determine σ_{xy}.

5. Consider a random experiment in which a fair coin is tossed three times. Let x be the total number of heads obtained and y the number of heads obtained on the second toss. Determine σ_{xy}.

6. Determine the variance of the sum that the petroleum company will have to pay out in bids for the situation described in Problem 4 for Section 2–8. Assume that there is no connection of any sort between the various bids.

7. Consider some project which extends over several years. The profit $y\{j\}$ in each year j will be a random variable. It is typical when evaluating a project which extends over a considerable period of time to determine the discounted profit z which is

$$z = \sum_{j=1}^{n} \frac{y\{j\}}{(1+i)^j} = \frac{y\{1\}}{1+i} + \cdots + \frac{y\{n\}}{(1+i)^n}.$$

The number $1/(1+i)^j$ is called the discount factor for year j and $y\{j\}/(1+i)^j$ is referred to as the present worth or discounted profit for year j. Profits obtained in the future are worth less than profits available today, because what is available now can be invested to earn more. The number i can be looked on as the interest rate which invested funds earn. Determine μ_z in terms of the μ_j, μ_j being the expected value of $y\{j\}$. Also determine σ_z^2. Under what circumstances can σ_z^2 be computed from the variances of the $y\{j\}$?

8. Determine the variance and standard deviation of x when x has the binomial distribution:

(a) $b(x; 10, 0.6)$; (b) $b(x; 100, 0.1)$; (c) $b(x; 1000, 0.2)$.

Sections 2–12 and 2–13

1. A manufacturer of resistors produces one type whose nominal resistance is 1000 ohms. One hundred resistors were selected at random from a lot of the resistors produced by this process and the resistance of each was determined carefully. The values obtained were as follows (when arranged in order of increasing resistance):

937.2	981.9	1002.3	1023.6
943.6	982.3	1002.8	1024.2
947.2	983.0	1003.6	1025.4
951.7	983.6	1004.7	1026.3
953.2	984.5	1005.2	1027.6
956.4	986.1	1005.6	1028.2
958.7	987.2	1006.8	1029.1
960.5	987.9	1007.1	1030.2
962.2	988.6	1007.9	1031.3
964.5	989.3	1008.2	1032.5
967.8	990.1	1009.6	1033.7
969.3	990.8	1010.4	1035.1
971.0	991.4	1011.3	1037.2
972.1	992.3	1012.6	1038.3
972.8	993.4	1013.1	1039.4
972.9	994.2	1013.9	1041.2
973.3	994.6	1014.6	1042.5
974.5	995.2	1015.2	1043.1
975.6	996.3	1016.7	1045.2
977.1	996.8	1017.3	1047.3
977.8	997.5	1018.6	1049.1
978.3	998.2	1019.2	1051.4
979.6	1000.1	1021.1	1054.3
980.2	1000.8	1022.3	1057.8
981.4	1001.4	1022.8	1065.3

Construct a relative frequency diagram and a histogram for these data using a class interval of length 20 ohms. Take $\gamma_0 = 930$. Repeat the procedure using a class

interval of length 10 ohms. On the same figure show the outline of the two histograms so obtained.

2. Repeat Problem 1 using class intervals of length 5 and 3 ohms respectively.

3. Repeat Problem 1 using $\gamma_0 = 935$.

4. Repeat Problem 2 using $\gamma_0 = 935$.

5. A department store selects at random the accounts of 75 of its credit-card customers to determine the amounts of their purchases over the past year. The sales in dollars to these customers were:

50.50	76.20	206.00
100.00	142.10	421.32
62.30	206.50	386.40
245.00	654.00	210.17
967.40	81.25	85.11
25.30	40.60	164.00
300.00	0.00	74.30
125.75	150.30	96.13
600.40	303.18	541.16
237.40	22.40	289.16
196.00	70.16	38.40
2500.00	100.41	121.60
10.00	263.70	333.11
97.40	216.10	716.50
296.00	15.80	5.41
400.00	161.30	206.00
345.50	47.00	346.70
35.00	86.00	47.92
86.25	118.50	2430.00
800.91	8700.00	613.32
76.40	315.10	81.40
99.00	416.50	212.16
187.30	58.40	113.98
391.00	93.10	187.42
18.00	15.00	56.66

Construct a relative frequency diagram and a histogram to represent these data. Use what you feel to be a suitable class interval.

6. For the situation described in Problem 1, what is your estimate of the probability that a resistor coming off the production line will have a resistance less than 950 ohms?

7. For the situation described in Problem 5, what is your estimate of the probability that a customer selected at random will purchase more than $300 worth of merchandise in a year?

8. Show that if x is going to be treated as a continuous random variable which can take on values in the interval $a \leq x \leq b$, then the area for $x = a$ to $x = b$ under the curve representing the density function $f(x)$ must be 1.

9. If a model which treats x as a continuous random variable is used, what is the probability that $x = \xi$? How can you explain this result intuitively?

10. Let x be a random variable which can take on values in the interval $a \le x \le b$. Suppose that all the values of x in this interval are equally likely, and the separation between two adjacent values of x is a constant γ. Show that if one constructs a model for this situation in which x is treated as a continuous random variable, then the density function for x is $f(x) = 1/(b - a)$. Represent $f(x)$ graphically. A continuous random variable whose density function is a constant is said to be uniformly distributed. The notion of a uniform distribution is the continuous analog of equally likely simple events. What is the probability that x lies in the interval $\alpha \le x \le \beta$, where $a \le \alpha < \beta \le b$? Note that this depends only on the length of the interval, so the probability that x lies in any interval of length λ is the same, regardless of where the interval lies in the event set.

11. Determine the cumulative function for the random variable x introduced in Problem 10. Represent the cumulative function graphically.

12. Often the quality of parts that a manufacturer receives from a supplier can be characterized by the behavior of some random variable which can be conveniently thought of as being continuous. The variable may be a length, diameter, weight or resistance. The manufacturer can gain a great deal of information about what his supplier is doing, without ever asking the supplier, by occasionally taking a sizable sample of the parts and measuring the value of the basic random variable for each unit in the sample. Then a histogram for the random variable is constructed and studied. As an example, suppose that a manufacturer purchases brass bushings from a supplier. The nominal inside diameter of the bushing is 1 inch, and the manufacturer cannot use it if the diameter deviates by more than 0.006 inch on the large side or 0.003 inch on the small side. He has indicated this to the supplier. From a given lot the manufacturer selects at random 1100 bushings and measures carefully the inside diameter. The results are given in Table 2–13 where the numbers of bushings

TABLE 2–13

DATA FOR BUSHING DIAMETERS

Interval	Number	Interval	Number
(−4, −3)	1	(2, 3)	100
(−3, −2)	142	(3, 4)	50
(−2, −1)	200	(4, 5)	25
(−1, 0)	275	(5, 6)	12
(0, 1)	160	(6, 7)	7
(1, 2)	125	(7, 10)	3

whose diameters lie in various intervals are listed. The interval (2, 3) means that the diameter ξ was in the interval $1.002 \le \xi < 1.003$. Construct a histogram for the bushing diameters. What can you conclude about the supplier's manufacturing process and the action he is taking? What would you estimate the density function representing the diameters of bushings coming off the supplier's production line

would look like? Hint: Not all the supplier's customers will necessarily have precisely the same limitations on the inside diameter of the bushing.

Section 2–14

1. Use the data in Table F at the end of the text to draw carefully the curve representing $\varphi(t)$.

2. Use the data in Table F at the end of the text to draw carefully the curves representing $n(x; 3, 1)$ and $n(x; 3, 3)$. Place both curves on the same figure.

3. Use the data in Table F at the end of the text to draw carefully the curves representing $n(x; 4, 2)$ and $n(x; 8, 2)$. Place both curves on the same figure.

4. Use the data in Table F at the end of the text to draw carefully the curves representing $n(x; -3, 2)$ and $n(x; 3, 4)$. Place both curves on the same figure.

5. Use the data in Table E at the end of the text to draw carefully the curve representing $\Phi(x)$.

6. Use the data in Table E at the end of the text to draw carefully the curve representing $N(x; 4, 2)$.

7. A manufacturer believes that the life of any particular light bulb he produces can be imagined to be a random variable which is normally distributed with mean 1100 hours and standard deviation 200 hours. What is the probability that a bulb will last for 900 hours or less?

8. For the situation described in Problem 7, what is the probability that a bulb will last for at least 1000 hours?

9. For the situation described in Problem 7, what is the probability that the bulb will last between 1000 and 1200 hours?

10. The weight of a box of detergent produced by a given manufacturer can be looked upon as a random variable which is normally distributed with mean 34 ounces and standard deviation 1 ounce. What is the probability that the box will weigh 32 ounces or less?

11. For the situation described in Problem 10, what is the probability that the weight of the box will be at least 33 ounces?

12. For the situation described in Problem 10, what is the probability that the weight of the box will be between 33 and 35 ounces?

13. Let x be a random variable with the normal distribution $n(x; \mu, \sigma)$. Show that the probability of the event $\mu - \sigma \leq x \leq \mu + \sigma$, that is, the event that x lies within one standard deviation of μ is 0.683. Note that this is independent of μ and σ. This is one reason that the standard deviation is an especially useful measure of fluctuations about the mean. Many distributions are close to normal, and hence the probability that x will be within one standard deviation of the mean for such is approximately 0.68. Determine the probability that x lies within two standard deviations of the mean and also the probability that it lies within three standard deviations of the mean.

14. Suppose that the diameter of Mikimoto cultured pearls formed in a given location is a random variable having a normal distribution with $\mu_x = 0.20$ inches and

$\sigma_x = 0.07$ inches. What is the long-run fraction of pearls which will have a diameter 0.35 or greater?

15. Suppose that the final examination grades for a class of 200 students are essentially normally distributed with mean 65 and standard deviation 20. About how many students will receive a grade of 90 or more? How many will receive a grade of 50 or less?

Sections 2–15, 2–16 and 2–17

1. Let x have the binomial distribution $b(x; 20, 0.5)$. Suppose that the random variable t is defined in terms of x by (2–68). What values does t take on? Construct carefully the histogram for t. On the same figure, draw the curve for $\varphi(t)$.

2. Let x have the binomial distribution $b(x; 10, 0.4)$. Suppose that the random variable t is defined in terms of x by (2–68). What values does t take on? Construct carefully the histogram for t. On the same figure, draw the curve for $\varphi(t)$.

3. Let x have the binomial distribution $b(x; 10, 0.2)$. Suppose that the random variable t is defined in terms of x by (2–68). What values does t take on? Construct carefully the histogram for t. On the same figure, draw the curve for $\varphi(t)$.

4. Let x have the binomial distribution $b(x; 10, 0.05)$. Suppose that the random variable t is defined in terms of x by (2–68). What values does t take on? Construct carefully the histogram for t. On the same figure, draw the curve for $\varphi(t)$.

5. Use the normal approximation to estimate
$$b(2; 20, 0.5); \quad b(10; 20, 0.5); \quad b(15; 20, 0.5); \quad b(20; 20, 0.5)$$
and compare the approximate values so obtained with the correct values.

6. Use the normal approximation to estimate
$$B(4; 20, 0.5); \quad B(10; 20, 0.5); \quad B(16; 20, 0.5)$$
and compare the approximate results with the correct values.

7. Use the normal approximation to estimate
$$b(2; 9, 0.2); \quad b(4; 9, 0.2); \quad b(8; 9, 0.2)$$
and compare the approximate values with the correct values.

8. Use the normal approximation to estimate the probability that $55 < x < 65$ when x has the distribution $b(x; 200, 0.3)$.

9. Use the normal approximation to estimate the probability that in a production run of 1000 units, the number of defectives will exceed 50 if the probability that any given unit will be defective is 0.04.

10. Use the normal approximation to estimate the probability that if 100 students are selected at random from a group of 1000 students containing 300 girls and 700 boys, at least 40 girls will be in the 100 students selected.

11. Consider the binomial distribution $b(x; 10, 0.5)$. Construct a histogram for this distribution, and on the same figure show the curve for the normal distribution having the same mean and standard deviation as $b(x; 10, 0.5)$.

12. Consider the binomial distribution $b(x; 20, 0.5)$. Construct a histogram for this distribution, and on the same figure show the curve for the normal distribution having the same mean and standard deviation as $b(x; 20, 0.5)$.

13. Consider the binomial distribution $b(x; 10, 0.2)$. Construct a histogram for this distribution, and on the same figure show the curve for the normal distribution having the same mean and standard deviation as $b(x; 10, 0.2)$.

14. A manufacturer produces shafts for electric motors. The diameter of the shaft can be conveniently imagined to be a continuous random variable, and when the process is in control, it behaves as if each shaft produced is the result of an independent trial of a random experiment in which the diameter x has the normal distribution $n(x; 0.75, 0.002)$. Suppose that to be classified as satisfactory, the diameter of the shaft cannot be less than 0.746 or greater than 0.753. What is the long-run fraction defective produced by the process?

15. For the situation described in Problem 14, suppose that the manufacturer can select the expected value of the diameter of the shaft by suitably adjusting the lathes and that this has no influence on the standard deviation. Construct a graph showing the fraction defective produced by the process as a function of μ, the expected diameter of the shaft. What value of μ gives the minimum fraction defective?

16. Suppose that x has a normal distribution. What must the standard deviation of this distribution be if the probability that the value of x is more than two units away from the mean (on either side) is 0.40?

17. A manufacturer produces resistors. Consider a particular resistor he produces whose nominal resistance is 1000 ohms. The actual resistance of any unit produced can be thought of as a random variable which has the normal distribution $n(x; 1000, 45)$. The manufacturer sells both one-percent and ten-percent resistors, one percent meaning that the resistance should be within one percent of the nominal value. To obtain these two classifications, he proceeds as follows. He inspects all resistors coming off the line, and those whose resistance does not differ from 1000 ohms by more than 10 ohms he selects as one-percent resistors. The rest he sells as ten-percent resistors. Thus, someone buying a ten-percent resistor will never get one having the nominal resistance. Determine the density function for ten-percent resistors, and sketch the graph of this function. If a ten-percent resistor is considered defective when the resistance lies outside the ten-percent limits, what is the fraction defective produced of ten-percent resistors?

18. Re-solve Problem 17 if the resistance x has the normal distribution $n(x; 990, 45)$.

19. Suppose that x and y are independent random variables with distributions $n(x; \mu_x, \sigma_x)$ and $n(y; \mu_y, \sigma_y)$. Then $z = y - x$ has a normal distribution $n(z; \mu, \sigma)$ where

$$\mu = \mu_y - \mu_x; \quad \sigma^2 = \sigma_y^2 + \sigma_x^2.$$

This is a useful result. As one illustration of its usefulness, suppose that a manufacturer makes shafts for electric motors, the diameter x being a normally distributed random variable with expected value 1 inch and standard deviation 0.0010 inch. Another manufacturer makes the bushings which the shaft passes through. The diameter of the hole in the bushing is a normally distributed random variable y with mean 1.002 inch and standard deviation 0.0010 inch. If we select a shaft and a bushing, $z = y - x$ is the clearance when the shaft is put in the bushing, a negative value of z meaning that the diameter of the shaft is greater than the diameter of the

bushing. What is the probability that for a shaft and bushing selected at random, the shaft will not fit into the bushing? To operate properly the clearance should be between 0.001 and 0.003 inch. What is the probability that the clearance will be in this interval?

20. For the situation described in Problem 19, suppose that nothing can be done about the variances. What should be the expected diameter of the hole in the bushing to maximize the probability that the clearance will be in the desired interval? Assume that the expected shaft diameter remains unchanged.

21. A company selects at random 1000 individuals from the names listed in the telephone book to be surveyed for their opinions of the company's products. Suppose that twenty percent of the persons in the phone book use the company's products. What is the probability that 300 or more of those selected use the company's products?

22. Use the results of Problem 4 for Section 2–6 to explain why the hypergeometric distribution may be approximated by the normal.

23. Consider once again the South African medical team referred to in Problems 1 and 2 for Section 2–6. Suppose that instead of selecting 100 natives for an examination they select 1000 natives. What is the probability that the fraction of natives in the sample with TB differs by more than 0.02 from the actual fraction of the population which has this disease. Does your answer here depend in any way on a knowledge of what fraction of the population has TB? Is the situation in this regard different from that of Problem 2 for Section 2–6? If the population is large with respect to the sample size, does your answer depend in any way on precisely how large the population happens to be?

24. In the Far East area there has been some concern about the infection of a particular type of fish with parasitic worms. These worms may be transferred to humans who have eaten infected fish which have not been cooked thoroughly. To investigate the problem the Japanese government sends inspectors to examine a sizable sample of the fish. A total of five thousand fish are examined over a period of several months. These are taken from catches made on different dates and in different parts of the Pacific. What is the probability that the fraction of inspected fish in the sample differs by more than 0.01 from the fraction of the entire population of this type of fish which is infected. Is it necessary to know the size of the fish population or the actual percentage of this population which is infected?

25. A polling organization wishes to determine the size of a random sample which should be taken from the voting population in the U.S. if the probability is 0.01 that the fraction of those in the sample who will vote for the Democratic candidate in the next election differs from the fraction of the entire population who vote Democratic by more than 0.005. What is the sample size? What might occur in practice which could increase the possibility of large errors?

26. In a box are 1000 seeds, 400 of which are of genotype AA and 600 of which are of genotype aa. One hundred seeds are selected at random from the box. What is the probability that there will be at least twice as many seeds of genotype aa as there are of type AA?

Section 2–18

1. Use numbered pennies to generate 50 two-digit numbers from the uniform distribution over the unit interval.

2. Use the results of Problem 1 and Table E to generate 50 random numbers from the standardized normal distribution. Use these numbers to construct a histogram. Does this histogram resemble $\varphi(t)$?

Section 2–19

1. Suppose it is desired to use simulation to study the way the inventory of a given item would behave if the rules for controlling the inventory of this item are: (1) Reorder stock when the amount on hand plus on order reaches 50 units; (2) always place an order for 200 units. Explain how simulation could be used here. Indicate what data would be needed to make the simulation.

2. How might simulation be useful to study the impact of converting from small branch warehouses to the use of a large centralized warehouse?

3. How might simulation be used to study various designs for a highway system?

4. How might simulation be used as an aid for design of an elevator system for a large building?

CHAPTER 3

Estimation

3-1 INTRODUCTION

In our discussion of the procedures for constructing probability models, we noted that the model builder must supply the probabilities of the simple events. Often, the only way to obtain these is by an empirical study of the nature of the real-world situation. We have also noted previously that we cannot, in general, expect to determine these probabilities exactly. The best that can be done is to *estimate* their values. Depending on the nature of the situation, the estimates may be quite accurate or there may exist considerable errors in the estimates.

We have indicated in Chapter 1 how the probabilities of the simple events might be estimated. As we progressed in Chapter 2, it became apparent that for many practical problems we are only interested in the value of some random variable. Furthermore, the entire probability model will be complete if one or two parameters associated with this random variable can be estimated. The reason for this is that the random variable x will often have a binomial, hypergeometric or normal distribution. If x has the distribution $b(x; n, p)$, the model for x will be determined completely once p is estimated, since n will be prespecified. Similarly, if x has the distribution $h(x; n, p, N)$, the model for x will be determined once p is specified. Finally, if x has the distribution $n(x; \mu, \sigma)$, the model for x will be complete once μ and σ are determined. For many practical problems then, it is not necessary to estimate the probabilities of a large number of simple events, but instead merely to estimate the values of a relatively small number of parameters associated with some random variable.

In this chapter we wish to study ways in which some parameter associated with a given random variable can be estimated. In the overwhelming

majority of cases, the parameter to be estimated can be interpreted as the expected value or variance (equivalently, standard deviation) of some random variable. We shall restrict our attention to estimating means and standard deviations. Conceivably, interest might center on estimating a variety of other parameters, such as the mode of a distribution, but these do not arise often enough for us to study them in detail here.

3–2 ESTIMATION OF A MEAN

Suppose that we would like to estimate the mean μ_x for some random-variable x. A logical procedure for doing this is suggested by the intuitive interpretation of μ_x as the long-run average value of x when the experiment $\mathfrak{R}$ with which x is associated is repeated unendingly. If n independent trials of $\mathfrak{R}$ are performed, a value of x will be observed on each trial of $\mathfrak{R}$ and thus n values of x, which we shall denote by $\xi_1, \ldots, \xi_n$, will be determined. If n is large, we expect intuitively that the average of these n values of x should be close to μ_x. In other words, if

$$\bar{\xi} = \frac{1}{n} \sum_{j=1}^{n} \xi_j = \frac{1}{n}(\xi_1 + \cdots + \xi_n) , \qquad (3\text{–}1)$$

then $\bar{\xi}$ should be a good estimate of μ_x which we can use in our model for μ_x.

The procedure for estimating μ_x is thus a very simple one conceptually. Several questions immediately arise in our minds, however, which need to be examined and which introduce some complications. First, it seems reasonable to ask how large n needs to be, that is, how many trials of $\mathfrak{R}$ must be performed if our estimate $\bar{\xi}$ of μ_x is to be sufficiently accurate. Second, for a specific value of n, how large might the error in our estimate of μ_x be? In order to answer these questions, insofar as they can be answered, we shall have to examine the procedure for estimating μ_x which we have just described in somewhat greater detail.

To estimate μ_x we perform n independent trials of $\mathfrak{R}$. This performance of n independent trials of $\mathfrak{R}$ constitutes another random experiment which we shall denote by $\mathfrak{R}_n$. Now associated with $\mathfrak{R}_n$ we can imagine that there are n random variables $x\{1\}, \ldots, x\{n\}$, where $x\{j\}$ is the random variable representing the value of x which is determined on the jth trial of $\mathfrak{R}$. If $x\{j\} = \xi_j$, then ξ_j is the value of x observed on the jth trial of $\mathfrak{R}$. Consider now the random variable

$$\bar{x} = \frac{1}{n}[x\{1\} + \cdots + x\{n\}] = \frac{1}{n}x\{1\} + \cdots + \frac{1}{n}x\{n\} . \qquad (3\text{–}2)$$

If $x\{j\} = \xi_j$, we then see from (3–1) that $\bar{x} = \bar{\xi}$. We can now make the interesting and important observation that our estimate of μ_x will merely be

the observed value for a random variable associated with $\mathfrak{R}_n$, that is, the value of the random variable $\bar{x}$.

What we have just noted here happens to be a universal characteristic of estimating procedures. We can describe in general terms the procedure by which any parameter θ associated with a random variable x is estimated as follows. Some random experiment $\mathfrak{R}^*$ is performed, and the value of a random variable w associated with $\mathfrak{R}^*$ is used as the estimate of θ. The random variable w whose observed value is used as the estimate of the parameter is called the *statistic* being used to estimate the parameter. To estimate μ_x we have suggested using $\bar{x}$ as a statistic. In this case $\mathfrak{R}^* = \mathfrak{R}_n$. In general, there is no unique statistic to be used in estimating a parameter, although as in estimating μ_x, there will often be some natural one to use. For example, to estimate μ_x we could use instead of $\bar{x}$ the average of the values of x obtained on the first and last trials of $\mathfrak{R}$ in $\mathfrak{R}_n$, that is $[x\{1\} + x\{n\}]/2$. The reader will see immediately that it would not be sensible to use this statistic because no use is made of the information obtained on the other trials. Nonetheless, it is a possible statistic.

We shall not be concerned in any detail with the problems involved in selecting the most desirable one of several possible statistics, since for the simple problems of interest to us there will be a more or less natural statistic to use. In passing, however, it is worthwhile to point out some principles which would influence a choice. Generally speaking, it is desirable, insofar as possible, to use statistics w whose expected value is the value θ of the parameter of interest, that is, $\mu_w = \theta$. A statistic with this property is said to be *unbiased*. If a statistic is biased, so that $\mu_w \neq \theta$, then on repeating $\mathfrak{R}^*$ a large number of times, the estimates of θ would tend to be high or low, depending on whether $\mu_w > \theta$ or $\mu_w < \theta$. For an unbiased statistic, the estimates would be spread about θ, and their deviations from θ would average to 0. In Figure 3–1 we have shown the density function for a biased statistic and in Figure 3–2 the density function for an unbiased statistic.

If we are given two unbiased statistics, clearly the one to be preferred is the one having the smaller standard deviation, because on repeating $\mathfrak{R}^*$, the estimates of θ will be more closely grouped about θ for the statistic

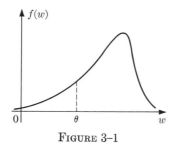

FIGURE 3–1

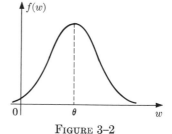

FIGURE 3–2

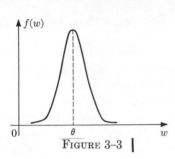

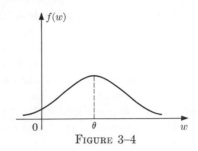

Figure 3–3 | Figure 3–4

with the smaller standard deviation than for the one with the larger stand-ard deviation. In Figures 3–3 and 3–4 we have shown the density functions for two different unbiased statistics. Clearly the one represented in Figure 3–3 would be preferable. If we are given the choice between a biased statis-tic with a small standard deviation and an unbiased one with a larger stand-ard deviation, the choice becomes more complicated, and we shall not con-sider it in any detail. However, it is clear that a slight amount of bias will do no harm, especially if the standard deviation can be reduced significantly by accepting a slight bias.

To answer the two questions that were posed originally for any parameter θ whose value we are interested in estimating, we must know something about the distribution of the statistic w that is being used to estimate θ. Let us now return to the problem of estimating μ_x, and suppose that $\bar{x}$ is the statistic to be used. We shall now see what can be said about the distribu-tion of $\bar{x}$. Note first that since $x\{j\}$ in (3–2) represents x for the jth trial of $\mathfrak{R}$, μ_j, the expected value of $x\{j\}$, is μ_x, the expected value of x, that is, $\mu_j = \mu_x$. By (2–33) we can express $\mu_{\bar{x}}$, the expected value of $\bar{x}$, in terms of the μ_j as follows:

$$\mu_{\bar{x}} = \frac{1}{n}\mu_1 + \cdots + \frac{1}{n}\mu_n . \tag{3–3}$$

Here $a_j = 1/n$. However, since $\mu_j = \mu_x$,

$$\mu_{\bar{x}} = \frac{1}{n}\mu_x + \cdots + \frac{1}{n}\mu_x = \mu_x\left(\frac{1}{n} + \cdots + \frac{1}{n}\right) = \mu_x\left(\frac{n}{n}\right) = \mu_x . \tag{3–4}$$

Thus

$$\mu_{\bar{x}} = \mu_x , \tag{3–5}$$

and we see that $\bar{x}$ is an unbiased statistic for estimating μ_x.

We can also determine the standard deviation of $\bar{x}$. In (3–2) the random variables $x\{j\}$ are independent random variables, since they refer to dif-ferent independent trials of $\mathfrak{R}$. Thus, by (2–53)

$$\sigma_{\bar{x}}^2 = \frac{1}{n^2}\sigma_1^2 + \cdots + \frac{1}{n^2}\sigma_n^2 , \tag{3–6}$$

where σ_j^2 is the variance of $x\{j\}$. However, since $x\{j\}$ is x for the jth trial of $\mathfrak{R}$, the variance of $x\{j\}$ is equal to the variance of x, that is $\sigma_j^2 = \sigma_x^2$. Consequently

$$\sigma_{\bar{x}}^2 = \frac{1}{n^2}\sigma_x^2 + \cdots + \frac{1}{n^2}\sigma_x^2 = \frac{n}{n^2}\sigma_x^2 = \frac{\sigma_x^2}{n}. \qquad (3\text{–}7)$$

Therefore

$$\sigma_{\bar{x}} = \frac{\sigma_x}{\sqrt{n}}. \qquad (3\text{–}8)$$

The standard deviation of $\bar{x}$ is thus proportional to the standard deviation of x, the factor of proportionality being $1/\sqrt{n}$, and, when $n > 1$, $\sigma_{\bar{x}}$ is less than σ_x.

The mean and standard deviation of $\bar{x}$ have now been determined. By use of the central limit theorem, we can even say something about the precise nature of the distribution of $\bar{x}$. Recall that in (3–2) the $x\{j\}$ are independent random variables. Thus if $y\{j\} = x\{j\}/n$, the $y\{j\}$ are also independent random variables, and furthermore, each $y\{j\}$ will have the same distribution. Furthermore $\bar{x} = \Sigma_{j=1}^n y\{j\}$. Therefore we can apply the central limit theorem to conclude that when n is large, the distribution of $\bar{x}$ will be essentially normal, regardless of what the distribution of x happens to be. This is an important and useful fact. To estimate μ_x with reasonable accuracy, n will usually need to be reasonably large. Since the sum of random variables often approaches a normal distribution very rapidly, it follows that in many practical problems we can treat $\bar{x}$ as being normally distributed with mean μ_x and standard deviation $\sigma_x/\sqrt{n}$. It also happens to be true, although we shall not attempt to prove it, that if x is normally distributed, $\bar{x}$ is exactly normally distributed for all values of n.

It is very important to understand the connection between x and $\bar{x}$, and for this reason we shall review the relationship again. We wish to estimate μ_x, the mean of a random variable x. We imagine that x is associated with some random experiment $\mathfrak{R}$. To estimate μ_x we make n independent trials of $\mathfrak{R}$, thus generating n values $\xi_1, \ldots, \xi_n$ of x. Then as our estimate of μ_x, we use $\bar{\xi}$ given by (3–1). Now the performance of n trials of $\mathfrak{R}$ can be looked upon as a random experiment which we have called $\mathfrak{R}_n$. We can now imagine a situation where $\mathfrak{R}_n$ is repeated a number of times (each trial of $\mathfrak{R}_n$ representing n trials of $\mathfrak{R}$). Each time $\mathfrak{R}_n$ is performed a value $\bar{\xi}$ will be determined. However, different values $\bar{\xi}$ can, in general, be obtained on different trials of $\mathfrak{R}_n$. Thus the value $\bar{\xi}$ obtained on any trial of $\mathfrak{R}_n$ can be looked upon as the value of a random variable which we have called $\bar{x}$. If $\mathfrak{R}_n$ is repeated a number of times, we can use the values of $\bar{x}$ so obtained to construct a histogram. If the number of trials of $\mathfrak{R}_n$ is very large, then the outline of this histogram will be essentially the same as the curve representing the density function for $\bar{x}$. We have shown above that this curve will be

essentially that of a normal distribution having a mean μ_x and standard deviation $\sigma_x/\sqrt{n}$. Note carefully that σ_x is the standard deviation of x and n is the number of trials of $\Re$ which are used in $\Re_n$.

We can illustrate in concrete form what we have been discussing by making a small-scale simulation. Suppose that x has a uniform distribution over the unit interval. To perform a random experiment which generates a value of x, we only need to select a random number from a table of random numbers from the uniform distribution over the unit interval. Such a table is given in Table J at the end of the text. Suppose now that we wish to estimate μ_x, which has the value 0.50 in this case. To do this let us generate four values of x and then average these. Thus $\Re_n$ will be $\Re_4$ in this case, since four trials of $\Re$ are used to estimate μ_x. Instead of performing $\Re_4$ only once, we shall perform $\Re_4$ forty times, thus obtained forty estimates of μ_x. Each of these estimates will be a value of $\bar{x}$, where

$$\bar{x} = \frac{1}{4}\left[x\{1\} + x\{2\} + x\{3\} + x\{4\}\right], \qquad (3\text{-}9)$$

and $x\{j\}$ is the value of x obtained on the jth trial of $\Re$ in performing $\Re_4$. The forty values of $\bar{x}$ can then be used to construct a histogram which will indicate roughly what the distribution of $\bar{x}$ is like.

Instead of representing each number using five digits as is done in Table J, we shall use only the first two digits. To perform one trial of $\Re_4$ we merely read off the first two digits of four numbers in Table J, place a decimal point in front of each of the pairs of digits and use the resulting four numbers as the values of x obtained on four trials of $\Re$. This is done forty times to obtain forty trials of $\Re_4$. For the first trial of $\Re_4$ we used the first four numbers in the first row of Table J and for the second trial the next four numbers in row 1. For the third trial we used the first four numbers in row 2 and so forth. The results are shown in Table 3–1, as are the values of $\bar{x}$. The resulting histogram, constructed using class intervals of length 0.10 is shown in Figure 3–5. This histogram looks like $\bar{x}$ might very well be treated as normally distributed, even though n is only 4 here and the distribution of x, shown in Figure 2–33, is quite different from a normal distribution. Thus we can see from this example that $\bar{x}$ may be quite close to being normally distributed for small n, even though the distribution of x is far from normal.

3–3 SAMPLE SIZE

In attempting to estimate μ_x, how can we decide how many values of x should be averaged to obtain μ_x to the desired accuracy, that is, what value of n should be used? We would like to investigate this problem now. The first important thing to note is that no matter how large n is made, it is

TABLE 3–1

SIMULATION OF THE DETERMINATION OF $\bar{x}$

Trial of $\mathcal{R}_4$	$x\{1\}$	$x\{2\}$	$x\{3\}$	$x\{4\}$	$\bar{x}$
1	.10	.32	.76	.13	0.33
2	.34	.54	.80	.09	0.44
3	.37	.04	.64	.74	0.45
4	.24	.24	.20	.10	0.19
5	.08	.68	.19	.09	0.26
6	.23	.02	.15	.34	0.18
7	.99	.02	.09	.70	0.45
8	.38	.31	.88	.74	0.58
9	.12	.99	.80	.36	0.57
10	.64	.36	.98	.16	0.53
11	.66	.74	.34	.76	0.62
12	.36	.36	.65	.39	0.44
13	.31	.10	.45	.82	0.42
14	.35	.42	.86	.07	0.43
15	.85	.77	.02	.65	0.57
16	.68	.74	.73	.85	0.75
17	.63	.32	.05	.47	0.37
18	.90	.57	.28	.28	0.51
19	.73	.45	.03	.64	0.46
20	.35	.34	.60	.20	0.37
21	.98	.17	.14	.68	0.49
22	.22	.40	.60	.93	0.54
23	.11	.05	.39	.27	0.21
24	.50	.68	.29	.24	0.43
25	.83	.99	.06	.98	0.72
26	.13	.70	.18	.40	0.35
27	.88	.40	.86	.58	0.68
28	.36	.67	.90	.76	0.67
29	.99	.67	.87	.64	0.79
30	.91	.08	.93	.61	0.63
31	.65	.17	.17	.50	0.37
32	.58	.76	.73	.57	0.66
33	.80	.35	.17	.08	0.35
34	.45	.22	.21	.78	0.42
35	.74	.99	.77	.77	0.82
36	.43	.00	.45	.64	0.38
37	.69	.26	.66	.29	0.48
38	.36	.87	.76	.13	0.53
39	.09	.20	.14	.68	0.28
40	.46	.56	.96	.78	0.69

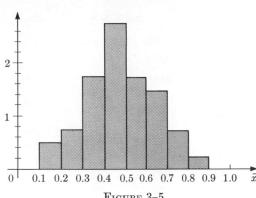

FIGURE 3–5

not possible to be absolutely certain that the observed value of $\bar{x}$ will not deviate by more than a specified amount from μ_x; $\bar{x}$ is a random variable, and there is usually always a positive probability that the observed value of $\bar{x}$ will deviate by more than a given amount from μ_x. Thus, if we interpret the above question as asking how large must n be if we wish to be absolutely certain that the value of $\bar{x}$ will not deviate by more than a specified amount from μ_x, the answer is that there is no such n. The best we can do is to determine an n such that the probability that $\bar{x}$ deviates by more than a specified amount from μ_x is not greater than a given value. Thus if a production manager asked us to determine an n such that if the diameters of n ball bearings were averaged the resulting number would not deviate by more than 0.001 inches from the expected diameter, we could not determine any such n for him. However, if he asked for an n such that the probability that the average of n diameters would deviate by more than 0.001 inches from the expected diameter should be less than or equal to 0.02, there exists the possibility of determining such an n.

The value of n used in estimating μ_x is often referred to as the *sample size*. The performance of $\mathcal{R}_n$ generates a sample, in this case n values of x, which is used to estimate μ_x. The sample size is the number of values of x in the sample. Let us now show how to determine an n such that the probability that the observed value of $\bar{x}$ deviates by more than δ from μ_x is less than or equal to some number α. Now $\bar{x}$ will deviate by more than δ from μ_x if $\bar{x} < \mu_x - \delta$ or $\bar{x} > \mu_x + \delta$. Call this the event A. The probability of this event is the area of the shaded regions shown in Figure 3–6, where the density function of $\bar{x}$ for a given n is shown.

We wish to determine an n such that $p(A) \le \alpha$. The value of n determines the standard deviation of the density function for $\bar{x}$. In Figure 3–7 we have shown the density function for $\bar{x}$ for a larger n than that used in Figure 3–6, and the probability of A in Figure 3–7 is smaller than in Figure 3–6. We see at once that if $p(A) \le \alpha$ for $n = n_1$, then $p(A) \le \alpha$ for any $n > n_1$. The smallest n for which $p(A) \le \alpha$ is that n for which $p(A) = \alpha$

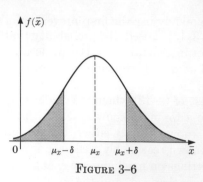

FIGURE 3–6

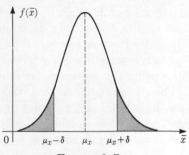

FIGURE 3–7

(or is as close to α as possible). What we really want to do then is to determine this smallest value of n.

Recall that $\bar{x}$ is essentially normally distributed with mean μ_x and standard deviation $\sigma_x/\sqrt{n}$. If we then treat $\bar{x}$ as normally distributed, the probability of A is

$$p(A) = \Phi\left(\frac{\mu_x - \delta - \mu_x}{\sigma_x/\sqrt{n}}\right) + 1 - \Phi\left(\frac{\mu_x + \delta - \mu_x}{\sigma_x/\sqrt{n}}\right)$$

$$= \Phi\left(\frac{-\delta\sqrt{n}}{\sigma_x}\right) + 1 - \Phi\left(\frac{\delta\sqrt{n}}{\sigma_x}\right)$$

$$= 2\left[1 - \Phi\left(\frac{\delta\sqrt{n}}{\sigma_x}\right)\right]. \tag{3-10}$$

Now we want $p(A) = \alpha$, so

$$2\left[1 - \Phi\left(\frac{\delta\sqrt{n}}{\sigma_x}\right)\right] = \alpha$$

or

$$1 - \Phi\left(\frac{\delta\sqrt{n}}{\sigma_x}\right) = \frac{\alpha}{2}$$

or

$$\Phi\left(\frac{\delta\sqrt{n}}{\sigma_x}\right) = 1 - \frac{\alpha}{2}. \tag{3-11}$$

From Table E, we can determine a value of t, call it λ, such that $\Phi(\lambda) = 1 - (\alpha/2)$. Then from (3–11), $\delta\sqrt{n}/\sigma_x$ must be equal to λ, so

$$\frac{\delta\sqrt{n}}{\sigma_x} = \lambda \quad \text{or} \quad \sqrt{n} = \frac{\lambda\sigma_x}{\delta}. \tag{3-12}$$

Therefore

$$n = \left(\frac{\lambda\sigma_x}{\delta}\right)^2, \tag{3-13}$$

and the value of n has been determined. The number computed from (3–13)

may not be an integer. In such a case we merely use the first integer larger than (3–13). When the n determined by (3–13) is used, the probability that $\bar{x}$ deviates from μ_x by more than δ will not in general be exactly α, because $\bar{x}$ is only approximately normally distributed and we made the computations using the normal distribution.

There is one problem involved in the use of (3–13) which the reader may have noted already. In order to compute n we must know σ_x. Now if μ_x is not known, then σ_x will normally not be known exactly either. For certain types of problems, however, σ_x may be known reasonably well even if μ_x is not. For example, problems where the settings on machines are changed to change μ_x will often leave σ_x essentially unchanged, and the σ_x obtained from the process in the past can be used to determine the size of n needed to estimate μ_x. In other cases, σ_x may not be known exactly, but one will be quite confident that σ_x is not greater than some value σ. In such cases σ might be used in place of σ_x in (3–13). When absolutely nothing is known about σ_x, it is not possible to determine n without first estimating σ_x. What might be done in this case is to first perform a small-scale experiment to estimate σ_x (we shall explain later how σ_x may be estimated). Then using this estimate of σ_x, an n, call it n_1, is determined from (3–13). Next n_1 trials of $\mathcal{R}$ are performed and the value of $\bar{x}$ determined. The n_1 values of x so generated are used to obtain a new estimate of σ_x. If this is not significantly greater than the original estimate, the value of $\bar{x}$ will be used as the estimate of μ_x. However, if the new estimate of σ_x is significantly larger than the previous one, a new n will be determined, call it n_2. Then an additional $n_2 - n_1$ trials of $\mathcal{R}$ will be performed, and μ_x will be estimated from the n_2 values of x obtained in total. One might even make a new estimate of σ_x and perform more trials if this seemed necessary.

Before going on to give some examples, it might be worthwhile to make an additional observation. If one individual estimates μ_x using n_1 trials of $\mathcal{R}$ and another estimates μ_x using $n_2 > n_1$ trials of $\mathcal{R}$, there is no guarantee that the estimate obtained from the larger number of trials will be better than that from the smaller number. Due to chance, the estimate using n_1 trials may be much better than that using n_2 trials. All we can claim is that if $\mathcal{R}_{n_1}$ and $\mathcal{R}_{n_2}$ are repeated a large number of times, the estimates from $\mathcal{R}_{n_2}$ will be more closely grouped about μ_x than the estimates from $\mathcal{R}_{n_1}$. The reason for this is that the $\bar{x}$ for $\mathcal{R}_{n_2}$ has a smaller standard deviation than the $\bar{x}$ for $\mathcal{R}_{n_1}$.

3–4 EXAMPLES

1. A manufacturer of light bulbs is planning to produce a bulb using a newly designed filament. The life of any particular bulb can be considered to be a random variable x, and the manufacturer desires to estimate the ex-

pected life of a bulb. More precisely, the manufacturer would like to test a sufficient number of bulbs so that the probability that the average of the lifetimes for these bulbs differs by more than 10 hours from the expected lifetime is equal to 0.05. A preliminary study in the research laboratory has shown that σ_x for the new filament is essentially 200 hours, which is the same as for the filament already in use.

For this situation $\alpha = 0.05$, $\alpha/2 = 0.025$ and $1 - (\alpha/2) = 0.975$. The number λ such that $\Phi(\lambda) = 0.975$ is, from Table E, $\lambda = 1.96$. If we use $\sigma_x = 200$ hours, then since $\delta = 10$ hours, (3–13) yields

$$n = \left[\frac{1.96(200)}{10}\right]^2 = 1538 ,$$

so that by testing about 1540 bulbs the desired result could be achieved. Note that there is no guarantee that the estimate will deviate by no more than 10 hours from the expected value. However, if $\mathfrak{R}_n$ was repeated many times, only about five percent of the estimates would deviate by more than 10 hours from μ_x. It is important for the reader to note that this does not depend on what value μ_x has. It is also worth noting that n can be determined without a knowledge of μ_x; σ_x must be used but not μ_x.

Suppose now that the manufacturer wishes to know the sample size such that the probability that the estimate deviates by more than 1 hour from μ_x is 0.05. The only difference between this and the previous case is that δ has been reduced from 10 to 1. Here $\lambda = 1.96$ again and

$$n = \left[\frac{1.96(200)}{1}\right]^2 = 153{,}800 .$$

Reducing δ by a factor of 10 has increased the sample size by a factor of 100.

Imagine now that δ is kept at 10 hours, but it is desired that the probability that $\bar{x}$ deviates from μ_x by more than δ be 0.01. Now $\alpha = 0.01$, $\alpha/2 = 0.005$ and $1 - (\alpha/2) = 0.995$. Then from Table E, $\lambda = 2.58$ and

$$n = \left[\frac{2.58(200)}{10}\right]^2 = 2660 .$$

Here again the sample size has increased significantly, but not so drastically as on reducing δ.

This example has pointed out clearly that the sample size needed increases rapidly when an effort is made to improve the accuracy of an estimate either by reducing δ or reducing the probability α. Since the cost of performing an experiment may be quite high, one should have the sort of accuracy needed clearly in mind in order to avoid increasing costs by making the sample size larger than needed. Unfortunately, it is not always easy to decide what reasonable values for δ and α are.

2. Suppose that a new process for producing injection-molded plastic parts

has been installed in a plant and is functioning properly. The production manager then desires to estimate the long-run fraction of the units produced that will be defective. This long-run fraction defective p is the probability that any given unit produced will be defective if we imagine that each unit produced represents an independent trial of a random experiment $\mathcal{R}$. As we know from (2–37), p is the expected value of a random variable y, where $y = 1$ if a unit is good and $y = 0$ if it is defective. The experiment $\mathcal{R}$ here consists of producing one unit, and n independent trials of $\mathcal{R}$ correspond to producing n units. If n units are produced and each is examined to determine whether or not it is defective, then $\bar{y}$, the estimate of p, is

$$\bar{y} = \frac{1}{n}\,[y\{1\} + \cdots + y\{n\}]\,, \tag{3-14}$$

where $y\{j\} = 1$ if the jth unit produced is good and 0 otherwise. Now $\sum_{j=1}^{n} y\{j\} = x$ is simply the number of defectives in the sample of size n, and $\bar{y} = x/n$. Thus the estimate for p is merely the number of defectives in the sample of size n divided by n.

Suppose that it is desired to determine n so that the probability that $\bar{y}$ deviates by more than 0.01 from p is 0.05. Here, then, $\alpha = 0.05$ and $\alpha/2 = 0.025$. We have shown in (2–55) that $\sigma_y^2 = p(1 - p)$. Now we encounter a difficulty; σ_y depends on the value of p which we are attempting to estimate. How do we proceed in this case? In Figure 3–8, we have shown how σ_y^2 varies

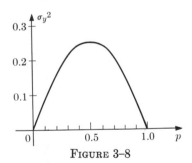

FIGURE 3–8

with p. For all p, $\sigma_y^2 \leq 0.25$, or $\sigma_y \leq 0.5$. One way to proceed would be to use $\sigma_0 = 0.50$ instead of σ_y. This is conservative, since we are sure that $\sigma_y \leq 0.50$. If this is done then, since $1 - (\alpha/2) = 0.975$ and $\lambda = 1.96$,

$$n = \left[\frac{1.96(0.50)}{0.01}\right]^2 = 9600\,.$$

The procedure just suggested may be quite inefficient, in that a much larger sample will be used than is really necessary. Suppose, for example, that the plant manager is quite sure that $p \leq 0.10$. Then we see that

$\sigma_y^2 \leq 0.1(0.9) = 0.09$ and $\sigma_y \leq \sqrt{0.09} = 0.30$. In this case then we can use $\sigma_1 = 0.30$ in place of σ_y and

$$ n = \left[\frac{1.96(0.30)}{0.01} \right]^2 = 3460 , $$

which is about one-third the number of observations required if nothing whatever is known about p.

A still more efficient procedure might be to first conduct a pilot study and check perhaps 500 units. This will give a rough estimate of p from which σ_y can be estimated. This estimate of σ_y can then be used to determine n.

If the production manager feels that p is quite small, say 0.01 or less, then the value $\delta = 0.01$ specified initially may be too large, because the interval determined by $p - \delta$ to $p + \delta$ extends from 0 to 0.02. A much narrower interval may be desirable and thus $\delta = 0.001$ might be used. Here $\sigma_y^2 \leq 0.01(0.99) \doteq 0.01$ and $\sigma_y \leq 0.1$. If $\delta = 0.001$, $\alpha = 0.05$ are used in (3–13), along with 0.1 for σ_y, then

$$ n = \left[\frac{1.96(0.1)}{0.001} \right]^2 = 38,400 , $$

and the sample size has become rather large. When p is small, a large sample will be needed to estimate p accurately because defectives appear only rarely.

3–5 ESTIMATING A POPULATION PROPORTION

The theory we have developed above can be applied to another type of estimation problem which arises frequently. Consider a population which consists of N members. The members may be people, resistors or animals, for example. Suppose that each member of the population can be characterized either as having a property or not having this property. The property may be that an individual is a Republican or that a unit is defective, for example. Assume that a fraction p of the members have this property. Imagine now that we wish to estimate p. The value of p could be determined exactly by examining each member of the population. Often, however, it is impractical to do this. Instead a random sample of size n will be selected from the population and the number r of the members in this sample having the property will be determined. The fraction r/n of the members of the sample which have the property is then used as the estimate of the fraction or proportion of the population which has the property.

Let us now study this estimation procedure in a little more detail and see how to determine the sample size to be used. The statistic being used here is $w = x/n$, where the random variable x is the number of members of the sample which have the property of interest. We have shown in Section 2–6

that x has the hypergeometric distribution $h(x; n, p, N)$. Furthermore, we noted in (2–72) but did not prove that

$$\mu_x = np; \quad \sigma_x^2 = npq\left(\frac{N-n}{N-1}\right). \tag{3-15}$$

Now since $w = x/n$, $\mu_w = \mu_x/n$ and $\sigma_w^2 = \sigma_x^2/n^2$. Thus

$$\mu_w = \frac{np}{n} = p; \quad \sigma_w = \sqrt{\frac{pq}{n}\left(\frac{N-n}{N-1}\right)}. \tag{3-16}$$

We see, therefore, that w is an unbiased statistic for estimating p. We cannot use the central limit theorem here to claim that w is essentially normally distributed, since w is not the sum of independent identically distributed random variables. Recall, however, that the hypergeometric distribution can be approximated by the normal distribution when np is fairly large and n/N is not too large. It usually turns out that these conditions are satisfied and thus x is essentially normally distributed. However, w is merely a constant times x and thus w is essentially normally distributed with mean and standard deviation given by (3–16). Thus we shall treat w as being normally distributed.

Suppose now that we wish to determine the value of n to use if the probability that the observed value of w will deviate from p by more than δ is to be α. Then precisely the same arguments which led to (3–11) show that

$$\Phi\left(\frac{\delta}{\sigma_w}\right) = 1 - \frac{\alpha}{2}, \tag{3-17}$$

and if λ is the number such that $\Phi(\lambda) = 1 - (\alpha/2)$, then $\delta/\sigma_w = \lambda$, so $\sigma_w^2 = (\delta/\lambda)^2$, or on using (3–16) for σ_w

$$\frac{pq}{n}\left(\frac{N-n}{N-1}\right) = \left(\frac{\delta}{\lambda}\right)^2. \tag{3-18}$$

It is not quite so easy to determine n here, but it is not difficult. On multiplying both sides of (3–18) by n we have

$$pq\left(\frac{N}{N-1}\right) - n\left(\frac{pq}{N-1}\right) = n\left(\frac{\delta}{\lambda}\right)^2 \tag{3-19}$$

or

$$n\left[\frac{pq}{N-1} + \left(\frac{\delta}{\lambda}\right)^2\right] = pq\left(\frac{N}{N-1}\right), \tag{3-20}$$

so

$$n = \frac{pq\left(\frac{N}{N-1}\right)}{\frac{pq}{N-1} + \left(\frac{\delta}{\lambda}\right)^2}, \tag{3-21}$$

where, of course, in (3–21), $q = 1 - p$. Precisely the same difficulty is encountered here as in the second example of the preceding section; p appears in σ_w and thus in (3–21). The difficulty can be resolved in the same way; $pq \leq 0.25$, so that instead of using the actual σ_w, one can use the σ_w obtained by replacing pq by 0.25. Thus pq in (3–21) would be replaced by 0.25. If something were known about p, then instead of 0.25, the largest possible value of pq would be used.

3–6 EXAMPLES

1. A newspaper in Berkeley, California, would like to estimate the proportion of the population who will vote for the school bond issue coming up in the next election. To do so it plans to select a random sample from the list of 40,000 registered voters. The proportion of those in the sample favoring the bonds will be used as the estimate of the proportion of all voters which favor them. The newspaper would like to determine the value of n such that the probability that the estimate differs by more than one percentage point from the true proportion is 0.05.

The paper believes that p is close to 0.5, so that to compute n, pq in (3–21) will be replaced by 0.25. A one percentage point deviation of w from p corresponds to setting $\delta = 0.01$, since a percentage is divided by 100 to convert it to a fraction. The value of α is 0.05, and hence from what we obtained in Section 3–4, $\lambda = 1.96$. Also, we see that $N/(N-1) = 1$ to a good approximation since $N - 1 \doteq N = 40{,}000$. Hence from (3–21),

$$n = \frac{0.25}{0.000006 + (0.0051)^2} = \frac{0.25}{0.000032} = 7820 .$$

Thus about 8000 voters should be questioned, or about one-fifth of all voters. The above computations assume, of course, that a voter will actually tell the truth about how he is going to vote. Voters do not always do this, and this is what sometimes causes such polls to yield very misleading results.

2. A manufacturer buys capacitors from a supplier in lots of 10,000. A lot has just been received and it is desired to estimate the fraction defective in the lot in such a way that the probability that the estimate differs from the true value p by more than 0.01 is equal to 0.10. The manufacturer is quite sure that the fraction defective is not greater than 0.05 so that the largest possible value of pq is $0.05(0.95) = 0.0475$. Here $\alpha = 0.10$, so $1 - (\alpha/2) = 0.95$ and from Table E, $\lambda = 1.65$. Since $\delta = 0.01$, the required sample size, if 0.0475 is used in place of pq, is

$$n = \frac{0.0475}{0.00000475 + (0.00606)^2} = \frac{0.0475}{0.0000415} = 1145 ,$$

so that a random sample of about 1150 capacitors should be checked.

3. Japanese inspectors have discovered that in recent months a number of the specimens inspected of a particular fish were infected with a parasitic worm. This worm can be transferred to humans if fish which have not been cooked thoroughly are eaten. To decide what measures should be taken, the Japanese authorities would like first to estimate the fraction of the population of this type of fish which is infected. They plan to collect a number of fish from various parts of the Pacific fishing grounds and use these as a random sample from the population. They would like the sample size to be such that the probability that the fraction infected in the sample differs from that of the population as a whole by more than 0.02 is 0.05. What should be the size of the sample selected?

Here we do not know the size of the population, but it is certainly very large; there will exist millions of fish of this type in the Pacific fishing grounds. The fraction infected is not known. Thus we shall use 0.25 in place of pq. It is desired that $\alpha = 0.05$. Consequently, as we noted previously, $\lambda = 1.96$ in this case. Hence $\delta/\lambda = 0.02/1.96 = 0.0102$ and $(\delta/\lambda)^2 = 0.000104$. We can now note that if N is one million or more $pq/(N - 1)$ is negligible with respect to $(\delta/\lambda)^2$, and of course $N/(N - 1)$ is essentially 1. Thus for the case we are considering (3–21) reduces to

$$n = pq \left(\frac{\lambda}{\delta}\right)^2 = \frac{0.25}{0.000104} = 2410 \, ,$$

and a sample size of about 2400 fish should be selected.

3–7 ESTIMATION OF STANDARD DEVIATION

Recall that σ_x^2, the variance of a random variable x, is the expected value of the random variable $(x - \mu_x)^2$. The expected value of $(x - \mu_x)^2$ can be interpreted intuitively as the long-run average value of $(x - \mu_x)^2$. If n independent trials of the random experiment $\Re$ with which x is associated are made, n values $\xi_1, \ldots, \xi_n$ of x are generated, and the average value of $(x - \mu_x)^2$ for these n trials is

$$\frac{1}{n} \sum_{j=1}^{n} (\xi_j - \mu_x)^2 \, . \tag{3–22}$$

For a very large n, this number should be close to σ_x^2. Thus it might seem that (3–22) could be used as an estimate of σ_x^2.

There is one major difficulty in attempting to use (3–22) to estimate σ_x^2. The difficulty is that μ_x is not known. We can estimate μ_x without a knowledge of σ_x, in the sense that $\bar{x}$ does not involve σ_x. However, to determine the number of trials needed to estimate μ_x to some prespecified accuracy, it is necessary to know something about σ_x. The situation is more complicated with σ_x, because μ_x appears directly in (3–22). The logical thing to do here

is to use $\bar{\xi}$, our estimate of μ_x, in (3–22) when estimating σ_x^2. If this is done our estimate s_x^2 of σ_x^2 then becomes

$$s_x^2 = \frac{1}{n} \sum_{j=1}^{n} (\xi_j - \bar{\xi})^2 , \tag{3–23}$$

and our estimate of σ_x is then simply s_x. Equation (3–23) is essentially the one normally used to estimate σ_x^2.

Now the s_x^2 computed from (3–23) can be looked upon as the value taken on by a random variable z, where

$$z = \frac{1}{n} \sum_{j=1}^{n} [x\{j\} - \bar{x}]^2 . \tag{3–24}$$

In (3–24), $x\{j\}$ is the random variable representing x on the jth trial of $\mathfrak{R}$ and $\bar{x}$ is the random variable defined by (3–2). Thus z is a statistic which could be used to estimate σ_x^2, that is, the value of z is used as the estimate of σ_x^2.

Let us now determine the expected value of z to see if it is an unbiased statistic. To do this we first add and subtract μ_x in (3–24) to yield

$$z = \frac{1}{n} \sum_{j=1}^{n} [(x\{j\} - \mu_x) - (\bar{x} - \mu_x)]^2 . \tag{3–25}$$

Next the square term is expanded to yield

$$z = \frac{1}{n} \sum_{j=1}^{n} [(x\{j\} - \mu_x)^2 - 2(x\{j\} - \mu_x)(\bar{x} - \mu_x) + (\bar{x} - \mu_x)^2] . \tag{3–26}$$

Then, by use of properties (2–23), (2–24) and (2–22) for summations,

$$z = \frac{1}{n} \sum_{j=1}^{n} (x\{j\} - \mu_x)^2 - \frac{2}{n} (\bar{x} - \mu_x) \sum_{j=1}^{n} (x\{j\} - \mu_x) + \frac{n}{n} (\bar{x} - \mu_x)^2 . \tag{3–27}$$

However,

$$\frac{1}{n} \sum_{j=1}^{n} (x\{j\} - \mu_x) = \frac{1}{n} \sum_{j=1}^{n} x\{j\} - \frac{n}{n} \mu_x = \bar{x} - \mu_x \tag{3–28}$$

by the definition of $\bar{x}$. Hence

$$z = \frac{1}{n} \sum_{j=1}^{n} (x\{j\} - \mu_x)^2 - (\bar{x} - \mu_x)^2 . \tag{3–29}$$

If we now introduce the random variables

$$y\{j\} = (x\{j\} - \mu_x)^2; \quad y = (\bar{x} - \mu_x)^2 , \tag{3–30}$$

then

$$z = \frac{1}{n} y\{1\} + \cdots + \frac{1}{n} y\{n\} - y , \tag{3–31}$$

and by (2–33)

$$\mu_z = \frac{1}{n}\mu_1 + \cdots + \frac{1}{n}\mu_n - \mu_y, \tag{3-32}$$

where μ_j is the expected value of $y\{j\}$ and μ_y is the expected value of y. The expected value of $y\{j\}$ is the expected value of $(x\{j\} - \mu_x)^2$. Since $x\{j\}$ is x for the jth trial of $\mathcal{R}$, the expected value of $(x\{j\} - \mu_x)^2$ is precisely σ_x^2. The expected value of y is the expected value of $(\bar{x} - \mu_x)^2$, which is the variance of $\bar{x}$. We have shown previously that the variance of $\bar{x}$ is σ_x^2/n. Therefore $\mu_j = \sigma_x^2$ and $\mu_y = \sigma_x^2/n$, so by (3–32)

$$\mu_z = \frac{1}{n}[\sigma_x^2 + \cdots + \sigma_x^2] - \frac{\sigma_x^2}{n} = \sigma_x^2 - \frac{1}{n}\sigma_x^2 = \frac{n-1}{n}\sigma_x^2. \tag{3-33}$$

We have reached the interesting conclusion that z is not an unbiased statistic, since μ_z is not σ_x^2. However, for large n, $(n-1)/n$ is close to 1 and z is essentially unbiased. It is very easy to obtain from z an unbiased statistic. All we need to do is divide by $n-1$ rather than by n in (3–24). Thus

$$w = \frac{1}{n-1}\sum_{j=1}^{n}[x\{j\} - \bar{x}]^2 \tag{3-34}$$

is an unbiased statistic for estimating σ_x^2, and is the one that statisticians normally use. For large n, there is essentially no difference between the statistics z and w. If n independent trials of $\mathcal{R}$ are performed, generating the n values $\xi_1, \ldots, \xi_n$ of x, the estimate s_x^2 of σ_x^2 obtained using the statistic w would be

$$s_x^2 = \frac{1}{n-1}\sum_{j=1}^{n}(\xi_j - \bar{\xi})^2, \tag{3-35}$$

and the estimate s_x of σ_x would be the square root of (3–35).

If n is large, it is tedious to compute (3–35) by hand. It is especially clumsy to have to subtract $\bar{\xi}$ from each ξ_j and then square the resulting number. This can be avoided by converting (3–35) to a slightly different form. On expanding the square in (3–35) we can write

$$s_x^2 = \frac{1}{n-1}\sum_{j=1}^{n}(\xi_j^2 - 2\xi_j\bar{\xi} + \bar{\xi}^2)$$

$$= \frac{1}{n-1}\sum_{j=1}^{n}\xi_j^2 - \frac{2\bar{\xi}}{n-1}\sum_{j=1}^{n}\xi_j + \frac{n}{n-1}\bar{\xi}^2 \tag{3-36}$$

$$= \frac{1}{n-1}\sum_{j=1}^{n}\xi_j^2 - 2\left(\frac{n}{n-1}\right)\bar{\xi}^2 + \frac{n}{n-1}\bar{\xi}^2,$$

so

$$s_x^2 = \frac{1}{n-1}\left[\sum_{j=1}^{n} \xi_j^2 - n\bar{\xi}^2\right] = \frac{1}{n(n-1)}\left[n\sum_{j=1}^{n}\xi_j^2 - \left(\sum_{j=1}^{n}\xi_j\right)^2\right]. \quad (3\text{–}37)$$

In (3–36) we used $n\bar{\xi} = \Sigma_{j=1}^{n}\xi_j$ to eliminate the latter sum, and in (3–37) we used $\bar{\xi} = (\Sigma_{j=1}^{n}\xi_j)/n$ to replace $\bar{\xi}$. It is the right-hand side of (3–37) that would normally be used to compute s_x^2. The computations are tedious, however, even with this simplified formula when n is large. They could, of course, easily be carried out on a computer.

EXAMPLE. Suppose that we wish to estimate the standard deviation of a random variable which has a uniform distribution over the unit interval. Let us do this using a sample size of twelve. In Table 3–1 are given a large number of values of x obtained from a random number table. Let us use the first twelve of these as the values of x obtained in collecting the sample. The details of computing s_x^2 are given in Table 3–2. The first column lists the trials, the second the values of x and the third the squares of these values. The last row gives the sum of the numbers in each column. Then by (3–37) it follows that

$$s_x^2 = \frac{1}{12(11)}[12(2.860) - (4.87)^2]$$

$$= \frac{1}{132}[34.3 - 23.7] = \frac{10.6}{132} = 0.0803 ,$$

so $s_x = \sqrt{0.0803} = 0.28$. The correct value of σ_x is 0.29. Thus σ_x has been estimated quite accurately here even though the estimate 0.40 of μ_x obtained from these twelve observations is rather poor.

TABLE 3–2

COMPUTATION OF s_x^2

Trial	ξ_j	ξ_j^2
1	.10	.010
2	.32	.102
3	.76	.578
4	.13	.017
5	.34	.116
6	.54	.292
7	.80	.640
8	.09	.008
9	.37	.137
10	.04	.002
11	.64	.410
12	.74	.548
	4.87	2.860

3–8 USE OF ESTIMATES IN MODELS

In the previous sections we have discussed how the mean and standard deviation of some random variable may be estimated. The procedure for estimating a population proportion was also considered. We showed how to determine the sample size needed to estimate μ_x or a proportion p in such a way that the probability that the estimate deviates by more than a specified amount δ from the true value is not greater than α. In order to compute this sample size it is necessary to have available an estimate of σ_x. We did not show how to determine the sample size needed to estimate σ_x to some specified accuracy. There were two reasons for this. First, it is not quite so easy to determine the sample size in this case. The second and more important reason is that the sample size used to determine μ_x is almost always satisfactory for obtaining an adequate estimate of σ_x simultaneously.

If the estimates of μ_x and σ_x are to be used in the construction of a probability model, then the estimates will be used in the model as μ_x and σ_x, that is, in the model one will use $\mu_x = \bar{\xi}$ and $\sigma_x = s_x$. For example, suppose that we wished to construct a model for the diameter of a ball bearing produced by the process considered in Section 2–12. Data on the nature of this process were generated by measuring the diameters of 10,000 ball bearings produced by this process, and the results are summarized in Table 2–11. A histogram for these data is shown in Figure 2–13. The outline of this histogram looks very much like a curve representing the normal distribution, and thus it would appear that as a model for the random variable x representing the diameter of a ball bearing we could use one in which x is assumed to be normally distributed.

To complete the model it is necessary to specify μ_x and σ_x in the normal distribution. The data can be used to estimate μ_x and σ_x, since the 10,000 diameters can be thought of as 10,000 values of x generated by performing 10,000 independent trials of the random experiment $\Re$ which corresponds to the manufacture of a ball bearing. The values of $\bar{\xi}$ and s_x determined from these 10,000 diameters are

$$\xi = 0.2500; \quad s_x = 0.0019 .$$

Thus it appears reasonable to take as the model for the diameter x of a ball bearing one in which x is normally distributed with $\mu_x = 0.2500$ and $\sigma_x = 0.0019$. To check this we have drawn in Figure 3–9 the curve representing $n(x; 0.2500, 0.0019)$, as well as the outline of the histogram from Figure 2–15. It would appear that the model being proposed is a reasonably good one in this case.

3–9 CONFIDENCE INTERVALS

Estimates of population proportions or of the expected value of a random variable will not always be used directly in constructing a probability

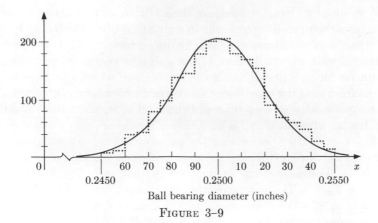

Ball bearing diameter (inches)

FIGURE 3–9

model. Sometimes these estimates are of great interest in themselves and are obtained for their intrinsic interest rather than with the intention of constructing a probability model. For example, the Colombian Government may be interested in the fraction of workers living in Bogotá who are unemployed, or a firm may be interested in the proportion of its customers who are women or *Time* may be interested in the proportion of voters who will vote for a given candidate in a coming election.

To obtain an estimate of the fraction of unemployed workers in Bogotá, the Colombian Government might commission an economist at the Universidad de los Andes to do a study to estimate this fraction. The question then arises as to how the economist should present his estimate after it is obtained. He could merely say that his study suggests that eighteen percent of the workers are unemployed. This gives no hint, however, as to what the accuracy of his estimate is. To be more precise, he could say that the probability that his estimate deviates by more than three percentage points from the true value is 0.05. This is a quite precise and suitable method for indicating the accuracy of the estimate. It is not the method normally used in scientific work, however, and it might be a little confusing to the government officials. When estimating some parameter such as the speed of light, a scientist will present his results in a form something like

$$299,792,500 \pm 300 \text{ meters per second.}$$

This has the form $f \pm \epsilon$ and it is normally taken to mean that the true value should lie between $f - \epsilon$ and $f + \epsilon$. In other words, the interval whose end points are $f - \epsilon$ and $f + \epsilon$ should contain the true value, and ϵ is an indication of the experimental error.

Statisticians have developed a procedure for presenting estimates of parameters, called the *method of confidence intervals*, which is quite similar to the scientific method just described. Suppose that μ_x is the parameter being estimated, and that n was chosen so that the probability that the ob-

served value of $\bar{x}$ deviates by more than δ from μ_x is α. After $\mathcal{R}_n$ is performed, it is then possible to obtain an estimate s_x of σ_x which may be much better than that which was available before performing $\mathcal{R}_n$. If the probability that $\bar{x}$ deviates by more than δ from μ_x is now computed using this better estimate for σ_x, a probability β will be obtained which may be somewhat different from α. If the experiment was carried out properly, however, β will not be significantly greater than α, because if it is, more trials should be made. It may turn out that β is less than α, however.

In any event, after performing the experiment, we can compute for an arbitrarily specified δ the approximate probability β that $\bar{x} > \mu_x + \delta$ or $\bar{x} < \mu_x - \delta$. The probability is then $1 - \beta$ that $\mu_x - \delta < \bar{x} < \mu_x + \delta$ or $-\delta < \bar{x} - \mu_x < \delta$. This is true regardless of what value μ_x has. If $\bar{x}$ satisfies $-\delta < \bar{x} - \mu_x < \delta$, then the inequality $-\delta < \mu_x - \bar{x} < \delta$ is also satisfied, as we see by multiplying by -1. Thus we see that if $\bar{x}$ satisfies $-\delta < \bar{x} - \mu_x < \delta$, the following inequality is also satisfied:

$$\bar{x} - \delta < \mu_x < \bar{x} + \delta .   \tag{3-38}$$

Furthermore (3-38) is satisfied only when $-\delta < \bar{x} - \mu_x < \delta$ is satisfied.

Let us now examine (3-38) more closely. Suppose that $\mathcal{R}_n$ is performed a large number of times. Each trial of $\mathcal{R}_n$ (each of which involves n trials of $\mathcal{R}$) generates a value $\bar{\xi}$. For each value $\bar{\xi}$ of $\bar{x}$ we can determine an interval whose end points are $\bar{\xi} - \delta$ and $\bar{\xi} + \delta$. In general, the interval so determined will vary from one trial of $\mathcal{R}_n$ to another because $\bar{\xi}$ will change. Now if it is true that $-\delta < \bar{\xi} - \mu_x < \delta$, then it will be true that

$$\bar{\xi} - \delta < \mu_x < \bar{\xi} + \delta ,   \tag{3-39}$$

that is, the interval whose end points are $\bar{\xi} - \delta$ and $\bar{\xi} + \delta$ contains μ_x, the parameter we are trying to estimate. The probability that $-\delta < \bar{\xi} - \mu_x < \delta$ is essentially $1 - \beta$, so that the long-run fraction of the trials of $\mathcal{R}_n$ for which $-\delta < \bar{\xi} - \mu_x < \delta$ holds is $1 - \beta$. Hence the long-run fraction of the intervals whose end points are $\bar{\xi} - \delta$ and $\bar{\xi} + \delta$ which contain μ_x is essentially $1 - \beta$.

An interval whose end points are $\bar{x} - \delta$ and $\bar{x} + \delta$ is called a *confidence interval for* μ_x. This interval may be thought of as a random interval, since its end points are determined by the value of the random variable $\bar{x}$. The probability that the random interval or confidence interval with end points $\bar{x} - \delta$ and $\bar{x} + \delta$ contains μ_x is approximately $1 - \beta$. In other words, if $\mathcal{R}_n$ is performed a large number of times, the fraction of the random intervals or confidence intervals so generated which contain μ_x is $1 - \beta$.

The situation can be illustrated graphically as shown in Figure 3-10. Each trial of $\mathcal{R}_n$ determines a confidence interval. The confidence interval determined by the value of $\bar{x}$ will change from one trial of $\mathcal{R}_n$ to the next, but the long-run fraction of them which contain μ_x will be essentially $1 - \beta$.

The length of each confidence interval is 2δ, but the location of its end points changes. A statistician would present his estimate of μ_x by giving $\bar{\xi}$ and then giving the statement that the probability is $1 - \beta$ that the confidence interval whose end points are $\bar{\xi} - \delta$ and $\bar{\xi} + \delta$ contains μ_x. Alternatively, he might write

$$\mu_x = \bar{\xi} \pm \delta, \tag{3-40}$$

holding with probability $1 - \beta$, and this is to be interpreted as meaning that the probability is $1 - \beta$ that the interval with end points $\bar{\xi} - \delta$ and $\bar{\xi} + \delta$ contains μ_x. Now if the economist at the Universidad de los Andes reported his estimate to the government officials in the form (3–40), they would believe that they understood exactly what it meant. They would ignore the probability statement and interpret δ as the maximum experimental error. This, of course, is wrong. However, if β is chosen to be small, then they will seldom be wrong in assuming that μ_x lies between $\bar{\xi} - \delta$ and $\bar{\xi} + \delta$. In any event, they obtain some feeling as to the possible experimental error.

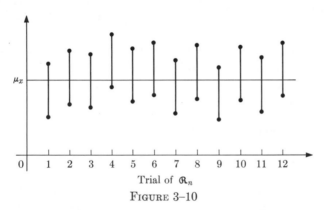

Trial of $\mathcal{R}_n$

FIGURE 3–10

We have introduced in the above paragraphs the notion of a confidence interval and have shown how it can be usefully employed in reporting an estimate of a parameter such as the mean. In the above discussion δ can be imagined to be selected arbitrarily and β is then determined as the probability that $\bar{x}$ does not deviate by more than δ from μ_x. In constructing confidence intervals, it is typical to go in the reverse direction and specify $1 - \beta$, the probability that the confidence interval will contain μ_x. Then from this δ is determined. Let us now see how to determine δ when β is specified. Since $\bar{x}$ is essentially normally distributed with mean μ_x and standard deviation $s_x/\sqrt{n}$ (we have used s_x, our estimate of σ_x), the random variable

$$t = \frac{(\bar{x} - \mu_x)\sqrt{n}}{s_x} \tag{3-41}$$

has approximately the distribution $\varphi(t)$. Now β is the probability that $\bar{x} > \mu_x + \delta$ or $\bar{x} < \mu_x - \delta$. This is the probability that $t > \delta\sqrt{n}/s_x$ or $t < -\delta\sqrt{n}/s_x$. The probability of this is $2[1 - \Phi(\delta\sqrt{n}/s_x)]$, and this is to be β. Thus

$$\Phi\left(\frac{\delta\sqrt{n}}{s_x}\right) = 1 - \frac{\beta}{2}. \tag{3-42}$$

Let λ be the value of t for which $\Phi(t) = 1 - (\beta/2)$. Then

$$\frac{\delta\sqrt{n}}{s_x} = \lambda \tag{3-43}$$

or

$$\delta = \frac{\lambda s_x}{\sqrt{n}}. \tag{3-44}$$

Here, then, we have shown how to compute δ given β. The end points of the confidence interval are then

$$\bar{\xi} - \frac{\lambda s_x}{\sqrt{n}} \quad \text{and} \quad \bar{\xi} + \frac{\lambda s_x}{\sqrt{n}}. \tag{3-45}$$

Frequently statisticians use $\beta = 0.05$ or 0.01, that is, $1 - \beta = 0.95$ or 0.99. A confidence interval with $\beta = 0.05$ is called a 95 percent confidence interval, since the probability that the interval contains μ_x is 0.95. Similarly, the confidence interval for $\beta = 0.01$ is called a 99 percent confidence interval.

EXAMPLES. *1.* A domestic airline is interested in determining the expected weight of baggage carried by non first class passengers on the Los Angeles to Honolulu run. As on other domestic flights, there is now no weight limitation, although only two bags of specified sizes may be checked. Over a period of one month the airline records the weight of the baggage carried by 5000 different passengers. The average weight is 36.7 pounds with $s_x = 15.0$ pounds. Let us determine a 99 percent confidence interval for the expected weight. Here $\beta = 0.01$ and $1 - (\beta/2) = 0.995$. Then from Table E, $\lambda_{.99} = 2.58$. Since $\sqrt{5000} = 70.7$, we see from (3–44) that

$$\delta = \frac{2.58(15)}{70.7} = 0.55.$$

Thus the confidence interval has end points 36.1 and 37.3. Therefore, one might report that $\mu_x = 36.7 \pm 0.6$ pounds with probability 0.99, and this means that the probability is approximately 0.99 that the confidence interval with end points 36.1 and 37.3 contains μ_x.

2. Confidence intervals on estimates of population proportions can, of course, be constructed in the same way as for estimates of μ_x. For example,

suppose that a company with its head office in New York City was interested in estimating the proportion of employees working there who would like to alter the working hours so as to begin work one hour early and finish one hour early. To do this it selected at random a sample of 500 of the 9000 employees who work at the head office and determined their preference. The result showed that a fraction 0.22 of these interviewed preferred the earlier hours. Let us determine a 99 percent confidence interval for the proportion of all workers who prefer the earlier hours.

We noted in Section 3–5 that the random variable w representing the proportion of the same having the characteristic of interest was essentially normally distributed with mean p and standard deviation given by (3–16). Thus $t = (w - p)/s_w$ has a distribution which is essentially $\varphi(t)$, s_w here being our estimate of σ_w. To determine s_w we use the sample proportion in place of p in (3–16). Hence

$$s_w = \sqrt{\frac{0.22(0.78)}{500}\left(\frac{10{,}000 - 500}{10{,}000 - 1}\right)} = \sqrt{0.000326} = 0.0181 \ .$$

Here $\beta = 0.01$ and $1 - (\beta/2) = 0.995$. The value of t, call it $\lambda_{.99}$, for which $\Phi(\lambda_{.99}) = 0.995$ is $\lambda_{.99} = 2.58$. In this case s_w corresponds to $s_{\bar{x}} = s_x/\sqrt{n}$ in (3–41). Thus

$$\delta = \lambda_{.99}s_w = 2.58(0.0181) = 0.047 \ .$$

The end points for a 99 percent confidence interval are then $0.22 - 0.047 = 0.173$ and $0.22 + 0.047 = 0.267$. Thus one might report here that $p = 0.22 \pm 0.047$ with probability 0.99.

The reader will note that the length 2δ of a confidence interval increases as $1 - \beta$, the probability that the interval contains the parameter of interest, is increased. To be very sure that the interval contains the parameter, a very broad interval must be used. The more narrow the interval the greater the probability that it does not contain the parameter of interest. It is therefore necessary to use a little care in selecting β in order not to mislead an untrained individual any more than necessary when reporting an estimate in the form $\bar{\xi} \pm \delta$.

3–10 THE STUDENT t DISTRIBUTION

We have frequently made use of the fact that when n is fairly large, the distribution of the random variable

$$t = \frac{(\bar{x} - \mu_x)\sqrt{n}}{\sigma_x} \tag{3–46}$$

is approximately $\varphi(t)$, regardless of what the distribution of x happens to be. In practice, σ_x is typically not known and σ_x is replaced by s_x, our estimate of σ_x, so that

$$t = \frac{(\bar{x} - \mu_x)\sqrt{n}}{\sqrt{w}}, \qquad (3\text{-}47)$$

where w is the random variable (3–34) whose value is s_x^2. In other words, t is a function not only of the random variable $\bar{x}$ but also of the random variable w. Each time $\mathcal{R}_n$ is performed, a value $\bar{\xi}$ of $\bar{x}$ and a value s_x^2 of w is determined. From this we then determine a value λ of t which is $\lambda = (\bar{\xi} - \mu_x)\sqrt{n}/s_x$. We cannot actually compute λ, because μ_x is not known; nonetheless, a value λ of t is implicitly determined in this way.

In the past we have ignored the fact that the random variable t is a function of w as well as $\bar{x}$, since we proceeded as if s_x were indeed σ_x rather than the value of a random variable. Suppose now that the distribution of t could be determined when t is given by (3–47), that is, when t is taken to be a function of two random variables $\bar{x}$ and w. Of what value would this be to us? The value would lie in the fact that we could then determine confidence intervals somewhat more accurately. By more accurately, we mean that if it were desired to determine a confidence interval which would cover μ_x with probability $1 - \beta$, the end points of this interval could be determined more precisely. The end points of the interval determined by treating t as having a standardized normal distribution do not make the probability that the random interval contains μ_x precisely $1 - \beta$. It is only approximately $1 - \beta$. However, if the exact distribution of t were known, then the end points could be determined so that the probability that the confidence interval contains μ_x is exactly $1 - \beta$.

Before turning to the question of whether it is possible to determine the exact distribution of t, let us show how this distribution could be used if it were known. Let λ be the value of t such that the probability of the event $-\lambda \leq t \leq \lambda$ is $1 - \beta$. The number λ could be determined from the distribution for t. If $-\lambda \leq t \leq \lambda$, then

$$-\lambda \leq \frac{(\bar{x} - \mu_x)\sqrt{n}}{\sqrt{w}} \leq \lambda$$

or

$$\frac{-\lambda\sqrt{w}}{\sqrt{n}} \leq \bar{x} - \mu_x \leq \frac{\lambda\sqrt{w}}{\sqrt{n}}$$

or

$$\bar{x} - \frac{\lambda\sqrt{w}}{\sqrt{n}} \leq \mu_x \leq \bar{x} + \frac{\lambda\sqrt{w}}{\sqrt{n}}. \qquad (3\text{-}48)$$

The event $-\lambda \leq t \leq \lambda$ occurs with probability $1 - \beta$. Thus the probability is $1 - \beta$ that the random interval or confidence interval whose end points

are $\bar{x} - \lambda\sqrt{w/n}$ and $\bar{x} + \lambda\sqrt{w/n}$ contains μ_x. Thus if $\mathcal{R}_n$ is performed and the values $\bar{\xi}$ of $\bar{x}$ and s_x of $\sqrt{w}$ are determined, the end points of the confidence interval are

$$\bar{\xi} - \frac{\lambda s_x}{\sqrt{n}} \quad \text{and} \quad \bar{\xi} + \frac{\lambda s_x}{\sqrt{n}}. \tag{3-49}$$

Now equation (3–49) is identical in appearance to (3–45). The only difference lies in the way λ was determined. In (3–45), λ was determined from $\varphi(t)$, while in (3–49) λ was determined using the exact distribution for t.

In general it is extremely difficult to determine the exact distribution for t. It depends on the distribution for x. There is one case where the exact distribution for t can be conveniently determined. This is the case where x itself is normally distributed. Recall that $\bar{x}$ is always approximately normally distributed; it is not by any means true that x is always approximately normally distributed. However, if x is normally distributed, the exact distribution for t can be found. This distribution is called the *Student t distribution* or often simply the *t distribution*. It was not developed by someone named Student, but rather by a man named W. S. Gosset who published his work under the pseudonym Student. As with the normal distribution, the most convenient way to work with the t distribution is through the use of tables. Such a table is given in Table G. This table does not give the cumulative function for t, but instead gives, for different α, values λ of t such that the area to the right of λ is α. From this table, the value of λ needed in (3–49) can be obtained directly for a specified β. If the probability that $-\lambda < t < \lambda$ is $1 - \beta$, then since the t distribution is symmetric, just as is the normal distribution, the area to the right of λ is $\beta/2$. Thus we merely read from the table the λ corresponding to $\beta/2$.

The distribution of t depends on n. For each n there is a different density function for t. Instead of tabulating the values of λ for different values of n, the values of λ are tabulated for various values of what is referred to as the degrees of freedom (abbreviated ν in Table G). The degrees of freedom is simply $n - 1$ when t is given by (3–47). The reason for using the degrees of freedom, that is, $n - 1$, rather than n is a technical one which we shall not attempt to justify. However, it is necessary to keep in mind that if it is desired to find λ for $\beta = 0.05$ when, for example, a sample of size 20 is used, this value is read in the column for $\alpha = \beta/2 = 0.025$ in the row for $\nu = 19$ (not 20). The value of λ is 2.09 in this case.

For large values of n the curve for the t distribution is essentially identical with that of $\varphi(t)$. In Figure 3–11 we have shown the curve for $\varphi(t)$ and that for t distribution when $n = 5$. There would normally be little point to using the t distribution for $n \geq 25$. It is only for small n that the differences become significant. For most estimation problems n will be greater than 25 and thus the methods introduced earlier based on the normal distribution are quite adequate in such cases. It should be kept in mind that the Student

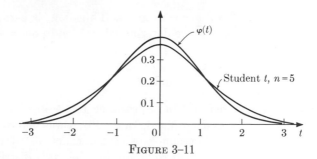

FIGURE 3–11

t distribution is the exact distribution for t only when x is normally distributed. It cannot be used when n is small unless there is reason to believe that x is moderately close to being normally distributed. The random variable does not need to be precisely normally distributed in order that the t distribution can be usefully employed for small n, but it cannot be employed in all cases.

EXAMPLE. A firm tests fifteen very expensive batteries used in space work to estimate the mean operating life for a battery of this type. The average life was 94.5 days and the value of s_x was 26 days. Let us determine a 95 percent confidence interval for the mean battery life if it is assumed that the life of any battery is normally distributed. The t distribution can be used in this case. The desired value of β is 0.05 and the number of degrees of freedom is 14(15 − 1 = 14). From Table G, for $\alpha = 0.025$ and $\nu = 14$ we read $\lambda_{.95} = 2.14$. Then

$$\delta = \frac{\lambda_{.95}s_x}{\sqrt{n}} = \frac{2.14(26)}{3.87} = 14.4 \ .$$

Thus the 95 percent confidence interval has end points 94.5 − 14.4 = 80.1 and 94.5 + 14.4 = 108.9. Equally well we can write

$$\mu_x = 94.5 \pm 14.4$$

with probability 0.95.

If the normal distribution were used instead of the t distribution, $\lambda_{.95} = 1.96$ and

$$\delta = \frac{1.96(26)}{3.87} = 13.2 \ ,$$

so that the end points of the confidence interval determined in this way are 81.3 and 107.7. This interval is slightly narrower than that obtained from the t distribution (this is always true), but the differences are very slight.

3–11 ERRORS OF MEASUREMENT

In all of our previous work we have assumed implicitly that when a random experiment was performed no errors or mistakes were made in determining

the outcome of this experiment. For example, if a sample of n items was selected at random from a lot and the number of defectives determined, we assumed that the inspector actually determined the number of defectives and did not make any errors by classifying a defective unit as good or a good unit as defective. In the real world, of course, inspectors will occasionally make mistakes. As another example, suppose that a biologist is attempting to estimate the expected weight of adult fruit flies. The weight of any given fruit fly can be considered to be a value ξ of a random variable x. However, when the biologist weighs the fly he may not obtain the value ξ due to various errors in the weighing process. Instead, the value he records as the weight may be ξ_1, which is different from ξ. Still another way in which errors of measurement can occur is in situations where a survey is being conducted and the person surveyed either intentionally or through a misunderstanding gives an incorrect answer to the question or questions. This sort of error arises frequently in market surveys and opinion polls.

The examples just given represent what may be called measurement errors, because errors were introduced as a result of measuring the value of x determined on each trial of $\Re$. Let us analyze measurement errors in a little more detail. We can now think of the process of performing $\Re$ and measuring the value of x so generated as a two-stage random experiment. The first stage consists in performing $\Re$ and generating a value ξ of x. At the second stage a random experiment is performed in which the value of x generated at the first stage is measured, and the result ξ_1 is obtained, which will in general be different from ξ. Normally, ξ_1 will be one of the allowable values of x (for otherwise it would presumably be clear that a mistake had been made). It is convenient to imagine that ξ_1 is the observed value of a random variable y, where y is the sum of the random variable x and the random variable ϵ, ϵ being the error introduced at the second stage, so that $y = x + \epsilon$.

Suppose now that we are trying to estimate μ_x, and that n independent trials of $\Re$ are performed to do so. We would like to use the value of $\bar{x}$ so generated as our estimate of μ_x, but because of the errors introduced we shall use the observed value of $\bar{y}$ instead, where $\bar{y} = \Sigma_{j=1}^{n} y\{j\}/n$, $y\{j\}$ being the random variable representing y for the jth trial of $\Re$. Now if $\epsilon\{j\}$ is the random variable representing ϵ on the jth trial of $\Re$, then

$$\bar{y} = \bar{x} + \frac{1}{n}\sum_{j=1}^{n} \epsilon\{j\} = \frac{1}{n}\sum_{j=1}^{n} y\{j\}. \tag{3–50}$$

Let us denote the expected value of ϵ by η. Then each $\epsilon\{j\}$ has the expected value η. Since the expected value of $\bar{x}$ is μ_x, we see that $\mu_{\bar{y}}$, the expected value of $\bar{y}$, is

$$\mu_{\bar{y}} = \mu_x + \eta. \tag{3–51}$$

If $\eta \neq 0$, then the statistic is $\bar{y}$ biased and the bias is η. The bias is introduced by the measuring process. Note that the statistic $\bar{y} - \eta$ is unbiased.

Thus if the expected value η of ϵ is known, one would use as the estimate of μ_x the value $\bar{\zeta} - \eta$, where $\bar{\zeta}$ is the value obtained for $\bar{y}$. Frequently the measuring process does not introduce any bias so that $\eta = 0$, and in this case $\bar{y}$ is an unbiased estimator of μ_x.

Consider next the variance of $\bar{y}$, where $\bar{y}$ is given by (3–50). The $2n$ variables $x\{j\}$ and $\epsilon\{j\}$ are not necessarily independent random variables. If the n variables $\epsilon\{j\}$ are not independent, so that the measurement error on one trial of $\Re$ may be dependent on the measurement error on a previous trial, then the errors are said to be *correlated*. Very frequently it is reasonable to assume that the errors are uncorrelated, so that the n random variables $\epsilon\{j\}$ are independent. Let us then assume that the errors are uncorrelated.

Even though the random variables $\epsilon\{j\}$ are independent and the random variables $x\{j\}$ are independent, it need not be true that the $2n$ random variables $x\{j\}$ and $\epsilon\{j\}$ are independent. The probability function or probability density for $\epsilon\{j\}$ may depend on the value of $x\{j\}$. In this case, the covariance of $\epsilon\{j\}$ and $x\{j\}$ need not be zero. Generally speaking, however, one would expect $\epsilon\{j\}$ to be independent of the $x\{i\}$, $i \neq j$. This, then, is equivalent to saying that we frequently expect the $y\{j\}$ to be independent random variables. When this is true, then $\sigma_{\bar{y}}^2 = n^{-1}\sigma_y^2$, and since $y = x + \epsilon$,

$$\sigma_y^2 = \sigma_x^2 + \sigma_\epsilon^2 + 2\Delta ,$$

where Δ is the covariance of x and ϵ. Hence

$$\sigma_{\bar{y}}^2 = n^{-1}(\sigma_x^2 + \sigma_\epsilon^2 + 2\Delta) = \sigma_{\bar{x}}^2 + n^{-1}(\sigma_\epsilon^2 + 2\Delta) . \tag{3–52}$$

When the $y\{j\}$ are independent, it follows from the central limit theorem that since the $y\{j\}$ are identically distributed, the distribution of $\bar{y}$ for large n will be essentially normal with mean $\mu_x + \eta$ and variance given by (3–52). This is a very useful result.

From what we have just shown, it follows at once that if $\eta = 0$ (so that the measurements are unbiased) and the $y\{j\}$ are independent random variables, then for large n, $\bar{y}$ is essentially normally distributed with mean μ_x, just as $\bar{x}$ is. However, the variance of $\bar{y}$ is greater than the variance of $\bar{x}$, as we see from (3–52). In such a case, we can use $\bar{y}$ to estimate μ_x, but in general, more observations will be needed, on account of the larger variance of $\bar{y}$, than would be the case when no measurement errors are made. If $\eta \neq 0$, but is known, $\bar{y} - \eta$ can be used as the statistic to estimate μ_x, and this statistic will be roughly normally distributed with mean μ_x and variance $\sigma_{\bar{y}}^2$. Thus we see that for a large number of important cases, measurement errors can be accounted for and the net result of them is that more observations may be required.

3–12 **ERRORS OF SELECTION**

There is another type of error that can arise frequently in performing an experiment for the purpose of estimating some parameter. This may be called a selection error. In constructing our models we have often assumed that the experiment was carried out in some specific way, that is, independent trials of $\Re$ were made, or a sample was selected at random from a population. The results obtained depend critically on these assumptions, and if they are not satisfied the model may yield very misleading results.

The greatest problems are usually encountered in selecting a random sample from a population. The basic problem lies in trying to make the sample random, that is, giving each set of n members of the population an equal chance of being chosen. Thus, for example, if one wished to select a random sample of 100 resistors from a lot it would be very unwise to assume that taking 100 resistors off the top layers of the lot would yield a random sample. The supplier might suspect this was being done and hence place only good units on the top layers. The number of defectives in the sample would then give no true indication of the quality of the lot.

One must be especially careful when attempting to draw random samples from populations whose members are people. Mistakes are often made which lead to very costly inappropriate decisions. Great care must be taken to be sure that the sample is random and that there do not exist factors which will automatically make the sample not representative of the population as a whole. Hence, if it were of interest to estimate the proportion of the population that will vote for the Republican presidential candidate in the next election, it would not be wise to select a random sample from the list of subscribers to the *Wall Street Journal*. The subset of the population which consists of subscribers to the *Wall Street Journal* is not representative of the population as a whole, and thus the sample could be extremely biased (probably in the direction of a higher-than-average percentage of Republicans). Although it may appear to the reader that such a ridiculous mistake would never be made, essentially this type of mistake is made frequently, in perhaps a little less obvious form.

A famous illustration of such a mistake is the poll conducted by the *Literary Digest* in 1936. Over two million voters filled in and returned sample ballots sent out by the *Digest*. Of these, 54.6 percent voted for Landon rather than Roosevelt. However, in the actual election, only 36.7 percent of the population voted for Landon, so that an error of almost 18 percentage points was made in the estimate. If the sample was indeed random, the probability of an error of only 0.1 percentage points was extremely small, and the probability of an error of 18 percentage points incredibly small. Thus the error was mainly due to bias. In this case the *Digest* did not poll

only its own subscribers, but instead selected persons from lists of individuals who owned telephones or automobiles. The sample was further biased in that only those who bothered to return a ballot were counted.

Before going on, it might be worthwhile to provide some other simple illustrations of how bias can be introduced. If a company were interested in test marketing a product to estimate the expected number of units to be sold per month per customer in stores, it might be dangerous to test market it only in stores in the East, since tastes in the Midwest or Far West might be different. Equally well, it might be very misleading to test market it only in suburban stores, since the tastes of those patronizing the central city stores might be quite different. Of course if the company were only interested in suburban stores, this would be another matter. Once again, although these mistakes seem fairly obvious, similar sorts of errors are made with great frequency.

How does one select a random sample from a population? In principle, the way to do this is to obtain a list of every member in the population, assign each a number, and then use a table of random numbers to decide which numbers and hence which members are to be included in the sample. In practice, however, for a variety of reasons, it may be very difficult if not impossible to select the sample in this manner. One simple reason might be that a list of the members of the population is not available, and hence even the first step in the process cannot be carried out. For example, if it is desired to select a random sample from a lot of resistors, the above process cannot be used because the resistors are not individually identified. We could identify them by painting a number on each one, but it would be easier to test each one and then no sample would be needed. Similarly, one could not use the above procedure to select a random sample from the population of the United States because there is no list giving the name of every inhabitant of the U.S. Even when a list of the members of the population is available, the procedure suggested above may be extremely costly or cumbersome to use.

Often, based on judgment concerning the physical situation, it is possible to obtain a random sample without use of any elaborate procedures. For example, if an oil company wishes to select a random sample of its credit card customers, it would probably be quite adequate to take something like every tenth or every fifteenth name from the file of customers, assuming they were arranged alphabetically. Similarly, if a supplier is known to be reliable and would not use a trick such as placing all good units on the top layers of a lot, then it might be satisfactory merely to use the units from the top of the lot as a random sample. However, this can lead to bias even if the supplier is reliable. For example, if he filled out the lot as units were produced, then only the last ones produced would be on top. If a change took place in the production process during the time the lot was being filled out,

then a sample from the top layers would be biased. Sometimes, the simplified sorts of procedures we have been discussing are referred to as *judgment sampling*. If the judgment used is sound, the sample will be unbiased and equivalent to a random sample.

3–13 EXPERIMENTAL DESIGN

In designing an effective experiment for estimating some parameter it is necessary to account for and control as needed the errors of measurement and selection that were discussed in the previous two sections. Furthermore, it is of interest to design the experiment so as to obtain the best estimate possible for the money expanded. By best estimate we essentially mean that the statistic with the smallest standard deviation should be used. The problem of designing the experiment to obtain the best estimate can be a complicated one. We shall in this section consider some very elementary notions concerning the design of an experiment.

In certain types of problems increased accuracy in an estimate can be obtained by using a technique known as *stratified sampling*. To illustrate this technique, consider a manufacturer of canned soups. During the canning season, ten different machines are working on tomato soup. In each machine the tomato mix is fed into the can and the top is put on while simultaneously the contents is sterilized. All the machines feed the cans onto a single conveyor system where they move on to be crated. The cans, of course, become thoroughly mixed on the conveyor belt, so that the output of the various machines is mixed. Now if any can is selected at random from a carton of tomato soup, the weight of tomato soup in this can is a random variable x with expected value μ_x.

Let us examine the problem of estimating μ_x. The various canning machines are not identical in their operation, that is, the mean weight of tomato soup in the cans produced by a given machine will vary somewhat from machine to machine, no matter how carefully they are adjusted. Similarly, the standard deviations will vary from machine to machine. One way that μ_x could be estimated would be merely to select a sample of n cans at random as they come off the conveyor just before crating. This sample would then normally contain a mixture of cans produced by the various machines. Although this is a perfectly satisfactory way to estimate μ_x, increased accuracy can be obtained for the same value of n by using a somewhat different procedure.

Suppose that instead of taking a single sample coming off the end of the conveyor, we instead take ten different samples, one for the output of each of the ten canning machines. In other words, a sample of size n_j cans is selected directly from the production of machine j. Assume that the n_j have the property that $\Sigma_{j=1}^{10} n_j = n$. Now μ_j, the expected weight of tomato soup

in a can produced on machine j, is estimated as the average of the weights of the n_j cans in the sample. Denote this estimate by $\bar{\xi}_j$. The production rates of each of the machines may not be precisely the same, since, for example, some machines may be newer than others. Let α_j be the fraction of the total production made on machine j. Consider the number

$$\bar{\xi} = \alpha_1 \bar{\xi}_1 + \cdots + \alpha_{10} \bar{\xi}_{10} . \tag{3-53}$$

This number $\bar{\xi}$ is a value of the random variable $\bar{y}$, where

$$\bar{y} = \alpha_1 \bar{y}\{1\} + \cdots + \alpha_{10} \bar{y}\{10\} , \tag{3-54}$$

and $\bar{y}\{j\}$ is the random variable representing the average weight of a sample of n_j cans from the production of machine j, that is, $\bar{y}\{j\}$ is the statistic being used to estimate μ_j. We now claim that it is possible to select the n_j in such a way that $\bar{y}$ is a better statistic to use for estimating μ_x than is $\bar{x}$, $\bar{x}$ being the random variable representing the average weight of a sample of n cans taken from the conveyor just before packaging.

To show that $\bar{y}$ is better than $\bar{x}$ when the n_j are chosen properly we shall show that both are unbiased statistics and that there exist n_j, $\Sigma_{j=1}^{10} n_j = n$ such that the variance of $\bar{y}$ is less than $\bar{x}$. Note first of all that the expected value of $\bar{y}\{j\}$ is μ_j, from what we have proved previously. Hence by (2–33) and (3–54), it follows that

$$\mu_{\bar{y}} = \alpha_1 \mu_1 + \cdots + \alpha_{10} \mu_{10} . \tag{3-55}$$

Let us now show that μ_x is just the right-hand side of (3–55). Suppose that x can take on the values $x_1, \ldots, x_k$. Consider the event that $x = x_i$ and the can was made on machine j. The probability of this event is $p(x_i|j)p(j)$, where $p(x_i|j)$ is the conditional probability that $x = x_i$, given that the can was made on machine j and $p(j)$ is the probability that the can was made on machine j. Now $p(j)$ is the long-run fraction of the cans which come from machine j and this is α_j. Thus $p(j) = \alpha_j$, and the probability that $x = x_i$ and the can was filled on machine j is $p(x_i|j)\alpha_j$. The events corresponding to different machines filling the cans are mutually exclusive. Therefore $p(x_i)$, the probability that $x = x_i$, is

$$p(x_i) = \sum_{j=1}^{10} p(x_i|j)\alpha_j = p(x_i|1)\alpha_1 + \cdots + p(x_i|10)\alpha_{10} . \tag{3-56}$$

Now recall that

$$\mu_x = \sum_{i=1}^{k} x_i p(x_i) ,$$

or by (3–56)

$$\mu_x = \alpha_1 \sum_{i=1}^{k} x_i p(x_i|1) + \cdots + \alpha_{10} \sum_{i=1}^{k} x_i p(x_i|10) . \tag{3-57}$$

However $\Sigma_{i=1}^{k} x_i p(x_i|j)$ is μ_j, the expected weight of tomato soup in cans filled on machine j. Thus

$$\mu_x = \alpha_1 \mu_1 + \cdots + \alpha_{10} \mu_{10} , \qquad (3\text{–}58)$$

which is what we wished to prove. Consequently, from (3–55) $\mu_{\bar{y}} = \mu_x$ and $\bar{y}$ is an unbiased statistic. Of course, $\bar{x}$ is also unbiased, as we know from our previous work.

Let us now determine the variance of $\bar{y}$ and $\bar{x}$ and show that there exist n_j for which $\sigma_{\bar{y}}^2 < \sigma_{\bar{x}}^2$. The random variables $\bar{y}\{j\}$ appearing in (3–54) are independent random variables since they refer to the output of different machines. Thus by (2–53), we see immediately that the variance of $\bar{y}$ is

$$\sigma_{\bar{y}}^2 = \frac{\alpha_1^2}{n_1} \sigma_1^2 + \cdots + \frac{\alpha_{10}^2}{n_{10}} \sigma_{10}^2 , \qquad (3\text{–}59)$$

where σ_j^2/n_j is the variance of $\bar{y}\{j\}$, σ_j^2 being the variance of the weight of the cans produced on machine j. Now $\sigma_{\bar{x}}^2 = \sigma_x^2/n$, where σ_x^2 is the variance of x. It remains then to determine σ_x^2. By definition

$$\sigma_x^2 = \sum_{i=1}^{k} (x_i - \mu_x)^2 p(x_i) . \qquad (3\text{–}60)$$

When (3–56) is used (3–60) becomes

$$\sigma_x^2 = \alpha_1 \sum_{i=1}^{k} (x_i - \mu_x)^2 p(x_i|1) + \cdots + \alpha_{10} \sum_{i=1}^{k} (x_i - \mu_x)^2 p(x_i|10) . \quad (3\text{–}61)$$

Consider next

$$\sum_{i=1}^{k} (x_i - \mu_x)^2 p(x_i|j) . \qquad (3\text{–}62)$$

If μ_x were replaced by μ_j, this would be σ_j^2, the variance of the weight produced on machine j. We can introduce μ_j by writing

$$(x_i - \mu_x)^2 = (x_i - \mu_j + \mu_j - \mu_x)^2 = [(x_i - \mu_j) + (\mu_j - \mu_x)]^2 \qquad (3\text{–}63)$$
$$= (x_i - \mu_j)^2 + 2(\mu_j - \mu_x)(x_i - \mu_j) + (\mu_j - \mu_x)^2 .$$

Use of (3–63) in (3–62) yields

$$\sum_{i=1}^{k} (x_i - \mu_x)^2 p(x_i|j) = \sum_{i=1}^{k} (x_i - \mu_j)^2 p(x_i|j)$$

$$+ 2(\mu_j - \mu_x) \sum_{i=1}^{k} (x_i - \mu_j) p(x_i|j) + (\mu_j - \mu_x)^2 \sum_{i=1}^{k} p(x_i|j) . \quad (3\text{–}64)$$

However,

$$\sum_{i=1}^{k} p(x_i|j) = 1 ,$$

since $p(x_i|j)$ is the probability that the weight is x_i for a can filled on machine j. Then

$$\sum_{i=1}^{k} (x_i - \mu_j)p(x_i|j) = \sum_{i=1}^{k} x_i p(x_i|j) - \mu_j = \mu_j - \mu_j = 0 ,$$

since $\Sigma_{i=1}^{k} x_i p(x_i|j) = \mu_j$. Therefore

$$\sum_{i=1}^{k} (x_i - \mu_x)^2 p(x_i|j) = \sigma_j^2 + (\mu_j - \mu_x)^2 , \qquad (3\text{-}65)$$

and by (3–61),

$$\sigma_x^2 = \sum_{j=1}^{10} \alpha_j \sigma_j^2 + \sum_{j=1}^{10} \alpha_j (\mu_j - \mu_x)^2 , \qquad (3\text{-}66)$$

and hence since $\sigma_{\bar{x}}^2 = \sigma_x^2/n$,

$$\sigma_{\bar{x}}^2 = \sum_{j=1}^{10} \frac{\alpha_j}{n} \sigma_j^2 + \sum_{j=1}^{10} \frac{\alpha_j}{n} (\mu_j - \mu_x)^2 . \qquad (3\text{-}67)$$

We are at last ready to compare (3–59) and (3–67). Equation (3–59) holds for any set of n_j whatever, provided each $n_j \geq 1$. We are interested in cases where $n = \Sigma_{j=1}^{10} n_j$. There yet remains the question of how the values of the individual n_j are selected, that is, how the total number of cans n to be inspected is to be allocated among the different machines. It would appear reasonable that larger samples should be taken from the machines that produce the larger fractions of the total output. Indeed, it appears reasonable to allocate n in proportion to the output of the machines, that is, to take (to the nearest integers)

$$n_j = \alpha_j n . \qquad (3\text{-}68)$$

This is referred to as proportional allocation. If (3–68) is used in (3–59), $\sigma_{\bar{y}}^2$ reduces to

$$\sigma_{\bar{y}}^2 = \sum_{j=1}^{10} \frac{\alpha_j}{n} \sigma_j^2 , \qquad (3\text{-}69)$$

so that from (3–67)

$$\sigma_{\bar{x}}^2 = \sigma_{\bar{y}}^2 + \sum_{j=1}^{10} \frac{\alpha_j}{n} (\mu_j - \mu_x)^2 . \qquad (3\text{-}70)$$

We assumed originally that the μ_j were not all the same. Thus $\mu_j - \mu_x \neq 0$ for at least one j and therefore $(\mu_j - \mu_x)^2 > 0$ for at least one j. Consequently,

$$\sigma_{\bar{x}}^2 > \sigma_{\bar{y}}^2 , \tag{3-71}$$

when the n_j are determined according to (3–68), and thus $\bar{y}$ is a better statistic to use than is $\bar{x}$.

It is not necessarily true that the proportional allocation rule (3–68) yields the smallest possible variance of $\bar{y}$. The rule (3–68) did not take into account the variances σ_j^2. It can be shown, although we shall not give the proof, that the allocation which minimizes $\sigma_{\bar{y}}^2$ is

$$n_j = \left(\frac{\alpha_j \sigma_j}{\beta}\right) n; \quad \beta = \sum_{j=1}^{10} \alpha_j \sigma_j . \tag{3-72}$$

The reader familiar with the method of Lagrange multipliers will be able to prove (3–72) very easily. Of course (3–72) cannot be used unless the σ_j are known at least approximately. After some mathematical analysis, we have shown that the stratified sampling method yields a better estimator of μ_x than the direct sampling of the production coming off the conveyor.

The general conditions under which stratified sampling may be useful can be described as follows. It is of interest to estimate some parameter associated with a random variable x, say μ_x. The experiment $\mathfrak{R}$ which generates a value of x does so by generating an object which has associated with it a value of x. The object may be a can of soup or a person. These objects have associated with them not only a value of x but some other characteristic K. The random experiment $\mathfrak{R}$ which generates an object can be looked upon as a two-stage random experiment where at the first stage the characteristic K of the object is selected (in our example the characteristic is the machine on which a can is filled) and then a particular object with this characteristic and having a given value of x is generated. The expected value of x and its variance will depend on K. To estimate μ_x it will be advantageous under these circumstances to use stratified sampling. First, for each category K, the expected value of x if the object has characteristic K is estimated. Each of these estimates is multiplied by the probability that the object will have characteristic K and the resulting numbers are added to obtain the estimate for μ_x.

EXAMPLE. We shall now provide an example illustrating the use of stratified sampling in a type of situation which arises frequently—that in which the objects are humans. Consider the experiment in which a child is selected at random from the grade-school students in a given city, and let x be the weight of the child selected. Suppose that it is desired to estimate μ_x. Assume that μ_x is to be estimated by selecting a sample of 100 students. There

are six grades in the elementary schools which contain 1000, 1250, 900, 1160, 1000 and 950 students, so that there are a total of 6260 students in the elementary schools. When a student is selected, he may be classified according to what grade he is in. Furthermore, the mean weight of students in the different grades would be expected to vary significantly from the first to sixth grades. Thus stratified sampling could be usefully employed here to obtain a good statistic. The procedure would be to select a random sample of students from each grade. This essentially corresponds to making a number of independent trials of $\Re$ for each grade (see Problem 7 for Section 2–6). Then the estimate of the expected weight for each grade would be obtained. Each of these numbers would be multiplied by the fraction of the school system's population which is in that grade and the results added.

For the case under consideration the weight variances in any grade would probably not be known, and thus the logical way to select the sample sizes for each grade would be to use proportional sampling. If n_j is the sample size to use for grade j, and α_j is the fraction of the students in grade j, then

$$\alpha_1 = \frac{1000}{6260} = 0.16 \; ; \quad \alpha_2 = \frac{1250}{6260} = 0.20 \; ; \quad \alpha_3 = \frac{900}{6260} = 0.14 \; ;$$

$$\alpha_4 = \frac{1160}{6260} = 0.19 \; ; \quad \alpha_5 = \frac{1000}{6260} = 0.16 \; ; \quad \alpha_6 = \frac{950}{6260} = 0.15 \; ,$$

and since $n = 100$,

$$n_1 = 16, \quad n_2 = 20, \quad n_3 = 14, \quad n_4 = 19, \quad n_5 = 16 \text{ and } n_6 = 15 \; .$$

To obtain a feeling as to how much the standard deviation of the estimate could be reduced by using stratified sampling, suppose that the mean weights in the grades are $\mu_1 = 55$, $\mu_2 = 65$, $\mu_3 = 73$, $\mu_4 = 84$, $\mu_5 = 92$ and $\mu_6 = 102$ pounds. Then

$$\mu_x = 0.16(55) + 0.20(65) + 0.14(73) + 0.19(84) + 0.16(92) + 0.15(102)$$
$$= 8.8 + 13.0 + 10.2 + 16.0 + 14.7 + 15.3 = 78.0.$$

Suppose that the variances of the weights are $\sigma_1^2 = 36$, $\sigma_2^2 = 28$, $\sigma_3^2 = 41$, $\sigma_4^2 = 48$, $\sigma_5^2 = 79$ and $\sigma_6^2 = 102$ pounds squared. Then by (3–69),

$$\sigma_{\bar{y}}^2 = 0.16(0.36) + 0.20(0.28) + 0.14(0.41) + 0.19(0.48) + 0.16(0.79)$$
$$\quad + 0.15(1.02)$$
$$= 0.058 + 0.056 + 0.057 + 0.091 + 0.126 + 0.153$$
$$= 0.541 \; ,$$

so $\sigma_{\bar{y}} = 0.74$ pounds.

Now the variance of the estimate that would be obtained simply by selecting at random 100 students is given by (3–70). Here

$$\sum_{j=1}^{10} \frac{\alpha_j}{n} (\mu_j - \mu_x)^2 = 0.14(5.29) + 0.20(1.69) + 0.14(0.25)$$
$$+ 0.19(0.36) + 0.16(1.96) + 0.15(5.76)$$
$$= 0.74 + 0.34 + 0.03 + 0.07 + 0.31 + 0.86$$
$$= 2.35.$$

Thus by (3–70)

$$\sigma_{\bar{x}}^2 = 0.54 + 2.35 = 2.89 ,$$

so $\sigma_{\bar{x}} = 1.70$, and $\sigma_{\bar{y}}$ is less than half $\sigma_{\bar{x}}$. We see then that stratified sampling can very significantly reduce the standard deviation of the estimate.

We have discussed stratified sampling in considerable detail. Let us now review very briefly one other technique which is often useful. This is referred to as *multistage sampling*. Suppose that it is desired to select a random sample from the population of the United States. We have already noted that this cannot be done directly, because no list of the residents of the U.S. is available. However, such a method would in many cases be unsatisfactory even if it could be carried out, because the persons in the sample would be widely scattered. This causes a great deal of difficulty if it is necessary to see the individuals in person, as would be true in some sort of interview survey. The time and expense required for the interviewers to travel to each individual's home would be prohibitive. Instead a procedure would be followed where first a number of cities and towns would be selected. Then several areas in each of the cities would be selected at random. Next a number of blocks in each area would be selected at random, and finally some residents of each block would be selected. In this way the sample would be obtained by a multistage selection process. This method has the advantage that it makes an orderly interview procedure possible, since in any one city the interviewers will cover one block and then move on to the next block in the same area of the city, until the entire sample from that city has been accounted for.

REFERENCES

1. Freund, J. E., *Modern Elementary Statistics, 3rd ed.* Prentice-Hall: Englewood Cliffs, N.J., 1967.
Gives an elementary discussion of estimation problems.
2. Hodges, J. L. Jr., and E. L. Lehmann, *Basic Concepts of Probability and Statistics.* Holden-Day: San Francisco, 1964.
Contains a good elementary discussion of estimation.
3. Hoel, P. G., *Elementary Statistics, 2nd ed.* Wiley: New York, 1962.
Gives a slightly more elementary discussion of estimation than that presented here

4. Hoel, P. G., *Introduction to Mathematical Statistics*, Wiley, New York, 1962. Gives a slightly more advanced and theoretical treatment of estimation than the one given here.

PROBLEMS

Section 3–2

1. Assume that the random variable x is normally distributed with $\mu_x = 10$ and $\sigma_x = 3$. Consider the random experiment $\Re_{100}$ in which 100 independent trials of the random experiment that generates a value of x are performed. Consider the random variable $\bar{x}$ given by (3–2), which is the average of the values of x obtained on the 100 trials. On the same figure draw the curve representing the density function for x and that representing the density function for $\bar{x}$.

2. Suppose that x is a discrete random variable which can only take on the values 1, 3 or 7 with the probabilities 0.2, 0.5 and 0.3 respectively. Let $\Re$ be the experiment with which x is associated. Let $\bar{x}$ be the random variable representing the average of the values of x obtained on making 100 independent trials of $\Re$. On the same figure draw a histogram for x and the curve representing the density function for $\bar{x}$.

3. Construct a diagram like that shown in Figure 2–8 for the data in Table 3–1. However, show on the figure the values of $\bar{x}$, not the values of x. Also draw two horizontal lines through the points $\mu_x + 3\sigma_{\bar{x}}$ and $\mu_x - 3\sigma_{\bar{x}}$ on the vertical axis. If $\bar{x}$ is indeed normally distributed, what is the probability that a value of $\bar{x}$ will be outside the band determined by these two lines? Do any of the values of $\bar{x}$ obtained lie outside this band?

4. Repeat the experiment which led to Table 3–1 using the last two digits of each random number instead of the first two. Place a decimal point before the first of the two digits so obtained. Construct a histogram for $\bar{x}$ using these new data. How does it compare with that obtained from Table 3–1?

5. Suppose that instead of using four trials of $\Re$ to estimate μ_x, as was done in Table 3–1, eight trials are used. Obtain data representing twenty repetitions of $\Re_8$ by taking the first eight numbers in Table 3–1 to represent the first trial of $\Re_8$, the second eight numbers to represent the second trial and so on. Construct a histogram for the values of $\bar{x}$ obtained in this way. Plot the outline of this histogram and the outline of the histogram obtained from Problem 4 on the same figure. How do they compare?

6. Show that, according to the criteria introduced in the text, to estimate μ_x, a statistic $\bar{x}$ based on a sample size of n_2 is preferred to the $\bar{x}$ for a sample size of n_1 when $n_2 > n_1$. Is it necessarily true that the estimate based on the sample size n_2 will be better than that obtained from the sample size n_1? Discuss.

7. Show that the statistic $y = \frac{1}{2}[x\{1\} + x\{n\}]$ introduced in the text is an unbiased estimator of μ_x. Why is $\bar{x}$ given by (3–2) a better statistic than y? Hint: What is σ_y^2.

8. Consider the statistic
$$y = \alpha_1 x\{1\} + \cdots + \alpha_n x\{n\}; \quad \alpha_1 + \cdots + \alpha_n = 1 .$$
Show that y is an unbiased estimator of μ_x. What is σ_y^2?

9. Consider the statistic
$$y = 0.01\ x\{1\} + 0.05\ x\{2\} + 0.25\ x\{3\} + 0.50\ x\{4\} + 0.19\ x\{5\} .$$
According to Problem 8, y is an unbiased estimator of μ_x. Compute σ_y^2 and compare with the variance of $\bar{x}$ for $n = 5$. Thus show that $\bar{x}$ is a better statistic than y for estimating μ_x.

10. Mark the numbers 1, 3 and 5 on three coins, one number per coin. Place the coins in a container. Consider the random experiment in which a coin is selected at random and the number on the coin noted. The number on the coin drawn can be imagined to be the value of a random variable x. What is μ_x? Suppose that the numbers on the coins were not known and it was desired to estimate μ_x by making 10 trials of $\mathcal{R}$. Denote this experiment by $\mathcal{R}_{10}$. Perform $\mathcal{R}_{10}$ a total of ten times, thus obtaining ten estimates for μ_x. Compare these estimates with μ_x, and represent the results graphically in the manner suggested in Problem 1.

Sections 3–3 and 3–4

1. A manufacturer of vacuum tubes would like to estimate the expected life of a new model. He wishes to test enough tubes so that the probability that the estimate of the life differs from the true mean by more than 10 hours is 0.05. He believes that the standard deviation of the life will be 150 hours, which is the same as on the current model. How many tubes should he test?

2. A large tire manufacturer wishes to estimate the expected mileage that a particular model of tire will give under turnpike conditions when used on a given model of automobile. It is desired that the probability that the estimate differ from the expected mileage by more than 100 miles be 0.01. The standard deviation of the mileage is believed to be 1200 miles. How many tires should be tested?

3. A manufacturer of color T.V. sets wishes to estimate the mean time until a set first suffers a failure. It is desired that the probability that the estimate differ from the true mean by more than 10 hours be 0.05. It is believed that the standard deviation of the time to failure is 100 hours. How many sets should be tested?

4. A chemist is interested in determining the sulfur content of a given organic chemical. The experiment performed to do so is really a random experiment, and the estimate of the sulfur content obtained can be looked upon as the value taken on by a random variable x. Studies on substances having a known sulfur content have shown that x can be considered to be normally distributed with a mean equal to the sulfur content (expressed as percent by weight) and a standard deviation of 0.5 (expressed as percent by weight). To improve the estimate of the sulfur content, the experiment is repeated on several different samples. How many times should the experiment be repeated if it is desired that the probability that the estimate deviates from the true sulfur content by more than 0.1 (percent by weight) is not greater than 0.05?

5. A company which produces stereo records is considering the purchase of a new high-speed, continuous-feed molding machine. There is some concern in management, however, about the fraction of records produced that will be defective. It is decided to make a test to determine this number. How many records should be produced if it is desired that the probability that the estimate of the fraction defective differs from the true value by more than 0.003 be less than or equal to 0.01? Answer this under the assumption that nothing is known about the fraction defective that will be produced, and then under the assumption that it is known that the fraction defective will not be greater than 0.05.

6. A manufacturer of canned soups would like to estimate the mean weight of cans of tomato soup. It is desired that the probability that the estimate of the mean differ from more than 0.10 ounces be 0.01. It is believed that the standard deviation of the weight is 0.30 ounces. How many cans should be weighed to obtain the estimate?

7. A manufacturer of X-ray film would like to estimate the expected weight of silver halide per sheet of film. It is desired that the probability that the estimate deviate by more than 0.01 grams from the true value be less than or equal to 0.05. The standard deviation of the weight per sheet is believed to be 0.25 grams. How many sheets of film should be checked?

8. What is the answer to Problem 7 if the deviation of 0.01 grams is replaced by 0.001 grams?

9. A processor of frozen orange juice would like to estimate the expected weight of juice obtained per orange from a given supplier in a given season. It is desired that the probability that the estimate deviate by more than 0.5 ounces from the true value be 0.01. The standard deviation is not known exactly, but is certainly not greater than 3 ounces. How many oranges should be checked?

10. A manufacturer of pressurized tanks for use in storing liquefied gases tests 50 of them to estimate the mean bursting pressure. The standard deviation of the bursting pressure is believed to be 500 pounds. What is the probability that the estimate will differ from the true mean bursting pressure by more than 50 pounds? How many tanks would need to be tested to reduce this probability to 0.05?

11. Suppose it is desired to estimate the probability p that a woman giving birth to her first child will, in fact, have twins. It is desired that the probability that the estimate deviate from the true value by more than 0.01 be 0.02. How large should the sample be if nothing is known about p? How large should the sample be if it is known that $p < 0.01$?

12. A biologist is studying fruit flies. The lifespan of any particular fruit fly can be considered to be a random variable x. The biologist desires to estimate μ_x so that the probability that the actual estimate differs from μ_x by more than 8 hours is 0.05. The standard deviation of the lifespan is unknown, but the biologist feels that it is not more than 5 days. What should the sample size be? What experimental difficulties might be encountered in trying to carry out the experiment to measure the lifespans of the flies?

13. The U.S. Army purchases bars of soap from a particular supplier. All the bars are supposed to weigh at least six ounces. A government inspector wishes to estimate the expected weight of a bar. He wishes to have the probability that the estimate differ from μ_x by more than 0.1 ounces be 0.01. The standard deviation of the weight is not known, but it is not greater than 1.0 ounce. How many bars of soap should be weighed?

14. A biologist wishes to estimate the expected weight of a salmon selected from a particular Alaskan stream during July, when the fish are returning to spawn. He wishes to have the probability that his estimate differ from the true mean by more than 0.2 pound be 0.05. The standard deviation of the weight is not known, but is not greater than 7 pounds. How many fish should be caught and weighed to achieve his goal?

15. For the situation described in Problem 14, what is the probability that the estimate deviates by one pound or more from the true value?

16. A metallurgist wishes to determine accurately the melting point of a new alloy. He first conducts a pilot study by measuring the melting point on ten samples of the alloy and obtains these results: 1013, 1005, 1025, 1001, 1009, 1020, 1017, 1010, 1014 and 1002, all the numbers being degrees centegrade. Experimental error has caused most of the variation in the results, although there may be very slight differences in melting point from one piece of alloy to another. If it is desired that the probability that the estimate differs by more than one degree from the true value be 0.01, how many repetitions of the experiment will be needed?

Sections 3–5 and 3–6

1. Prove that if (3–15) holds then (3–16) must hold.

2. Under the assumption that voters would tell the truth about how they were going to vote, how large a sample would be needed to have the probability that the estimate of the fraction of votes to be received by a given candidate in a presidential election would differ from the true fraction by more than 0.5 percentage points be 0.005?

3. A firm with over 100,000 hourly employees wishes to estimate the proportion of the employees who favor a proposed change in the pension plan. How many employees should be queried if it is desired that the probability that the estimate deviates from the true proportion by more than one percentage point is 0.05?

4. The United States Army has received a shipment of 10,000 lightweight rifles. It is desired to estimate the fraction of the rifles which are defective. How many rifles should be tested if it is desired that the probability that the estimate deviates from the true proportion by more than three percentage points be 0.10? Solve this under the assumption that nothing is known about the fraction defective and under the assumption that it is fairly certain that the fraction defective will be less than 0.1.

5. A manufacturer of rotary pumps is contemplating the introduction of a new model. The pump, like the others in the line, would be sold mainly to industrial customers. It is desired to estimate the proportion of customers currently using other models in the line who would also find the new pump useful. How many

customers should be questioned if it is desired that the probability that the estimate will deviate from the true proportion by more than five percentage points be 0.01? The manufacturer has about 2000 customers in total.

6. To estimate the proportion of defective resistors in a lot of 10,000 resistors, the quality-control inspector for a radio manufacturer selects at random a sample of 100 resistors, and determines the fraction of the resistors in the sample that are defective. What is the probability that the fraction defective in the sample will deviate by more than 0.01 from the lot fraction defective? Solve the problem under the assumption that nothing is known about the lot fraction defective and under the assumption that the lot fraction defective is almost certain to be less than 0.1.

7. A department store wishes to estimate the proportion of its 50,000 credit-card customers who spend more than $200 per year at the store. It is desired that the probability that the estimate deviates by more than one percentage point from the true proportion should be 0.05. How many accounts should be checked in order to obtain the estimate?

8. A bank wishes to estimate the proportion of its 71,000 checking account customers who maintained a minimum balance of $1000 during the past month. It is desired that the probability that the estimate deviates by more than 0.5 percentage points from the true value be 0.10. How many accounts should be checked?

9. A national magazine wishes to estimate the proportion of its subscribers who earn more than $40,000 per year. It desires that the probability that the estimate differ from the true value by as much as 0.01 be 0.05. How many subscribers should be checked? Assume that individuals will tell the truth when answering.

Section 3–7

1. Select a set of twelve numbers from Table 3–1, different from those used in the example for this section, and estimate the standard deviation of x. How does the estimate compare with that obtained in Table 3–2?

2. Suppose that x is a random variable which takes on the value 1 with probability p and the value 0 with the probability $q = 1 - p$. What does (3–37) reduce to in this case?

3. Assume that the numbers 1, 3 and 9 are marked on three coins, one number per coin. Consider the random experiment in which the three coins are placed in a container and one of them is selected at random. Let x represent the number on the coin selected. Determine μ_x and σ_x. Suppose now that μ_x and σ_x were not known and twenty independent trials of $\Re$ are performed to generate a sample of twenty values of x. These are used to estimate μ_x and σ_x. Actually perform twenty trials of $\Re$ and estimate μ_x and σ_x. How do the estimates compare with the correct values?

4. A sample of ten bushings was selected from a day's production and the inside diameter of each was measured. This yielded the following results in inches: 1.207, 1.196, 1.202, 1.208, 1.185, 1.200, 1.199, 1.203, 1.195 and 1.201. Estimate the expected value and standard deviation of the random variable x representing the inside diameter of a bushing.

5. Suppose that the values of x obtained in a sample of size n all have the form $f + e_j$, where f is the same for every observed value of x. For example, f might be 2, so that the numbers might then read something like 2.013, 1.996, 2.014 and so forth. Show how the computation of the estimate of the standard deviation and mean of x can be simplified in such cases. Apply these results to the data of Problem 4.

6. Imagine that a sample of size n has been taken by making n independent trials of $\mathfrak{R}$ to generate n values of x. Suppose, however, we are not given the actual values of x but are instead given a table like Table 2–11, where the number of values of x lying in different class intervals is given. Suggest a way to estimate μ_x and σ_x in this case and use the method suggested to estimate the expected diameter of a ball bearing and its standard deviation from the data given in Table 2–11.

7. Repeat the computations of Problem 6 for Table 2–12.

8. An individual takes a bus every morning which is supposed to arrive at 8 A.M. For 15 days he determines the actual time of arrival of the bus and obtains the following results: 8:05, 8:10, 7:59, 8:15, 7:50, 8:02, 8:07, 8:01, 7:55, 7:59, 8:12, 8:05, 8:16, 7:56 and 8:03. Estimate the expected time of arrival and the standard deviation of the arrival time.

Section 3–8

1. Consider the data given in Problem 1 for Sections 2–12 and 2–13. A histogram was constructed for these data in the process of solving the problem. Does it appear reasonable from this histogram that one can use a normal distribution as a model for the resistance of any unit produced? What μ_x and σ_x should be used? On the same figure which contains the histogram plot the graph of $n(x; \mu_x, \sigma_x)$.

2. Re-solve Problem 1 for the data given in Problem 5 for Sections 2–12 and 2–13

Section 3–9

1. For each estimate of μ_x determined in Table 3–1 construct a confidence interval with $\delta = 0.1$. Represent each of these confidence intervals geometrically in the manner shown in Figure 3–9.

2. Re-solve Problem 1 using $\delta = 0.05$.

3. Construct a 95 percent confidence interval for each of the first five estimates of μ_x given in Table 3–1. Use the correct value of σ_x given in Section 3–7.

4. Re-solve Problem 3 using in each case the estimate of σ_x obtained from the same four numbers used to estimate μ_x.

5. Obtain a 99 percent confidence interval for the expected diameter of a ball bearing when the estimated diameter obtained from measurements on 10,000 ball bearings is 0.2500 and the estimate of the standard deviation is 0.00019.

6. The adjustment of machines which cut tin plate for cans is very critical. A machine has just been adjusted to produce the bottoms for a number 10 can. The diameters of the first 100 bottoms produced are carefully measured. The average diameter was 4.027 inches and $s_x = 0.034$ inches. Determine a 95 and 99 percent

confidence interval for the expected diameter of a can. To meet standards, the diameter of a bottom must lie between 3.978 and 4.091 inches. What is your estimate of the fraction of the bottoms produced that will be defective? What is a reasonable estimate as to what range of values this fraction defective may have? What assumption did you make?

7. A manufacturer of batteries for transistor radios tests 2000 of these to estimate the mean life for a battery. The estimate turns out to be 200 hours with $s_x = 50$ hours. Determine a 95 percent confidence interval for the expected life of a battery.

8. A producer of liquid detergents carefully determines the weight of liquid in 100 plastic bottles coming off the production line. The average weight in the 100 bottles is 17.43 ounces and $s_x = 0.21$ ounces. Determine a 99 percent confidence interval for the expected weight of liquid in a bottle. To meet government standards the weight of liquid in a bottle must be 16.95 ounces. What is your estimate of the probability that a government inspector, selecting a bottle at random, will get one that weighs less than 16.95 ounces? What assumption did you make?

9. To estimate the number of unemployed workers in Bogotá, an economist selected at random 1000 families in the working-class residential areas of Bogotá. He found that these families contained 1521 persons who claimed to be in the working class. Of these 284 were unemployed. Determine a 95 percent confidence interval for the proportion of the Bogotá work force which is unemployed.

10. A manufacturer of molded plastic parts inspects 1000 plastic knobs being produced on a given machine and finds that 87 are defective. Determine a 99 percent confidence interval for the long-run fraction defective.

11. The personnel manager of a firm surveys 150 of the 500 employees to determine their preference for reducing the length of the lunch hour and stopping work earlier. A total of 92 of those questioned were in favor of the change. Determine a 99 percent confidence interval for the proportion of all employees who favor the change.

12. A national magazine interviews 10,000 voters and finds that 5472 of them are going to vote for the Democratic candidate in the forthcoming presidential election. Determine a 95 percent confidence interval for the actual proportion of the voting population who will cast their votes for the Democratic candidate. What factors could make this confidence interval meaningless?

13. A manufacturer of radar tubes tests 50 of them to estimate the expected tube life. The average life of these tubes is 890 hours with $s_x = 120$ hours. Determine a confidence of length 50 hours for the expected life. What is the probability that this confidence interval contains the expected value?

14. A biologist weighs 10 salmon taken from an Alaskan stream in July and finds that the average weight is 10.2 pounds with $s_x = 3.1$ pounds. Determine a 95 and a 99 percent confidence interval for the expected weight of salmon in the stream during July.

15. A medical researcher tests a new drug on a total of 500 patients. He finds that 30 of them develop adverse side effects as a result of using the drug. Determine a 95 and a 99 percent confidence interval for the fraction of patients which will be adversely affected by the drug.

16. A firm wishes to estimate the fraction of the adult population which saw and could remember at least one of its commercials which were given on three different programs in the past week. One thousand households were selected at random. Of these 150 could remember at least one of the commercials. Determine a 95 and a 99 percent confidence interval for the fraction of the population which saw and had some recollection of the commercials.

17. What difficulties would be encountered in determining a confidence interval for the fraction of patients adversely affected by a new drug if the normal distribution could not be used to approximate the binomial?

Section 3–10

1. Simulate an experiment, call it $\mathfrak{R}_4$, which generates four two-digit random numbers from $n(x; 0, 1)$. Do this by first generating random numbers from the uniform distribution over the unit interval, and then, using Table E, convert them to random numbers from $n(x; 0, 1)$. Repeat $\mathfrak{R}_4$ a total of ten times and construct a table like Table 3–1. Use each trial of $\mathfrak{R}_4$ to obtain an estimate for the expected value of x. Determine s_x for each trial of $\mathfrak{R}_4$ and construct a 95 percent confidence interval for each estimate treating $t = (\bar{x} - \mu_x)\sqrt{n}/\sqrt{w}$ as normally distributed, that is, having the distribution $\varphi(t)$.

2. Determine the confidence intervals for Problem 1 treating $t = (\bar{x} - \mu_x)\sqrt{n}/\sqrt{w}$ as having a Student t distribution.

3. To estimate the expected life of very expensive nickel-cadmium batteries, the life of 15 of them is measured. The results of the experiment are $\bar{x} = 2005$ hours and $s_x = 215$ hours. Determine a 99 percent confidence interval for the expected life using the Student t distribution. What assumption is being made here?

4. A naturalist is interested in estimating the expected weight of black bears in Yellowstone Park during the fall months. Occasionally bears are caught to be tagged, and then it is sometimes possible to weigh them. This cannot be done frequently, however. Over a period of three years, seven bears were weighed in the fall yielding the following weights in pounds: 250, 400, 325, 375, 490, 275 and 295. Assume that the weight of the bears is normally distributed. Then determine a 95 percent confidence interval for the expected weight of the bears.

5. An anthropologist is interested in estimating the expected height of male adult bushmen. The bushmen appear only rarely in the Kalahari reserve of South Africa and in only very small numbers. The anthropologist succeeds in measuring the heights of eight different, apparently unrelated men during a two-year visit to the reserve. The heights in inches are: 52.0, 56.4, 53.1, 50.4, 51.2, 58.0, 55.0 and 53.7. Determine a 95 percent confidence interval for the expected height of a bushman. What assumptions need to be made in making the computations?

Section 3–11

1. It is desired to determine the mean resistance of the resistors produced by a production process. These are to be sold as 1000-ohm resistors. Previous experience indicates that the random variable x, representing the resistance of any given unit, is normally distributed with a standard deviation of 25 ohms. The resistance is

measured with an ohm-meter. The measurement process can be looked upon as a random experiment whose outcome yields a reading which is the value of a random variable y. Assume that y is normally distributed with the mean being the actual resistance and the standard deviation being 10 ohms. It is believed that the measurement error is completely independent of the value of the resistance being measured. It is desired that the probability that the estimate of μ_x will differ from μ_x by more than 10 ohms be 0.01 or less. How many resistors should be included in the sample? How should μ_x be estimated? How large would the sample need to be if there were no measurement errors?

2. Suppose that a voter, when asked what candidate he will vote for, will tell the correct candidate with probability 0.8 and the other candidate with probability 0.2. This is independent of which of the candidates he plans to vote for. If a sample of n individuals is selected at random and asked which candidate he will vote for, is it true that the portion in the sample who will vote for a given candidate is an unbiased estimator of the proportion of all voters who cast their ballot for that candidate? What is an unbiased estimator? Suppose that 10,000 individuals are interviewed and 5600 say they will vote for candidate A. What is your estimate of the proportion of the population that will vote for A?

3. For the situation discussed in Problem 2, it is desired to interview enough voters so that the probability that the estimate of the proportion of the voters who will vote for A differs by more than one percentage point from the true proportion is equal to 0.05. What should n be? What would n need to be if every voter told the truth? By how much has the sample size been increased?

4. The biologist weighing fruit flies believes that the standard deviation of fruit fly weight is about 0.05 grams. The balance used for weighing gives an unbiased estimate of the actual weight of any fly, but the standard deviation of the weights on repeated weighings is 0.005 grams. How many flies should be weighed if it is desired that the probability that the estimate differ from the true mean by more than 0.002 grams be 0.01?

Section 3–12

1. Can you offer any explanation as to why the *Literary Digest* estimate was so grossly in error? Try to analyze the various factors which might have had an effect.

2. Suppose that a firm wished to gain information by sending out questionnaires to customers. What sort of bias might be introduced by using data only on questionnaires that were returned?

3. It is desired to estimate the fraction defective produced by a process. What sort of bias could be introduced by selecting the entire sample at one point in time, taking all units as they are produced?

4. It is desired to estimate the output per man hour in assembly work for a company which works three shifts. The industrial engineering department sets up an elaborate procedure to estimate the output by observing at random the workers on the first shift. What sort of bias could be introduced by this procedure?

Section 3–13

1. The producer of canned soups referred to in the text plans to estimate the mean weight of tomato soup in a can by carefully weighing the contents of 1000 cans. The fraction of the total output produced by each machine is $\alpha_1 = 0.08$, $\alpha_2 = 0.12$, $\alpha_3 = 0.05$, $\alpha_4 = 0.15$, $\alpha_5 = 0.10$, $\alpha_6 = 0.10$, $\alpha_7 = 0.12$, $\alpha_8 = 0.08$, $\alpha_9 = 0.16$ and $\alpha_{10} = 0.04$. If nothing is known about the standard deviation of the weights on the various machines, how many cans should be selected from each machine? Suppose that it was known that the standard deviations are $\sigma_1 = 0.25$, $\sigma_2 = 0.10$, $\sigma_3 = 0.28$, $\sigma_4 = 0.21$, $\sigma_5 = 0.15$, $\sigma_6 = 0.18$, $\sigma_7 = 0.09$, $\sigma_8 = 0.14$, $\sigma_9 = 0.22$ and $\sigma_{10} = 0.17$ ounces. What should be the sample sizes? Suppose that the estimates of the expected weight per can for each of the machines are: $\bar{\xi}_1 = 10.72$, $\bar{\xi}_2 = 10.81$, $\bar{\xi}_3 = 10.68$; $\bar{\xi}_4 = 10.84$, $\bar{\xi}_5 = 10.75$, $\bar{\xi}_6 = 10.69$, $\bar{\xi}_7 = 10.83$, $\bar{\xi}_8 = 10.65$, $\bar{\xi}_9 = 10.77$, $\bar{\xi}_{10} = 10.79$ ounces. What then is the estimate of μ_x?

2. For the situation discussed in Problem 1 use the estimates of the means for the true means to determine approximately how much the standard deviation of the estimate of μ_x is reduced using stratified sampling rather than merely selecting 1000 cans coming off the conveyor. Make the computation both for the case where proportional allocation is used and where the theoretically best allocation is used. Is there a significant reduction in the standard deviation using the best allocation rather than the proportional allocation?

CHAPTER 4

Hypothesis Testing

4–1 INTRODUCTION

In this chapter we wish to discuss a subject referred to as hypothesis testing. Our interest in hypothesis testing will be in its usefulness as a tool for decision making. The types of decision problems that we shall be studying will be of a relatively simple type, but they nonetheless arise with considerable frequency. Decision theory is currently undergoing extensive changes. Hypothesis testing represents what is now referred to as the classical approach for applying statistics to decision problems. In the next chapter we shall study some of the newer techniques for treating decision problems.

4–2 HYPOTHESES

Very frequently, one must deal with random variables whose distribution is not known in complete detail. For example, suppose that a coin is tossed ten times and the number of heads is noted. If x is the random variable representing the number of heads obtained, then x should have the binomial distribution $b(x; 10, p)$, where p is the probability of a head. However, we can never really know what p is, and therefore the distribution of x is not completely known. As another example, consider a process which turns out shafts for electric motors. Let x be the random variable representing the diameter of any shaft produced. It is never possible to know the distribution of x; the best one can do is determine approximately what it is. Thus one might assume that to an adequate approximation x has a normal distribution with a given mean and variance. Any conjecture about the distribution of a random variable x is referred to in statistics as a

hypothesis. For example, one might make the hypothesis that x has a normal distribution, or the hypothesis that the value of p in $b(x; n, p)$ is 0.15.

Is there any way one can prove that a hypothesis about some distribution is correct or false? It is important to note that in general there is no way in which this can be done. By performing random experiments which in one way or another involve the distribution, data can be gathered which may strongly indicate that the hypothesis is false or true, but normally no procedure exists by which one could ever demonstrate incontrovertibly that the hypothesis is true or false. There are, however, exceptions to this statement. For example, let x be the random variable representing the number of giraffes in a sample of size n selected at random from the N giraffes in an African game reserve which are afflicted with a certain disease. Then we expect that x has the hypergeometric distribution $h(x; n, p, N)$, where p is the fraction of the giraffes in the reserve which have the disease. A hypothesis concerning the distribution of x might be that p has some specified value. This hypothesis could be proved or disproved by actually checking each giraffe in the reserve and determining the fraction which have the disease. This situation is unusual, however. There is no way, for example, to prove that the mean of the distribution for the diameter of motor shafts turned out by some production process is indeed one inch, or that the probability of a human male birth is exactly 0.514.

Generally, the procedure by which one attempts to check or "test" a hypothesis is to perform a random experiment which will provide some information concerning the validity of the hypothesis. Let us now illustrate by a simple example how a random experiment can be used to test a hypothesis. This example will point up an asymmetry in the testing procedure which is very frequently encountered. The experiment will have the characteristic that it may very well convince us that the hypothesis is false, but it cannot convince us that the hypothesis is true. What we have in mind here will become clear as the example is studied.

Suppose we make the hypothesis that a given coin is fair. We can think of this as making a hypothesis about the distribution of a random variable, since the number of heads x obtained on n tosses is a random variable having the binomial distribution $b(x; n, p)$, and we are making the hypothesis that $p = \frac{1}{2}$. Assume that this coin is tossed 100 times, and 98 heads are obtained. Although this could happen with a fair coin (the event has a positive probability of occurring), most individuals would feel that the outcome of this experiment provided sufficient evidence to abandon the hypothesis that the coin is fair. What is the justification for reaching this conclusion? It is not based on the fact that the probability of obtaining 98 heads is very small if the coin is fair, since the probability of obtaining 50 heads is also small. Rather it is based on the following type of reasoning. The expected value of x when $n = 100$ and $p = \frac{1}{2}$ is $\mu = 50$, and the

standard deviation of x is $\sigma = \sqrt{25} = 5$. For a μ of 50, $b(x; n, p)$ can be approximated reasonably well by the normal distribution, and we know that the probability that the value of x differs by more than σ from μ is about 0.3173 and that the value of x differs by more than 2σ from μ is about 0.0455 and that the value of x differs by more than 3σ from μ is about 0.003. Now $3\sigma = 15$. Hence, only about 3 times in 1000 would x be greater than 65 or less than 35. On the other hand, if $p = 0.80$, for example, the probability that $x > 65$ is very large. This type of reasoning leads us to suspect very strongly that $p > \frac{1}{2}$. The above discussion has thus illustrated how the outcome of an experiment could well lead an individual to abandon a particular hypothesis.

We might now ask if the outcome of an experiment of the type considered above could ever convince us that the hypothesis was correct. The answer here is generally no. If we obtained 50 heads on 100 tosses, does this convince us that $p = \frac{1}{2}$? There is no reason why it should, since obtaining 50 heads is quite consistent with having $p = 0.49$ or $p = 0.51$ or $p = 0.5052$ or $p = 0.47823$, for example. By performing more and more trials, we could progressively narrow down the range of p values which we felt were consistent with the outcome of the experiment, but it would never be possible to decide with reasonable assurance that one particular p value was indeed the probability of a head. This points up the asymmetrical nature of the experiment in giving information about or *testing* the hypothesis $p = \frac{1}{2}$. The outcome of the experiment may well convince us that the hypothesis is not true, but it cannot convince us that it is true. This is the sort of situation normally encountered. The reader may wonder at this point what is to be gained by formulating a hypothesis and then testing it by performing an experiment. We shall return to this question shortly, after discussing the procedure for testing hypotheses in a little more detail.

4–3 TESTING HYPOTHESES

It is convenient to subdivide hypotheses into several different categories. An hypothesis concerning some parameter of a distribution, such as the mean or variance, is called a *parametric hypothesis*. A parametric hypothesis which states that the parameter has a specific value, say $\mu = 2$, is called a *simple hypothesis*. A parametric hypothesis which states that the parameter may have one of several possible specified values, or lies in some given interval, say $2 \leq \mu \leq 4$, is called a *composite hypothesis*. A hypothesis which makes a statement about the form of the distribution rather than about a parameter is called a *nonparametric hypothesis*. Thus the hypothesis that x is normally distributed would be what is called a nonparametric hypothesis.

A procedure whereby a random experiment is performed, and on the

basis of the outcome of this experiment we decide whether or not to reject some given hypothesis, is referred to as a *statistical test* of the hypothesis. The hypothesis being tested is frequently referred to as the *null hypothesis*. The reason for this is that one is frequently trying to decide which of two or more hypotheses seems the most reasonable, and the one being tested is given a special name to distinguish it from the others. We noted in the previous section that an experiment may allow us to reject a hypothesis, but it will seldom be of such a nature that it will convince us that the hypothesis is true. In spite of this observation, it is usually the case in practice that if we do not reject a hypothesis as a result of a statistical test, then we proceed as if we accepted the hypothesis as true, even though we may suspect or even know that it is not true. The rationale behind this will appear below. Thus one frequently finds in statistics books the statement that as a result of testing a hypothesis we either accept or reject the null hypothesis. This is not true at all. As a result of the experiment, we may reject the hypothesis, or we may decide that the outcome of the experiment does not contradict the hypothesis. In the latter case we may then take the same action as if the hypothesis were true. This is not the same as accepting the hypothesis to be true.

Let us now formalize the procedure of testing an hypothesis, call it H_0. The procedure is basically the same regardless of the type of hypothesis being tested. To test H_0 we perform a random experiment $\Re$, which yields a value ζ of some random variable y. Based on the value ζ we either reject H_0 or we do not reject H_0 (and this is frequently referred to as accepting H_0). The random variable y is referred to as the *statistic* being used to test the hypothesis. Generally speaking, y is selected so that as the value of y gets larger and larger (or smaller and smaller), the truth of the hypothesis becomes more and more unlikely. The basic decision to be made is how large must ζ, the observed value of y, be if it is to be considered to be large enough to reject the hypothesis.

This decision is usually made as follows. It will be assumed that if H_0 is true the distribution for y is determined. The problem then reduces to one of deciding whether or not ζ, the observed value of y, is consistent with the distribution for y determined by H_0. If ζ is too large, it will be rather unlikely that y does have the distribution determined by H_0, and in this case H_0 is rejected. Denote the probability that $y \geq \zeta$ when H_0 is true by $Q(\zeta)$. The procedure for determining whether ζ is large enough to reject H_0 is then to select a number α, and to reject H_0 if $Q(\zeta) \leq \alpha$. In other words, if the probability that $y \geq \zeta$ is less than or equal to α, then the hypothesis is rejected. Otherwise the hypothesis is not rejected. The critical probability α is called the *significance level* of the test. How is α to be selected? The theory provides no rules for doing this. Generally, and arbitrarily, α is often selected either to have the value 0.01 or 0.05. If it turns out that ζ, the

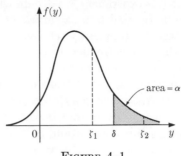

FIGURE 4–1

observed value of y, is such that $Q(\zeta) \leq \alpha$, then the test is said to be *significant*. When the test is significant, the null hypothesis is rejected.

The situation just described is illustrated geometrically in Figure 4–1, where the graph of the density function for y when H_0 is true is shown. The shaded region will be imagined to have area α. If the value of y observed is ζ_1, the area to the right of ζ_1 is greater than α, and in this case the hypothesis would not be rejected. However, if $y = \zeta_2$, the area to the right of ζ_2 is less than α. In this case the test is significant, and the null hypothesis is rejected. We can now note that when y is a continuous random variable, there is a unique value of y, call it δ, such that $Q(\delta) = \alpha$. Given α, the value of δ is determined. However, once δ is known, it is very simple to decide whether the test is significant or not. If $\zeta \geq \delta$, the test is significant, and H_0 is rejected. If $\zeta < \delta$, H_0 is not rejected. If the distribution of y is discrete, there may not be any value δ of y such that $Q(\delta) = \alpha$. There will, however, always be a smallest value of y, call it δ_1, such that $Q(\delta_1) \leq \alpha$. Thus if $\zeta \geq \delta_1$, H_0 is rejected and if $\zeta < \delta_1$, H_0 is not rejected. The number δ or δ_1 is frequently referred to as the *critical value* of the statistic y.

This explains how one proceeds to test an hypothesis using some specified experiment $\Re$. Whether or not the test yields a significant result depends, of course, on the arbitrarily chosen value for α. As we have indicated above, α is generally chosen to have the value 0.01 or 0.05. We might also observe that when y is a continuous random variable, then α is the probability that H_0 will be rejected if, in fact, H_0 is true. If y is discrete, $Q(\delta_1)$ is the probability that H_0 will be rejected if H_0 is true. Normally, this will be very close to α, so that in either case α can usually be looked upon as the probability of rejecting H_0 when H_0 is true.

We have in this section given a general discussion of the way in which an hypothesis is tested. In the next several sections we shall expand on this by considering a variety of specific cases which arise rather frequently. Before turning to these specific cases, let us pause to ask once again the question: Why would one be interested in testing an hypothesis? Con-

ceivably it could be merely scientific curiosity, or someone may have constructed a model for the random variable x and we wish to test experimentally some hypothesis made in constructing the model. For problems that will be of interest to us, however, there will be a very practical reason for carrying out a test of a hypothesis. Hypothesis testing will provide an approach to treating certain types of *decision problems*. In general terms we shall be considering a situation where a decision maker must choose one of precisely two alternative courses of action. The appropriate course of action will depend on the value of some parameter associated with a random variable. The possibility exists of performing a random experiment to obtain information about the value of the parameter. It is to this type of situation that the theory of hypothesis testing can be applied to aid the decision maker in selecting an action. The problem is thought of as one in which a hypothesis is being tested. If as a result of the experiment the hypothesis is rejected, then one of the actions will be taken. If the hypothesis is not rejected the other action will be selected. This general description will no doubt seem a little vague to the reader, but it will become much clearer when concrete situations are studied.

4–4 TESTING A MEAN

Let us begin with an example. The management of a firm which produces electric light bulbs is currently studying the claim of an engineer that he has developed a new filament which will increase the expected life of the bulb. The management would like to decide whether it seems worthwhile to consider manufacturing the new filament or whether the filament does not seem to represent a significant improvement over the one currently in use, thus implying that the project will be dropped or more research will be needed. To reach a decision, a number of light bulbs using the new filament will be tested, and the average life of these bulbs will be compared with the known expected life of bulbs already in existence. This procedure, whereby one of two alternative actions is selected depending on the outcome of a random experiment, can be looked upon as a situation in which a hypothesis is being tested. If the hypothesis is rejected, one action is taken, and if it is not rejected the other action is taken. Furthermore, the hypothesis-testing concept can be used to provide a clearly defined decision rule for relating the action to be taken to the outcome of the experiment. Let us now see how this can be done.

To be specific, suppose the expected life μ_0 of bulbs using the current filament is known, as is σ, the standard deviation of the life. In fact, $\mu_0 = 1000$ hours and $\sigma = 100$ hours. The precise claim that the engineer makes is that μ_1, the expected bulb life for the new filament, is greater than μ_0, and, in addition, the standard deviation of the life for the new filament is

essentially the same as that for the current filament. What is the hypothesis to be tested? There exists a whole array of hypotheses which might conceivably be tested. For example, the management might try to test the hypothesis that $\mu_1 = 1100$ or $\mu_1 = 1200$ or $\mu_1 = 1250$. The important thing to note here is that what the management wishes to ascertain is whether the new filament represents any significant improvement over the old one. The way this would typically be done would be to take H_0 to be the hypothesis that there is no difference between the new filament and the one currently in use, that is, the distribution of the bulb life for the new filament is precisely the same as for the current filament. In particular, the hypothesis H_0 implies that $\mu_1 = \mu_0 = 1000$. When H_0 is selected in this way, there is no need to attempt to guess what a reasonable value for μ_1 might be. If on performing the experiment the test is significant and H_0 is rejected, then the management will presumably agree that the new filament looks sufficiently good that the possibility of using it should be studied. If the test is not significant, it will then be concluded that the new filament does not appear to be a sufficiently significant improvement over the current one to justify consideration of any change at the present time.

Having decided what H_0 should be, let us now proceed to consider the experiment to be used in testing H_0. Suppose that n bulbs containing the new filament are tested and the time until burn-out of each is determined. Since we are concerned here with the expected life of the bulbs using the new filament, the logical statistic to use to test H_0 is the random variable $\bar{x}$, the average life of the n bulbs tested. Recall that $\bar{x}$ can be written as in (3–2), and $\bar{x}$ is an unbiased estimator of μ_1. It now seems intuitively obvious that if the observed value of $\bar{x}$ is large enough, then the new filament definitely seems better than the current one and H_0 should be rejected, while if $\bar{x}$ is close to 1000 then H_0 should not be rejected. In other words, there should be a number δ such that if $\bar{x} \geq \delta$, H_0 is rejected, while if $\bar{x} < \delta$ then H_0 is not rejected (and this is frequently referred to as accepting H_0). Here, then, management has a decision rule which tells it which action to take for every possible outcome of the experiment. The situation is illustrated graphically in Figure 4–2.

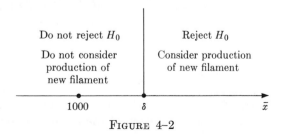

FIGURE 4–2

The task which yet remains is that of determining the numerical value of δ. The method for doing this was described in general terms in the previous section. Suppose that H_0 is true, so that $\mu_1 = \mu_0 = 1000$. Furthermore, the standard deviation of the life of the new filament will be the same as that for the current filament, that is, 100 hours. Thus if the number of bulbs tested with the new filament is reasonably large, then when H_0 is true, $\bar{x}$ will be essentially normally distributed with mean 1000 and standard deviation $100/\sqrt{n}$. Now the management is rather conservative, and does not want to move toward production of the new filament unless the evidence indicating that it is better is very convincing. This is expressed in the way the significance level α is chosen. Recall from the previous section that the test of a hypothesis is normally carried out in such a way that if the probability that y is larger than the observed value ζ of y is less than α then the hypothesis is rejected. Otherwise the hypothesis is not rejected. When α is made small, then the hypothesis will not be rejected unless the evidence is quite convincing that the observed value of y is not consistent with H_0 being true. As was noted in the previous section, α is often selected to be 0.05 or 0.01.

Suppose that the management selects $\alpha = 0.01$. Let us determine the value δ of $\bar{x}$ such that the probability that $\bar{x} \geq \delta$ is α when H_0 is true. Since $\bar{x}$ is essentially normally distributed, the probability that $\bar{x} \geq \delta$ is

$$1 - \Phi\left(\frac{\delta - 1000}{100/\sqrt{n}}\right)$$

when H_0 is true. Thus we want

$$1 - \Phi\left(\frac{\delta - 1000}{100/\sqrt{n}}\right) = 0.01 \quad \text{or} \quad \Phi\left(\frac{\delta - 1000}{100/\sqrt{n}}\right) = 0.99 . \qquad (4\text{--}1)$$

The value of t, call it λ, such that $\Phi(\lambda) = 0.99$ is, from Table E, $\lambda = 2.33$. Therefore

$$\frac{\delta - 1000}{100/\sqrt{n}} = 2.33 \quad \text{or} \quad \delta = 1000 + \frac{233}{\sqrt{n}} . \qquad (4\text{--}2)$$

We have thus shown how δ can be determined for every value of n, the number of bulbs tested, when a significance level of 0.01 is used. The manner in which δ varies with n is shown in Figure 4–3. Suppose, for example, $n = 1000$. Then

$$\delta = 1000 + \frac{233}{\sqrt{1000}} = 1007.4 .$$

Thus if 1000 bulbs are tested, then when using a one percent significance level ($\alpha = 0.01$), the null hypothesis is rejected if $\bar{x} \geq 1007.4$ and is not

rejected otherwise. Note that to determine the decision rule, that is, the value of δ, it was necessary to specify the significance level α for the test and the sample size n.

Imagine now that 1000 bulbs are tested and $\bar{x} = 1020$ hours. In this case $\bar{x} > \delta$ and hence the null hypothesis would be rejected. What does it mean to reject the null hypothesis? It means that the management would feel that the distribution of the bulb life for the new filament is not one with a mean of 1000 and a standard deviation of 100. Thus the management would presumably conclude that $\mu_1 > 1000$. Note, however, this is not necessarily a justifiable conclusion. If it happened that $\mu_0 = \mu_1$, but the standard deviation of the life of the bulbs using the new filaments has been increased greatly, then it is indeed correct to reject H_0, but not to conclude that $\mu_1 > \mu_0$. The rejection of H_0 implies that it is believed that something has changed. To conclude that $\mu_1 > \mu_0$ requires an addi-

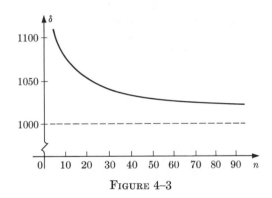

FIGURE 4–3

tional assumption about the sort of changes that can occur. In this case the data generated from the experiment could be used to estimate the standard deviation for the new filament. If this turns out to be essentially 100 hours, then it certainly seems likely that $\mu_1 > \mu_0$. However, if the estimate of the standard deviation is much greater than 100 hours, there is much less certainty that $\mu_1 > \mu_0$.

Let us now look at the test procedure just outlined from another point of view. Suppose that the management tells the engineer that they will test 1000 bulbs using the new filament, and if $\bar{x} \geq 1007.4$ they will agree that the new filament is better, while if $\bar{x} < 1007.4$ they will feel that the improvement, if any, is not significant enough to bother about. The engineer is now interested in knowing, given management's decision rule, the probability that $\bar{x} < 1007.4$ if the expected life is μ_1 and the standard deviation 100 hours, that is, the probability that management will reject

the new filament in this case. The probability that $\bar{x} < 1007.4$ when the mean is μ_1 is

$$p(\mu_1) = \Phi\left(\frac{1007.4 - \mu_1}{3.16}\right), \tag{4-3}$$

and $p(\mu_1)$ is the probability that management will reject the new filament and act as if H_0 is true, when in fact the mean for the new filament is $\mu_1 > \mu_0$. A curve showing how $p(\mu_1)$ varies with μ_1 is shown in Figure 4–4. This curve is called the *operating characteristic curve* for the test. This curve gives the probability of accepting H_0 when the expected filament life is μ_1. It will be noted that if $\mu_1 > 1014$ is almost certain that the management will agree that the new filament is an improvement over the one currently in use.

The example just discussed in some detail is illustrative of a type of

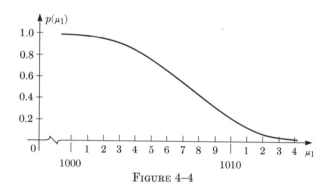

FIGURE 4–4

problem which arises with considerable frequency. A decision maker wishes to select one of two alternative actions, call them a_1 and a_2. The proper action to take will depend on whether or not the expected value of some random variable x is significantly greater than a value μ_0. If μ_x is not significantly greater than μ_0, then a_1 is the action to take, whereas if μ_x is significantly greater than μ_0, then a_2 is the action to take. The decision maker has the opportunity to perform an experiment $\Re$ to gain information about μ_x before making a decision. To decide what action to take the decision maker will use a decision rule of the following type. If $\bar{x} < \delta$, take action a_1. If $\bar{x} \geq \delta$, take action a_2. The basic problem, then, is to determine the critical value δ. There are many ways that δ could be selected. The procedure which we have been studying here is that where $\Re$ is looked upon as testing the hypothesis H_0 that $\mu_x = \mu_0$.

To test H_0, one must be able to determine the distribution of the statistic being used when H_0 is true. The logical statistic to use here is $\bar{x}$. If σ_0, the

standard deviation of x, is assumed known, then $\bar{x}$ is essentially normally distributed with mean μ_0 and standard deviation $\sigma_0/\sqrt{n}$ when H_0 is true and the sample size n is reasonably large. To determine δ for a given n, the significance level α for the test is first selected; α is the probability that H_0 will be rejected if, in fact, H_0 is true. Frequently, α is chosen to be 0.05 or 0.01. With α so small, this implies that the decision maker definitely wishes to avoid taking action a_2 when H_0 is true. Given α, δ is determined from

$$\Phi\left(\frac{\delta - \mu_0}{\sigma_0/\sqrt{n}}\right) = 1 - \alpha. \tag{4-4}$$

Here then, we have described in detail how a decision rule can be selected using the theory of hypothesis testing.

4-5 TWO-SIDED TEST OF A MEAN

The type of decision problem we considered in the last section was one where action a_1 should be selected if μ_x, the expected value of x, is close to μ_0, and a_2 should be selected if μ_x is significantly greater than μ_0. Precisely the same sort of procedure can be used, of course, if a_1 should be selected when μ_x is close to μ_0 and a_2 when μ_x is significantly less than μ_0. In this case the decision rule would have the form: take a_2 if $\bar{x} \leq \delta$ and take a_1 if $\bar{x} > \delta$. We now wish to consider a slightly different sort of situation. Assume once again that a decision maker wishes to select one of two possible actions a_1 and a_2, and that the appropriate action to be selected depends on the value of μ_x, the expected value of x. Suppose now, however, that action a_2 should be selected whenever μ_x deviates significantly from a value μ_0. Thus action a_2 should be taken if μ_x is significantly larger or significantly smaller than μ_0. Action a_1 should be taken when μ_x is essentially equal to μ_0. Once again, an experiment $\Re$ can be performed before making a decision to obtain information about μ_x. As before, it seems logical that the decision maker will use the value of $\bar{x}$ to determine the appropriate action to take. Now, however, the decision rule will have this form: Take action a_2 if $\bar{x} \geq \delta_2$ or $\bar{x} \leq \delta_1$ and take action a_1 if $\delta_1 < \bar{x} < \delta_2$. This is illustrated graphically in Figure 4-5.

One way to determine the critical values δ_1 and δ_2 is to visualize the

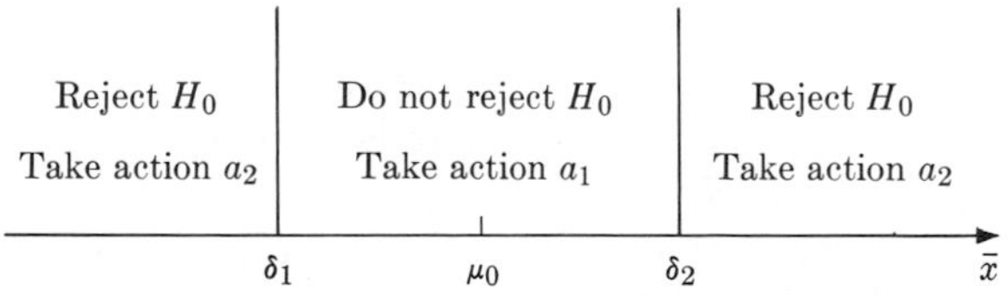

Reject H_0	Do not reject H_0	Reject H_0
Take action a_2	Take action a_1	Take action a_2

$$\delta_1 \qquad \mu_0 \qquad \delta_2 \qquad \bar{x}$$

FIGURE 4-5

problem as one in which a hypothesis H_0 that $\mu_x = \mu_0$ is being tested. Suppose that the standard deviation of x is σ_0 when $\mu_x = \mu_0$. Then when H_0 is true, $\bar{x}$ has a normal distribution (approximately) with mean μ_0 and standard deviation $\sigma_0/\sqrt{n}$. We wish to reject H_0 when $\bar{x}$ deviates too much from μ_0. A reasonable way to make the decision is to reject H_0 when $\bar{x}$ deviates by more than d from μ_0, that is, when $\bar{x} \leq \mu_0 - d$ or $\bar{x} \geq \mu_0 + d$. Then $\delta_1 = \mu_0 - d$ and $\delta_2 = \mu_0 + d$. To determine d, the probability α of rejecting H_0 when H_0 is in fact true is specified; α is again called the significance level. The situation then reduces to that shown in Figure 4–6. We have noted previously that the area to the left of $\mu_0 - d$ is the same as the area to the right of $\mu_0 + d$ for the normal distribution. Since these areas must sum to α, each has the value $\alpha/2$. Thus, when considering the area lying to the right of $\mu_0 + d$ we see that d must satisfy the equation

$$1 - \Phi\left(\frac{\mu_0 + d - \mu_0}{\sigma_0/\sqrt{n}}\right) = \frac{\alpha}{2} \tag{4-5}$$

or

$$\Phi\left(\frac{d\sqrt{n}}{\sigma_0}\right) = 1 - \frac{\alpha}{2}. \tag{4-6}$$

If λ is the value of $\Phi(t)$ such that $\Phi(\lambda) = 1 - (\alpha/2)$, then

$$\frac{d\sqrt{n}}{\sigma_0} = \lambda \quad \text{or} \quad d = \frac{\lambda\sigma_0}{\sqrt{n}} \tag{4-7}$$

and

$$\delta_1 = \mu_0 - \frac{\lambda\sigma_0}{\sqrt{n}}; \quad \delta_2 = \mu_0 + \frac{\lambda\sigma_0}{\sqrt{n}}. \tag{4-8}$$

The reader should note that the decision rule developed here is symmetric in that δ_1 and δ_2 are equally spaced about μ_0. If it was much more important to take action a_2 when $\mu_x < \mu_0$ than when $\mu_x > \mu_0$, then it might be desirable to have δ_1 closer to μ_0 than δ_2. Then δ_1 and δ_2 would be determined separately by specifying separately the area to the left of δ_1 and that to the right of δ_2, as shown in Figure 4–7. We shall not consider the case in

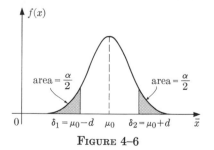

FIGURE 4–6

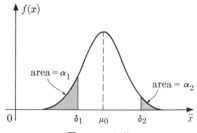

FIGURE 4–7

detail since it is seldom used when the hypothesis-testing approach to decision problems is employed.

EXAMPLE. Let us now give a practical example illustrating how the theory developed in this section might be applied. A manufacturer of cans is about to start a long run on a particular type of sanitary can on one of the lines. The machine which rolls the side of the can into a cylindrical shape, forms the seam on the side of the can and solders the seam has now been adjusted. It is essential that the expected diameter of the can be controlled very carefully at the value 2.63 inches, since an expected diameter which is significantly larger or smaller than this value will cause problems in the machines that place the bottoms on the cans. The plant manager would like to know whether to accept the set-up and proceed with the run or whether additional adjustments are needed. To check on the adjustment, 25 cans are run through the machine and the diameter of each is measured. The average diameter of the 100 cans is 2.66 inches and $s_x = 0.05$ inches. This estimate of the standard deviation is very close to the standard deviation that has been obtained in the past on cans of this sort.

To derive a decision rule which will allow him to select the action to take in this case, the plant manager decides to look at the problem as one of testing the hypothesis that the expected diameter of a can is 2.63. The value of n is 25 here. All that remains for him to do is to select the significance level α. It is very time-consuming to adjust the machine, and thus the plant manager does not want to reject H_0 if it is true. Furthermore, the man who did the set-up is very reliable and thus the manager is reasonably confident that it is satisfactory. After some introspection, the manager decides to use $\alpha = 0.05$. Then $1 - (\alpha/2) = 0.975$. If $\Phi(\lambda) = 0.975$, then from Table E, $\lambda = 1.96$, so from (4–7)

$$d = \frac{1.96(0.05)}{5} = 0.02 ,$$

so that

$$\delta_1 = 2.63 - 0.02 = 2.61 \quad \text{and} \quad \delta_2 = 2.63 + 0.02 = 2.65 .$$

The value $\bar{x} = 2.66 > \delta_2$, so the test is significant, and hence according to his decision rule the plant manager should require that the machine be readjusted.

In order to investigate in a little more detail the nature of his decision rule, the plant manager computes for various values of μ_1, $p(\mu_1)$, the probability that H_0 will be accepted (not rejected) when the expected diameter of a can is μ_1. This is the probability that $2.61 \leq \bar{x} \leq 2.65$, which is

$$p(\mu_1) = \Phi\left(\frac{2.65 - \mu_1}{0.01}\right) - \Phi\left(\frac{2.61 - \mu_1}{0.01}\right). \tag{4–9}$$

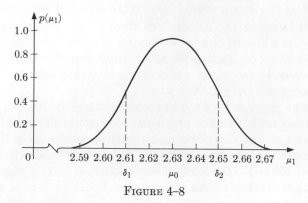

FIGURE 4–8

The curve showing how $p(\mu_1)$ varies with μ_1 is shown in Figure 4–8. This curve is called the *operating characteristic curve* for the test. After examining this curve the production manager feels that it exhibits roughly the type of behavior he wants, and thus he is satisfied that the decision rule should be satisfactory.

4–6 USE OF STUDENT t DISTRIBUTION

In testing an hypothesis H_0 that $\mu_x = \mu_0$, it is frequently the case that σ_0, the standard deviation of x, is not known and that the estimate s_x of σ_0 obtained from the experimental data must be used for σ_0. When the sample size n is fairly large, as is often the case, it is quite satisfactory to treat $\bar{x}$ as being normally distributed with mean μ_0 and standard deviation $s_x/\sqrt{n}$ if H_0 is true, so that $t = (\bar{x} - \mu_0)\sqrt{n}/s_x$ has the distribution $\phi(t)$ in this case. In reality, t is a function of two random variables $\bar{x}$ and w, where w is the statistic (3–34) used to estimate σ_0, so that $t = (\bar{x} - \mu_0)\sqrt{n}/\sqrt{w}$. When n is small, t will not have a distribution which is very close to $\phi(t)$, and in general it is difficult to determine the distribution of t. However, if it is known that x is normally distributed, then t has a Student t distribution with $n - 1$ degrees of freedom, as we noted in Section 3–10. This result can be conveniently made use of when the sample size n is small and it is believed that x (not $\bar{x}$) is close to being normally distributed. It is only necessary to use the table for the t distribution rather than the table for the normal distribution when making the computations. Let us illustrate the procedure with an example.

EXAMPLE. The National Aeronautics and Space Administration (NASA) is trying to determine if the expected life of a receiving unit used in satellites has met contract specifications. If it appears that the contract specification has been met then a new contract will be given to the subcontractor. If it does not seem to have been met, then NASA will consider selecting a new

subcontractor. The contract required that the expected life of the unit be at least 10,000 hours in the actual operating environment of outer space. NASA has only a limited amount of data on which to base its decision. The lifetimes of the unit during five different satellite flights have been measured and the average length of operation of the receiving unit on these flights was 8000 hours with $s_x = 2000$ hours.

The decision problem here can be formulated as a problem in hypothesis testing, just as we have done previously. The problem of deciding if the minimum contract specification of an expected operating life of 10,000 hours has been met can be viewed as a problem of testing the null hypothesis that the expected life is 10,000 hours at some specified level of significance α. Since n is quite small here and since s_x must be used for σ_x, it appears appropriate to treat $t = (\bar{x} - 10,000)\sqrt{n}/\sqrt{w}$, where w is the random variable used to estimate the variance, as having a Student t distribution. The NASA scientists do not know what the precise distribution of x is like, but they feel that it is essentially normal. If this is true, then it is justifiable to use the t distribution. We shall do this.

We wish here to reject the null hypothesis if t is small enough (small values of t are significant here—not large values). Let λ be the value of t such that the probability of the event $t \leq \lambda$ is α. Then the null hypothesis will be rejected if $t \leq \lambda$, that is, if

$$\frac{(\bar{x} - 10,000)\sqrt{n}}{\sqrt{w}} \leq \lambda$$

or

$$\bar{x} \leq 10,000 + \frac{\lambda\sqrt{w}}{\sqrt{n}}.$$

Whether or not H_0 is rejected depends not only on the value of $\bar{x}$ but on w as well. In this case there is no number δ such that H_0 is rejected if $\bar{x} \leq \delta$ and not otherwise.

For the case under consideration $n = 5$ and $\sqrt{w} = s_x = 2000$. The NASA official does not want to consider a new subcontractor unless it is quite clear that the contract has not been fulfilled. Thus, on examining the α values available in the tables, he suggests using $\alpha = 0.025$. The value of λ such that the probability that $t \leq \lambda$ is 0.025 will be negative. Recall that the area to the left of $-t$ is the same as the area to the right of t for the t distribution. Therefore if we determine the number λ_1 such that the area to the right of λ_1 is 0.025 then $\lambda = -\lambda_1$. The degrees of freedom here is $\nu = 5 - 1 = 4$. From Table G, for $\nu = 4$ and $\alpha = 0.025$ we read $\lambda_1 = 2.78$. Hence $\lambda = -2.78$ and

$$10,000 + \frac{\lambda\sqrt{w}}{\sqrt{n}} = 10,000 - 2.78(817) = 7730.$$

Now $\bar{x} = 8000$ is not less than or equal to 7730, and so the test is not significant in spite of the fact that the observed average life of 8000 hours is considerably less than the minimum expected life of 10,000 hours specified in the contract. Hence the decision rule would indicate that the subcontractor should be given a follow-on contract. The reader might check that if $\alpha = 0.05$, then $\lambda = -2.13$ and $10,000 + (\lambda\sqrt{w}/\sqrt{n}) = 8260$. In this case the test would be significant. This example is one where the performance of the equipment appears to be borderline with respect to the contract specification and the action to take depends critically on the value of α selected, for α in the range under consideration.

4–7 TESTING A DIFFERENCE OF TWO MEANS

A type of problem slightly different from those we have considered thus far is one where it is of interest to decide if there is any significant difference in quality between two products. We shall restrict our attention to cases where the quality of each of the products can be characterized by the expected value of some random variable. For example, a consumer testing organization is interested in evaluating two competing brands of transistor batteries used in radios and clocks. The quality of such a battery can be conveniently measured by its expected life. The consumer testing organization would like to determine whether the expected life of one of the brands appears to be significantly greater than the other. Let us now show how the theory of hypothesis testing can be applied here to help answer the question of interest.

Suppose that to test the quality of the two competing batteries, call them 1 and 2, n_1 of type 1 and n_2 of type 2 are installed in a given type of transistor radio, the radios are all operated in a specified manner (a given number of operating hours per day) and the useful lifetime of each battery is determined. These are averaged for each brand to yield the average number of operating hours for each. To be completely general we shall not necessarily assume that the same number of each brand is tested. Let μ_1 and μ_2 be the expected lifetimes of batteries 1 and 2 in operating hours, and σ_1 and σ_2 the corresponding standard deviations of the operating life. Presumably, σ_1 and σ_2 will not be known before performing the experiment. Denote by $\bar{x}$ the random variable representing the average life of the n_1 batteries of type 1 tested and by $\bar{y}$ the random variable representing the average life of the n_2 batteries of type 2 tested. Then $\mu_{\bar{x}} = \mu_1$ and $\mu_{\bar{y}} = \mu_2$, $\sigma_{\bar{x}} = \sigma_1/\sqrt{n_1}$ and $\sigma_{\bar{y}} = \sigma_2/\sqrt{n_2}$. Furthermore, if n_1 and n_2 are reasonably large, $\bar{x}$ and $\bar{y}$ are essentially normally distributed.

Intuitively, it is clear that if on performing the experiment just described it turns out that the value of $\bar{x}$ is significantly greater than that of $\bar{y}$, then it should be concluded that brand 1 is better than brand 2, whereas if the

there is no improvement is significant at the one percent level they should plan on a large-scale test. Otherwise, they should be more cautious and continue on a small scale for awhile.

If there is no improvement over the current drug then the probability that any particular affected individual should respond is 0.50 and the probability that r individuals out of n treated will respond favorably is the binomial probability $b(r; n, 0.50)$. For $n = 10$, the random variable x representing the number of favorable responses can be moderately well approximated by the normal distribution since $p = 0.50$. However, we can also make the computation exactly using binomial distribution since it is tabulated for $n = 10$ and $p = 0.50$ in Table C. We shall carry out the computations in both ways and then compare the results.

Consider the normal approximation in which x is treated as a continuous random variable first. When H_0 is true

$$\mu_x = np = 5; \quad \sigma_x = \sqrt{2.5} = 1.58 \ .$$

H_0 is to be rejected if x is too large. If $1 - \Phi(\lambda) = \alpha$ or $\Phi(\lambda) = 0.99$, then $\lambda = 2.33$. The critical number δ is then determined from

$$\frac{\delta - 5}{1.58} = 2.33 \quad \text{or} \quad \delta = 5 + 1.58(2.33) = 8.68 \ .$$

Now $x = 9 > 8.68$ and thus, based on the normal approximation, it is to be concluded that the test is significant and the move toward a large-scale test seems justified according to the criterion selected.

Let us now use the binomial distribution. In this case there is no number δ such that the probability that $x \geq \delta$ is precisely 0.01. However, there is a smallest value of x, call it δ_1, such that the probability that $x \geq \delta_1$ is less than or equal to 0.01. From Table C we see that the probability that $x \geq 9$ is 0.0108 and that $x \geq 10$ is 0.0010. Thus, strictly speaking $\delta_1 = 10$ and the test is not significant. However, 0.0108 is so close to 0.01 that the firm would no doubt be willing to use $\delta_1 = 9$ and then the test again becomes significant. It is just on the borderline, however.

2. Figure 1–3 shows the results obtained from tossing a quarter 100 times. A total of 59 heads was obtained in the 100 tosses, and we suggested earlier that most people would not feel this coin was fair, although some might not feel that obtaining as many as 59 heads was serious enough to abandon the notion that the coin was fair. Let us now look at the problem of deciding whether to reject the notion that the coin is fair as one of hypothesis testing. Let H_0 be the hypothesis that the coin is fair. Then when H_0 is true, the random variable x representing the number of heads obtained will have a binomial distribution with

$$\mu_x = 50 \quad \text{and} \quad \sigma_x = \sqrt{25} = 5 \ .$$

The normal approximation to the binomial should be quite good here, and hence we shall treat x as a continuous random variable which is normally distributed with $\mu_x = 50$ and $\sigma_x = 5$.

In this case we wish to reject H_0 is x is either too large or too small. Let us test the hypothesis at the five percent significance level. We shall then determine the number d such that the probability that x deviates by more than d from 50 is 0.05. Thus

$$\Phi\left(\frac{d}{5}\right) = 1 - \frac{\alpha}{2} = 0.975 \,,$$

and from Table E

$$\frac{d}{5} = 1.96 \quad \text{or} \quad d = 9.80 \,,$$

so that $\delta_1 = 50 - 9.8 = 40.2$ and $\delta_2 = 50 + 9.8 = 59.8$. Now $x = 59 < \delta_2$, and the test is not significant at the five percent level, although it is very close to being significant. Thus if a person does not want to reject the hypothesis of a fair coin unless there is very substantial evidence that it is not fair, he would not reject H_0 in this case. On the other hand, anyone who was willing to use a significance level of any value slightly greater than 0.05 would reject the null hypothesis H_0.

3. Suppose that a fertilizer manufacturer has brought out a new fertilizer which he claims gives an increased yield of corn as compared to that of his standard fertilizer. A farmer wishes to perform an experiment to see if this claim appears to be valid. One way to look at the farmer's problem is as one of testing an hypothesis. The null hypothesis is made that there is no difference between the new and standard fertilizers. As a result of the experiment, the farmer either rejects the null hypothesis (and hence presumably concludes that the new fertilizer is better) or he does not reject the null hypothesis (and in this case presumably concludes that there is no substantial evidence indicating that the new fertilizer is better). Let us imagine then that the farmer does proceed in this manner and decides to test the null hypothesis at the five percent significance level.

We shall now examine the situation in more detail. Imagine that based on his previous experience, the farmer believes that the yield per acre for the standard fertilizer can be looked upon as a random variable x with a normal distribution having mean μ and standard deviation σ. The mean μ varies from year to year depending on weather conditions, but σ seems to remain essentially constant at 111 bushels per acre. According to the null hypothesis then, the random variable y representing the yield per acre when the new fertilizer is used will have the normal distribution $n(y; \mu, 111)$. The important thing to note at this point is that the farmer cannot

proceed by simply raising corn on several acres treated with the new fertilizer, computing $\bar{\zeta}$ the average yield per acre and rejecting the hypothesis if $\bar{\zeta}$ is too large. This could be done if μ were known and never changed from year to year. However, it has been assumed that μ does change. The farmer runs into a certain difficulty here which is typical of problems which involve comparison of two treatments. He would like to test the fertilizers under identical conditions, which among other things would imply using the same piece of land in each case and the same growing conditions. Essentially the same growing conditions can be obtained if he makes a comparison of the regular and new fertilizer in the same growing season. This automatically rules out using the same piece of land for both, however. Since land is more uniform from one acre to another than are growing conditions from one year to the next, he decides to conduct the test by trying out both fertilizers in the same year.

The farmer selected 16 acres of rather uniform land and treats eight of them with the new fertilizer and eight with the regular fertilizer. Let $\bar{x}$ and $\bar{y}$ be the random variables representing the average yield per acre from the eight acres treated with the regular fertilizer and those treated with the new fertilizer respectively. Now $\bar{x}$ will be essentially normally distributed (it will be exactly normally distributed if x does indeed have a normal distribution) with standard deviation $\sigma_{\bar{x}} = 111/\sqrt{8} = 39.3$ and mean μ; μ will depend on what growing conditions are like. If H_0 is true, then y will have the same distribution as x and $\bar{y}$ will therefore be normally distributed with mean μ and standard deviation $\sigma_{\bar{y}} = \sigma_{\bar{x}}$. Thus when H_0 is true, the random variable $v = \bar{y} - \bar{x}$ will be essentially normally distributed with mean 0 and standard deviation

$$\sigma_v = \sqrt{\sigma_{\bar{x}}^2 + \sigma_{\bar{y}}^2} = \sqrt{2\sigma_{\bar{x}}^2} = \sigma_{\bar{x}}\sqrt{2} = 1.414(39.3) = 55.5 .$$

This follows from the theory of Section 4–7.

Now it is desired to reject H_0 if $\bar{y}$ is significantly greater than $\bar{x}$, that is, if v is significantly greater than 0, and not to reject H_0 otherwise. Let δ be the value such that the probability that $v \geq \delta$ when H_0 is true $\alpha = 0.05$. Then δ is determined from the equation

$$\Phi\left(\frac{\delta}{55.5}\right) = 1 - \alpha = 0.95 .$$

Hence from Table E,

$$\frac{\delta}{55.5} = 1.65 \quad \text{or} \quad \delta = 1.65(55.5) = 91 .$$

Thus H_0 is rejected when $v \geq 91$, that is, $\bar{y} \geq \bar{x} + 91$, and is not rejected otherwise. When H_0 is rejected, the farmer will presumably conclude that

the new fertilizer is significantly better than the standard one and will consider using it. If H_0 is not rejected, the farmer will not be interested in using
the new fertilizer.

It should be noted that in order for the test the farmer conducted to be
meaningful several things must be true. First of all the relative effectiveness
of the new fertilizer as compared to the standard one should not depend
critically on growing conditions. If the new fertilizer did well for some types
of weather but not others, the farmer would need to repeat the experiment
for different types of growing conditions. Similarly, if the relative effectiveness of the new fertilizer depended strongly on the type of land, then the
farmer might need to perform the experiment on different types of land if
there were significant variations on his property.

4–9 TESTING A POPULATION PROPORTION

The theory of hypothesis testing can be applied in a straightforward way to
obtain decision rules in situations where a decision maker wishes to choose
one of two possible actions and where the appropriate action to select depends upon the proportion of some population which has a given characteristic. Before making a decision, n members of the population are selected
at random and the number x having the characteristic of interest is determined. Then from the theory of Section 3–5, we know that the random
variable $w = x/n$ is essentially normally distributed with

$$\mu_w = p; \quad \sigma_w = \sqrt{\frac{pq}{n}\left(\frac{N-n}{N-1}\right)}, \tag{4–12}$$

provided p is not too small and n is reasonably large but is small compared
with the population size N. With this information it is possible to test hypotheses concerning p.

Let us now give a simple example. A firm is contemplating the purchase
of a choice piece of land in the Diamond Head area of Honolulu for the
construction of a hotel complex. However, there exists the possibility that
the land will be condemned by the City, for use as a park, before work can
begin. Whether or not this will be done depends on the outcome of an election to be held two years from now. If the voters favor a park, then it
would be a mistake to purchase the land now. If they do not favor a park
then it would be a very sound investment to have the land.

The company decides to have a polling firm select 1000 voters at random
and question them about how they will vote on this issue. Based on the
fraction w of voters in the sample who favor the park, the company will
decide what action to take. There are really three actions that can be taken.
They are: a_1—purchase the land now; a_2—take an option on the land now

for purchase at a later date; and a_3—give up the idea of purchasing the land. If w is significantly less than 0.5, then a_1 is the action to take. If w is close to 0.5, a_2 is the action to take, and if w is significantly greater than 0.5, a_3 is the action to take, since then the chances of a park are extremely good. The use of a_2 will involve paying a higher price for the land at a later date, but it does guarantee the possibility of getting it; a_2 should be used, however, only if there is considerable doubt about the outcome of the election. The desired decision rule will then have the form: Take a_1 if $w \leq \delta_1$; take a_2 if $\delta_1 < w < \delta_2$; and take a_3 if $w \geq \delta_2$.

The values of δ_1 and δ_2 can be determined by testing the hypothesis that $\mu_w = 0.5$. The company decides that it would like to use a significance level $\alpha = 0.01$. The theory developed in Section 4–5 can now be applied. Assume that there are 100,000 voters in Honolulu. Then since $n = 1000$, we see from (4–12) that

$$\sigma_w = \sqrt{\frac{0.5(0.5)}{1000} \left(\frac{99{,}000}{100{,}000} \right)} = \sqrt{0.00025} = 0.016 \,,$$

when H_0 is true, that is, when $p = 0.5$. To determine δ_1 and δ_2, d is first determined from (4–6), which becomes

$$\Phi \left(\frac{d}{0.016} \right) = 0.995 \,,$$

so that we see from Table E that $d/0.016 = 2.58$, or $d = 0.041$. Hence

$$\delta_1 = 0.5 - 0.041 = 0.459; \quad \delta_2 = 0.5 + 0.041 = 0.541 \,.$$

The decision rule is then as shown in Figure 4–10.

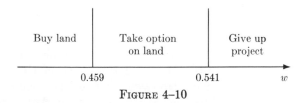

FIGURE 4–10

4–10 DETERMINATION OF SAMPLE SIZE

In our discussion of hypothesis testing we have always assumed that the experiment $\mathfrak{R}$ to be performed in testing the hypothesis was specified. In other words, we have always assumed that the sample size n was given in advance. The question now arises as to how the sample size is determined. To make it possible to determine n, something new must be introduced. This is done as follows. Note that in testing a null hypothesis H_0 we merely attempt to decide whether or not to reject H_0. We do not attempt to select

between two or more alternative hypotheses. Suppose that instead of considering H_0 only, we consider two alternative hypotheses H_0 and H_1, and we look upon the experiment to be performed as serving to reject one of these two hypotheses, thus implying that we shall act as if the one not rejected is true. For example, in Example 3 for Section 4–8, H_0 might be the hypothesis that the new fertilizer is the same as the regular one, while H_1 is the hypothesis that the new fertilizer improves the average yield per acre by 50 bushels, everything else remaining unchanged.

Let us study for the moment the following sort of situation. Suppose that the expected value μ_x of a random variable x will either have the value μ_0 or the value μ_1. No other values are possible. We would like to decide whether to use μ_0 or μ_1 in the probability model for x. Suppose also that if $\mu_x = \mu_0$ then $\sigma_x = \sigma_0$ and if $\mu_x = \mu_1$, then $\sigma_x = \sigma_1$. To be specific imagine that $\mu_1 > \mu_0$. Let H_0 be the hypothesis that $\mu_x = \mu_1$. In order to decide between H_0 and H_1, a random experiment $\Re$ is performed which involves generating n values of x and computing the resulting value of $\bar{x}$. The value of $\bar{x}$ will be used to decide whether to accept H_0 or H_1. A reasonable sort of decision rule would have the form: Select H_0 if $\bar{x} < \delta$ and select H_1 if $\bar{x} \geq \delta$. If H_0 is true, then provided n is reasonably large, $\bar{x}$ is normally distributed with mean μ_0 and standard deviation $\sigma_0/\sqrt{n}$. The curve representing this normal distribution is labeled I in Figure 4–11. When H_1 is true, $\bar{x}$ is normally distributed with mean μ_1 and standard deviation $\sigma_1/\sqrt{n}$. The curve representing this normal distribution is labeled II in Figure 4–11.

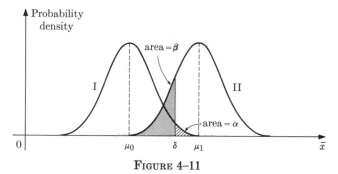

FIGURE 4–11

Now imagine that a δ is selected arbitrarily as shown in Figure 4–11. The value of δ completely specifies the decision rule. The probability of rejecting H_0 when H_0 is true is the area of the crosshatched region in Figure 4–11. This probability is what we called α, the significance level of the test, when only H_0 was being tested. However, we are now considering an alternative hypothesis H_1. The probability that H_1 will be rejected when H_1 is in fact true is the probability that $\bar{x} < \delta$ when H_1 is true. This is the area of the shaded region in Figure 4–11. This area will be denoted by β. When we deal

with two hypotheses, there are now two types of errors that are of interest. We may reject H_0 when H_0 is true and thus erroneously assume H_1 is true. This is called an *error of type 1*, and we have denoted the probability of this error by α. Alternatively we may reject H_1 when H_1 is true and thus erroneously assume H_0 is true. This is called an *error of type 2*, and we have denoted the probability of an error of this type by β. The situation can be summarized as shown in Table 4–1.

TABLE 4–1

ERRORS

	H_0 true	H_1 true
Accept H_0	No error	Type 2 error Probability $= \beta$
Accept H_1	Type 1 error Probability $= \alpha$	No error

In our previous work, δ was determined by specifying α. Once α is specified, then β is also determined when n is fixed, since α determines δ, and β is determined once δ is. Now α was always chosen to be very small in our previous work. When this is done it may turn out that β is quite large, as in the case in Figure 4–11. When considering two hypotheses it is often desirable to control the magnitude β, and, indeed, it is often of interest to make β as small as α. How can this be done? The way to reduce β for a fixed α is to increase the sample size n. As n is increased both $\sigma_0/\sqrt{n}$ and $\sigma_1/\sqrt{n}$ decrease, and, in fact, it is possible to reduce β to as small a value as desired by making n large enough.

We have now discovered a basis for selecting the experiment to be used in making the test. If the values of α and β are both specified, this will determine not only the critical value δ, but also the sample size n. Let us now see how this can be done. Suppose that we wish to perform an experiment to choose between H_0, which is $\mu_x = \mu_0$, and H_1, which is $\mu_x = \mu_1$. It is desired to have the probability of rejecting H_0 when it is true be α and the probability of rejecting H_1 when it is true be β. We wish to determine the sample size n to use and the critical value δ.

The probability of rejecting H_0 when H_0 is true is the probability that $\bar{x} \geq \delta$, that is,

$$1 - \Phi\left(\frac{\delta - \mu_0}{\sigma_0/\sqrt{n}}\right).$$

We wish this to have the value α. Thus

$$1 - \Phi\left(\frac{\delta - \mu_0}{\sigma_0/\sqrt{n}}\right) = \alpha \quad \text{or} \quad \Phi\left(\frac{\delta - \mu_0}{\sigma_0/\sqrt{n}}\right) = 1 - \alpha .$$

Let λ_0 be the number such that $\Phi(\lambda_0) = 1 - \alpha$. Then

$$\frac{(\delta - \mu_0)\sqrt{n}}{\sigma_0} = \lambda_0 . \tag{4–13}$$

The probability of rejecting H_1 when H_1 is true is the probability that $\bar{x} < \delta$, that is,

$$\Phi\left(\frac{\delta - \mu_1}{\sigma_1/\sqrt{n}}\right) .$$

It is desired that this probability be β. Thus

$$\Phi\left(\frac{\delta - \mu_1}{\sigma_1/\sqrt{n}}\right) = \beta .$$

Let λ_1 be the number such that $\Phi(\lambda_1) = \beta$. Then

$$\frac{(\delta - \mu_1)\sqrt{n}}{\sigma_1} = \lambda_1 . \tag{4–14}$$

Equations (4–13) and (4–14) can now be solved for δ and n. To do this write (4–13) and (4–14) as

$$\delta = \mu_0 + \frac{\lambda_0 \sigma_0}{\sqrt{n}} \tag{4–15}$$

$$\delta = \mu_1 + \frac{\lambda_1 \sigma_1}{\sqrt{n}} . \tag{4–16}$$

Equating the two expressions for δ in (4–15) and (4–16), we obtain

$$\mu_1 + \frac{\lambda_1 \sigma_1}{\sqrt{n}} = \mu_0 + \frac{\lambda_0 \sigma_0}{\sqrt{n}}$$

$$\sqrt{n}(\mu_1 - \mu_0) = \lambda_0 \sigma_0 - \lambda_1 \sigma_1 .$$

Hence

$$n = \left(\frac{\lambda_0 \sigma_0 - \lambda_1 \sigma_1}{\mu_1 - \mu_0}\right)^2 . \tag{4–17}$$

Now n has been expressed in terms of known quantities. Given n, δ can be immediately obtained from (4–15) or (4–16). It is possible that the n determined from (4–17) will not be an integer. In this case one merely uses for n the smallest integer greater than the number determined from (4–17).

We have been studying a hypothetical situation where μ_x could take on only two different values and it was of interest to perform an experiment to try to decide which was the actual value of μ_x. Let us now explain how what we have just studied can be applied to decision problems. Suppose that a decision maker would like to select one of two possible actions a_1 and a_2. The appropriate action to take depends on μ_x, the expected value of some random variable x; a_1 is the action to take if μ_x is sufficiently small and a_2 otherwise. Before making a decision it is possible to perform a random experiment to gain information about μ_x. The experiment, that is, the sample size, can be chosen freely. The decision maker then wishes to determine what sample size should be used when performing the experiment and what decision rule to use in selecting an action. One way he can proceed to do this is as follows.

He selects a value μ_0 such that if $\mu_x = \mu_0$ then action a_1 should be taken, and a value μ_1 such that if $\mu_x = \mu_1$ then action a_2 should be taken. The experiment is now looked at as one in which a choice is being made between the hypothesis H_0 that $\mu_x = \mu_0$ and the hypothesis H_1 that $\mu_x = \mu_1$. To determine n and δ he then simply specifies the probabilities α and β of type 1 and type 2 errors respectively. It is very important to note that this procedure does not imply that the decision maker believes μ_0 and μ_1 are the only possible values which μ_x can take on. The reasoning is that, so far as selecting the appropriate action to take, it is possible to proceed as if μ_0 and μ_1 are the only two values which μ_x can take on. In other words, a model which assumes that μ_x can take on only the values μ_0 or μ_1 is quite adequate to determine the appropriate action to take.

Let us now illustrate what we have been discussing by returning to the farmer considered in Example 3 of Section 4–8. Recall that the farmer used the new fertilizer on eight acres and the regular one on eight acres. The farmer was testing the hypothesis that $\mu_v = 0$, that is, that there is no difference between the fertilizers. Suppose now that the farmer decides that if $\mu_v = 50$, so that the expected increase in yield per acre is 50 bushels, he should definitely use the new fertilizer. He feels that σ_v will be independent of μ_v, since σ_y should always be equal to σ_x. Thus if $\mu_v = 50$, v will be essentially normally distributed with mean 50 and standard deviation 55.5. The farmer is interested in knowing the probability that if $\mu_v = 50$ (we shall refer to this as the hypothesis H_1 in the following), H_0 will be accepted as true and the new fertilizer will not be adopted even though it should be. This probability is

$$\Phi\left(\frac{91 - 50}{55.5}\right) = \Phi\left(\frac{41}{55.5}\right) = \Phi(0.74) = 0.77 = \beta .$$

Thus when eight acres are treated with each fertilizer, the probability of rejecting H_1 when it is true is very large. The situation is represented

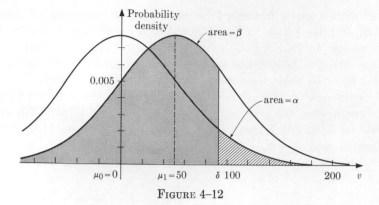

FIGURE 4–12

graphically in Figure 4–12, where the curves representing the density function for $\bar{x}$ under the alternative hypothesis H_0 and H_1 are shown.

The farmer is somewhat disturbed by the revelation that there is a good
chance that he will reject the new fertilizer if it turns out that $\mu_v = 50$. He
is interested in knowing what will happen if he uses larger plots of ground
to test the fertilizers. The situation when 72 acres are treated with the new
fertilizer and 72 acres with the standard fertilizer is shown in Figure 4–13.
The value of δ was again determined so that the probability of rejecting H_0
when it is true is 0.05. The value of δ obtained now is considerably smaller
than when eight acres were used. Furthermore, the probability of rejecting
the new fertilizer when H_1 is true has been reduced tremendously. The
farmer decides that he would much prefer the situation represented in Figure 4–13 to that shown in Figure 4–12.

After some thought, the farmer decides that what he would really like is
to have the probability of rejecting H_0 when it is true be 0.05 and of rejecting H_1 when it is true be 0.10. In other words, he wants to have $\alpha = 0.05$
and $\beta = 0.10$. Let us then determine the number of acres that should be

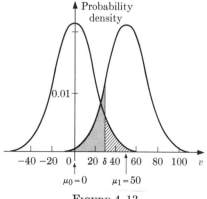

FIGURE 4–13

treated with the new fertilizer (the same number being treated with the standard fertilizer) in order to achieve this. We shall also determine δ. To do so, we use (4–17). For the case under consideration, $\mu_0 = 0$ and $\mu_1 = 50$. Furthermore, the farmer has assumed that σ_v is the same regardless of whether H_0 or H_1 holds, the reason being that the standard deviation of the yield for the new fertilizer should be the same as for the standard fertilizer, regardless of what the expected improvement in yield is. Recall from Example 3 for Section 4–8 that $\sigma_x = 111$ bushels. Thus $\sigma_y = 111$ bushels, by assumption, and $\sigma_{\bar{x}} = \sigma_{\bar{y}} = 111/\sqrt{n}$. Hence

$$\sigma_v^2 = \sigma_{\bar{x}}^2 + \sigma_{\bar{y}}^2 = \frac{2\sigma_x^2}{n} = \frac{2(111)^2}{n} .$$

Thus in (4–17), $\sigma_0 = \sigma_1 = (111)\sqrt{2}$, since in the current context σ_0 and σ_1 represent σ_v when $n = 1$. Hence (4–17) becomes

$$n = \frac{2(111)^2(\lambda_0 - \lambda_1)^2}{(50)^2} .$$

It remains to determine λ_0 and λ_1. Now λ_0 is the value such that $\Phi(\lambda_0) = 1 - \alpha = 0.95$. From Table E, $\lambda_0 = 1.65$. Similarly, λ_1 is the number such that $\Phi(\lambda_1) = \beta = 0.10$. Now $\lambda_1 < 0$, and hence $1 - \Phi(-\lambda_1) = 0.10$ or $\Phi(-\lambda_1) = 0.90$, so $-\lambda_1 = 1.28$ and $\lambda_1 = -1.28$. Then $\lambda_0 - \lambda_1 = 1.65 + 1.28 = 2.93$, and

$$n = 2\left[\frac{(111)(2.93)}{50}\right]^2 = 84 .$$

Hence 84 acres should be treated with the new fertilizer and the same number with the regular fertilizer. By treating 84 acres instead of 8 with each type of fertilizer, the probability of a type 2 error has been reduced from 0.77 to 0.10 while maintaining the probability of a type 1 error at 0.05. The critical value δ, when $n = 84$, can now be found from (4–15) and is

$$\delta = 1.65(111)(1.414)/\sqrt{84} = 28.3 ,$$

which is considerably smaller than the value of 91 obtained when $n = 8$ acres.

The sample size to use for a two-sided test can be determined in precisely the same manner as in the case considered above. Consider the sort of situation studied in Section 4–5, where action a_1 should be taken in μ_x is close to μ_0 and action a_2 should be taken otherwise. Imagine now that a_2 should definitely be taken if μ_x deviates by as much as d from μ_0, that is, if $\mu_x = \mu_0 + d = \mu_1$ or $\mu_x = \mu_0 - d = \mu_2$. Suppose that H_0 is the hypothesis that $\mu_x = \mu_0$ and H_1 is the hypothesis that $\mu_x = \mu_1$ or $\mu_x = \mu_2$. Imagine that, as in Section 4–5, it is desired to have the probability of rejecting H_0 when it is true be α. Now assume, in addition, that the probability of rejecting

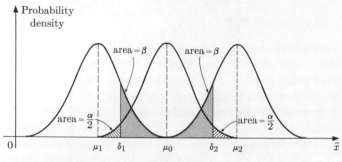

FIGURE 4–14

H_1 when $\mu_x = \mu_1$ is true to be β, and the probability of rejecting H_1 when $\mu_x = \mu_2$ is also to be β. Then the critical values δ_1 and δ_2 are symmetrically placed about μ_0. Finally assume that $\sigma_x = \sigma_0$ at $\mu_x = \mu_0$ and $\sigma_x = \sigma_1$ at $\mu_x = \mu_1$ or at $\mu_x = \mu_2$. Then, the density function for $\bar{x}$ is determined given $\mu_x = \mu_0$ or μ_1 or μ_2 and given n, and the situation looks like that shown in Figure 4–14. By analogy with Figure 4–11, δ_2 must satisfy the equations

$$1 - \Phi\left(\frac{\delta_2 - \mu_0}{\sigma_0/\sqrt{n}}\right) = \frac{\alpha}{2} \; ; \quad \Phi\left(\frac{\delta_2 - \mu_2}{\sigma_1/\sqrt{n}}\right) = \beta \, . \qquad (4\text{–}18)$$

Hence if λ_0 and λ_1 are the values of t for which $\Phi(\lambda_0) = 1 - (\alpha/2)$, $\Phi(\lambda_1) = \beta$, then once again n is given by (4–17) with μ_1 replaced by μ_2 and

$$\delta_2 \;=\; \mu_0 + \frac{\lambda_0 \sigma_0}{\sqrt{n}} \, . \qquad (4\text{–}19)$$

Of course, once δ_2 is determined, then if $d = \delta_2 - \mu_0$, it follows that $\delta_1 = \mu_0 - d$. Thus the critical values and the sample size are determined.

The careful reader will note that an approximation was made in writing

$$\Phi\left(\frac{\delta_2 - \mu_2}{\sigma_1/\sqrt{n}}\right) = \beta \, .$$

In reality we want the area under the curve representing the density function for $\bar{x}$ when $\mu_x = \mu_2$ to be β in the interval $\delta_1 \leq \bar{x} \leq \delta_2$, not the interval $\bar{x} \leq \delta_2$, since if $\bar{x} < \delta_1$, H_1 will not be rejected. Thus we should have written

$$\Phi\left(\frac{\delta_2 - \mu_2}{\sigma_1/\sqrt{n}}\right) - \Phi\left(\frac{\delta_1 - \mu_2}{\sigma_1/\sqrt{n}}\right) = \beta \, .$$

However, it is essentially always true, as is the case in Figure 4–14, that

$$\Phi\left(\frac{\delta_1 - \mu_2}{\sigma_1/\sqrt{n}}\right)$$

is extremely close to 0, and hence the approximation used in (4–18) is valid.

4–11 GOODNESS OF FIT

Suppose that we have asked a statistician to construct a probability model for a random experiment $\mathcal{R}$. The model $\mathcal{E}$ constructed by the statistician has the event set $E = \{e_1, \ldots, e_m\}$, with the probability of the simple event e_j being p_j. We now wish to determine whether or not $\mathcal{E}$ is a realistic model. To do so, suppose we perform n independent trials of $\mathcal{R}$ and find that e_j occurs precisely n_j times, $j = 1, \ldots, m$. How can we use these data to determine whether the model is realistic or not? The obvious thing to do is to compute the relative frequency n_j/n of each e_j and to note whether this is close to p_j or not. The value n_j/n will not normally be precisely equal to p_j for any j. The question then arises as to how much n_j/n can differ from p_j without making us feel that the model is invalid.

To make a decision it is necessary to consider $(n_j/n) - p_j$ for each $j = 1, \ldots, m$, and then weight the various deviations in some manner. One way in which this could be done would be to base the decision on the sums of the squares of the deviations, that is, on

$$\sum_{j=1}^{m} \left(\frac{n_j}{n} - p_j\right)^2 = \left(\frac{n_1}{n} - p_1\right)^2 + \cdots + \left(\frac{n_m}{n} - p_m\right)^2. \qquad (4\text{--}20)$$

However, if $(n_j/n) - p_j = 0.01$ when $p_j = 0.001$, this would be considered a serious error, while a deviation of 0.01 if $p_j = 0.50$ would be considered to be a small error. Thus what the decision maker is really interested in, for most cases, is the relative error

$$\frac{\frac{n_j}{n} - p_j}{p_j} = \frac{n_j - np_j}{np_j}. \qquad (4\text{--}21)$$

It would then seem more appropriate to consider the sum of the squares of the relative errors, that is,

$$\sum_{j=1}^{m} \left(\frac{n_j - np_j}{np_j}\right)^2. \qquad (4\text{--}22)$$

It turns out to be somewhat more convenient to use

$$\epsilon = \sum_{j=1}^{m} \frac{(n_j - np_j)^2}{np_j}, \qquad (4\text{--}23)$$

and it is this latter expression which is normally used. Now if $x\{j\}$ is the random variable representing the number of times e_j occurs on n trials of $\mathcal{R}$, we have assumed above that $x\{j\} = n_j$. We can then see that ϵ is the observed value of a random variable χ^2, which is related to the $x\{j\}$ by

$$\chi^2 = \sum_{j=1}^{m} \frac{(x\{j\} - np_i)^2}{np_i}.$$ (4–24)

The use of χ^2 to represent the random variable (4–24) may seem a little odd to the reader. It is, however, the symbol typically used.

It now seems reasonable that if the value of χ^2 is too large, the decision maker should conclude that the model is not realistic, whereas if χ^2 is sufficiently close to 0, he has no grounds for rejecting the model based on the results of the n trials of $\Re$. In other words, a decision rule which could be used to accept or reject the model has this form: Reject the model of $\chi^2 \geq \delta$ and accept the model if $\chi^2 < \delta$. One way to determine δ is to think of the experiment which involves performing n trials of $\Re$ as serving to test a null hypothesis, the null hypothesis H_0 being that the model is indeed correct. It is possible to determine the exact distribution of χ^2 when H_0 is true. This distribution cannot be used conveniently because of its complexity, however.

Now it is a remarkable fact that for reasonably large n, the distribution of χ^2 is essentially independent of the values of the p_j and n, and is closely approximated by a distribution called the *chi-squared distribution* which depends on m only. Since n is normally fairly large, probability computations for the random variable χ^2 can be carried out using the chi-squared distribution. Now it can be shown (the reader is asked to do so in the problems) that the expected value of χ^2 given by (4–24) is $m - 1$; $m - 1$ is called the degrees of freedom for χ^2. The chi-squared distribution involves a parameter which is called the degrees of freedom, the degrees of freedom being the expected value of the chi-squared distribution. In approximating the actual distribution of χ^2, we use a chi-squared distribution with the same number or degrees of freedom as χ^2. Thus to make probability computations with χ^2 we use a chi-squared distribution with $m - 1$ degrees of freedom.

As with the normal and Student t distributions, the only convenient way to make computations by hand using the chi-squared distribution is through the use of a table. Such a table is presented in Table H at the end of the text. This table does not give the cumulative chi-squared distribution, but instead gives the value δ, for various values of α, such that the probability that $\chi^2 \geq \delta$ is α. This is an especially convenient form because δ, the critical value in the test of H_0, is determined directly. This is done for different values of the degrees of freedom ν (ν is used to represent the degrees of freedom in Table H). For example, if χ^2 has 13 degrees of freedom, the value of δ such that the probability that $\chi^2 \geq \delta$ is 0.05 is 22.36. The graphs of the chi-squared distribution for various values of the degrees of freedom are shown in Figure 4–15.

Let us now give a practical illustration of the theory just developed. A

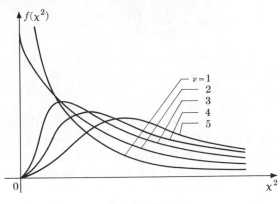

FIGURE 4–15

test of the type we have been considering is typically referred to as a *chi-squared test*.

EXAMPLE. An industrial engineer has constructed a model for the daily demand for a particular type of spare part in the maintenance shop of an automated factory. The model is summarized in Table 4–2. To check the

TABLE 4–2

PROBABILITY MODEL

Units demanded	0	1	2	3	4
Probability	0.4	0.3	0.1	0.1	0.1

model, the plant manager keeps a record of the number of units demanded in each day for 30 working days. He finds that on 14 days no units were demanded, on 10 days one unit was demanded, on 2 days two units were demanded, on 3 days three units were demanded and on one day four units were demanded. He wishes to determine whether he should accept or reject the model if a significance level $\alpha = 0.05$ is used. Here $m = 5$, so the degrees of freedom $\nu = 4$. It is seen from Table H that for $\nu = 4$ and $\alpha = 0.05$, the critical value is $\delta = 9.49$. Thus the model should be rejected if $\chi^2 \geq 9.49$ and not otherwise. The details of the computation of χ^2 are given in Table 4–3. Here $\chi^2 = 2.10$ is less than 9.49 and hence the plant manager will presumably accept the model.

In the above discussion we have not indicated how large n must be in order that the chi-squared test may be applied. As with other approximations it is difficult to give a precise answer. A rule of thumb often used is that if each $np_i > 5$, then n is large enough. However, the chi-squared

TABLE 4–3

COMPUTATION OF χ^2

Demand	p_i	np_i	$x\{j\}$	$(x\{j\} - np_i)^2$	$\dfrac{(x\{j\} - np_i)^2}{np_i}$
0	0.4	12	14	4	0.33
1	0.3	9	10	1	0.11
2	0.1	3	2	1	0.33
3	0.1	3	3	0	0.00
4	0.1	3	1	4	1.33

$$\chi^2 = 2.10$$

distribution is often an adequate approximation to the exact distribution, even if some of the np_i are considerably less than 5. In the above example, some of the np_i are less than 5, but the chi-squared test is adequate here, especially in view of the fact that the value of χ^2 is not close to the critical value.

4–12 FURTHER OBSERVATIONS CONCERNING HYPOTHESIS TESTING

We have presented the theory of hypothesis testing as a theory which can be used for solving certain simple types of decision problems. The reader, on reflecting on some of the examples, may well have had certain reservations as to whether the hypothesis testing procedure introduced really yielded a sound decision rule. To illustrate what we have in mind, let us return to the example given in Section 4–4 concerning the new filament for light bulbs.

Recall that the decision rule there was obtained by testing at a given level of significance the null hypothesis that there was no difference between the new filament and the one currently in use. The decision rule reduced to a form: reject H_0 if $\bar{x} \geq \delta$ and do not reject H_0 otherwise. Now we might note that δ depends on the sample size n, δ decreasing as n increases. For a very large value of n, δ will be only slightly greater than 1000, and hence H_0 will be rejected if $\bar{x}$ is only slightly greater than 1000. What this says is that for large n even very slight improvements in the filament quality will be detected. We can now ask: is this actually what we want? If interest centers on merely detecting any improvement in the filament mean life, no matter how small, then this is indeed the sort of behavior we wish the decision rule to exhibit. Is this, however, really what the management wants?

The management presumably wants to decide whether to start making the new filament or not. Will they wish to start making the new filament

if the increase in the expected life is only 0.01 hours? Unquestionably they would not. The improvement is too insignificant to be of any interest. Instead, there will be a value μ^* such that if $\mu_x > \mu^*$ it is desirable to produce the new filament, while if $\mu_x < \mu^*$, it will not be worthwhile to do so. How is μ^* determined? To do so, it is necessary to consider the economics of the situation. Without a knowledge of μ^* it is not possible to construct a suitable test that will give the sort of results desired. The test suggested in Section 4–4 may be satisfactory for detecting an improvement, but not for deciding whether production of the new filament should be undertaken.

Let us now explain how a knowledge of μ^* can be used in constructing a meaningful test. We shall use the theory of Section 4–10 to determine both a critical number δ and a sample size n. Recall that the operating characteristic curve for a test gives $p(\mu_x)$ the probability of accepting H_0 as a function of μ_x. The type of operating characteristic curve the decision maker would like is the step-type solid curve shown in Figure 4–16. It is

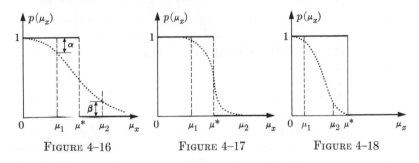

FIGURE 4–16   FIGURE 4–17   FIGURE 4–18

desired to have $p(\mu_x) = 1$ for $\mu_x < \mu^*$ and $p(\mu_x) = 0$ for $\mu_x > \mu^*$. In the present case, it is not possible to achieve this curve using a finite sample size, although we can come arbitrarily close to it by making n large enough. To obtain an operating characteristic curve which approximates the ideal curve, it is only necessary to select a value of $\mu_0 < \mu^*$ and a value of $\mu_1 > \mu^*$. Then H_0 is taken to be the hypothesis that $\mu_x = \mu_0$ and H_1 the hypothesis that $\mu_x = \mu$. Once α and β are selected, the values of δ and n are determined. The operating characteristic curve is also determined and will look something like the dotted curve shown in Figure 4–16. If smaller values of α and β are selected a larger value of n will result and the operating characteristic curve obtained will approximate more closely the ideal curve as shown in Figure 4–17.

We can now see that if the decision maker can sketch in about μ^* the sort of operating characteristic curve he would like, he can construct a test which will give essentially this operating curve by selecting a convenient μ_0 and μ_1 and reading the values of α and β from the curve as shown in Figure 4–16. Use of μ_0, μ_1, α and β to construct a test will then yield a test

which gives essentially (but not necessarily exactly) the operating characteristic curve drawn originally. By drawing an operating characteristic curve which is very close to the ideal one, the decision maker can construct a test which will make the chances of making a wrong decision rather small. However, the resulting value of n may be very large and the resulting experiment very expensive.

How, then, can the decision maker decide what sort of operating characteristic curve he desires? This is not easy to do because he must take into account somehow the cost of making the experiment and the costs of making errors, and the procedure for doing this lies outside the theory we have developed. Thus we see that while it is essential to know at least roughly what μ^* is in order to develop a good test, it is in general necessary to know even more, because one must be able to draw in approximately the sort of operating characteristic curve desired. We might note before going on what happens if μ^* is not known, and μ_0 and μ_1 happen to both be selected to be on the same side of μ^*. This is shown in Figure 4–18. The operating characteristic curve will look something like that shown and will not approximate the ideal curve at all. In the case illustrated, the hypothesis will normally be rejected for a whole range of μ_x where it should not be rejected.

In this section we have shown that in order to use hypothesis-testing methods in a way that will yield useful results, it is often necessary to perform some economic analyses on the side. The theory we have developed gives no rules for carrying out such economic analyses. The next chapter will be devoted to introducing the modern theory of decision making. It can be applied to a much broader class of problems than the classical theory, and it has the great advantage that economic considerations are integrated into the decision model directly. This makes it more appealing in many ways than the classical theory, and it also provides a more complete analysis of the problem.

REFERENCES

The references for Chapter 3 are also appropriate here.

PROBLEMS

Sections 4–2 and 4–3

1. If the hypothesis is made that the mean μ_x of a random variable x has the value μ_0, it seems logical to test this hypothesis using the statistic $\bar{x}$ that would be used if

μ_x were to be estimated, and to reject the hypothesis if $\bar{x}$ deviates too far from μ_0. What is the logic in rejecting the hypothesis if the observed value of $\bar{x}$ deviates by more than a given amount from μ_0?

2. The experiment referred to in Problem 1 might convince us that $\mu_x \neq \mu_0$. Could it ever serve to convince us that $\mu_x = \mu_0$? Why is this normally impossible?

3. Suppose that the null hypothesis is made that $\mu_x = 10$. Assume that it is known that $\sigma_x = 3$. Imagine now that a sample of 25 values of x is obtained, and let $\bar{x}$ be the random variable representing the average value of x for the sample. What is the distribution of $\bar{x}$ (approximately)? Use Table F to illustrate graphically the curve representing this distribution. Suppose now that the observed value of $\bar{x}$ is 10.8. Illustrate this value on the figure just obtained and shade the region whose area is the probability that $\bar{x} \geq 10.8$. What is this probability? Does it seem reasonable to reject the hypothesis in this case?

4. What factors can you think of which might influence the choice of the significance level α?

5. For the situation described in Problem 3, suppose that if $\mu_x \neq 10$ then $\mu_x > 10$, that is, μ_x cannot be less than 10. If $\alpha = 0.05$, what is the critical value δ? If $\bar{x} = 10.8$, should H_0 be rejected?

6. Generally speaking, the way an hypothesis concerning some parameter is tested is that one first estimates the value of the parameter, using the methods introduced in Chapter 3. After the estimate is obtained, a decision must then be made as to whether this estimate is consistent with the hypothesis. It is at this point that something new must be introduced in order to reach a decision. Why must something new be introduced and what is the nature of the additional theoretical framework which is needed?

Section 4–4

1. An engineer claims that a process which turns out ball bearings is such that the diameter of any given ball bearing can be looked upon as the value of a random variable x which is normally distributed with mean 0.2510 and standard deviation 0.0020. Suppose that we believe that he is correct in stating that x is normally distributed with standard deviation 0.0020, but we are not sure about the mean. The mean may be greater than 0.2510. Imagine that we make the null hypothesis $\mu_x = 0.2510$, and to test this hypothesis, the diameters of ten ball bearings are measured. The mean diameter obtained is 0.2530. Would you reject the hypothesis at the five percent significance level? What statistic did you use? What is the critical number? For what range of α values would the hypothesis be rejected? Explain carefully the procedure used for testing the hypothesis and the reasoning behind it.

2. A brewer believes that he has developed a beer which can in taste be clearly distinguished from his competitor's brand. To test this he selects 25 individuals and allows them to sample his beer. He then gives each two unlabeled glasses, one containing his beer and one containing his competitor's beer, and asks them to select the glass which contains his beer. It turns out that 17 of them correctly identify his beer. At the five percent significance level should one agree that it is possible to

distinguish his beer by its taste? What is the null hypothesis to be tested in this case? What is the statistic? For what range of α values would the null hypothesis be rejected? What is the critical number when $\alpha = 0.05$?

3. A chemist has developed for a tire company a new synthetic rubber which he feels is better than that being used currently. Extensive testing over the years on the synthetic rubber currently in use has shown that the mileage which can be put on a tire until wearout can be looked upon as a random variable x which is normally distributed with mean 15,000 miles and standard deviation 3000 miles. Fifty tires are tested using the new synthetic rubber, and the average mileage until wearout turns out to be 16,200 miles. At the one percent significance level, would it appear that the new synthetic rubber should be considered to be superior to that currently in use? What is the critical number?

4. Determine the operating characteristic curve for the test of Problem 1 if σ_x does not vary with μ_x.

5. Determine the operating characteristic curve for the test of Problem 3 if σ_x does not vary with μ_x.

6. A new air-pollution control law has been enacted in the county where a large steel mill operates. To comply with this law an electrostatic precipitator has been installed in one of the stacks. The new law requires that the mean solids content per 1000 cubic feet of gas emitted from the stack be no more than 3 grams. To determine whether the new unit will meet these standards, the company selects a 1000 cubic foot sample of gas from the stack on each of 30 different days. The mean solids content per 1000 cubic feet for the 30 samples is 3.6 grams and $s_x = 2$ grams. Determine the decision rule which the firm should use to either reject or not reject the null hypothesis that the mean solids content per 1000 cubic feet is indeed 3 grams at a significance level of 0.10. Given the outcome of the experiment, should the null hypothesis be rejected?

7. Construct an operating characteristic curve for the test in Problem 6 if s_x can be used as the standard deviation regardless of what value μ_x has. Obtain the operating characteristic curve if $\sigma_x = 2 + 1.5(\mu_x - 3)$.

8. A radio station tells a manufacturer of a consumer product that his sales will be increased in the city served by the station if he advertises on that station. The manufacturer decides to try the advertising program suggested by the station for a period of four weeks. In the recent past, sales in that city have averaged $200,000 per week with a standard deviation of $40,000. Sales over the four-week period during which advertising was carried out over the radio station averaged $230,000 per week. At the five percent significance level, should the manufacturer reject the null hypothesis that the advertising campaign produced no change in expected sales?

9. An industrial engineer has suggested to management that productivity on the third shift could be increased by introducing certain changes in the work methods. To check on this, management decides to try out the proposed changes for 20 working days. Normal productivity in the past has averaged 100 units, with a standard deviation of 17. Determine the decision rule which management should use if it wishes to test the null hypothesis that the changed methods produce no improve-

ment on the average at a significance level of $\alpha = 0.01$. If the average output over the 20-day period is 120 units, should the null hypothesis be rejected? Determine the operating characteristic curve for the test if the standard deviation is always 17.

10. A psychologist claims that the expected I.Q. of underprivileged children in a certain city is 62. A social worker believes the value is somewhat lower. To test this claim, 25 underprivileged children are selected at random by the social worker, and their I.Q.'s are measured. The average I.Q. for those tested is 54 with $s_x = 15$. Should the psychologist's claim be rejected at the five percent level of significance level?

11. The U.S. Navy purchases candy bars in large numbers for use on ships and at bases. These bars are specially packaged so as to hold up under rather rigorous conditions. The expected weight of a bar is supposed to be six ounces. Occasionally, either by accident or by intention, the expected weight is somewhat less than this. A government inspector goes to one supplier's plant to check bars coming off the production line. He selects at random 50 bars and weighs each carefully. The average weight is 5.75 ounces, with $s_x = 0.30$ ounces. At the five percent level of significance, should the government inspector reject the null hypothesis that the expected weight of a bar is six ounces?

12. An anthropologist claims that the expected number of children a woman will have in a certain South Pacific tribe is 2.6. To check on this, another anthropologist, who believes the mean to be greater than this, selects at random from the tribe 25 women who have been married for many years and finds that the average number of children these women had was 3.2 with $s_x = 1.6$. At the five percent level of significance, should the null hypothesis that the expected number of children is 2.6 be rejected? For what range of α values will the null hypothesis be rejected?

13. A white hunter claims that the expected weight of elephant tusks from elephants living in the Belgian Congo is 150 pounds. A buyer at the ivory center in Mombassa, Kenya, believes this is a little high. To check on it he selects 15 tusks from different elephants that came from the Congo. The average weight is 125 pounds with $s_x = 50$ pounds. Should the buyer reject the white hunter's claim at the five percent level of significance? For what range of α values would the white hunter's claim be rejected?

14. A company executive who frequently rents cars for driving from an airport in a given city to one of his company's plants has become convinced that the odometers on the cars are set in such a way that they always give a mileage reading which is too high. To check on this he has one of the plant staff measure the distance accurately. He then keeps a record of the distance indicated on the odometers of the cars he rents. Over a period of six months he rents 20 different cars, and the average odometer reading is 5 miles too high, with $s_x = 7$ miles. At the one percent level of significance should the executive reject the null hypothesis that on the average the odometers are correct, that is, that the expected deviation from the true mileage is 0? For what range of α values will the null hypothesis be rejected?

15. A drug company has developed a diet pill which it claims will result in an average weight reduction of 20 pounds in a period of one month. A doctor believes

this value to be too high. To check on this he selects 25 patients to use the pill for a month. The average weight reduction turns out to be 10 pounds with s_x, the estimate of the standard deviation of the random variable x representing the weight reduction, being 15 pounds. Should the doctor reject the null hypothesis that the expected weight reduction is 20 pounds if a significance level of $\alpha = 0.05$ is used?

16. Imagine that a random variable x has the normal distribution $n(x; \mu, 2)$, and it is desired to test the hypothesis that $\mu \leq 4$ by performing n independent trials of an experiment, where on each trial a value of x will be observed. Let $\bar{x}$ be the random variable used to test the hypothesis. Show that the hypothesis does not uniquely determine the distribution of $\bar{x}$. For any given value of $\mu \leq 4$, what is the distribution of $\bar{x}$? Compute the probability of rejecting the hypothesis as a function of μ, $\mu \leq 4$, when the critical value used for $\bar{x}$ is $\delta = 6$. Plot a graph of this probability as a function of μ, and show that the probability of rejection is largest when $\mu = 4$. Show that by specifying the maximum probability of rejection, call it α, one can then determine a unique critical value for $\bar{x}$, and that the value of δ so obtained is precisely the same as that which one would obtain for testing the null hypothesis $\mu = 4$ at the level of significance α. This is the way that the hypothesis $\mu \leq 4$ would normally be tested, that is, it would be tested as the hypothesis $\mu = 4$ at some specified significance level, using $\bar{x}$ as a statistic and assuming that only large values of $\bar{x}$ are unfavorable to the hypothesis.

Section 4–5

1. An electrical components manufacturer has set up for a long run of 1000-ohm resistors. It is important that the expected resistance of any unit produced not deviate significantly from 1000 ohms. To check on the set-up the industrial engineering department determines the resistance of 200 units coming off the line. The average resistance is 985 ohms, with $s_x = 100$ ohms. Determine the decision rule that will result if the null hypothesis that $\mu_x = 1000$ is tested at the five percent significance level. Should the null hypothesis be rejected in this case?

2. Determine the operating characteristic curve for the test in Problem 1 under the assumption that σ_x is independent of μ_x. What is the probability that H_0 will not be rejected if $\mu_x = 972$?

3. A manufacturer of detergents is about to make a run on one particular type. It is of interest to know whether the filling machine has been adjusted correctly. The expected weight placed in a box should be 35 ounces. If the actual mean weight is significantly greater than this, there will be product giveaway, and if it is significantly less there may be trouble with government inspectors. To check on this, the weight of each of 100 boxes coming off the line is determined carefully. The average weight turns out to be 35.4 ounces, with $s_x = 0.8$ ounces. Determine the decision rule that will result if the null hypothesis that $\mu_x = 35$ is tested at a significance level of $\alpha = 0.10$. Should the null hypothesis be rejected?

4. Determine the operating characteristic curve for the test in Problem 3 under the assumption that σ_x is independent of μ_x.

5. A manufacturer of ball bearings has just installed a new automated line to produce ball bearings whose nominal diameter is $\frac{1}{4}$ inch. It is important that the

expected diameter of any bearing not deviate significantly from 0.25 inches. To check on the machine settings the diameters of 500 ball bearings are measured carefully. The average diameter of these is 0.2507 inches with $s_x = 0.008$ inches. Determine the decision rule implied by testing the null hypothesis that $\mu_x = 0.2500$ at a significance level of $\alpha = 0.01$. Should the null hypothesis be rejected?

6. Determine the operating characteristic curve for the test of Problem 5 under the assumption that σ_x is independent of μ_x.

7. The manufacturer of cans referred to in the text is also preparing the machines that produce the bottoms of the cans. One operation involves placing a small amount of sealant around the outer edge of the bottom. This sealant serves to make the joint with the side of the can leakproof (the bottom is not soldered to the sides). The quantity of sealant applied is rather critical. If two little is applied the can may leak and if too much is applied some of the sealant will seep into the interior of the can. It is desired that the expected amount applied be 0.8 grams. It is difficult to scrape the sealant off the bottoms to determine how much has been applied. However, this is done on 25 bottoms and it is found that the average amount applied is 0.88 grams with $s_x = 0.15$ grams. Determine the decision rule that is implied if the null hypothesis that $\mu_x = 0.80$ is tested at a significance level of $\alpha = 0.05$. Should the null hypothesis be rejected?

8. Determine the operating characteristic curve for the test of Problem 7 under the assumption that δ_x is independent of μ_x.

9. A doctor claims to have the ability to determine very frequently the sex of a child just before birth from x-rays taken of the baby in the womb. To check on this the administrator of a hospital takes x-rays of 30 maternity patients and asks the doctor to determine which will have male and which female children. He makes the correct choice in 20 of the 30 cases. Should the administrator reject, at the five percent level of significance, the hypothesis that the doctor merely tosses a fair coin to decide whether it will be a boy or a girl? Does the result depend on what the probability of a male birth is?

10. A drug manufacturer is about to start making a run of the vaccine used for cholera immunization. The machine which fills the glass vials has been set up to fill the vials used for the first of the two injections needed for cholera. The quantity of liquid that will be placed in any vial will be a random variable. The expected amount, however, must be controlled very closely at 0.55 cubic centimeters. If this is indeed the mean, then essentially all vials will be satisfactory. However, if it falls below this value a certain fraction of the vials will contain too little vaccine to give a proper immunization. If it goes above this value, a certain number of vials will contain too much vaccine and will give an undesirable side effect. To check on the adjustment of the machine, 50 vials are filled and the volume of vaccine in each is carefully measured. The average volume in these is 0.49 cubic centimeters with $s_x = 0.13$ cubic centimeters. Should the null hypothesis that $\mu_x = 0.55$ be rejected at the five percent level? For what range of α values will the null hypothesis be rejected?

11. A gambler claims he is using a fair die. An individual who is somewhat dubious about this claim tosses the die 100 times and notes that the face with one dot on it

turns up 25 times. Should the null hypothesis that the die is fair be rejected at the five percent level of significance?

12. Out of 200,000 births in a given city, a fraction 0.514 of them are male. Is this evidence enough to reject at the one percent level the null hypothesis that the probability of a male birth is $\frac{1}{2}$?

13. A total of 1000 seeds are obtained from crossing an AA type plant with an Aa type. The genes here refer to the color of the flowers. It turns out that 475 of the plants have pink flowers and the remainder have red flowers. Should the null hypothesis that the Mendelian theory is correct be rejected here at the five percent level?

14. A watch manufacturer claims that the settings on his watches are such that the expected gain or loss in time over a given period is 0. One hundred of these watches are tested for a month and the gains and losses are averaged. The average result is a gain of 10 minutes with the estimate of the standard deviation of the gains and losses being $s_x = 35$ minutes. Should the null hypothesis that there is no expected gain or loss be rejected at the five percent level of significance?

Section 4–6

1. A firm which produces mechanical cotton pickers has received five samples of a new metal alloy gear which the supplier claims has a longer life than the gear currently in use. The gear currently in use has an expected life of about 1000 hours of operation in the field. The firm installs one of these gears in each of five machines and tests them under simulated field operating conditions. The average life of the five gears is 1100 hours with $s_x = 125$ hours. If it is desired to test the null hypothesis that the new gears have an expected life of 1000 hours at the five percent significance level using the t distribution, should the null hypothesis be rejected?

2. Solve Problem 7 for Section 4–5 using the t distribution.

3. Solve Problem 8 for Section 4–4 using the t distribution.

4. A firm has developed a new type of expensive radar tube which is being considered for use in the Australian Navy. Eight of these tubes are shipped to Australia for testing. The current tube used has a mean life of 300 hours. On testing the new tubes it is found that the mean life of the eight tubes is 425 hours with $s_x = 150$ hours. If the null hypothesis that the new tubes represent no improvement over the current one is to be tested at the five percent significance level using the t distribution, should the null hypothesis be rejected?

5. A firm has received a sample of five typewriter ribbons of a new type which the manufacturer claims last longer than the standard ribbons. The average life of a standard ribbon is 40 days. The new ribbons are tried out in the office on randomly selected typewriters and the average life of these ribbons is 51 days, with $s_x = 15$ days. If the null hypothesis that the new ribbons are no improvement over the standard ribbon is to be tested at the five percent significance level using the t distribution, should the null hypothesis be rejected?

6. An anthropologist has remains of the skulls of five ancient men taken from a particular level in the Olduvai Gorge in Tanzania. He finds that the average weight

of the jawbones is 885 grams with $s_x = 75$ grams. The average jawbone weight for modern males is 820 grams. Should the null hypothesis that the expected jawbone weight for these prehistoric men is the same as for modern man be rejected at the five percent level of significance? What assumptions are needed in making the computations?

Section 4–7

1. A manufacturer wishes to compare two different brands of batteries produced by foreign firms for possible use in the battery-operated clocks he produces. He tests 100 of each brand of battery in a given make of clock. The average operating life for brand 1 is 2500 hours with $s_x = 500$ hours, while the average operating life for brand 2 is 2300 hours with $s_x = 200$ hours. If he tests the null hypothesis that there is no significant difference between the batteries at the five percent significance level, what is the decision rule that is implied by this test? What should the manufacturer conclude as a result of the experiment performed?

2. Determine the operating characteristic curve for the test in Problem 1 under the assumption that the standard deviations are independent of the means.

3. A manufacturer of fluoride toothpaste selects at random 500 sixth-grade school children to use its toothpaste for one year and another 500 to use a competing brand of toothpaste for the same period of time. The average number of cavities that developed in the set of students who used the given manufacturer's toothpaste was 2.1 with $s_x = 1.5$ and with the competitor's toothpaste was 2.4 with $s_x = 2.0$. As a result of this, the manufacturer launches an advertising campaign claiming that his toothpaste reduces cavities by 25 percent as compared to its leading competitor. Is there any justification for this claim? What would be the conclusion based on testing the null hypothesis that there is no significant difference between the brands at the five percent level?

4. A department store chain with stores in several cities wishes to determine if there is any difference in expected sales per credit-card customer in their large downtown stores in two different cities. To gain information on this, the accounts of 200 customers are selected at random from each store. The average sales per customer over the period of one year in store 1 are \$211 with $s_x = \$86$, and in store 2 are \$184 with $s_x = \$95$. As a result of this experiment the financial vice president states that average annual sales to credit card customers are 15 percent higher in store 1 than in store 2. Is this claim justified? Would the null hypothesis that there is no difference between average annual sales to credit card customers be rejected at the five percent significance level?

5. A firm is interested in determining if the long-run fraction defective generated on the third shift is any different from that on the first shift. To check on this, the industrial engineering department checks 2000 units produced on the first shift and 2000 units produced on the third shift. The fraction defective of those produced on the first shift is 0.03 and that on the third shift is 0.035. The industrial engineering department then prepares a report which claims that about 15 percent more defectives are turned out on the third shift than on the first shift. Is this claim justified? Would the null hypothesis that the long-run fraction defective is the same on both shifts be rejected at the five percent significance level?

6. A large company cafeteria is trying to decide if the expected sales per day of main dish A are greater than those of main dish B. Over a period of three months, A is served on 7 days and B is served on 7 different days. The average sales of A per day are 572 with $s_x = 100$ and the average sales of B are 495 with $s_x = 111$. The cafeteria manager claims on the basis of these data that expected daily sales of A are 20 percent higher than B. Is this claim justified? Would the null hypothesis that there is no difference in average sales be rejected at the five percent level?

7. A land developer claims that there is more rain in the Diamond Head area of Honolulu than there is in the Waikiki Beach area. To test this claim the amount of rainfall at selected spots in each area was determined each day for 365 days. The average rainfall in the Diamond Head location was 0.084 inches per day with $s_x = 0.07$ inches, and was 0.073 inches per day in the Waikiki area with $s_x = 0.065$. At a significance level of $\alpha = 0.01$, should the hypothesis that there is no difference in average daily rainfall between the two areas be rejected?

8. A farmer is trying to decide whether one of two brands of feed is better than the other. To do this he selects 200 young pigs and feeds 100 of them brand A for the year and the other 100 brand B. At the end of the year the average weight of a hog fed with A is 151 pounds with $s_x = 18$ pounds and the average weight of a hog fed with B is 158 pounds with $s_x = 13$ pounds. If the null hypothesis that there is no difference between the feeds is tested at the significance level $\alpha = 0.05$, what decision rule is obtained from this? What conclusion should the farmer draw based on this decision rule?

9. A professor makes the claim that students learn better using textbook A than they do textbook B. To test this, 350 students in a class are broken up into two groups of 175 each. One group is taught using text A and the other using text B. At the end of the semester they are all given the same final exam. The average grade for those taught with text A is 71 with $s_x = 18$ and is 65 with $s_x = 21$ for those taught with B. Should the null hypothesis that both texts are equally good be rejected at the five percent significance level?

10. A naturalist claims that the expected weight of salmon caught in streams in Washington state is the same as that for salmon caught in Alaskan streams. To check this claim a sportsman catches 100 salmon in Washington. Their average weight is 8.5 pounds with $s_x = 2.4$ pounds. He also catches 200 salmon in Alaskan waters. The average weight of these is 9.1 pounds with $s_x = 3.1$. At the five percent significance level, should the sportsman reject the null hypothesis that the expected weights are the same in Washington and Alaska?

Section 4–8

1. A textile mill purchases nylon yarn to be used in synthetic fibers. If the expected breaking strength of the yarn is less than 50 pounds, difficulties will be encountered in the spinning and weaving operations because of an excessively high number of breaks. A lot of this yarn consists of thousands of spools. To test the quality of a lot, 25 spools are selected at random, and then a one-inch piece of yarn is selected at random from the yarn on each of these spools to yield 25 one-inch pieces of yarn. To determine the breaking strength, one end of the yarn is clamped,

and the force which must be applied at the other end to break the yarn is noted. The average breaking strength for the 25 pieces is 48.2 pounds, with $s_x = 5.4$ pounds. If the null hypothesis that the mean breaking strength is 50 pounds is tested at a significance level of $\alpha = 0.05$, what is the decision rule which should be used to decide whether or not to accept the lot? Should the lot under consideration be accepted?

2. Re-solve Problem 1 using the t distribution.

3. A large manufacturer of fertilizers uses paper bags to package the fertilizer. It is important that the expected bursting strength of a bag be at least 110 pounds. A lot of 100,000 bags has just been received, and it is desired to test the quality of these. To do so, 100 bags are selected at random and the bursting strength of each is measured. The average bursting strength for these is 104 pounds with $s_x = 12.2$ pounds. Suppose the null hypothesis is made that the mean bursting strength is 110 pounds and this is tested at the one percent significance level. Should the null hypothesis be rejected in this case?

4. A grower of fruit wishes to test out a new spray which a manufacturer claims will reduce the loss due to damage by a certain insect. To test the claim the grower sprays 200 trees with the new spray and 200 with the standard one. The average yield per tree for those treated with the new spray is 240 pounds with $s_x = 47$ pounds while the average yield per tree for those treated with the standard spray is 227 pounds with $s_x = 30$ pounds. Should the null hypothesis that there is no difference between the sprays be rejected at the five percent significance level?

5. A pharmaceutical firm wishes to test two new cold vaccines. To do so, 1000 students are selected in a given city and are divided into two groups of 500 each. One group is innoculated with vaccine 1 and the other with vaccine 2. The average number of days during the season that each student treated with 1 suffered with a cold was 6 with $s_x = 4$, while the average number of days for students treated with 2 was 4.8 with $s_x = 3.5$. Should the null hypothesis that there is no difference between the vaccines be rejected at the one percent level? What is the decision rule implied by the test?

6. The pharmaceutical firm referred to in Problem 5 also wishes to determine whether either vaccine seems to be better than no vaccine at all. For this purpose a control group of 500 students which are not treated with any vaccine is used. The average number of days that each student in this group suffered with a cold during the season was 7.1 with $s_x = 4.8$. Should the null hypothesis that vaccine 1 is of no value be rejected at the one percent level? Should the null hypothesis that vaccine 2 is of no value be rejected at the one percent level?

7. A firm produces shafts and bushings for use on small generators. It is desired to have the expected clearance when a shaft is placed in a bushing be 0.004 inches. If the value is significantly greater than this there will be too much play, and if it is significantly less, the shafts will frequently either not fit into the bushings or will bind. To check on this the diameters of 100 shafts are measured carefully and the average diameter is 1.002 inches with $s_x = 0.003$ inches. The inside diameters of 200 bushings are measured and the average diameter is 1.009 inches with $s_x = $

0.0045. Should the null hypothesis that the expected clearance is 0.004 be rejected at the five percent significance level?

8. A coin is tossed 15 times. For what values of x, the number of heads obtained, would the null hypothesis that the coin is fair be rejected at the five percent significance level? What is the answer if the coin is tossed 10 times instead of 15 times?

Section 4–9

1. A government official has recently been involved in a scandal and he is interested in knowing whether this has influenced his popularity with the voters in his home state. A total of 54 percent voted for him in the previous election. To gain some information he has a polling organization interview 1000 voters selected at random to determine how many would vote for him. The results indicate that 49 percent of those interviewed would still vote for him. There are a total of 700,000 voters in his state. Should the null hypothesis that the fraction of the voters who will vote for the individual is still 54 percent be rejected at the five percent significance level?

2. A University of Chicago professor claims that 70 percent of the faculty live within walking distance of the University and actually walk to work. A new faculty member, desiring to check on this selects at random 50 of the 1000 faculty members. It turns out that 28 live near the University and walk to classes. At the five percent level of significance, should he reject the null hypothesis that 70 percent of the faculty members live near the University and walk to classes? For what range of values of α would the null hypothesis be rejected?

3. An official of a state university claims that the families of at least 50 percent of the students earn over $12,000 per year, and it seems inappropriate that such students should receive free tuition. To check on this a newspaper reporter selects at random 100 of the 12,000 students (all from different families) and determines from each student if the family income is greater than $12,000. It turns out that 38 students indicate this to be the case. Assume the students tell the truth in answering the reporter's question. Should the null hypothesis that at least 50 percent of the student's families earn more than $12,000 per year be rejected at the five percent level of significance? Hint: Can this problem be treated as one where the null hypothesis states that precisely 50 percent of the families earn more than $12,000?

4. A dairy firm is considering the marketing of an imitation milk in a certain area. Whether it is sound to do so depends on the outcome of the forthcoming city elections. One candidate, A, has indicated that he will make an effort to block the sale of such a product in the city, while the other candidate, B, has no objections to its sale. There are only two candidates. The firm has a polling organization conduct a poll of how voters are going to cast their ballots in the coming election. A total of 2000 of the 100,000 registered voters are selected and 1094 indicate that they are going to vote for B. The firm decides that it will test the null hypothesis that 50 percent of the votes will be for B and will reject the hypothesis only if it appears that the percentage who will vote for B is significantly greater than 50. The significance level $\alpha = 0.01$ will be used. The new product will be introduced if the null hypothesis is rejected. Should the new product be introduced?

5. The operator of two large apartment building complexes, one for middle-income families and another for high-income families wishes to determine if there is any difference in the fraction of families which have dogs in the two complexes. To gain some information, he selects at random 100 of the 500 families living in the middle-income development and finds that 42 percent of those checked have dogs. He selects at random 80 of the 400 families living in the high-priced units and finds that 38 percent of those interviewed have dogs. Should the null hypothesis that the fraction of families having dogs is essentially the same be rejected at the five percent significance level? Hint: To estimate the standard deviation, it is necessary to use some value for p. What is a reasonable way to estimate p?

6. A firm which is nonunionized has two plants which are located in different cities. It is interested in determining whether there is any difference in the percentage of employees at the two plants which favor unionization. To do this it selects a random sample of 100 of the 600 employees at one plant and finds that 28 percent favor a union. A random sample of 250 of the 1200 employees at the other plant are selected and it is found that 32 percent favor a union. Should the null hypothesis that there is no difference between the proportion of the employees favoring a union at the two plants be rejected if the test is made at the ten percent significance level?

Section 4–10

1. Determine the sample size and critical number that should be used in Problem 1 for Section 4–4 if the alternative hypothesis H_1 that $\mu_x = 0.2550$ is introduced and it is desired that $\beta = 0.10$. Use $\alpha = 0.05$. Assume that $\sigma_x = 0.0020$ in all cases.

2. Determine the sample size and critical number that should be used in Problem 3 for Section 4–4 if the alternative hypothesis H_1 that $\mu_x = 16,000$ is introduced and it is desired to have $\beta = 0.01$. Use $\alpha = 0.01$. Assume that $\sigma_x = 3000$ miles in all cases.

3. Determine the number of days for which the air should be sampled for the situation discussed in Problem 6 for Section 4–4 if the alternative hypothesis H_1 that $\mu_x = 3.5$ grams is introduced and it is desired that $\beta = 0.10$. Use $\alpha = 0.10$. Assume that $\sigma_x = 2$ grams in all cases. What is the critical number?

4. Determine the number of weeks that the advertising program should be carried out in Problem 8 for Section 4–4 if the alternative hypothesis H_1 that the expected weekly sales will be increased to $220,000 is introduced and it is desired that $\beta = 0.10$. Use $\alpha = 0.05$. Assume that $\sigma_x = \$40,000$ in all cases. What is the critical number?

5. For how many days should production be checked in Problem 9 for Section 4–4 if the alternative hypothesis H_1 is introduced that productivity will be increased to 130 with $\sigma_x = 17$, and it is desired that $\beta = 0.05$. Use $\alpha = 0.01$.

6. Consider a situation where a given sample size is being used to test the null hypothesis H_0 that $\mu_x = \mu_0$. Consider the alternative hypothesis H_1 that $\mu_x = \mu_1$. For each μ_1 there is determined a value of β. The curve which gives $1 - \beta$ as a function of μ_1 is called the power curve for the test. Note that $1 - \beta$ is the probability

of rejecting H_0 when H_1 is true. Explain the relation between the power curve and the operating characteristic curve. Determine the power curve for Problem 3 of Section 4–4. Assume $\sigma_x = 0.0020$ always.

7. Determine the sample size and critical values that should be used in Problem 1 for Section 4–5 if the alternative hypothesis H_1 that $\mu_x = 1025$ or 975 is introduced and it is desired that $\beta = 0.05$. Use $\alpha = 0.05$. Assume that $\sigma_x = 100$ ohms in all cases.

8. Determine the sample size that should be used in Problem 3 for Section 4–5 if the alternative hypothesis H_1 that $\mu_x = 35.5$ or 34.5 is introduced and it is desired that $\beta = 0.10$. Assume $\sigma_x = 0.8$ ounces always.

9. Determine the sample size and critical values that should be used in Problem 5 for Section 4–5 if the alternative hypothesis H_1 that $\mu_x = 0.2505$ or 0.2495 is introduced and it is desired that $\beta = 0.01$. Use $\alpha = 0.01$ and assume that $\sigma_x = 0.008$ always.

10. For the situation described in Problem 9, draw the curves representing the distribution of $\bar{x}$ when H_0 is true and also when H_1 is true.

11. Determine the sample size and critical values that should be used in Problem 7 for Section 4–5 if the alternative hypothesis H_1 that $\mu_x = 0.85$ or 0.75 is introduced and it is desired that $\beta = 0.05$. Use $\alpha = 0.05$ and assume $\sigma_x = 0.15$ grams always.

12. Determine the sample size and critical value that should be used in Problem 1 for Section 4–7 if the alternative hypothesis that the mean life of brand 1 is 200 hours greater than that of brand 2 is introduced, and it is desired that $\beta = 0.05$. Use $\alpha = 0.05$ and assume $\sigma_1 = 500$ and $\sigma_2 = 200$ always. Assume that the same number of batteries of each brand will be tested.

13. Determine the sample size and critical value that should be used in Problem 3 for Section 4–7 if the alternative hypothesis is introduced that the manufacturer's toothpaste should produce 0.5 fewer cavities than the competitor's, and it is desired that $\beta = 0.05$. Use $\alpha = 0.05$ and assume that σ is the same for both toothpastes and has the value 1.85. Assume that the same number of children will be tested with each brand.

14. Determine the sample size and critical value that the chain store of Problem 4 for Section 4–7 should use if the alternative hypothesis is introduced that expected sales in store 1 are $50 greater than in store 2, and it is desired that $\beta = 0.10$. Use $\alpha = 0.05$ and assume $\sigma_1 = \$86$ and $\sigma_2 = \$95$ always. Assume that the same number of accounts will be checked in each store.

15. Determine the sample size and critical value that should be used in Problem 5 for Section 4–7 if the alternative hypothesis is made that the expected fraction defective on the third shift is 0.01 greater than on the first shift, and it is desired that $\beta = 0.05$. Use $\alpha = 0.05$. Assume that the same number of units will be examined from the output of each shift.

16. What sample size and critical value should the farmer use in Problem 8 for Section 4–7 if the alternative hypothesis is made that B results in an expected

weight increase of 10 pounds over A and it is desired that $\beta = 0.05$? Use $\alpha = 0.05$ and assume that the standard deviation of the weight for both brands is 15 pounds.

17. What sample size and critical value should be used in Problem 1 for Section 4-8 if the alternative hypothesis that $\mu_x = 48$ pounds is introduced and it is desired that $\beta = 0.10$. Use $\alpha = 0.05$ and assume $\sigma_x = 5$ in all cases.

18. What sample size and critical value should be used in Problem 3 for Section 4-8 if the alternative hypothesis that $\mu_x = 100$ is made and it is desired to have $\beta = 0.01$? Use $\alpha = 0.01$ and assume $\sigma_x = 12$ always.

19. What sample size and critical value should be used in Problem 4 for Section 4-8 if the same sample size is to be used for trees treated with the new spray and those treated with the standard spray, and the alternative hypothesis that the new spray increases the expected yield by 25 pounds is introduced. It is desired that $\alpha = \beta = 0.05$. Assume that the standard deviation for the new spray is 50 and is 35 for the standard one.

20. What sample size and critical value should be used in Problem 7 for Section 4-8 if the alternative hypothesis is introduced that the expected clearance is 0.003 and it is desired that $\beta = 0.05$. Use $\alpha = 0.05$ and assume the standard deviation of the shaft diameters is 0.003 while that of the bushings is 0.0045. These values are independent of the means. Also assume that the same number of bushings will be examined as shafts.

21. Just as in estimation problems, it is not possible to determine the sample size to be used in testing hypotheses without a knowledge of the standard deviations. Why is this so? How might one proceed if the standard deviations were not known?

22. An electronics manufacturer has received a lot of 100,000 resistors to be used in a particular production run. They have a nominal resistance of 1000 ohms. For the type of circuits in which they are to be used, it is much more serious if the resistance is less than 1000 ohms than if it is greater than this value. The manufacturer believes that if the expected resistance μ_x of any unit is 1000 ohms he should accept the lot, but if it is 950 ohms he should reject the lot. Let H_0 be the hypothesis that $\mu_x = 1000$ and H_1 the hypothesis that $\mu_x = 950$. Suppose that it is desired that $\alpha = \beta = 0.05$. What sample size should be used, and what decision rule should be employed to decide whether to accept or reject the lot? Assume $\sigma_x = 50$ always.

23. For the situation described in Problem 1 for Section 4-4, draw on the same figure the curve representing the distribution of $\bar{x}$ when H_0 is true and that representing the distribution of $\bar{x}$ when the alternative hypothesis H_1 that $\mu_x = 0.2520$ is true. Assume $\sigma_x = 0.0020$ in both cases. What is the probability β of a type 2 error here?

24. For the situation described in Problem 3 for Section 4-4, draw on the same figure the curve representing the distribution of $\bar{x}$ when H_0 is true and that representing the distribution of $\bar{x}$ when the alternative hypothesis H_1 that $\mu_x = 16,000$ is true. Assume $\sigma = 3000$ in both cases. What is the probability β of a type 2 error here?

25. For the situation described in Problem 1 for Section 4–8, draw on the same figure the curves representing the distribution of $\bar{x}$ when H_0 is true and that representing the distribution of $\bar{x}$ when the alternative hypothesis H_1 that $\mu_x = 48$ is true. Assume $\sigma_x = 5$ in both cases. What is the probability β of a type 2 error here?

26. Determine the operating characteristic curve for the test obtained in Problem 1.

27. Determine the operating characteristic curve for the test obtained in Problem 2.

28. Determine the operating characteristic curve for the test obtained in Problem 7.

29. A cement manufacturer purchases bags in lots of 10,000 from a local supplier. The mean bursting strength of the bag determines its quality; the higher the mean bursting strength the better the quality. To monitor the quality a sample of n bags is selected from every lot and the bursting strength of each is measured. Based on the average bursting strength a decision is made either to accept or to reject the lot. It is desired to accept lots having an expected bursting strength of 250 pounds about 90 percent of the time and to reject those having a mean bursting strength of 200 pounds about 90 percent of the time. Experience indicates that the standard deviation of the bursting strength is about 50 pounds. What sample size and decision rule should be used?

30. The U.S. Defense Department buys packages of dried milk in lots of 100,000. The principal quality determinant of the milk itself is the butterfat content, which is measured on a special scale designed for this purpose. The higher the mean butterfat content, the higher the quality of the milk. It is desired to accept lots for which the expected butterfat content is 7 about 90 percent of the time and to reject those for which it is 5 about 90 percent of the time. The standard deviation of the butterfat content is about 0.70. Determine the number of packages which should be inspected and the decision rule to be used.

31. The U.S. Army buys bars of soap whose nominal weight is 6 ounces in lots of 25,000 bars. Occasionally a supplier will attempt to cheat on the weight of the bars. To monitor this a sample of n bars is selected from each lot and the weight of each is carefully determined. Based on the average weight of the bars in the sample a decision is made either to accept or to reject the lot. It is desired to use a sampling plan which accepts lots having an expected weight of 5.95 ounces about 95 percent of the time and rejects those with a mean weight of 5.40 ounces about 90 percent of the time. Determine the sample size and the decision rule which should be used if $\sigma = 0.30$ always. Illustrate the operating characteristic curve for this sampling plan.

32. A large utility firm buys coal by the trainload for one of its plants. The plant operates in an area which restricts the quantities of certain sulfur compounds which can be ejected into the atmosphere. If the sulfur content of the coal is sufficiently low, no treatment of the stack gases from the boiler is necessary, but if the sulfur content is too high, the gases must be subjected to a fairly expensive treatment. To decide what can be done with any given trainload of coal, a sample of n pieces of coal is selected at random from various parts of the train and the sulfur content of

each is determined. Based on the average sulfur content of the sample, a decision is made either to accept the load (which means that no special treatment of the gases will be used) or to reject the load (which means that the special treatment process must be used). It is desired to accept loads having a sulfur content of 0.5 percent by weight about 90 percent of the time and to reject those having a sulfur content of 1.5 percent about 90 percent of the time. Determine the sample size and decision rule to be used in this case if the standard deviation of the sulfur content is 0.5 percent by weight.

33. Consider a coin and imagine that we make the null hypothesis H_0 that the probability p of a head satisfies $p \leq \frac{1}{2}$. Suppose that we take as the alternative hypothesis H_1, which says $p > \frac{1}{2}$. To test the hypothesis H_0, we toss the coin n times and use the random variable x, the number of heads obtained, as the statistic. Show that the distribution of x, given that H_0 is true, is not uniquely determined, and that neither is the distribution of x when H_1 is true. To determine the distribution of x, p must be specified. Suppose now that a value of n and a critical value δ for x are also selected. Given n and δ, let $q_1(p)$ be the probability of rejecting H_0 when H_0 is true and the probability of a head is p, $p \leq \frac{1}{2}$. Also let $q_2(p)$ be the probability of accepting H_0 when H_1 is true and the probability of a head is p, $p > \frac{1}{2}$. Consider the function

$$
q(p) = \begin{cases} q_1(p), & p \leq \dfrac{1}{2} \\ q_2(p), & p > \dfrac{1}{2}. \end{cases}
$$

This is the probability of making an error when the probability of a head is p. Show that $q_1(p)$ and $q_2(p)$ decrease as p gets farther and farther away from $\frac{1}{2}$. Also show that $q(p)$ can be discontinuous at $p = \frac{1}{2}$. Plot $q(p)$ for $n = 10$ and for $\delta = 4, 5, 6, 7$. Is it possible in this case to specify arbitrarily the maximum probability of a type 1 error and of a type 2 error? What is the difficulty here?

34. Let x be a random variable which has the distribution $n(x; \mu, 2)$. Suppose that the null hypothesis H_0 is made that $\mu \leq 4$, and the alternative hypothesis H_1 is taken to be $\mu > 4$. To test H_0, n independent trials of a random experiment are performed, where on each trial a value of x is observed. Let $\bar{x}$ be the statistic used to test H_0. Let δ be the critical value of $\bar{x}$ and $q(\mu)$ be the probability of making a wrong choice when the expected value of x is μ. The function $q(\mu)$ can be determined as in Problem 33. For $n = 100$, plot $q(\mu)$ in the cases where $\delta = 3.8, 4, 4.2$. Is it possible in this case to specify arbitrarily the maximum probability of a type 1 error and a type 2 error? What is the difficulty here?

35. It was shown in Problems 33 and 34 that difficulties can be encountered in testing an hypothesis such as $\mu \leq \mu$ against the alternative $\mu > \mu_0$. A typical way to resolve this difficulty is to reformulate the problem by selecting a $\mu_1 < \mu_0$ and a $\mu_2 > \mu_0$ and testing the hypothesis $\mu = \mu_1$ against $\mu = \mu_2$, the probabilities α and β being specified. This now reduces to a problem of the type studied in the text. For the situation outlined in Problem 29, suppose that H_0 is taken to be $p = 0.45$ and H_1 to be $p = 0.55$ with $\alpha = \beta = 0.1$. Determine the experiment to be performed and the critical value of x in this case.

36. Let x be a random variable with the distribution $n(x; \mu, \sigma)$, where σ is known. Discuss the difficulties encountered in testing the hypothesis H_0, which is $\mu = \mu_0$, against H_1, which is $\mu \neq \mu_0$. Can one specify arbitrarily the probabilities α and β in this case? What is the difficulty here, and how might it be avoided?

Section 4–11

1. The null hypothesis is made that a coin is fair. To test this, the coin is tossed 25 times. A total of 17 heads is obtained. According to the chi-squared test, should the null hypothesis be rejected at a five percent significance level?

2. The null hypothesis is made that a die is fair. To test this the die is tossed 51 times. The number of times that the six different faces appeared are, respectively, 13, 6, 10, 9, 5 and 8. According to the chi-squared test, should the null hypothesis be rejected at a five percent significance level?

3. The null hypothesis is made that the long-run fraction defective produced by a process is 0.10. To check this, 100 units coming off the line are checked, and it is found that 6 are defective. If a chi-squared test is applied using a significance level of one percent, should the null hypothesis be rejected?

4. The industrial engineering department of a firm has constructed a model to represent the demand for a given spare part on any day. The model has four simple events corresponding to having 0, 1, 2 or 3 units demanded, with the probabilities of these simple events being 0.30, 0.20, 0.35 and 0.15 respectively. To check this model, the demand for the spare is noted on each of 51 days. The number of times that 0, 1, 2 and 3 units were demanded turned out to be 18, 7, 20 and 6. If the null hypothesis is made that the model is correct and a chi-squared test is applied using a significance level of $\alpha = 0.05$, should the null hypothesis be rejected?

5. Consider the same situation described in Problem 4, but now suppose that the number of days on which 0, 1, 2 and 3 units were demanded turned out to be 12, 13, 22 and 3. Suppose also that on one day four units were demanded. Re-solve Problem 4 under these assumptions.

6. How might the chi-squared test be used to test the hypothesis that a continuous random variable has a specified distribution such as a normal distribution $n(x; \mu, \sigma)$? Hint: Subdivide the range of variation of x into class intervals.

7. Use the results of Problem 6 to test at the five percent significance level the null hypothesis that the diameter of a ball bearing made by the process which generated the data in Table 2–11 is normally distributed with mean $\mu = 0.2500$ and standard deviation $\sigma = 0.0019$.

8. Use the results of Problem 6 to test at the five percent significance level the null hypothesis that the random variable, values of which are given in Table 3–1, has a uniform distribution over the unit interval.

9. Show that the expected value of χ^2 in (4–23) is $m - 1$.

10. To test Mendel's theory, 1000 seeds obtained from crossing Aa plants with Aa plants are planted and the colors of the flowers on the resulting plants are noted. It turns out that 225 have red flowers, 285 have white flowers, and the remainder have pink flowers. Use the chi-square test to determine whether Mendel's theory should be rejected at the five percent level of significance.

CHAPTER 5

Modern Decision Theory

5-1 INTRODUCTION

In Chapter 4 we studied what is now referred to as the classical approach to decision making. The word classical is used because in the last fifteen or so years a new, quite different and much more comprehensive theory of decision making, which is sometimes called modern decision theory, has evolved. In this chapter we wish to provide the reader with an introduction to the modern theory of decision making.

The scope of decision problems which can be treated by the theory developed in Chapter 4 is extremely limited. It concentrated on situations where the decision maker was considering only two, or at most three, alternative courses of action. The appropriate action to take depended on the expected value of some random variable. This expected value was unknown, but the possibility existed of performing a random experiment, before making the decision, to obtain more information about the value of this parameter. The reader will be well aware that it is by no means true that all decision problems fall into the framework just described. However, even for problems which do fall into this framework, the reader may well have felt that something was lacking in the approach taken in the previous chapter, since economic considerations, that is, costs or profits, did not enter directly in determining the decision rule. The modern theory of decision making encompasses a much broader class of decision problems than does the classical theory. It also has the advantage that, among other things, it integrates the economic considerations directly into the determination of decision rules. The greatest interest in these new methods has been in the fields of business, economics and engineering, where costs and profits arise in a natural way. However, interest is also growing rapidly in many other

areas of the social sciences. Let us then proceed to a study of these newer developments.

5–2 THE STRUCTURE OF SINGLE-STAGE DECISION PROBLEMS

We shall begin with an example. Let us consider once again the problem faced by a supermarket manager in deciding how many loaves of a particular type of bread to stock on a given day of the week. The manager has the freedom to select the number of loaves to be stocked at the beginning of the day. However, his problem is complicated by the fact that he does not know precisely how many loaves will be demanded during the day. If he stocks too much there will be bread left over which will have to be sold at a loss on the stale-bread counter. If he stocks too little, profits will be lost on those loaves which he could have sold if they had been stocked. In addition, customers finding the store out of bread may become annoyed and buy less of other things, causing additional losses.

Let us now study the structure of this problem in a little more detail. We can describe the time sequence of events in a somewhat more abstract way as follows. First the decision maker selects the *action* to be taken, the action here referring to the number of loaves to be stocked. Next there follows the events of the day which determine the number of loaves that will be demanded. A convenient way to represent this is as the performance of a random experiment. We can imagine that after a decision is made on the number of loaves to be stocked, a random experiment is then performed which determines the number of loaves to be demanded. The number of loaves demanded can then be looked upon as a random variable x having a probability function $p(x)$. It will be assumed that on the basis of historical records the decision maker can build a model for this random experiment. After the decision is made on the number of loaves to be stocked and after the random experiment is performed which determines the demand, then the supermarket's profit on this type of bread for the day under consideration will be determined. Note that the profit will depend both on the action taken and on the outcome of the random experiment which is performed.

The problem we have just discussed is a special case of a general type of decision problem which we shall refer to as a *single-stage decision problem*. We now wish to describe the structure of the general single-stage decision problem. In formulating decision problems we shall find it convenient to imagine that each time a decision is made the decision maker selects precisely one of a finite number of alternatives. Things will be defined in such a way that *only one* of the alternatives will be selected. It is typical to refer to these alternatives as possible *actions;* the decision maker will then be selecting one action to take. The notion of an action will be used here in a

somewhat more general way than it normally is in everyday speech. By the term action we may be referring to something very complicated, involving a number of what we usually think of as actions, or it might be something as simple as deciding to carry one's umbrella on a given day. A relatively complicated action would be the building of five new plants of given sizes, in given locations, at given times, the introduction of two new products in a foreign market and the initiation of a profit sharing plan for employees. Note that carrying out an action may require an extended period of time. It is sometimes necessary to use relatively complicated actions in order that the decision maker will be selecting precisely one and not several actions simultaneously.

In general terms, a single-stage decision problem is one that can be viewed as a problem in which the decision maker makes precisely one decision. There then follows a random experiment, the outcome of which determines the *prize* to be received by the decision maker. Let us attempt to characterize in more detail what we mean by a prize. The term prize simply refers to all the consequences of the situation which may be relevant to the decision maker. The word prize does not necessarily imply something favorable and there is usually no implication that the prize is a monetary reward. For example, if the decision situation is one in which a general is contemplating a military coup, the prize may be the firing squad if things go wrong or the premiership of the country if things go well. The prize is not something which is necessarily obtained at a single point in time. It may be revealed over a period of many years. Thus if a decision situation involves the choice of a career, the results of this decision will develop over a considerable period of time. Thus when we refer to a prize we should have a mental picture of a written-out description, which may be very complicated and take up many pages, explaining precisely what the prize is. It is important to note that the action selected and the outcome of the random experiment determine completely the prize. There is no uncertainty about what the prize is once these two things are specified.

We shall refer to the simple events in the random experiment whose outcome influences critically the prize obtained by the decision maker as the *states of nature*. What are these states of nature? They are simply the various real-world events which determine the outcome of any action selected. The different states of nature correspond to the different real-world events which can occur, and which have the property that at the time the decision maker selects an action he does not know which set of real-world events will ultimately be realized. The state of nature in the bread-stocking problem is suitably described by the number of loaves demanded during the day. There is a different state of nature corresponding to each possible number of loaves which might be demanded. In this case the states of nature are quite simple conceptually and can easily be characterized by num-

bers. The situation, however, is not always this simple. Let us illustrate with an example.

Suppose that before leaving home in the morning an individual is trying to decide whether to carry his umbrella. The states of nature will here describe the possible types of weather that can occur, if weather is the only thing that influences his decision as to whether or not to carry the umbrella. If other factors enter in, then they too should be considered in specifying the states of nature. However, even when we restrict the states of nature to describing the behavior of the weather, we note immediately that we can think of an arbitrarily large number of different weather patterns which might take place during the day, since we could describe the precise temperature at each minute; precisely when it rains if it does, and how much; precisely when the sun is shining; whether the sky is cloudy when the sun is shining, and so on. It becomes clear at once that it is hopeless to try to describe the possible state of nature in great detail. Fortunately, it is also important to note that it is unnecessary to do this. All the individual is trying to do is decide whether or not to carry his umbrella. Therefore, all we need to do is represent enough about the various weather possibilities to let him make a sound decision. For example, it would probably be adequate to consider that there are four possible states of nature. These are: (1) the sun shines all day long; (2) the sun shines when the individual is out, but it rains at some other times; (3) it rains when he is out, but the sun shines part of the time; and (4) it rains all day.

We have characterized a single-stage decision problem as one in which the decision maker makes only one decision. Alternatively, a single-stage decision problem can be characterized as one in which precisely two decisions are made. First the decision maker selects an action and then nature selects the state of nature. The net result of these two decisions determines the prize to be obtained by the decision maker. At the time the decision maker selects an action, he does not know what state will be selected by nature. However, we are assuming that the decision maker is able to construct a probability model which gives the probability of occurrence of each state of nature. The use of the terminology that nature selects a state of nature has no metaphysical connotations here. It is simply a convenient terminology to use in describing the structure of the problem.

We are now ready to set up a model of a single-stage decision problem. Assume that there are r different actions which the decision maker is considering; denote these by $a_1, \ldots, a_r$. It will be supposed that there are m states of nature $e_1, \ldots, e_m$. We shall assume that the set E of states of nature is independent of which particular action happens to be selected. However, the probability p_{ij} that the state of nature will turn out to be e_j will, in general, depend on the action a_i selected. There is no real loss in generality in assuming that E is independent of the action selected, since

if some particular state of nature cannot occur for a given action we merely assign it a probability of 0. After the decision maker selects an action, say a_i, we imagine that a random experiment $\Re$ is performed. The model $\mathcal{E}_i$ used to represent $\Re$ depends on a_i. The event set $E = \{e_1, \ldots, e_m\}$ is always the same. However, the probabilities of the simple events will, in general, change with i; the probability of e_j in $\mathcal{E}_i$ we have denoted by p_{ij}. Once an action a_i is selected and $\Re$ is performed, the prize for the decision maker is determined. If a_i is the action and the state of nature turns out to be e_j, the prize will be denoted by Q_{ij}; Q_{ij} does not necessarily represent a number but is instead the symbolic representation for the detailed description of the prize.

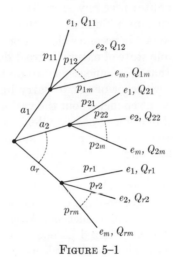

FIGURE 5–1

The nature of a single-stage decision problem can be conveniently represented by a tree similar to that used to represent two-stage experiments. Such a tree is shown in Figure 5–1. From a given node, r branches are drawn, one for each action. These are labeled by the symbol for the action. Then from the terminal point of each of these branches are drawn m branches, one for each state of nature. These are labeled with the corresponding probabilities of occurrence of the state of nature. At the ends of these branches are given the symbols for the states of nature and the prizes. The tree shown in Figure 5–1 is called a *decision tree*. It can be imagined that at each node a decision is made. At the first node the decision maker selects an action. At the remaining nodes nature selects the state of nature.

It is by no means true that all decision problems can be characterized as single-stage problems. Indeed, some of the most interesting and important decision problems are what we refer to as multi-stage or sequential decision problems. The basic characteristic of a single-stage decision prob-

lem is that only a single decision is made by the decision maker. After this decision and after nature selects the state of nature the prize received by the decision maker is determined.

The next type of problem in order of complexity is called a two-stage decision problem. A decision problem which is referred to as a two-stage decision problem has the characteristic that precisely two decisions must be made by the decision maker. Furthermore, after the first decision, but before the second, a random experiment is performed. The outcome of this random experiment has an important influence upon the second decision, and thus the second decision must be delayed as long as possible to learn as much as possible about the outcome of this random experiment. After the second decision, there follows another random experiment. Then the prize obtained by the decision maker is determined.

As a simple example of a two-stage decision problem, suppose that the supermarket we have been considering can call at noonday and order an additional supply of bread which will be delivered at 1 P.M. No additional bread can be obtained during the day after the 1 P.M. delivery. Now the manager must decide how much to order for morning delivery and how much for the afternoon delivery. The amount to be ordered for afternoon delivery will depend on how much bread was purchased during the morning. The morning demand can be imagined to be the outcome of a random experiment. This experiment influences the second decision. After the second decision another random experiment is performed which determines the demand for bread in the afternoon. Then the profit resulting from the day's decisions is determined.

Decision problems in which precisely n decisions must be made by the decision maker, with a random experiment taking place between each decision and after the last one, are called n-stage decision problems. Sometimes the number of decisions is not fixed in advance, but depends on the actions selected and the outcomes of the random experiments. Such problems are referred to in general as sequential decision problems. The term sequential decision problem is also used to refer to n-stage problems. Sequential decision problems are generally quite difficult to solve and we shall not study them. We shall study mainly simple varieties of single-stage problems. We shall, however, show how to solve one particular type of two-stage problem.

5–3 THE BASIC DECISION RULE

Let us now return to studying single-stage decision problems. Suppose that a model has been constructed for some particular problem, that is, a decision tree of the form of Figure 5–1 is available. It will be supposed that the decision maker believes this to be an accurate representation of the

situation. The fundamental question which now arises is that of how the decision maker should select the action to take. This is a surprisingly difficult question to answer in general, even when all the information contained in the model is available.

To investigate this in more detail, let us return to the problem of stocking bread in the supermarket. In this situation, it seems very reasonable to use the profit received by the store from selling the bread as a description of the prize received for stocking any given number of loaves when the demand turns out to be a specified value. In other words, in this case it is possible to use a numerical measure to characterize the prize, and this measure is simply the profit received.

Suppose now that the manager of the supermarket always stocks i loaves of bread on the particular day of the week under consideration. The profit received as a result of this will not always be the same, however, since the demand will fluctuate from one week to another, and the profit depends on the demand x as well as on i. Let U_{ij} be the profit for the day when i loaves are stocked and the demand turns out to be for j loaves. The probability that $x = j$ is $p(j)$. Although the profit on any particular day obtained from stocking i units is a random variable, the long-run average profit that will result for this day of the week if i loaves are always stocked can be determined; it is nothing but the expected profit. Denote by $\pi(i)$ the expected profit when i loaves are stocked. Then

$$\pi(i) = \sum_{j=1}^{m} U_{ij} p(j) . \qquad (5\text{--}1)$$

Inasmuch as the supermarket manager faces the same problem each week on the particular day under consideration, it seems reasonable that he should select the number of loaves to stock in such a way as to maximize the long-run average profit, that is, the expected profit (5–1). In this way things will work out as well as possible in the long run. Here we have obtained a very useful generalization of the generally accepted criterion of profit maximization for deterministic situations. When chance plays a role it appears appropriate to maximize the expected profit.

The rule that the action to select is the one which maximizes the expected profit is a very appealing one and is widely used. Indeed, it is the rule which we shall use here. It is important to recognize at the outset, however, that it is not a rule of universal validity. Let us now investigate why this is so. First of all, it is not in general true that the profit fully characterizes the prizes. There may be other important factors which must be taken into account. In some cases, the prizes are in no way connected with profits and hence cannot be characterized by profitability. The plausibility for maximizing expected profits was obtained by considering a situation which would be repeated over and over again. Not all decision problems

have this characteristic. Some problems occur only once and are not expected to occur again. For such problems it is not at all clear that it is appropriate to select that action which maximizes the expected profit. Here we have pointed out a number of situations where it is not clear that maximization of expected profits is appropriate. However, even in cases where the prizes can seemingly be described by the profit and where, in addition, the situation may possibly be repeated over and over again, it is still not necessarily true that profit maximization is the appropriate rule to use. We shall now illustrate by an example why this is the case.

Consider the following greatly simplified situation. A businessman is contemplating which of two contracts, *a* or *b*, he should undertake. The ultimate profit to be received from the contracts will be determined by a sequence of unpredictable events. To make the situation as simple as possible, let us suppose that for each contract the businessman decides that the profit received will have only two possible values, and in each case, one of these values will represent a loss. The decision maker has also estimated the probability of each of these outcomes for each contract. The results are summarized in Table 5–1. Thus the businessman believes that

TABLE 5–1

SUMMARY OF TWO CONTRACTS BEING CONSIDERED

Contract *a*		Contract *b*	
Profit	Probability	Profit	Probability
$100,000	0.60	$50,000	0.30
−$100,000	0.40	−$10,000	0.70

if he takes contract *a* he will either get $100,000 with probability 0.60 or lose $100,000 with probability 0.40, and if he takes contract *b*, he will either get $50,000 with probability 0.30 or lose $10,000 with probability 0.70. The expected value of the profit received for contract *a* is

$$100,000(0.60) - 100,000(0.40) = \$20,000 \ ,$$

and the expected value for contract *b* is

$$50,000(0.30) - 10,000(0.70) = \$8000 \ .$$

Thus if the businessman selected the contract with the largest expected value of the profit received, he would select contract *a*. It is by no means true, however, that every businessman would prefer contract *a* to contract *b*. Why is this? The reason is that some businessmen would find a loss as large as $100,000 horrifying to contemplate. Such a loss might wipe out such businessmen completely, and they simply could not afford to undertake any contract which held out the possibility of so large a loss, regard-

less of the fact that the expected value of the profit received is as large as \$20,000. Such businessmen would prefer contract *b* where the maximum loss of \$10,000 is much smaller. They would prefer contract *b* in spite of the fact that the expected profit received from contract *a* is greater than from contract *b*. Other businessmen, for whom the loss of \$100,000 would not be such a serious event, would probably prefer contract *a*. What this means is that expected monetary values do not always accurately reflect a decision maker's true feelings about situations. This is especially true when the sums of money involved are large compared to the decision maker's resources.

We have now seen that there are a variety of situations in which it may not be appropriate for a decision maker to select the action which maximizes the expected profit. What are suitable rules to use in these other cases? It is a remarkable fact that if the decision maker behaves in such a way as to satisfy a set of rather reasonable assumptions, then there exists a numerical function, often called the decision maker's *utility function*, such that the appropriate action to take is the one which maximizes the expected utility. This utility function makes it possible to handle each of the cases above where it seemed that maximization of expected profit was inappropriate. The utility function takes into account all of the decision maker's feelings about the prizes and thus is a very personal thing. Different decision makers may have different utility functions and hence may prefer different actions in the same situation. It also follows from this that maximization of expected utility will not necessarily lead to the choice of the same action as maximization of expected profit. Thus in maximizing his expected utility the businessman who preferred contract *b* in Table 5–1 would indeed be led to select this contract even though maximization of expected profit would lead to the selection of contract *a*.

We shall not attempt to study the theory of utility in this text. The theory of utility and its applications to decision making are considered in some detail in reference [2]. We shall restrict our attention to problems in which maximization of expected profits is equivalent to maximization of expected utility, that is, to situations where the appropriate rule to use in selecting the action to take is the maximization of expected profits. Most business problems in which the sums involved are not too large fall into this category. It is important to recognize, of course, that the procedures we shall be using applies to an even wider class of problems if the profits are replaced by utilities. We shall in our development make use of one result which follows from the theory of utility but which is not especially obvious from our above discussions. This is the fact that it is appropriate to maximize expected profits even in situations which will not be repeated, provided that the profits do adequately represent the decision maker's feelings. Thus even for situations which will never be repeated again we

shall use the criterion of maximizing expected profits to determine the most appropriate (often called optimal) action to take.

Before going on, it will be helpful to note that the profit-maximization criterion can be stated in some alternative forms which will be found useful later. Let $U\{i\}$ be the random variable representing the profit to be received when the decision maker selects action a_i. Now by the definition of profit

$$U\{i\} = T\{i\} - C\{i\} , \qquad (5\text{--}2)$$

where $T\{i\}$ is the revenue received and $C\{i\}$ is the cost incurred, both $T\{i\}$ and $C\{i\}$ will, in general, be random variables. Denote by $\psi(i)$ the expected revenue to be received and by $\gamma(i)$ the expected cost to be incurred if action a_i is selected. Then since the expected value of the difference of two random variables is the difference of the expected values (as we showed in Chapter 1), it follows that

$$\pi(i) = \psi(i) - \gamma(i) , \qquad (5\text{--}3)$$

$\pi(i)$ being the expected profit. Equation (5–3) sometimes provides a useful way to compute the expected profit. It also states that maximization of the expected profit is equivalent to maximization of the difference between expected revenues received and expected costs.

Let us now consider a case where the revenue received is not a random variable but is instead a constant T, which is independent both of the action selected and of the state of nature. In other words, the revenue to be received does not depend on the action the decision maker selects or on the final state of nature. In this case, the expected revenue is merely T and (5–3) becomes

$$\pi(i) = T - \gamma(i) . \qquad (5\text{--}4)$$

We can now note from (5–3) that the i which yields the largest value of the expected profit $\pi(i)$ must be the i which yields the smallest value of the expected cost $\gamma(i)$. Consequently, we have reached the important conclusion that when the revenues received are independent of the action selected or of the state of nature, the same action will be selected by minimizing the expected cost as would be obtained by maximizing the expected profit. The criterion of minimization of expected cost is often a convenient one to use when revenues are fixed. A special case of (5–4) is that where there are no revenues, so that $T = 0$. In such cases, minimization of expected cost is the criterion that we shall normally employ.

5–4 SOLUTION OF SINGLE-STAGE DECISION PROBLEMS

One way to represent the structure of the model of a single-stage decision problem is through the use of a decision tree, such as that shown in Figure

TABLE 5–2

NORMAL FORM FOR A SINGLE STAGE DECISION PROBLEM

	e_1	e_2	e_3	e_4	$\ldots$	e_m
a_1	p_{11} U_{11}	p_{12} U_{12}	p_{13} U_{13}	p_{14} U_{14}	$\ldots$	p_{1m} U_{1m}
a_2	p_{21} U_{21}	p_{22} U_{22}	p_{23} U_{23}	p_{24} U_{24}	$\ldots$	p_{2m} U_{2m}
. . .	. . .					
a_r	p_{r1} U_{r1}	p_{r2} U_{r2}	p_{r3} U_{r3}	p_{r4} U_{r4}	$\ldots$	p_{rm} U_{rm}

5–1. This is sometimes referred to as the *extensive form* for the decision problem. An alternative representation which can be especially useful when making numerical computations is referred to as the *normal form* and is illustrated in Table 5–2. The normal form is nothing but a table or tableau, as it is sometimes called. There is a row in this table for each action and a column for each state of nature. In the cell formed by row a_i and column e_j are written the numbers p_{ij}, the probability that the state of nature will be e_j when action a_i is selected, and the corresponding profit U_{ij}. The ith row of the table then describes the possible outcomes if action a_i is selected. Since the e_j form the simple events for the experiment which is performed, we see that the probabilities in each row must sum to 1. Table 5–2 is often referred to as the *payoff table* for the problem.

If the normal form for a single-stage decision problem is available, it is then a straightforward matter to determine the optimal action. The action which the decision maker should take is the one which yields the largest value of the expected profit $\pi(i)$,

$$\pi(i) = \sum_{j=1}^{m} U_{ij} p_{ij} . \tag{5–5}$$

To determine the largest expected profit, the expected profit can be determined for each action, that is, $\pi(i)$ is computed for $i = 1, \ldots, r$, and the largest of these numbers is determined. The action to take is the one

which yields the largest of these numbers. To compute $\pi(i)$ for a given i, we go to the ith row of Table 5–2. Then for each cell in this row we multiply together the two numbers appearing in this cell, and add the numbers so obtained for all of the cells in the ith row. This yields $\pi(i)$. The values $\pi(i)$ can be conveniently represented by annexing one additional column to Table 5–2: a $\pi(i)$ value for each row in the table.

From the description just given, it might seem a trivial matter to solve any single-stage decision problem. This is by no means true, however. First of all, there may be a very large number of different states of nature, and it may be difficult to list all those which should be included. For certain types of problems there can easily be millions of states of nature. Then, of course, it can also be quite an undertaking to determine the probabilities and profits involved. We shall only discuss relatively simple problems here. Some problems are sufficiently complicated that they can be more easily solved using simulation than by the methods we have just described.

5–5 **EXAMPLES**

1. The simplest of all single-stage decision problems are those where only two actions are open to the decision maker, and there are only two states of nature. To illustrate such a problem consider a businessman who is trying to decide which of two projects, call them 1 and 2, to undertake. The profitability of these projects will be strongly influenced by a tax law which Congress is considering. If the law is not passed, project 1 is considerably more favorable than project 2. However, if the law is passed, then project 2 is the more desirable one. The decision maker estimates that the probability of the law being passed is independent of what project he undertakes, and feels that the probability of its being passed is 0.3. If he undertakes project 1, call this action a_1, and if he undertakes project 2, call this action a_2. Denote by e_1 the state of nature that the law will be passed and by e_2 the state of nature that it will not be passed. If he undertakes project 1 and the law is passed, his profit will be \$20,000, whereas if the law is not passed, his profit will be \$70,000. If he undertakes project 2 and the law is passed, his profit will be \$40,000, whereas if the law is not passed, his profit will be \$60,000. The payoff table for this problem is shown in Table 5–3. In the last column of this table are given the expected profits for each action. For example, the expected profit for action a_1 is

$$0.3(20,000) + 0.7(70,000) = \$55,000 .$$

The expected profit for action a_1 is greater than for a_2, and therefore he should undertake project 1.

The decision problem illustrated here is one which will presumably occur only once and not be repeated. Nonetheless, the action to be used was de-

TABLE 5–3

PAYOFF TABLE FOR TAX LAW EXAMPLE

	e_1	e_2	Expected profit
a_1	0.3 20,000	0.7 70,000	55,000
a_2	0.3 40,000	0.7 60,000	54,000

termined by maximizing the expected profit. As was pointed out in Section 5–4, the justification for maximizing the expected profit in this case lies in the theory of utility, which we have not studied. However, this sort of problem brings up another interesting point. What do the probabilities mean here? We originally indicated that probabilities would normally be thought of intuitively as long-run relative frequencies. Here, however, it is not possible to use this interpretation. In this case the probabilities must be given the personal or subjective interpretation which we indicated at the outset is becoming more and more important in decision theory. We can now see in part why the subjective interpretation is important. To treat decision problems which will occur only once and will not be repeated, it is necessary to use personal probabilities. These probabilities are, of course, manipulated in precisely the same way as probabilities which can be interpreted as relative frequencies.

2. Every Saturday a bakery in a large shopping center stocks a few of a very special and expensive cake which appeals to some of its best customers. Unfortunately, these customers do not necessarily come in every Saturday, nor do they always buy one of these cakes when they are in the bakery. There is no way of finding out ahead of time how many cakes will be demanded. However, over the past two years the owner has kept a record of the number of these cakes demanded on a Saturday. He feels justified in using the historical relative frequencies for probabilities. The historical data show that the number of cakes demanded on any given Saturday was always less than or equal to five. The relative frequencies give the following probabilities of having j cakes demanded:

$$p(0) = 0.10, \quad p(1) = 0.20, \quad p(2) = 0.20,$$
$$p(3) = 0.30, \quad p(4) = 0.15, \quad p(5) = 0.05.$$

We can then think of the number of cakes demanded as a random variable x, which can take on the values 0, 1, 2, 3, 4 and 5. This random variable can be conveniently used to define a set of states of nature, each state of nature corresponding to the number of cakes demanded. There are, therefore, six different states of nature, call them $e_0, \ldots, e_5$. The different actions correspond to stocking different numbers of cakes at the beginning of the day. Since no more than five cakes have ever been demanded, the owner will not stock more than five. He may stock any number from 0 to 5, however. There are thus six possible actions, one for each number of cakes stocked, which we shall denote by $a_0, \ldots, a_5$.

For the sums of money involved here, the owner indicates that monetary values reflect his true preferences, and the proper number of cakes to stock can be determined by maximizing the expected profit. Let us then see how to determine the profit when he uses action a_i, and the state of nature turns out to be e_j. The cakes retail for $5.50 and cost the bakery $3.50. Any cakes left over on Saturday evening will on Monday morning be placed on a half-price counter. All cakes placed on this counter will be sold at half price, that is, at $2.75 on Monday. If a customer comes in and requests a cake when the bakery is out of stock, the owner has observed that the customer usually becomes somewhat annoyed, and as a consequence buys less than usual of other things. The owner feels that, on the average, each time a cake is demanded when he is out of stock costs him about $2.00. However, the inability to supply a customer's demand has no carry-over effect to future weeks, the owner has observed, since his is the only high-quality bakery in the area. Therefore, the only effect of not being able to meet a customer's request for a cake is the two-dollar loss in profit on other sales. We can now easily determine the profit if he stocks i cakes and the demand turns out to be j.

When $i > j$, all demands will be met, and he will have $i - j$ cakes left over at the end of the day. If $i = j$, all demands will be met, and there will be no leftovers. If $i < j$, there will be no leftovers, but $j - i$ demands will occur which he cannot fill. Recall that the profit is equal to the revenues received from all sources less all costs. The loss in profits due to customer annoyance if not enough cakes are stocked can be looked upon as a cost. The profit in terms of i and j is then

$$\begin{aligned}
&5.50j - 3.50i + 2.75(i - j) , && i > j \\
&(5.50 - 3.50)i = 2.00i , && i = j \\
&5.50i - 3.50i - 2.00(j - i) , && i < j .
\end{aligned} \qquad (5\text{–}6)$$

The reasoning behind these expressions is the following. If he stocks i cakes, these cost him $3.50i. If $i > j$, he sells j cakes, the number demanded, and the revenues received from these is $5.50j. In addition, he has $i - j$ left

over which bring in an additional revenue of $2.75(i - j)$ when they are sold at half price on Monday. The profit for $i > j$ is then given by the first expression in (5–6). If $i = j$, the demand is for precisely the number stocked, and in this case there are no leftovers and no unfilled demands. The revenues received are \$5.50i, while the cost of the cakes is \$3.50i. Thus the profit is given by the second expression in (5–6). Finally, if $j > i$, he sells all that he stocked, so that revenues received are \$5.50i. Furthermore, the cost of the cakes is still \$3.50i. Now, however, there is an additional cost to the bakery of \$2.00 for every cake demanded which cannot be supplied. This cost is then in total \$2.00$(j - i)$. Thus the profit is that given by the third expression in (5–6).

By use of (5–6), we can easily fill the entries in the payoff table. This is done in Table 5–4. Since the probabilities of the states of nature do not

TABLE 5–4

PAYOFF TABLE FOR BAKERY PROBLEM

	e_0	e_1	e_2	e_3	e_4	e_5	Expected profit
$p(e_j)$	0.10	0.20	0.20	0.30	0.15	0.05	
a_0	0	−2.00	−4.00	−6.00	−8.00	−10.00	−4.70
a_1	−0.75	2.00	0	−2.00	−4.00	− 6.00	−1.175
a_2	−1.50	1.25	4.00	2.00	0	− 2.00	1.40
a_3	−2.25	0.50	3.25	6.00	4.00	2.00	3.025
a_4	−3.00	−0.25	2.50	5.25	8.00	6.00	3.225
a_5	−3.75	−1.00	1.75	4.50	7.25	10.00	2.7125

depend on the action taken, we have simply listed these in the first row of the tableau. The task of filling out the payoff table here is somewhat arduous. The expected profits are shown in the final column. The expected profit for stocking two units is, for example,

$$-1.50(0.10) + 1.25(0.20) + 4.00(0.20) + 2.00(0.30) + 0(0.15)$$
$$-2.00(0.05) = -0.15 + 0.25 + 0.80 + 0.60 - 0.10 = \$1.40 .$$

The decision maker should select the action which maximizes his expected profit. This is action a_4. In other words, he should stock four cakes.

To determine the optimal action for a decision maker to use it is by no

means necessary always to go through the arduous task of constructing a payoff table and then determining the expected profit for each action. It is frequently possible to develop simplified procedures for finding an optimal action which can be used for certain types of problems. We shall not attempt to give any examples of the derivation of such simplified procedures. Some examples are given, however, in reference [2].

In the next four sections we shall digress to consider what can be done in cases where it is not possible to assign prior probabilities to the states of nature. The reader can, if he wishes, omit these sections and turn to Section 5–10 to continue the main thread of the development.

5–6* PROBABILITIES NOT KNOWN

The mathematical model of a single stage decision problem which we have been using assumes that the probabilities of the states of nature are known. The theory does not take into account how vague the decision maker may be about these probabilities, but rather treats the numbers assigned to the probabilities as if they were the correct numbers. One might ask the question, "What should the decision maker do if he knows absolutely nothing about the probabilities of the states of nature?" Clearly, in such a case the decision maker is in rather bad shape, and there is a limit to what mathematical analysis can do to help. There is some question as to whether situations are ever encountered in which the decision maker has no basis whatever for estimating the probabilities. Nonetheless, it is not unreasonable to study at least briefly such situations. As a matter of fact, the case where the probabilities of the states of nature are not known has received quite a bit of attention, and a variety of rules have been suggested for selecting an action in such cases. Unfortunately, none of these rules is very satisfactory. We shall very briefly survey in this section the five most important rules which have been proposed.

Suppose that we are considering a single-stage decision problem with m states of nature and r possible actions, and that the profit associated with each (a_i, e_j) pair is known. The probabilities of the states of nature will be assumed to be unknown. We can then construct a payoff table such as that in Table 5–2, except that no probabilities will appear. We would like to decide what action the decision maker should take in this case. Clearly the payoff table gives us information which should be useful, and it may make it possible to rule out some actions immediately. For example, if $U_{ij} \geq U_{kj}$ for each j, then action a_i will be at least as good as action a_k, since regardless of what the state of nature turns out to be, the profit received from action a_i is at least as great as that from action a_k.

* Starred sections contain material which may be omitted without loss of continuity.

In such a case we say that action a_i *dominates* action a_k. Let us then drop all actions which are dominated by other actions. Once this is done, the situation is more complicated, because one action will be preferred if one state of nature should occur while another will be preferred for a different state. We shall now consider five possible ways to select an action. To illustrate each of these as they are discussed, we shall use the decision problem whose payoff table is shown in Table 5–5.

TABLE 5–5

PAYOFF TABLE FOR EXAMPLE

	e_1	e_2	e_3	e_4
a_1	2	5	1	3
a_2	6	4	2	0
a_3	3	1	5	2

1. LAPLACE CRITERION. Laplace, in discussing other matters, suggested a procedure which can be used here. He said that if nothing is known one simply assumes that the states of nature are equally likely, so that if there are m states of nature, one sets $p_j = 1/m$. This is merely a subjective evaluation of the probabilities and reduces the problem to the case we have been considering. Thus one computes

$$\frac{1}{m} \sum_{j=1}^{m} U_{ij} \tag{5-7}$$

for each i and selects the action which yields the largest of these numbers. Now the largest of the numbers (5–7) can be found by selecting the largest of the numbers $\sum_{j=1}^{m} U_{ij}$. It is unnecessary to divide by m. For the example of Table 5–5, if we sum the numbers in each row, we obtain respectively 11, 12, and 11. The number 12 corresponding to row 2 is the largest, and thus according to the Laplace criterion action a_2 would be selected. This rule does not seem especially sound unless one actually has reason to believe that the states are equally likely.

2. MAX-MIN CRITERION. This criterion, which was suggested by the theory of two-person games, is very conservative or pessimistic in the sense that it assumes that nature will do the worst thing possible. Denote by U_i^+ the smallest profit in row i. Then the action to be selected is the one which yields the largest U_i^+. We can say therefore that we choose the action which maximizes the minimum profit. The U_i^+ for the example of Table 5–5 are respectively 1, 0, and 1. The largest of these numbers is 1, and this occurs

for either a_1 or a_3. Thus according to the max-min criterion the decision maker would select either a_1 or a_3. Note that the max-min criterion protects one against large losses, but takes no account whatever of possible large profits to be made.

3. MAX-MAX CRITERION. This is a very optimistic criterion. It assumes that the best imaginable outcome will occur. Let U_i^* be the largest of the U_{ij}, i.e., the action whose row in the payoff table contains the largest number U_{ij} in the payoff table. For the example in Table 5–5, the U_i^* are respectively 5, 6, and 5. Hence according to the max-max criterion action a_2 should be selected.

4. THE OPTIMISM INDEX. This criterion, suggested by the economist Hurwicz, is a combination of 2 and 3 above. Let U_i^+ be defined as in 2 and U_i^* as in 3. Then consider the numbers

$$\hat{U}_i(\alpha) = \alpha U_i^* + (1 - \alpha)U_i^+, \quad 0 \le \alpha \le 1 . \tag{5–8}$$

Now imagine that the decision maker selects a value of α, $0 \le \alpha \le 1$, and computes $\hat{U}_i(\alpha)$ for each i. The action he is to select is then the one corresponding to the largest of the $\hat{U}_i(\alpha)$ values. The number α is called the decision maker's *optimism index*. When $\alpha = 0$, he uses a max-min criterion, and when $\alpha = 1$, he uses a max-max criterion. Thus one can imagine that the larger the value of α he selects, the more optimistic he is. Rather than solving the example of Table 5–5 for a particular α, we shall illustrate geometrically what the situation is for all α. Note that $\hat{U}_i(\alpha)$ can be written

$$\hat{U}_i(\alpha) = U_i^+ + \alpha(U_i^* - U_i^+) ,$$

and if we plot $\hat{U}_i(\alpha)$ as a function of α, we obtain a straight line (see Section 7–3 for a discussion of straight lines). There will be generated a straight line for each value of i, that is, for each action. The three $\hat{U}_i(\alpha)$ functions are

$$\hat{U}_1(\alpha) = 1 + 4\alpha; \quad \hat{U}_2(\alpha) = 6\alpha; \quad \hat{U}_3(\alpha) = 1 + 4\alpha .$$

Note that $\hat{U}_1(\alpha)$ and $\hat{U}_3(\alpha)$ are the same. The resulting two lines are shown in Figure 5–2. The action to be selected is the one with the largest $\hat{U}_i(\alpha)$. From Figure 5–2 it is seen that for $\alpha < \alpha_1$, a_1 (or a_3) is the action to use. When $\alpha = \alpha_1$, $\hat{U}_1 = \hat{U}_2 = \hat{U}_3$, and any one of the actions can be used. When $\alpha > \alpha_1$, action a_2 is the one to select. We see from Figure 5–2 that the decision maker does not need to determine with great accuracy what he thinks his value of α is, since for any α such that $0 \le \alpha \le \alpha_1$, he will select action a_1 (or a_3) while for any α, $\alpha_1 \le \alpha \le 1$ he will select a_2. The same sort of thing will apply when there are more actions. There will be a range of α values for which a given action will be the one to be selected. This action will change from one range of α's to another, however.

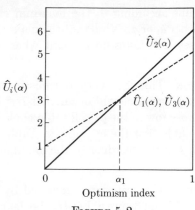

FIGURE 5–2

5. MIN-MAX REGRET. The *regret* on taking any action a_i when the state of nature turns out to be e_j is defined to be the difference between the maximum profit which could have been obtained when the state of nature is e_j and that actually obtained. Denote by M_j the largest number in column j of the payoff table. Then M_j is the largest profit which can be obtained if the state of nature turns out to be e_j. The regret L_{ij} when action a_i is taken and the state of nature turns out to be e_j is defined to be

$$L_{ij} = M_j - U_{ij}. \tag{5-9}$$

The statistician Savage has suggested that instead of using the max-min criterion it might be more appropriate to minimize the maximum regret. Let $L_i{}^+$ be the largest of the regrets for action i. Then one selects the action for which L_i^+ is the smallest. A table of regrets for the example of Table 5–5 is given in Table 5–6. We see that the values of the L_i^+ are respectively 4, 3, and 4. The smallest L_i^+ is 3 and occurs for a_2. Thus a_2 is the action to be selected.

We have now introduced the most important rules which have been suggested to cope with the case when nothing is known about the proba-

TABLE 5–6

REGRET TABLE FOR EXAMPLE

	e_1	e_2	e_3	e_4
a_1	4	0	4	0
a_2	0	1	3	3
a_3	3	4	0	1

bilities of the states of nature. It will be clear to the reader that the rules leave something to be desired. Some specific objections will be considered in the problems. For the simple example used to illustrate the rules, it is notable that each of the three actions seemed appropriate according to at least one of the rules.

5–7 MIXED STRATEGIES

For problems where the probabilities of the states of nature are not known, a decision maker might have considerable difficulty deciding which of two or more actions he really preferred. In desperation, to decide among two competing actions, he might toss a coin and use one action if the coin landed heads up and the other if it landed tails up. This suggests an interesting procedure which could be used to select an action. Instead of deciding what action to take, the decision maker could instead play a game of chance and let the outcome of the game determine the action to be used. It might seem ridiculous to suggest to a businessman that he flip a coin to decide whether to build a plant or not, and indeed normally it is. However, it is interesting to pursue this idea a little further. Suppose that to decide what action to take, the decision maker plays a game of chance. It will be imagined that this game of chance will select action a_i with probability q_i. We shall refer to this game of chance that is used for selecting an action as a *mixed strategy*. This terminology was first used in game theory and has been carried over to decision theory. When we say that a decision maker is using a mixed strategy, we mean that he does not decide what action to take but plays a game of chance or lottery whose outcome determines the action to take. The decision maker does not know himself what action will be used until after playing the game. We shall refer to a rule which simply says take action a_i as a *pure strategy*. There are r different pure strategies, one for each action, so that the pure strategy i says take action a_i.

To characterize a mixed strategy, it is necessary to specify for each i the probability q_i that a_i will be the action selected by the game of chance. For every different set of probabilities q_i, we obtain a different mixed strategy. There are thus an infinite number of different mixed strategies for the single stage decision problem under consideration. Note that a pure strategy can be looked upon as a special case of a mixed strategy, since the mixed strategy with $q_i = 0$, $i \neq k$, and $q_k = 1$ is equivalent to the pure strategy which says take action a_k. The reason for this is that if $q_k = 1$ we can imagine that the game will always say to use action a_k.

The decision maker was originally faced with the problem of selecting one of a finite number of actions. We then suggested that he might not be able to decide what action to take and would let a game of chance

decide this for him, that is, he would use a mixed strategy. However, there
are an infinite number of different mixed strategies he could use, that is, an
infinite number of different games he could play. How does he decide which
game is most suitable? This would appear to be a more difficult problem
than selecting an action. We can then ask the questions: How should he
select a mixed strategy, and what can possibly be gained by introducing
mixed strategies? Let us now examine these questions.

To select a mixed strategy, the decision maker could simply apply one
of the rules discussed in the previous section. We shall not bother to illustrate here how all of these can be generalized to apply to selecting mixed
strategies; the procedure will be illustrated only for the max-min criterion,
the generalization of the other rules being left to the problems. Note that
when a mixed strategy is used, the decision maker does not know what his
profit will turn out to be if the state of nature turns out to be e_j, since this
will depend on what action the game tells him to select. However, it is
possible for him to compute his expected profit if the state of nature turns
out to be e_j. His expected profit is then

$$\sum_{i=1}^{r} q_i U_{ij} = q_1 U_{1j} + \cdots + q_r U_{rj}, \tag{5-10}$$

since the probability that the profit will be U_{ij} is q_i. The expected profit
will depend on the mixed strategy chosen, in other words, on the values
of the q_i. Let us now explain how the max-min criterion can be generalized
to treat mixed strategies. For a given mixed strategy we have an expected
profit (5–10) for each state of nature. Denote by w the smallest of these
expected profits, that is, the minimum of (5–10) over j. Now w will change
as the mixed strategy changes. Suppose that we find that set of numbers
$q_i, i = 1, \ldots, r$ satisfying

$$q_i \geq 0, \quad i = 1, \ldots, r, \quad \sum_{i=1}^{r} q_i = 1,$$

which yields the largest possible value of w. This set of q_i (which may
not be unique) defines a mixed strategy which we shall call the max-min
mixed strategy. This is the way the max-min criterion is generalized to
handle mixed strategies.

Let us now see how a max-min strategy can be determined. When w is
the smallest of the m numbers (5–10), it must be true that

$$\sum_{i=1}^{r} q_i U_{ij} \geq w, \quad j = 1, \ldots, m,$$

with the strict equality holding for at least one j. We wish to find a set
of q_i such that w is as large as possible. In other words, we wish to determine a set of numbers $q_1, \ldots, q_r$ and a number w satisfying

$$\sum_{i=1}^{r} q_i U_{ij} - w \geq 0, \quad j = 1, \ldots, m,$$

$$\sum_{i=1}^{r} q_i = 1, \quad q_i \geq 0, \quad i = 1, \ldots, r, \tag{5-11}$$

such that w has the largest value possible. This is a type of problem referred to as a linear programming problem. Therefore, by solving this linear programming problem, one can determine a max-min strategy. We shall not attempt to illustrate the use of linear programming to determine max-min strategies. Instead, in the next section, we shall illustrate how such strategies can be determined by the use of an interesting type of geometrical analysis in the case where there are only two different states of nature. We shall also examine in the next section the question of what is to be gained by using a mixed strategy.

5-8* GEOMETRIC INTERPRETATION

Let us consider a single-stage decision problem in which there are only two states of nature, e_1 and e_2, and two actions a_1 and a_2. A mixed strategy is then characterized by specifying the probability q that action a_1 will be selected by the game of chance, since the probability that action a_2 will be selected is $1 - q$. The expected profit if the state of nature turns out to be e_1 or e_2 are respectively

$$qU_{11} + (1 - q)U_{21} = U_{21} + q(U_{11} - U_{21}) = x(q), \tag{5-12}$$

$$qU_{12} + (1 - q)U_{22} = U_{22} + q(U_{12} - U_{22}) = y(q). \tag{5-13}$$

The first of these we have denoted by $x(q)$ and the second by $y(q)$; $x(q)$ and $y(q)$ are, respectively, the expected profit when the state of nature turns out to be e_1 and a_1 is selected with probability q, and the expected profit if the state of nature turns out to be e_2 and a_1 is selected with probability q. Now observe that (x, y) can be plotted as a point in the xy-plane. We shall next show how to illustrate graphically the set of all points $(x(q), y(q))$ generated as q takes on all possible values $0 \leq q \leq 1$.

Note that for $q = 1$, $x(q) = U_{11}$ and $y(q) = U_{12}$. This mixed strategy is simply the pure strategy which says to take action a_1, that is, the point (U_{11}, U_{12}) in the plane corresponds to the pure strategy a_1. If $q = 0$, $1 - q = 1$ and $x(q) = U_{21}$ and $y(q) = U_{22}$. This mixed strategy is the pure strategy which says to take action a_2, that is, the point (U_{21}, U_{22}) in the plane corresponds to the pure strategy a_2. Next observe that when $0 < q < 1$, $x(q)$ lies between U_{11} and U_{21} and $y(q)$ lies between U_{12} and U_{22}. We can easily relate $x(q)$ and $y(q)$, since if we solve (5-12) for q, we obtain

$$q = \frac{x(q) - U_{21}}{U_{11} - U_{21}},$$

and if we substitute this into (5–13), we have

$$y(q) = \left(\frac{U_{12} - U_{22}}{U_{11} - U_{21}}\right) x(q) + U_{22} - U_{21} \left(\frac{U_{12} - U_{22}}{U_{11} - U_{21}}\right). \qquad (5\text{-}14)$$

Equation (5–14) is called a linear equation and it is represented graphically by a straight line (for more details on graphing linear equations see Section 3 of Chapter 7). Furthermore, since $x(q)$ lies between U_{11} and U_{21} and $y(q)$ lies between U_{12} and U_{22}, the set of points $(x(q), y(q))$ corresponding to all mixed strategies must just be that part of the line which joins (U_{11}, U_{12}) to (U_{21}, U_{22}). Let us illustrate this geometrically with an example. Suppose that the payoff table for the two action, two states of nature problem under consideration is that shown in Table 5–7.

TABLE 5–7

	e_1	e_2
a_1	1	3
a_2	5	2

The points corresponding to the pure strategies a_1 and a_2 are then $(1, 3)$ and $(5, 2)$, respectively. These are plotted in Figure 5–2. The set of all points corresponding to all mixed strategies is then the line segment joining these two points as shown in Figure 5–2. In particular, if we choose $q = 0.2$, then from (5–12) and (5–13)

$$x(q) = 5 + 0.2(1 - 5) = 5 - 0.8 = 4.2; \quad y(q) = 2 + 0.2(3 - 2) = 2.2 .$$

This point does indeed lie on the line segment and is indicated in Figure 5–3.

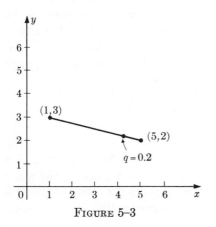

FIGURE 5–3

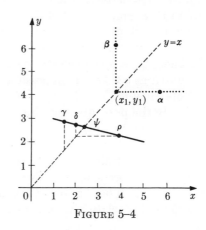

FIGURE 5–4

We shall now show how a max-min mixed strategy for the example of Table 5–7 can be determined with the aid of Figure 5–3. We wish to find that q for which the smaller of $x(q)$ and $y(q)$ is as large as possible. For any point $(x(q), y(q))$ with $y(q) \geq x(q)$, then $x(q)$ is the smaller number, whereas if $x(q) \geq y(q)$, then $y(q)$ is the smaller number. If $x(q) = y(q)$, either one can be looked upon as the smaller. If there is a q for which $x(q) = y(q)$, then this point lies on the line $y = x$ which makes a 45 degree angle with the x-axis. In Figure 5–3 we have reproduced the situation in Figure 5–2 and have also shown the line $y = x$. Consider a particular point (x_1, y_1) on the line $y = x$, and the horizontal and vertical lines emanating from the point as shown in Figure 5–4. Note that any point α on the horizontal line has the form (x, y_1), $x \geq y_1$, and the minimum of x and y_1 is thus y_1. Similarly, any point β on the vertical line can be written as (x_1, y), $y > x_1$ so that the smaller of x_1 and y is x_1. Consequently, for any point (x, y) on either the horizontal or vertical line emanating from (x_1, y_1), the minimum value of x and y is $x_1 = y_1$.

To determine the optimal max-min mixed strategy, we wish to find that point on the line segment in Figure 5–3 whose smallest coordinate is as large as possible. Let us select a point on this line segment, say the point γ shown in Figure 5–4. Its smallest coordinate is the x-coordinate. Is this point the one on the line segment with its smallest coordinate as large as possible? Clearly not, because the points δ and ρ both have a smallest coordinate which is larger than that of γ. Observe that the x-coordinate is the smallest one for δ and the y-coordinate is the smallest one for ρ. The smallest coordinates have the same value for both of these points, however. It is now clear that the point whose smallest coordinate is as large as possible is the point ψ where the 45° line cuts the line segment under consideration. We can easily find the mixed strategy which yields this point, since it is the point such that $x(q) = y(q)$. For the example under consideration,

$$x(q) = 5 - 4q, \quad y(q) = 2 + q \,.$$

Hence

$$5 - 4q = 2 + q \quad \text{or} \quad 5q = 3 \,,$$

that is,

$$q = \tfrac{3}{5} = 0.60 \,.$$

The max-min mixed strategy is then unique and is the one which selects action a_1 with probability 0.60.

Let us now ask what is gained by using a max-min mixed strategy rather than by simply selecting the max-min criterion action. If we applied the max-min criterion of the previous section to determine the action to take, it would be a_2 for the example of Table 5–7, and the worst that could result would be a profit of 2 if e_2 occurred. Now if the max-min mixed

strategy is used, the expected profit is the same regardless of which state of nature occurs and has the value $y(0.6) = 2 + 0.6 = 2.6$. This is larger than the worst that can happen if the decision maker simply uses a max-min action. Thus by introducing mixed strategies it is possible to increase the expected profit of the worst outcome. Observe carefully, however, that the worst profit the decision maker may actually encounter can be much worse using a max-min mixed strategy rather than a max-min action. Only the expected profit is increased. In particular, for the example under consideration, if the chance mechanism selects a_1 and the state of nature turns out to be e_1, the decision maker will have a profit of 1, whereas using a max-min action the worst he could do would be better than this, a profit of 2.

We have seen how to find the max-min mixed strategy for a particular example. Let us now generalize the procedure so that it can be used to find such a strategy for any two action, two states of nature problem. Denote by X the set of all points $(x(q), y(q))$ generated as q varies between 0 and 1 in (5–12) and (5–13). Suppose that we have represented this set geometrically, and imagine now that it is the line segment shown in Figure 5–5. Consider a collection of right angles with sides vertical and horizontal and vertices on the line $y = x$ as shown in Figure 5–5. What we wish to do is find that right angle whose vertex has an x-coordinate (or y-coordinate, since $y = x$ at the vertex) which is as large as possible, while the line segments forming the right angle still have at least one point in common with the set X. Thus in Figure 5–5 the right angle which has

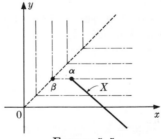

FIGURE 5–5

the largest x-coordinate, while still having a point in common with X, is the one whose vertex is β. Consequently, the point α in X is the point whose smallest coordinate is as large as possible. The point α, as we have shown above, corresponds to a pure strategy, and the max-min mixed strategy in this case is one of the two possible pure strategies.

We can now generalize the preceding geometric interpretation for two-action problems, to problems having r actions. We shall continue to assume, however, that there are only two states of nature. To specify

a mixed strategy, one must now specify more than a single probability q; one must specify r probabilities q_i which sum to 1, such that q_i is the probability that the lottery will select action a_i. If U_{i1} and U_{i2} are the utilities when action a_i is taken and the state of nature turns out to be e_1 and e_2, respectively, then the expected profit corresponding to a given mixed strategy when the state of nature turns out to be e_1 is

$$\sum_{i=1}^{r} q_i U_{i1} = x. \tag{5-15}$$

Similarly, the expected profit if the state of nature turns out to be e_2 is

$$\sum_{i=1}^{r} q_i U_{i2} = y. \tag{5-16}$$

Once again (x, y) can be looked upon as a point in the plane. Denote by X the set of all such points generated by all possible mixed strategies. We shall now show how to represent this set geometrically and also how to determine a max-min mixed strategy. This will be done by use of an example.

Consider the single-stage decision problem whose payoff table is that shown in Table 5–8. We shall illustrate for this example the set X referred to above. First let us illustrate the points corresponding to the seven pure strategies. These are the points $(1, 6)$, $(1.5, 5)$, $(7, 0.5)$, $(5, 3)$, $(3, 4)$, $(2.5, 4.5)$, and $(5.5, 2)$, corresponding to the actions a_1 through a_7, respectively. These points are plotted in Figure 5–6. Next observe that all points on the line segments joining any two of these points must be in the set X. The reason for this is that if we set all $q_i = 0$, except q_u and q_v, then $q_v = 1 - q_u$, and by what we proved for the case where there

TABLE 5–8

	e_1	e_2
a_1	1	6
a_2	1.5	5
a_3	7	0.5
a_4	5	3
a_5	3	4
a_6	2.5	4.5
a_7	5.5	2

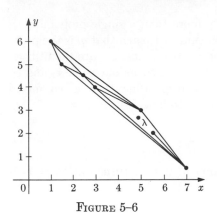

FIGURE 5–6

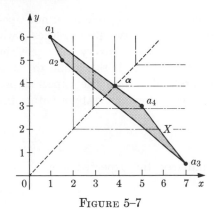

FIGURE 5–7

were only two actions, as q_u varies between 0 and 1 we trace out the line segment joining the point for pure strategy u to that for pure strategy v. Some but not all of these line segments are also shown in Figure 5–6.

It remains to consider what happens when more than two q_i are positive. Suppose that $q_1 = 0.3$, $q_3 = 0.5$, and $q_4 = 0.2$, with all other q_i being zero. The point (x,y) corresponding to this mixed strategy is then

$$x = 0.3(1) + 0.5(7) + 0.2(5) = 0.3 + 3.5 + 1.0 = 4.8,$$
$$y = 0.3(6) + 0.5(0.5) + 0.2(3) = 1.80 + 0.25 + 0.60 = 2.65 \,.$$

This is the point λ shown in Figure 5–6. We now claim that the set X is the shaded set of points shown in Figure 5–7, constituting a polygon and its interior. Every mixed strategy will yield one of these points, and each point will correspond to at least one mixed strategy. We shall not attempt to prove this in detail. The reader might convince himself of the truth of the statement by plotting points corresponding to several other mixed strategies. Note that the corners of the polygon correspond to pure strategies. These have been labeled by the action to which they correspond.

The procedure for finding the max-min mixed strategy is now precisely the same as for the case where there were only two actions. We find the right angle with vertex on the line $y = x$ having the coordinates of the vertex as large as possible while still having at least one point of X lying on the lines which form the right angle. From Figure 5–6 we see that the point α corresponds to the max-min mixed strategy. The point α lies on the line segment joining the points corresponding to the pure strategies a_1 and a_4. Hence the mixed strategy has only q_1 and q_4 positive. To determine q_1 (and hence $q_4 = 1 - q_1$), we note that for α, $x = y$, and by (5–15) and 5–16)

$$x = q_1 + 5(1 - q_1), \quad y = 6q_1 + 3(1 - q_1) \,,$$

or on setting $x = y$,

$$q_i + 5(1 - q_1) = 6q_1 + 3(1 - q_1) \,,$$

that is,

$$7q_1 = 2 \quad \text{or} \quad q_1 = \tfrac{2}{7}, \quad \text{and} \quad q_4 = \tfrac{5}{7} \,.$$

Hence the max-min mixed strategy is one with $q_1 = \tfrac{2}{7}$, $q_4 = \tfrac{5}{7}$, and all other $q_i = 0$.

Figures 5–8 and 5–9 indicate two other possibilities that could arise in determining the max-min mixed strategy. In Figure 5–8, the max-min mixed strategy is a pure strategy. In Figure 5–9, the max-min strategy is not unique, for any point on the line segment joining the points α and β serves to define a max-min strategy. This case is a little bit peculiar,

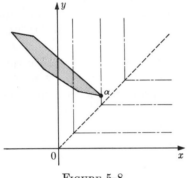

FIGURE 5–8

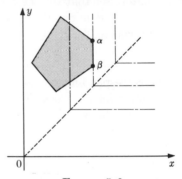

FIGURE 5–9

however, since the action corresponding to the pure strategy of the point α dominates the action corresponding to β, and hence there would be no reason to ever include this latter action. This observation brings up a new idea. Just as we could define dominance for actions, we can define dominance for mixed strategies. If we have two mixed strategies S_1 and S_2 which yield the points (x_1, y_1) and (x_2, y_2), and if $x_1 \leq x_2$ and $y_1 \leq y_2$ with strict inequality holding for at least one of the inequalities (so that we are dealing with two different points), then S_2 is said to dominate S_1, and hence S_1 and the point (x_1, y_1) never need be considered in studying mixed strategies. The set of mixed strategies having the property that no mixed strategy in the set is dominated by some other mixed strategy is called the set of *admissible mixed strategies*. The set of admissible strategies generates a set of points Y which is a subset of X. We can easily illustrate the set of points Y when the set X is that shown in Figure 5–7. To do this, it is convenient to note that the set of points corresponding to the mixed strategies which are dominated by the mixed strategy yielding the point

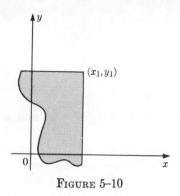

FIGURE 5–10

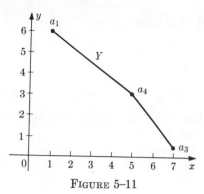

FIGURE 5–11

(x_1, y_1), is the shaded set in Figure 5–10. By use of this observation, we immediately see that the set Y of points corresponding to mixed strategies which are not dominated by other mixed strategies is that shown in Figure 5–11. Of course, the point corresponding to the max-min mixed strategy is in Y.

When there are three states of nature, one could illustrate the set of points corresponding to all mixed strategies geometrically by a set in three-dimensional space. The geometric representation is much more cumbersome here and not nearly so clear, and we shall not attempt to provide an example. When there are four or more states of nature, we cannot provide any convenient geometric interpretation. Nonetheless, a max-min strategy can always be determined by solving the linear programming problem introduced in the previous section.

5–9* BAYES STRATEGIES

We observed in the previous section that in the event that nothing is known about the probabilities of the states of nature, a decision maker might feel that it is better to use a max-min mixed strategy than a max-min action. Now, there is nothing to prevent a decision maker from using a mixed strategy even when the probabilities of the states of nature are known. The question then arises in our minds: "Could there be any possible advantage to using a mixed strategy in cases where the probabilities of the states of nature are known?" We now wish to investigate whether the decision maker can somehow improve his expected profit by use of mixed strategies.

Suppose the decision maker uses a mixed strategy in which the probability of selecting action a_i is q_i. We shall restrict our attention here to cases where the probabilities of the states of nature do not depend on the action selected. Then the probability that the state of nature turns out to be e_j can be denoted by p_j. The decision problem can now be looked

on as a two-stage random experiment. First a random experiment (a game of chance) is performed to determine what action will be taken; then another random experiment is performed (the same experiment regardless of what action is taken) to determine the state of nature. The profit, however, depends not only on the state of nature but also on the action taken. Thus the probability that the profit is U_{ij} is determined by the probability that at the first stage action i will be taken and at the second stage the state of nature will turn out to be e_j. We showed in Section 1–13 that the probability of this should be $q_i p_j$. We can think of the U_{ij} as being the values of a random variable for the two-stage random experiment. The expected profit for the entire random experiment, which we shall denote by Ω, is

$$\Omega = U_{11}q_1p_1 + U_{12}q_1p_2 + \cdots + U_{1m}q_1p_m + U_{21}q_2p_1 + \cdots + U_{2m}q_2p_m$$
$$+ \cdots + U_{r1}q_rp_1 + \cdots + U_{rm}q_rp_m$$
$$= \left(\sum_{j=1}^{m} U_{1j}p_j\right) q_1 + \cdots + \left(\sum_{j=1}^{m} U_{rj}p_j\right) q_r. \tag{5-17}$$

Let us imagine that the decision maker now chooses the mixed strategy so as to maximize Ω. The mixed strategy which maximizes Ω is called a *Bayes' strategy*.

We shall now prove that there is always a pure strategy which is a Bayes' strategy. In other words, the decision maker cannot, by the introduction of mixed strategies, increase his expected profit above that which he can obtain simply by selecting the action which yields the largest value of $\sum_{j=1}^{m} U_{ij}p_j$. Consequently, in cases where the probabilities of the states of nature are known, we need not concern ourselves with mixed strategies. There is no reason why a decision maker should ever consider using one. In other words, it would never be desirable in such cases to tell the decision maker to toss a coin to decide what action to take. The proof is easy to carry out. Let θ be the largest of the numbers $\sum_{j=1}^{m} U_{ij}p_j$. Then

$$\sum_{j=1}^{m} U_{ij}p_j \leq \theta, \quad i = 1, \ldots, r, \tag{5-18}$$

with the strict equality holding for at least one i. However, since $q_i \geq 0$,

$$\left(\sum_{j=1}^{m} U_{ij}p_j\right) q_i \leq \theta q_i, \quad i = 1, \ldots, r. \tag{5-19}$$

Summing (4–63) over i, we obtain

$$\Omega = \left(\sum_{j=1}^{m} U_{1j}p_j\right) q_1 + \cdots + \left(\sum_{j=1}^{m} U_{rj}p_j\right) q_r \leq \theta(q_1 + \cdots + q_r) = \theta, \tag{5-20}$$

since the q_i sum to 1. Thus for any mixed strategy, $\Omega \leq \theta$. Hence, if the strict equality holds in (5–18) for $i = k$, then it is impossible for the decision maker to increase his expected utility above that for taking action a_k by introducing mixed strategies. He can, of course, have $\Omega = \theta$ by using the mixed (and pure) strategy $q_i = 0$, $i \neq k$, $q_k = 1$. This is what we wished to prove, and it relieves us of the necessity of considering mixed strategies in cases where the probabilities of the states of nature are known.

When there are just two states of nature, we can illustrate geometrically the determination of a Bayes' strategy using the same diagrams used in the previous section. To do this, note that we can write

$$\Omega = U_{11}q_1p_1 + U_{21}q_2p_1 + \cdots + U_{r1}q_rp_1 + U_{12}q_1p_2 + \cdots + U_{r2}q_rp_2$$

$$= \left(\sum_{i=1}^{r} q_iU_{i1} \right) p_1 + \left(\sum_{i=1}^{r} q_iU_{i2} \right) p_2 = p_1x + p_2y, \qquad (5\text{–}21)$$

where x and y are defined by (5–15) and (5–16), respectively. Consider the set of points X of the form (x, y) generated by all possible mixed strategies. Now we wish to select that point in X which yields the largest value of Ω, when Ω is given in terms or x and y by (5–21). Note that in (5–21), p_1 and p_2 are the probabilities of the states of nature and are assumed to be known. Now if we select a particular value of Ω, say Ω_0, the set of points (x, y) in the plane which satisfy the equation

$$\Omega_0 = p_1x + p_2y \qquad (5\text{–}22)$$

is a straight line. For each different value of Ω we select we get a different straight line. These straight lines so generated will have the interesting characteristic that they are all parallel. We wish to determine the par-

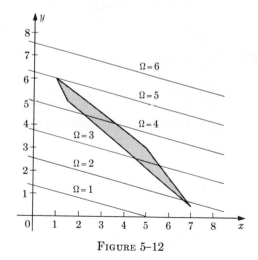

FIGURE 5–12

ticular line having the largest value of Ω which has at least one point in common with the set X. Let us illustrate the situation for the example of Table 5–8. Previously we determined the set X; it is shown in Figure 5–7. Suppose that the probabilities of the states of nature are $p_1 = 0.2$ and $p_2 = 0.8$. In Figure 5–12 we have reproduced the set X and have indicated several of the lines

$$\Omega = 0.2x + 0.8y \qquad (5\text{–}23)$$

corresponding to different values of Ω. It is seen that the line with the largest value of Ω which has at least one point in common with X is that for $\Omega = 5$. This line has just one point in common with X, and this is the point $(1, 6)$ which corresponds to the pure strategy a_1. Thus the Bayes' strategy in this case is a pure strategy, as expected.

5–10 USE OF EXPERIMENTS IN DECISION PROBLEMS

We shall now return to the main thread of development where it was left in Section 5–5. We have been visualizing a single-stage decision problem as one in which first the decision maker selects an action and then later nature selects a state of nature. Now there is nothing in the formulation of the problem which requires that the state of nature actually be determined *after* the decision maker selects an action. All that is implied is that at the time the decision maker selects the action he does not know what the state of nature is. It is irrelevant whether nature selected the state of nature millions of years before the decision maker selects an action or whether the state of nature is determined after the action is chosen, so long as in either case the decision maker does not know what the state of nature is, but can think of the state of nature as being the outcome of a random experiment.

Many interesting decision problems have the characteristic that the state of nature is determined prior to the time the decision maker selects an action. There is one very important distinction between problems of this sort and those where the state of nature is determined after an action is selected. When the state of nature has been determined before the decision maker selects an action, then there exists the possibility that the decision maker can gain additional information about what the state of nature is or, for a sufficiently high cost, even ascertain precisely what the state of nature is. We shall now concern ourselves with problems where the state of nature has been determined before the decision maker selects an action and where there exists the possibility of getting more information about the state of nature before making a decision. We shall be assuming that the way in which the decision maker can obtain additional information is to perform a random experiment. The experiment will yield some information about the state of nature, although in general it will not be able to determine pre-

cisely what the state of nature is. We shall begin by showing how the experimental results can be combined with other information to determine the action to select. Later we shall consider the problem of whether it is worthwhile to perform an experiment to obtain additional information, and if so, which of a variety of possible experiments is the one that should actually be performed.

Consider a single-stage decision problem for which there are m states of nature, which will be represented symbolically by $e_1, \ldots, e_m$, and r possible actions, symbolized by $a_1, \ldots, a_r$, one of which will be selected by the decision maker. Let U_{ij} be the profit if the decision maker selects action a_i and the state of nature turns out to be e_j. We shall assume that the state of nature has been determined before the decision maker selects the action to take, but that he does not know what the state of nature is. We shall suppose, however, that he has assigned a probability p_j that the state of nature is e_j. If the situation is one that is repeated, then p_j is the long-run fraction of the time that nature selects the state e_j. If the situation is not one that can be given a frequency interpretation, then the p_j are personal probabilities, that is, weights, which are assigned by the decision maker to reflect his feelings about the situation. Since we are assuming that the state of nature is already determined when the action is selected, it is clear that the probabilities assigned to the states of nature must be independent of the action selected. We shall be making this assumption throughout the remainder of this chapter.

Let us now imagine that before selecting an action, the decision maker decides to perform a random experiment $\Re$ to try to learn something more about the state of nature. We shall suppose that the event set for $\Re$ is $F = \{f_1, \ldots, f_s\}$, and that this is independent of what the state of nature happens to be. However, the probability of any particular simple event f_k will, in general, depend on the state of nature, and the probability of any particular simple event f_k when the state of nature is e_j will be denoted by q_{jk}. The appropriate mathematical model to use for $\Re$ will thus depend on the state of nature. After $\Re$ is performed and the simple event of $\Re$ which occurs is determined, the decision maker then uses this information in aiding him in selecting an action. We shall now explain how the outcome of $\Re$ can be incorporated into the model of a single-stage decision problem and thus be used in deciding what action should be taken.

The situation we are considering is one where at some time in the past a random experiment $\Re_N$ was performed which determined the state of nature. The outcome of $\Re_N$ is not known. However, the model $\mathcal{E}_N$ being used for $\Re_N$ has the event set $E = \{e_1, \ldots, e_m\}$, with the probability of e_j being p_j. The combined experiments $\Re_N$ and $\Re$ can be looked upon as a two-stage random experiment $\Re^*$. $\Re^*$ consists in first performing $\Re_N$ and then $\Re$. We studied in Chapter 1 the procedure by which a model for $\Re^*$

could be constructed from the models $\mathfrak{R}_N$ and $\mathfrak{R}$. The representation of $\mathfrak{R}^*$ by means of a tree is shown in Figure 1–6. We studied two-stage random experiments in some detail in Chapter 1. The novel thing about the random experiment $\mathfrak{R}^*$ we are now considering is that the purpose of performing the second stage is to learn something about the outcome of the first stage. What the performance of the second stage does is to give us some information about the outcome of $\mathfrak{R}^*$, although it does not, in general, tell us which simple event of $\mathfrak{R}^*$ has occurred, that is, which simple event occurred when $\mathfrak{R}_N$ was performed. However, a knowledge that f_k occurred on performing $\mathfrak{R}$ allows us to construct a conditional probability model which can be used as a new model for $\mathfrak{R}_N$.

From our study of conditional probabilities, we know that as a result of knowing the outcome of the second stage of $\mathfrak{R}^*$, our estimate of the probabilities of the possible outcomes $e_1, \ldots, e_m$ of the first stage will be modified, the probability of e_j changing from p_j to $p(e_j|f_k)$, the conditional probability of e_j given f_k. The probability $p(e_j|f_k)$ can be determined from the model for $\mathfrak{R}^*$, since by (1–22),

$$p(e_j|f_k) = \frac{p(e_j \cap f_k)}{p(f_k)}. \qquad (5\text{–}24)$$

However, $e_j \cap f_k$ is the simple event $\langle e_j, f_k \rangle$ and has the probability $p_j q_{jk}$ For the model $\mathfrak{R}^*$, the event that the second stage yields f_k is $\{\langle e_1, f_k \rangle \ldots, \langle e_m, f_k \rangle\}$ and hence, since $p(f_k)$ is the sum of the probabilities of the simple events in f_k, it follows that

$$p(f_k) = \sum_{u=1}^{m} p_u q_{uk}. \qquad (5\text{–}25)$$

Consequently

$$p(e_j|f_k) = \frac{p_j q_{jk}}{\displaystyle\sum_{u=1}^{m} p_u q_{uk}}. \qquad (5\text{–}26)$$

This formula for computing the conditional probabilities $p(e_j|f_k)$ is referred to as *Bayes' law* and is very important in decision theory. Indeed, the modern approach to decision theory which we are now studying is often referred to as *Bayesian decision theory*. Inasmuch as $q_{jk} = p(f_k|e_j)$, Bayes' law is often written

$$p(e_j|f_k) = \frac{p(f_k|e_j)p_j}{\displaystyle\sum_{u=1}^{m} p(f_k|e_u)p_u}. \qquad (5\text{–}27)$$

It is instructive to examine (5–26) or (5–27) in a little more detail. Recall that p_j is the probability that e_j will occur when $\Re_N$ (the first stage of $\Re^*$) is performed. However, $p(e_j|f_k)$ is also a probability of e_j. How can we have two different probabilities for the same event? The answer lies in the interpretation of these probabilities. By definition, p_j is the probability that we assign to the outcome e_j before $\Re^*$ is performed or before we know what the outcome of the second stage is. On the other hand, $p(e_j|f_k)$ is the probability we would assign to e_j after both stages of $\Re^*$ are completed and we know that the second stage yielded the event f_k. The number p_j is the long-run relative frequency with which the event e_j would be observed if $\Re_N$ or $\Re^*$ were repeated unendingly. The number $p(e_j|f_k)$ is the long-run relative frequency of e_j for those cases where the second stage yielded the event f_k. Observe that these will, in general, be different relative frequencies. It is convenient to use different names to refer to these two probabilities. We shall refer to p_j as the *prior* or *a priori probability*, since it refers to the probability that would be assigned to e_j prior to performing $\Re$. The conditional probability $p(e_j|f_k)$ will be referred to as the *posterior* or *a posteriori probability*, since it refers to the probability that should be assigned to e_j after $\Re$ has been performed and yielded the outcome f_k.

Once the posterior probabilities of the states of nature have been found, the action to be selected is determined by maximizing the expected profit, just as in the case when no experiment is performed. The only difference is that now the posterior probabilities are used as the probabilities of the states of nature, so that the expression to be maximized is

$$\sum_{j=1}^{m} U_{ij} p(e_j|f_k) \ . \qquad (5\text{–}28)$$

The only difference between the approach when no experiment is performed prior to selecting an action and that where an experiment is performed is that one additional step is inserted into the procedure. This step is the one in which the outcome of the experiment is used, by means of Bayes' law, to modify the prior probabilities to obtain the posterior probabilities. The procedure is thus very simple conceptually. First the actions and states of nature are defined. Then the profit to be received for each action-state of nature pair (a_i, e_j) is determined. Next the prior probabilities p_j are estimated. Before making a decision, an experiment is performed to obtain information about the state of nature. The outcome of this experiment is used to modify the prior probabilities, converting them by Bayes' law into posterior probabilities. Once the posterior probabilities are obtained, the action selected is the one which maximizes the expected profit, the posterior probabilities being used in computing the expected profit.

It is important to note that the posterior probabilities cannot be determined without some assignment of the prior probabilities. The posterior probabilities depend both on $\Re$ and on the prior probabilities. We feel

intuitively, however, that if the experiment $\mathcal{R}$ is very discriminating, then it is not important to know the prior probabilities with great accuracy. This is essentially correct, and we shall see more clearly in the next section how a discriminating experiment can "wash out" the prior probabilities.

5-11 EXAMPLES ILLUSTRATING THE USE OF BAYES' LAW

Before illustrating how to solve single-stage decision problems in which an experiment is performed, it will be helpful to provide some examples illustrating the use and properties of Bayes' law.

1. A manufacturer buys ball bearings from a supplier in lots of 10,000 ball bearings. Usually, the fraction of the ball bearings in the lot which are defective is about 0.001. However, occasionally something goes wrong with the supplier's production process and the fraction defective jumps to about 0.02. For simplicity, we shall suppose that every lot received has a fraction defective of 0.001 or 0.02. The supplier has indicated, and this is confirmed by the manufacturer's records, that the probability of receiving a lot having a fraction defective of 0.02 is 0.05. The manufacturer attempts to check on the quality of the supplier's lot by selecting at random a sample of 500 ball bearings from the lot and determining the number of defective bearings in the sample. On checking a particular lot, it is found that there are five defective ball bearings in the sample. What are the posterior probabilities that the lot fraction defective is 0.001 and 0.02?

We can think of the situation just described as a two-stage random experiment. The first stage was carried out at the supplier's factory where the lot fraction defective was determined. The second stage involves taking a sample of 500 ball bearings from the lot and noting the number of defectives. We wish to compute the conditional probabilities $p(0.001|5)$ and $p(0.02|5)$ that the lot fraction defective is 0.001 or 0.02, given that five defectives were found in the sample. By (5–27)

$$p(0.001|5) = \frac{p(5|0.001)p(0.001)}{p(5|0.001)p(0.001) + p(5|0.02)p(0.02)}.$$

Now by assumption, the prior probabilities are $p(0.001) = 0.95$ and $p(0.02) = 0.05$. Consider next the conditional probabilities $p(5|0.001)$ and $p(5|0.02)$. The probability $p(5|0.001)$ is the probability that 5 defective ball bearings will be found in a sample of 500 selected from a lot of 10,000 ball bearings having a fraction defective of 0.001. This probability is the hypergeometric probability $h(5; 500, 0.001, 10,000)$. Since the sample size is small with respect to the lot size and r is not large with respect to the total number of defectives, and p is very small, we can approximate this probability by the Poisson probability $p(5; np)$. Since $n = 500$ and $p = 0.001$, $np = 0.5$. From Table D, $p(5; 0.5) = 0.0002$. Thus $p(5|0.001) =$

0.0002. Similarly, $p(5|0.02)$ can be approximated by the Poisson probability $p(5; 10) = 0.0378$. Hence

$$p(0.001|5) = \frac{0.0002(0.95)}{0.0002(0.95) + 0.0378(0.05)}$$

$$= \frac{0.00019}{0.00019 + 0.00189} = \frac{0.00019}{0.00208} = 0.09 .$$

Thus as a result of taking the sample, the manufacturer's estimate that the lot fraction defective is 0.001 changed from 0.95 to 0.09. The probability $p(0.02|5) = 0.91$, since it is merely $1 - p(0.001|5)$.

2. We suggested at the end of the last section that if the experiment performed is quite discriminating then the posterior probabilities should be quite accurate, even if the prior probabilities are not known with great precision. Let us prove, in particular, that if $\mathcal{R}$ actually determines the state of nature, then, provided the decision maker did not assign a prior probability of 0 to this state of nature, call it e_t, the posterior probability of e_t computed from Bayes' law is 1. Imagine that f_k is the event which says that e_t is the state of nature when e_t is actually the state of nature. If $\mathcal{R}$ does indeed indicate what the state of nature is, we must have $q_{tk} = 1$, $q_{jk} = 0$, $j \neq t$. Then when e_t is the state of nature, $\mathcal{R}$ yields the outcome f_k and (5–8) becomes

$$p(e_t|f_k) = \frac{p_t}{p_t} = 1, \quad \text{if } p_t \neq 0 .$$

3. We have shown in the previous example that the prior probabilities are essentially irrelevant if the experiment $\mathcal{R}$ actually determines the state of nature. One can see by example what the situation is when the experiment is very discriminating but not perfect. Let us now do this. Suppose that there are three states of nature e_1, e_2 and e_3 to which the prior probabilities $p_1 = 0.2$, $p_2 = 0.5$, $p_3 = 0.3$ are assigned. Suppose that we perform an experiment $\mathcal{R}$ which has three outcomes f_1, f_2, f_3, where f_u says that the state of nature is e_u. Imagine that this experiment is quite discriminating, and $q_{11} = 0.90$, $q_{12} = 0.04$, $q_{13} = 0.06$, $q_{21} = 0.07$, $q_{22} = 0.92$, $q_{23} = 0.01$, $q_{31} = 0.02$, $q_{32} = 0.04$, $q_{33} = 0.94$. Consider the case where $\mathcal{R}$ yields the outcome f_1. Then the posterior probabilities are

$$p(e_1|f_1) = \frac{0.90(0.2)}{0.90(0.2) + 0.07(0.5) + 0.02(0.3)} = \frac{0.18}{0.221} = 0.815 ;$$

$$p(e_2|f_1) = \frac{0.07(0.5)}{0.221} = \frac{0.035}{0.221} = 0.158 ;$$

$$p(e_3|f_1) = \frac{0.02(0.3)}{0.221} = \frac{0.006}{0.221} = 0.027 .$$

Thus the prior probability $p_1 = 0.2$ is changed to a posterior probability of 0.815 as a result of the experiment.

Suppose that instead of the above prior probability assignment, the assignment $p_1 = 0.7$, $p_2 = 0.2$, $p_3 = 0.1$ is used. Let us again compute the posterior probabilities, given that f_1 is the outcome of $\mathcal{R}$. Now

$$p(e_1|f_1) = \frac{0.90(0.7)}{0.90(0.7) + 0.07(0.2) + 0.02(0.1)} = \frac{0.63}{0.646} = 0.975 ;$$

$$p(e_2|f_1) = \frac{0.07(0.2)}{0.646} = 0.022 ;$$

$$p(e_3|f_1) = \frac{0.02(0.1)}{0.646} = 0.003 .$$

Thus, although the prior probabilities are quite different in the two cases just considered, there is a relatively small difference in the posterior probabilities. We are now ready to illustrate how decision problems where an experiment is performed before making a decision can be solved. This will be done by giving two detailed examples.

5-12 THE INDEPENDENT OIL PRODUCER EXAMPLE

We shall begin with a grossly oversimplified version of a type of decision problem which is faced with great frequency by so-called independent operators in the oil business. An independent operator is trying to decide whether or not to buy a tract of land for the possibility that a reservoir of crude petroleum lies beneath the surface. The individual concerned is a trained geologist and has decided that, so far as making the right decision is concerned, it is sufficient to imagine that there are only three states of nature. These are: e_1, there is no oil; e_2, there are 500,000 barrels that can be produced; and e_3, there are two million barrels that can be produced. Based on intuition and prior experience, he assigns the following prior probabilities to these states of nature:

$$p_1 = 0.6; \quad p_2 = 0.3, \quad p_3 = 0.1 . \qquad (5\text{-}29)$$

Before deciding whether to purchase the tract, the independent operator decides to perform a seismic experiment. The seismic experiment will not tell him what the state of nature is, but will give him additional information on the matter. Imagine that the seismic experiment has four outcomes f_1, f_2, f_3 and f_4, which we might think of as low, medium, high and very high readings. (This is far from realistic, but a realistic description would be quite complicated and extended). The conditional probabilities $q_{ik} = p(f_k|e_j)$ are shown in Table 5-9. Note that the numbers in each column (not each row) must sum to 1. The numbers in Table 5-9 might, in practice, be obtained from the physics of the problem.

For the problem under consideration, the operator believes that if he

Table 5–9

Conditional probabilities for
seismic experiment

	e_1	e_2	e_3
f_1	0.50	0.40	0.10
f_2	0.30	0.40	0.30
f_3	0.15	0.15	0.35
f_4	0.05	0.05	0.25

buys the tract and no oil is present, he will incur a loss of $800,000. If 500,000 barrels are present, he will make a profit of $400,000, and if two million barrels are present, the profit will be $2 million. If no seismic experiment is performed, his expected profit on buying the tract is

$$0.6(-800,000) + 0.3(400,000) + 0.1(2,000,000) = -\$160,000 \,,$$

and therefore he would not buy the tract, since on not buying it his expected profit is 0.

Suppose now that the seismic experiment is performed and the outcome is f_3, a high reading. Let us determine the best action for the operator to take. First we compute the posterior probabilities of the states of nature using Bayes' law (5–27). From (5–29) and Table 5–9 we see that

$$p(e_1|f_3) = \frac{0.15(0.6)}{0.15(0.6) + 0.15(0.3) + 0.35(0.1)} = \frac{0.09}{0.09 + 0.045 + 0.035}$$

$$= \frac{0.09}{0.170} = 0.530 \;;$$

$$p(e_2|f_3) = \frac{0.045}{0.170} = 0.264; \quad p(e_3|f_3) = \frac{0.035}{0.170} = 0.206 \,.$$

Note that in computing the posterior probabilities, we needed only the third row in Table 5–9. Note also that the denominator in Bayes' formula is the same for each case, that is, $p(f_3)$.

Using the posterior probabilities, let us next compute the expected profit if he buys the tract. This gives

$$0.530(-800,000) + 0.264(400,000) + 0.206(2,000,000) = \$93,600 \,.$$

With f_3 being the outcome of the experiment, it becomes desirable to purchase the tract, since the expected profit if he does not purchase it remains 0. In the computations we did not include the cost of making the seismic experiment because it is unnecessary to do so. If the cost is α and we included this cost, then the expected profit if he purchases the tract is $93,600 - \alpha$ and is $-\alpha$ if he does not. Thus the same action is chosen. Through the simple example we see that when an experiment is performed, the optimal action to choose, in the light of the experimental evidence, may

be quite different from the one which would be selected if there were no opportunity to perform an experiment before selecting an action.

5–13 THE RADIO MANUFACTURER EXAMPLE

A small manufacturer of private brand radios has purchased a lot of 200,000 resistors from a foreign producer for use in an upcoming production run of a particular model. Let us suppose that the manufacturer is concerned about the fraction of defective resistors in the lot. Prior experience with the supplier indicates that the fraction defective will either be 0.001, 0.005 or 0.03; and he believes the probability that the fraction defective is 0.001 is 0.70, that it is 0.005 is 0.25 and that it is 0.03 is 0.05. To check on the quality, he selects at random from the lot a sample of 1000 resistors and determines the number of defectives in the sample. Given the outcome of the sample, the manufacturer must decide among three alternative courses of action. These are: a_1, use the lot in the production run without further inspection; a_2, inspect the entire lot and replace defective resistors with good ones purchased locally; a_3, return the lot and use only resistors purchased locally. The manufacturer does have the freedom to return the lot, but if he does so, he must use locally procured resistors in making the production run because there will not be sufficient time available for getting a new lot from the foreign supplier.

Before going on to discuss the costs involved, let us summarize what has been said thus far in the terminology of a decision problem. We imagine that at the supplier's plant a random experiment is performed which leads to one of three outcomes e_1, e_2 and e_3, e_1 being that the lot fraction defective $p = 0.001$, e_2 that $p = 0.005$ and e_3 that $p = 0.03$. The probabilities assigned to these three events are $p_1 = 0.70$, $p_2 = 0.25$ and $p_3 = 0.05$. There are three actions a_1, a_2 and a_3 open to the decision maker. Before actually selecting an action he performs the random experiment, which involves the selection of a sample of 1000 resistors from the lot and determining the number of defectives in the lot.

Let us now turn to the costs involved. For simplicity, we shall imagine that included in the manufacturing process for locally produced resistors there is an automatic checking device, so that all locally produced resistors will be good. However, each locally purchased resistor costs $0.06, which is $0.01 more than the foreign produced one. Each defective resistor used in the production process will be caught at a particular quality-control point which checks the partially completed chasis. However, to replace a defective part with a good one (purchased locally) costs $0.50. To inspect the entire lot remaining after the sample is taken costs $1600. If the manufacturer returns the lot, he incurs shipping charges of $150. In this case, of course, he does not pay for the lot.

The manufacturer feels that monetary values adequately represent his feelings in the situation, and therefore the action to take is the one which minimizes his expected cost. Given the above data, we can construct the payoff table shown in Table 5–10. In this table we have given only those costs incurred over and above the cost of purchasing the lot from the foreign supplier and the cost of making the experiment. These are the only costs needed, since to obtain the total cost it is only necessary to add some number δ to every entry in Table 5–10. The same action will be selected either way.

TABLE 5–10

PAYOFF TABLE FOR RADIO MANUFACTURER EXAMPLE

	e_1	e_2	e_3
a_1	$100 - 0.44r$	$500 - 0.44r$	$3000 - 0.44r$
a_2	1612	1660	1960
a_3	2150	2150	2150

Consider then the determination of the entry when a_1 is the action selected and e_1 turns out to be the state of nature. We recall that a_1 means that the lot is sent into production after the sample is taken, and e_1 means that $p = 0.001$, that is, there are 200 defective resistors. Each defective going into production costs \$0.50. We shall assume that before going into production any defectives in the sample are replaced by good ones procured locally. Thus if r defectives are found in the sample, the cost will be

$$0.5(200 - r) + 0.06r = 100 - 0.44r .$$

The other entries in the first row are determined in the same manner. For the second row consider the entry in the column corresponding to e_1. In this case the entire lot is inspected at a cost of \$1600 (over the cost of the experiment), and the 200 defectives are replaced by good units at a cost of \$12. The other two entries in the second row are obtained in the same manner. Finally, for a_3 the lot is returned to the supplier at a cost of \$150 and 200,000 resistors are procured locally at a cost of \$2000 above the cost of the lot received from the foreign supplier. On examining Table 5–10, we note immediately that regardless of what the state of nature turns out to be, the cost on taking action a_2 is less than that on taking a_3. Hence a_3 will never be the action selected and we can ignore it when computing the expected costs. In other words, it would never pay to ship back the lot and use only locally produced resistors.

Suppose now that on taking the sample of 1000 resistors, the manufacturer finds that 25 are defective. Let us determine whether he should take action a_1 or a_2 in this case. To do so we first determine the posterior probabilities using Bayes' law. To use Bayes' law we must determine the

conditional probabilities $p(25|0.001)$, $p(25|0.005)$, and $p(25|0.03)$, where, for example, by $p(25|0.001)$ we mean the probability of obtaining 25 defectives in a sample of 1000 when the lot fraction defective is 0.001. The precise computation of these probabilities would require the use of the hypergeometric distribution. However, since the sample size is small with respect to the lot size and the lot fraction defective is small in every case, the hypergeometric distribution can be approximated by the Poisson distribution with mean 1000 p, where p is the lot fraction defective. However, for $p = 0.03$, $np = 30$ and Table D does not go up to $\beta = 30$. Recall though that when $np > 25$, the normal approximation to the hypergeometric distribution can be used. Thus the normal approximation will be used for $p = 0.03$. Thus

$$p(25|0.001) = p(25; 1) \doteq 0.0000$$
$$p(25|0.005) = p(25; 5) \doteq 0.0000$$

$$p(25|0.03) \doteq \Phi\left(\frac{25.5 - 30}{5.47}\right) - \Phi\left(\frac{24.5 - 30}{5.47}\right)$$
$$= \Phi(-0.824) - \Phi(-1.01) = 0.0488 \,.$$

The probabilities $p(25|0.001)$ and $p(25|0.005)$ are 0 to four decimals. Hence we see immediately from Bayes' law that to four decimal places, the posterior probabilities are

$$p(e_1|25) = 0; \quad p(e_2|25) = 0; \quad p(e_3|25) = 1 \,,$$

and after making the experiment, the manufacturer concludes that it is almost certain that the lot fraction defective is 0.03. Note that he would reach this same conclusion regardless of what the prior probabilities of the states of nature were, provided that p_3 was not so small that more than four decimal places would be needed to reach a conclusion.

Given the above posterior probabilities we see at once that the expected cost of action a_1 is taken is $3000 - 0.44(25)$ and is 1960 if action a_2 is taken. The minimum expected cost occurs when action a_2 is selected. Thus the entire lot should be inspected and defective resistors replaced with good ones purchased locally.

In the next section we shall digress again to consider methods which do not make use of prior probabilities. The reader who wishes to continue with the main thread of development can skip to Section 5–15.

5–14* TECHNIQUES NOT INVOLVING PRIOR PROBABILITIES

In Sections 5–6 and 5–7 we discussed some procedures which could be used for selecting an action in cases where the probabilities of the states of

---

* Recall that sections marked with an asterisk contain material which may be omitted without loss of continuity.

nature were not known. We would now like to generalize those rules to problems in which an experiment is performed to gain information about the state of nature before an action is selected. The procedures introduced in Sections 5–6 and 5–7 will be generalized in such a way that they do not make any use of prior probabilities for the states of nature, but do make use of the experimental results. In particular, they make use not only of the outcome f_k of the experiment $\mathcal{R}$, but also of the conditional probabilities $p(f_k|e_j)$ that the outcome is f_k when the state of nature is e_j.

To proceed we shall introduce the notion of a *strategy*. The experiment $\mathcal{R}$ will yield one of the outcomes $f_1, \ldots, f_s$. A strategy will here be defined as a rule which tells the decision maker what action to take for each of the s possible outcomes of the experiment. A strategy S is then characterized by an ordered array of s symbols $\langle \alpha_1, \ldots, \alpha_s \rangle$, where each α_k denotes one of the actions a_i, and the ordered s-tuple of numbers is to be interpreted to mean that if f_k is the outcome of $\mathcal{R}$, then the action symbolized by α_k should be used. Thus one possible strategy would be $S = \langle a_1, a_1, a_2, \ldots, a_5 \rangle$, and it means that if $\mathcal{R}$ yields f_1 or f_2 take action a_1, if it yields f_3 take action a_2, etc., if it yields f_s take action a_5. Suppose that there are r different actions. Then α_1 can represent any one of these r actions and similarly for $\alpha_2, \ldots, \alpha_s$. Thus there are $h = r^s$ different strategies of the type which we have just introduced. What we now wish to do is select one of these strategies for the decision maker to use. Any strategy tells the decision maker what to do under every possible circumstance, since it tells him what action to take for every possible outcome of the experiment.

Suppose now that we have written down each of the h possible strategies and have numbered them from 1 to h. We shall denote the vth one by S_v. Let us imagine that the decision maker decides to use S_v and that S_v says to use action a_{i_k} when the experiment $\mathcal{R}$ yields the outcome f_k. As usual, U_{ij} will denote the profit received if the decision maker selects action a_i and the state of nature turns out to be e_j. If the decision maker selects strategy S_v and the state of nature turns out to be e_j, the profit for the decision maker is not determined because the action selected will depend upon the outcome of $\mathcal{R}$. In other words, once S_v is selected, and it is the case that the state of nature is e_j, the profit becomes a random variable defined with respect to $\mathcal{R}$. Denote by $\overline{U}_{vj}$ the expected profit when strategy S_v is selected and the state of nature is e_j. Then

$$\overline{U}_{vj} = \sum_{k=1}^{s} U_{i_k j} p(f_k|e_j), \qquad (5\text{-}30)$$

where $p(f_k|e_j)$ is the probability that $\mathcal{R}$ will yield f_k when the state of nature is e_j.

At this point we can construct a table similar to a payoff table in which the columns refer to the states of nature, but the rows refer to strategies

TABLE 5–11

STRATEGY PAYOFF TABLE

	e_1	e_2		e_m
S_1	$\overline{U}_{11}$	$\overline{U}_{12}$	. . .	$\overline{U}_{1m}$
S_2	$\overline{U}_{21}$	$\overline{U}_{22}$		$\overline{U}_{2m}$
$\vdots$	$\vdots$	$\vdots$		
S_h	$\overline{U}_{h1}$	$\overline{U}_{h2}$		$\overline{U}_{hm}$

rather than actions. The entries $\overline{U}_{vj}$ in the table are the expected profits when the strategy selected is S_v and the state of nature turns out to be e_j. The format for this table is shown in Table 5–11. We shall call this the strategy payoff table. It should be noted that it can be a staggering undertaking to construct a strategy payoff table even for decision problems of very modest size. The reason for this is that the number of strategies r^s can be very large even though r and s are not very large. Furthermore, each entry in the table requires the computation of an expected value.

Once one has the strategy payoff table, it is fairly obvious how the rules introduced in Section 5–6 should be generalized to the type of situation under consideration. One applies the rules to the strategy payoff table in precisely the same way that they were applied to the payoff table in Section 5–6. Before explaining the way in which these rules are applied here, it will be instructive to show the connection between the strategy payoff table and the payoff table in the case where the prior probabilities of the states of nature are known. Suppose that p_j is the prior probability for e_j. Recall that if $\mathcal{R}$ yields the outcome f_k, the way the decision maker determines the action to select is by computing the posterior probabilities of the states of nature

$$p(e_j|f_k) = \frac{p(f_k|e_j)p_j}{p(f_k)}, \tag{5-31}$$

and then determining the action which maximizes

$$\sum_{j=1}^{m} U_{ij}p(e_j|f_k) = \frac{1}{p(f_k)} \sum_{j=1}^{m} U_{ij}p(f_k|e_j)p_j. \tag{5-32}$$

Note, however, that since $p(f_k)$ does not depend on i, the value of i which maximizes (5–32) also maximizes

$$\sum_{j=1}^{m} U_{ij} p(f_k|e_j) p_j. \tag{5-33}$$

Conversely, the i which maximizes (5–33) maximizes (5–32). Thus the action to take is the one which maximizes (5–33). Suppose that we denote by β_k the action which maximizes (5–33) for f_k. There is then an optimal action to take for each value of k. Let $S^* = \langle \beta_1, \beta_2, \ldots, \beta_s \rangle$. Then S^* is a strategy and it is what we shall call an optimal strategy for the given prior probabilities, since regardless of what the outcome of $\mathfrak{R}$ is, it tells the decision maker to use the best action for that outcome. Note that S^* is one of the strategies listed in the strategy payoff table.

Suppose now that in the strategy payoff table we select the strategy (or one of the strategies) which maximizes

$$\sum_{j=1}^{m} U_{vj} p_j. \tag{5-34}$$

We then claim that S^* is such a strategy. To prove this, we use (5–30) and note that

$$\sum_{j=1}^{m} U_{vj} p_j = \sum_{j=1}^{m}\left[\sum_{k=1}^{s} U_{i_kj} p(f_k|e_j)\right] p_j = \sum_{k=1}^{s}\left[\sum_{j=1}^{m} U_{i_kj} p(f_k|e_j) p_j\right], \tag{5-35}$$

where a_{i_k} is the action which S_v says to select when $\mathfrak{R}$ yields the outcome f_k. We wish to determine $i_1, \ldots, i_s$ so that (5–35) is as large as possible. Now (5–35) will be maximized if and only if for each k, i_k is chosen so as to maximize

$$\sum_{j=1}^{m} U_{i_kj} p(f_k|e_j) p_j.$$

But this is precisely (5–33) and we know that it is maximized when the action is β_k. This is true for every k. Thus S^* maximizes (5–34), and this is what we wished to prove.

We have shown that there are two ways in which we can determine an optimal strategy S^* for a given set of prior probabilities. One way is to proceed in the manner discussed in Section 5–10. We compute for a given outcome of $\mathfrak{R}$ the posterior probabilities and then determine the action which maximizes the expected profit. This is done for each outcome of $\mathfrak{R}$, and the set of actions obtained (which may not be unique) yields S^* or an S^*. This procedure makes use of the payoff table. The alternative procedure is to construct a strategy payoff table. We then look at the numbers in this table as being profits and, using the prior probabilities to compute the expected profit, select the strategy which maximizes the expected profit. The S^*

obtained above will also be the strategy or one of the strategies obtained by this latter procedure.

Let us now return to showing how the rules of Section 5–6 can be applied here. To apply the Laplace criterion, one simply finds the strategy which maximizes (5–34) when $p_j = 1/m$. As before, this corresponds to assuming that a priori the states of nature are equally likely. A strategy determined in this way will be called a Laplace strategy. To apply the max-min criterion, let

$$U_v{}^+ = \min_j \overline{U}_{vj} .\qquad (5\text{–}36)$$

The notation on the right in (5–36) means the minimum of the $\overline{U}_{vj}$ in row v of the strategy payoff table. Then the strategy to use is the one which maximizes $U_v{}^+$. Such a strategy will be called a max-min strategy. It is the strategy for which the smallest entry in any row of the strategy payoff table is made as large as possible. To apply the max-max criterion, let

$$U_v{}^* = \max_j \overline{U}_{vj} .\qquad (5\text{–}37)$$

Then the strategy is selected for which $U_v{}^*$ is as large as possible. Such a strategy will be called a max-max strategy. The optimism index is applied as follows. A number $\alpha, 0 \le \alpha \le 1$ is selected, and the numbers

$$\hat{U}_v(\alpha) = \alpha U_v{}^* + (1 - \alpha)U_v{}^+ .\qquad (5\text{–}38)$$

are computed, and the strategy chosen is the one which yields the largest value of $\hat{U}_v(\alpha)$. Such a strategy will be called an optimism index strategy. Finally, to use the min-max regret criterion, one computes

$$M_j = \max_v \overline{U}_{vj} ,\qquad (5\text{–}39)$$

and then the regrets

$$L_{vj} = M_j - \overline{U}_{vj} .\qquad (5\text{–}40)$$

Next one computes

$$L_v{}^+ = \max_j L_{vj} ,\qquad (5\text{–}41)$$

and the strategy selected is the one yielding the smallest $L_v{}^+$. Such a strategy will be called a min-max regret strategy. Thus each of the criteria is used on the strategy payoff table in precisely the same way as on the payoff table in Section 5–6.

We shall now illustrate several of these rules by applying them to the independent oil producer example studied in Section 5–12. The reader should review that example at this point. There are two actions and four outcomes of the experiment, and therefore there are $2^4 = 16$ strategies.

TABLE 5–12

STRATEGY PAYOFF AND REGRET TABLE FOR OIL PRODUCER

	Payoff			Regret		
	e_1	e_2	e_3	e_1	e_2	e_3
$S_1 = \langle a_1, a_1, a_1, a_1 \rangle$	$-800{,}000$	$400{,}000$	$2{,}000{,}000$	$800{,}000$	0	0
$S_2 = \langle a_1, a_1, a_1, a_2 \rangle$	$-760{,}000$	$380{,}000$	$1{,}500{,}000$	$760{,}000$	$20{,}000$	$500{,}000$
$S_3 = \langle a_1, a_1, a_2, a_1 \rangle$	$-680{,}000$	$340{,}000$	$1{,}300{,}000$	$680{,}000$	$60{,}000$	$700{,}000$
$S_4 = \langle a_1, a_1, a_2, a_2 \rangle$	$-640{,}000$	$320{,}000$	$800{,}000$	$640{,}000$	$80{,}000$	$1{,}200{,}000$
$S_5 = \langle a_1, a_2, a_1, a_1 \rangle$	$-560{,}000$	$240{,}000$	$1{,}400{,}000$	$560{,}000$	$160{,}000$	$600{,}000$
$S_6 = \langle a_1, a_2, a_2, a_1 \rangle$	$-440{,}000$	$180{,}000$	$700{,}000$	$440{,}000$	$220{,}000$	$1{,}300{,}000$
$S_7 = \langle a_1, a_2, a_2, a_2 \rangle$	$-400{,}000$	$160{,}000$	$200{,}000$	$400{,}000$	$240{,}000$	$1{,}800{,}000$
$S_8 = \langle a_1, a_2, a_1, a_2 \rangle$	$-520{,}000$	$220{,}000$	$900{,}000$	$520{,}000$	$180{,}000$	$1{,}100{,}000$
$S_9 = \langle a_2, a_1, a_1, a_1 \rangle$	$-400{,}000$	$240{,}000$	$1{,}800{,}000$	$400{,}000$	$160{,}000$	$200{,}000$
$S_{10} = \langle a_2, a_2, a_1, a_1 \rangle$	$-160{,}000$	$80{,}000$	$1{,}200{,}000$	$160{,}000$	$320{,}000$	$800{,}000$
$S_{11} = \langle a_2, a_2, a_2, a_1 \rangle$	$-40{,}000$	$20{,}000$	$500{,}000$	$40{,}000$	$380{,}000$	$1{,}500{,}000$
$S_{12} = \langle a_2, a_2, a_2, a_2 \rangle$	0	0	0	0	$400{,}000$	$2{,}000{,}000$
$S_{13} = \langle a_2, a_2, a_1, a_2 \rangle$	$-120{,}000$	$60{,}000$	$700{,}000$	$120{,}000$	$340{,}000$	$1{,}300{,}000$
$S_{14} = \langle a_2, a_1, a_2, a_2 \rangle$	$-240{,}000$	$160{,}000$	$600{,}000$	$240{,}000$	$240{,}000$	$1{,}400{,}000$
$S_{15} = \langle a_2, a_1, a_1, a_2 \rangle$	$-360{,}000$	$220{,}000$	$1{,}300{,}000$	$360{,}000$	$180{,}000$	$700{,}000$
$S_{16} = \langle a_2, a_1, a_2, a_1 \rangle$	$-280{,}000$	$180{,}000$	$1{,}100{,}000$	$280{,}000$	$220{,}000$	$900{,}000$

In Table 5–12 we have listed these strategies and have constructed the strategy payoff table. In the table a_1 stands for the action which involves purchasing the property, while a_2 stands for the action of not purchasing it. The minimum $\overline{U}_{vj}$ always occurs in the first column. The largest number in the first column is 0, which occurs for S_{12}. Thus the max-min strategy is S_{12}. This strategy says that he should not buy the property regardless of what the outcome of the experiment is. Clearly, the max-min strategy is too conservative here and is not of much value to the oil producer, since in applying it he would essentially always get the same result and would be forced out of business. To find the max-max strategy, we look for the largest entry in the table. This is 2,000,000 and occurs for S_1. Thus S_1 is the max-max strategy. It says he should always buy the property regardless of what the outcome of the experiment is. This is not a very sound strategy either. In Table 5–12 we have also presented a regret table. The reader can easily check that the minimum value of the maximum regret is taken on at S_9, so that S_9 is the min-max regret strategy. This strategy borders on being a reasonable one. It says that he should not buy the property if the outcome of $\mathcal{R}$ is the lowest reading, while he should buy it for every other outcome. We leave to the problems the task of obtaining the Laplace strategy and the determination of the optimism index strategy for every possible value of α, $0 \leq \alpha \leq 1$. We also ask the reader to show that when the prior probabilities are given by (5–29), then $S^* = S_{10}$ is the optimal strategy.

It is possible to introduce mixed strategies for problems of the type being

considered here, just as we did in Section 5–7. The decision maker is using a mixed strategy if instead of selecting one of the strategies S_v considered above, which will now be referred to as pure strategies, he allows a game of chance to select the strategy he will use. A mixed strategy is then characterized by a set of probabilities $q_1, \ldots, q_h$, where q_v is the probability that the game of chance will select pure strategy S_v. The rules discussed above can be generalized to deal with mixed strategies in precisely the same way as in Section 5–7. When a mixed strategy is used, a two-stage experiment is performed before the decision maker selects an action. The first stage is the game of chance which selects the strategy, and the second stage is the random experiment $\mathfrak{R}$ whose outcome determines which action will be taken once the strategy is selected. The expected utility, computed with respect to this two-stage experiment given that the state of nature is e_j, is

$$\sum_{v=1}^{h} q_v \overline{U}_{vj}. \tag{5-42}$$

A max-min mixed strategy is one where the q_v are selected so that the minimum over j of (5–42) is made as large as possible. The max-min mixed strategy can be found by solving the linear programming problem

$$\sum_{v=1}^{h} q_v \overline{U}_{vj} - \rho \geq 0, \quad v = 1, \ldots, h,$$

$$\sum_{v=1}^{h} q_v = 1; \quad q_v \geq 0, \text{ all } v, \tag{5-43}$$

and the variable ρ is to be maximized.

When there are only two states of nature, the situation can be represented geometrically in precisely the same manner described in Section 5–8, the only difference being that the corners in Figure 5–7 will now refer to pure strategies rather than actions. The max-min strategy for the present case can be found in precisely the same manner that was discussed in Section 5–8.

If the prior probabilities of the states of nature are specified to be p_j, $j = 1, \ldots, m$, a Bayes' mixed strategy is one in which the q_v are determined so as to maximize

$$z = \sum_{j=1}^{m} \left[\sum_{v=1}^{h} q_v \overline{U}_{vj} \right] p_j = \sum_{v=1}^{h} \left[\sum_{j=1}^{m} \overline{U}_{vj} p_j \right] q_v. \tag{5-44}$$

Precisely the same proof as that given in Section 5–9 shows that at least one pure strategy will be a Bayes' mixed strategy, and in fact, S^* is a Bayes' mixed strategy.

5-15 SELECTION AMONG ALTERNATIVE EXPERIMENTS

In Section 5–10 through 5–13 we showed how the results obtained from an experiment may be used to aid the decision maker in selecting the appropriate action to take in some single-stage decision problem. We did not attempt to decide whether it was desirable to perform the experiment or which of several alternative experiments would be the best one to perform. We now wish to show how it can be determined whether it is desirable to perform an experiment and, if so, which of a number of alternative experiments should be performed. In order to make this analysis, it is necessary, of course, to know the cost of performing each of the random experiments. Note that the problem of determining the sample size which should be used in an experiment is a special case of what we are going to consider here, since each sample size can be viewed as a different random experiment. However, the theory we are about to consider is much more general than one which merely selects a sample size, since it can be used to choose between entirely different types of random experiments or the alternative of not performing any random experiment.

Let us suppose that it is desired either to select precisely one of the experiments $\mathcal{R}_1, \ldots, \mathcal{R}_v$ to be performed or to conclude that it is best not to perform an experiment. Each experiment will have some set of possible outcomes. Let $f_{h1}, \ldots, f_{hs_h}$ be the outcomes for experiment $\mathcal{R}_h$. Two subscripts are now used on the outcomes. The first refers to the experiment and the second to the particular outcome for that experiment. Thus f_{hj} is the jth simple event for $\mathcal{R}_h$. We shall denote the alternative of not performing an experiment by $\mathcal{R}_0$. We can, if we like, think of this as a free experiment which has only one outcome f_{01} (which symbolizes no information obtained). With this convenient terminology we can then look at the problem as one in which it is desired to select precisely one of the $v + 1$ experiments $\mathcal{R}_0$, $\mathcal{R}_1, \ldots, \mathcal{R}_v$.

Denote, as usual, by U_{ij} the profit that will be obtained, exclusive of the cost of any experiment, if the decision maker selects action a_i and the state of nature turns out to be e_j. Then if C_h is the cost of performing experiment $\mathcal{R}_h$, $(C_0 = 0)$, the profit that will be obtained if experiment $\mathcal{R}_h$ is performed, action a_i is selected and the state of nature turns out to be e_j is

$$U_{hij} = U_{ij} - C_h . \tag{5-45}$$

Three subscripts are now needed on the profit, since it depends on the experiment performed, the action selected and on the state of nature.

We shall now proceed to show how to determine which one of the $v + 1$ experiments should be performed. We might note that the problem we are considering is really a two-stage decision problem. First the decision maker

must select an experiment. Then, after the experiment is performed, he must select an action, this latter decision being based on the outcome of the experiment. We shall thus show how to solve this two-stage decision problem.

In order to compare the various experiments we must determine for each experiment the expected profit to be received if that experiment is performed and then the appropriate action is taken. Note carefully that what is needed is the expected profit *before* the experiment is performed. Let us then see how to determine the expected profit if $\Re_h$ is performed, this being the expected profit before the outcome of $\Re_h$ is known. Now $\Re_h$ will yield one of the simple events f_{hk}, $k = 1, \ldots, s_h$. Consider the situation where the outcome is f_{hk}. How does the decision maker decide what action to take in this case? Here we have a situation where a specified experiment $\Re_h$ is performed and yields the outcome f_{hk}. The way the decision maker selects the appropriate action is to use the theory developed in Section 5–10. He first computes the posterior probabilities $p(e_j|f_{hk})$ and then determines for each i

$$\overline{U}_{hki} = \sum_{j=1}^{m} U_{hij} p(e_j|f_{hk}) , \tag{5–46}$$

the expected profit if he takes action i. The action to take is the one for which $\overline{U}_{hki}$ is the largest. Thus if $\overline{U}_{hkg}$ is the largest of the $\overline{U}_{hki}$, then action a_g is the one to take. Denote by $\overline{U}_{hk}$ the largest of the $\overline{U}_{hki}$. Thus

$$\overline{U}_{hk} = \max_i \overline{U}_{hki} . \tag{5–47}$$

The notation on the right in (5–47) simply means the maximum with respect to i of the $\overline{U}_{hki}$.

Now $\overline{U}_{hk}$ is the expected profit to be received if $\Re_h$ yields the outcome f_{hk} and the decision maker takes the optimal action for this outcome. Note that the optimal action to take will, in general, differ for different outcomes of $\Re_h$. Suppose now that for each outcome f_{hk} of $\Re_h$, the optimal action to take and $\overline{U}_{hk}$, the expected profit to be received under these circumstances, is determined. Note that to do this really requires the solution of s_h single-stage decision problems. Up to this point we have determined the expected profit to be received for each possible outcome of $\Re_h$. It remains to combine these to obtain the expected profit before $\Re_h$ is performed.

Associated with the simple event f_{hk} of $\Re_h$ is the number $\overline{U}_{hk}$, which is the expected profit if $\Re_h$ yields f_{hk} and the decision maker selects the optimal action for this outcome. Thus the $\overline{U}_{hk}$ define a random variable with respect to $\Re_h$, call it $U\{h\}$, and the expected value $\overline{U}_h$ of this random variable is the expected profit before $\Re_h$ is performed. If $p_h(f_{hk})$ is the probability that $\Re_h$ yields f_{hk}, then the expected value of $U\{h\}$ is

$$\bar{U}_h = \sum_{k=1}^{s_h} \bar{U}_{hk} p_h(f_{hk}) \,, \tag{5-48}$$

and we have shown how $\bar{U}_h$ can be determined if $p_h(f_{hk})$ can be computed. Let us now see how to compute $p_h(f_{hk})$.

Recall that f_{hk} can be looked upon as the outcome of the second stage of a two-stage random experiment. The first stage is the experiment $\Re_N$ which determines the state of nature, and the second stage if $\Re_v$. Given this, we see immediately that (1–20) can be applied to compute $p_h(f_{hk})$, since all we are trying to do is to compute the probability of the simple event f_{hk} for the second stage. Thus

$$p_h(f_{hk}) = \sum_{j=1}^{m} p_h(f_{hk}|e_j) p_j \,, \tag{5-49}$$

where $p_h(f_{hk}|e_j)$ is the conditional probability that $\Re_h$ will yield the simple event f_{hk} if e_j is the state of nature, and p_j is the prior probability for the state of nature e_j.

Once $\bar{U}_h$ has been determined for each experiment, it is very easy to determine the experiment to be performed. The experiment to be performed is the one which yields the largest value of the expected profit, that is, the one for which $\bar{U}_h$ is largest. If $\bar{U}^*$ is the largest of the $\bar{U}_h$, so that

$$\bar{U}^* = \max_h \bar{U}_h \,, \tag{5-50}$$

then the experiment for which the expected profit is $\bar{U}^*$ is the one to perform. Observe that in the process of determining this experiment, we have also determined the action that the decision maker should take for each possible outcome of this experiment. The solution procedure can be summarized by the following four equations:

$$\bar{U}_{hki} = \sum_{j=1}^{m} U_{hij} p_h(e_j|f_{hk}) \,; \tag{5-51}$$

$$\bar{U}_{hk} = \max_i \bar{U}_{hki} \,; \tag{5-52}$$

$$\bar{U}_h = \sum_{k=1}^{s_h} \bar{U}_{hk} p_h(f_{hk}) \,; \tag{5-53}$$

$$\bar{U}^* = \max_h \bar{U}_h \,. \tag{5-54}$$

The procedure can be represented graphically by a decision tree like that shown in Figure 5–13. There is a $\bar{U}$ associated with each node. One begins at the end of the tree and works back towards the initial node. The entire tree is not shown in Figure 5–13 because of its complexity.

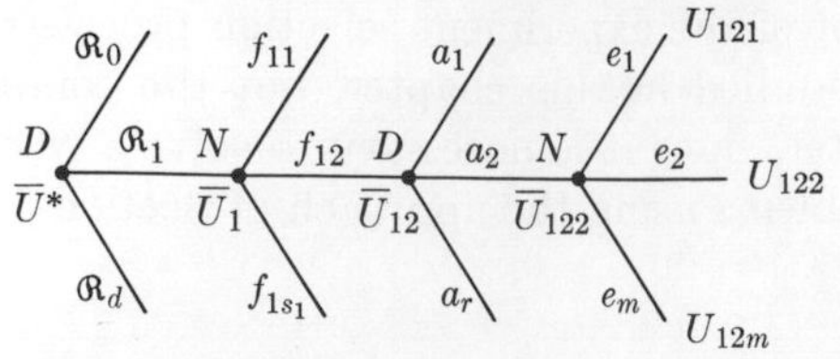

FIGURE 5–13

Frequently, instead of maximizing the expected profit it is more convenient to minimize the expected cost. If K_{ij} is the cost incurred, exclusive of the cost of performing any experiment, if action a_i is selected and the state of nature turns out to be e_j, the cost including the cost of performing the experiment is

$$K_{hij} = K_{ij} + C_h . \qquad (5\text{--}55)$$

The expected cost if experiment $\Re_h$ is performed and yields outcome f_{hk} and, in addition, action a_i is selected as

$$\bar{K}_{hki} = \sum_{j=1}^{m} K_{hij} p_h(e_j|f_{hk}) . \qquad (5\text{--}56)$$

The expected cost, given that $\Re_h$ is performed, yields the outcome f_{hk}, and the optimal action for this outcome is selected as

$$\bar{K}_{hk} = \min_i \bar{K}_{hki} . \qquad (5\text{--}57)$$

The expected cost before $\Re_h$ is performed, given that the optimal action is taken in light of the outcome of $\Re_h$, is

$$\bar{K}_h = \sum_{k=1}^{s_h} \bar{K}_{hk} p_h(f_{hk}) , \qquad (5\text{--}58)$$

and the appropriate experiment to perform is determined by computing

$$K^* = \min_h K_h . \qquad (5\text{--}59)$$

Thus if expected costs are minimized rather than expected profits maximized, equations (5–51), (5–52), (5–53) and (5–54) are replaced by (5–56), (5–57), (5–58) and (5–59) respectively.

We shall next illustrate the computational procedure just developed by working an example. The reader may well feel that the material in this section is somewhat difficult to understand. This is indeed true; the problem of selecting the experiment to undertake is conceptually more difficult than the other situations we have studied. The example will help clarify the approach, but two or more readings may reasonably be required. Not

only is the concept of the experiment selection process more difficult than the other topics studied in this chapter, but the computational effort is considerably greater, since it is necessary to solve a whole series of single-stage decision problems using the approach of Section 5–10.

5–16 **EXAMPLE**

A firm which designs and builds automatic electronic control devices markets one device which controls high temperature and pressure reactors for the chemical process industry. Part of the manufacturing process involves alignment of these devices under simulated operating conditions. The systems are then shipped to field offices where installation in the customer's plant is handled. For one reason or another, the systems occasionally get out of alignment in transportation. In the field, a simple piece of checkout equipment is used to determine whether the system seems to be in alignment. Unfortunately, the checkout equipment cannot simulate actual operating conditions too well, and thus it does not always indicate correctly whether the system is indeed in alignment. The checkout test can then be looked at as a random experiment performed to obtain information on the state of nature. Let us suppose that there are just two states of nature: e_1, the system is in alignment, and e_2, the system is not in alignment. The checkout equipment gives one of two readings: f_1, which says that the system is in alignment, and f_2, which says that the system is not in alignment. The conditional probabilities of these readings given the state of nature are

$$p(f_1|e_1) = 0.95 \; ; \quad p(f_2|e_1) = 0.05 \; ; \tag{5–60}$$
$$p(f_1|e_2) = 0.10 \; ; \quad p(f_2|e_2) = 0.90 \; . \tag{5–61}$$

At a particular field office, past experience has shown that ten percent of the time the systems get out of alignment during transportation from the plant. Each system is tested using the checkout equipment. Based on the outcome of the test, the system is either installed as is at the customer's plant, call this action a_1, or some engineers are sent out from the plant to align the system, call this action a_2. The cost of bringing engineers out from the plant to align the system is $500. If a system which is out of alignment is installed in the customer's plant, the cost of correcting things varies somewhat, but the value of this cost is estimated to be about $20,000. The cost of performing the test with the checkout equipment amounts to $50.

Let us show that it is definitely worthwhile to use the checkout equipment, and also determine the expected cost reduction per installation on using it. To do this, it is only necessary to determine the expected cost associated with alignment on installing a system when the checkout equipment is not used, and that when it is. The difference will be the expected savings per system installed as a result of using the checkout equipment.

TABLE 5–13

PAYOFF TABLE FOR EXAMPLE

	e_1	e_2
a_1	0	20,000
a_2	500	500

In Table 5–13 we have shown the payoff table using costs for the case where no checkout is made. When the checkout equipment is used, one adds the $50 cost of the checkout to each entry in the table.

The prior probabilities of e_1 and e_2 are $p_1 = 0.90$ and $p_2 = 0.10$. Given that $\mathfrak{R}_0$ is performed, that is, no checkout is made, the expected cost if action a_1 is taken is $0.10(20,000) = \$2000$, and if action a_2 is taken is $\$500$. Thus the optimal action when no checkout is made is a_2, that is, engineers from the plant should come out to align the system and the expected cost associated with this action is $\$500$. Thus $\bar{K}_0 = \bar{K}_{00} = \$500$ (since there is only one outcome of $\mathfrak{R}_0$, call it f_0, which can be imagined to say "no information").

Consider now the case where the checkout equipment is used. We first determine the optimal action to take and the corresponding expected cost for each of the two possible readings on the checkout equipment. Suppose that the reading is f_1, which says that the system is in alignment. The posterior probabilities of the states of nature are then

$$p(e_1|f_1) = \frac{p(f_1|e_1)p_1}{p(f_1)} \; ; \quad p(e_2|f_1) = \frac{p(f_1|e_2)p_2}{p(f_1)} , \qquad (5\text{–}62)$$

and from (5–45) and (5–46),

$$p(f_1) = p(f_1|e_1)p_1 + p(f_1|e_2)p_2 = 0.95(0.9) + 0.1(0.1) = 0.865 , \qquad (5\text{–}63)$$

so that

$$p(e_1|f_1) = \frac{0.855}{0.865} = 0.989; \quad p(e_2|f_1) = \frac{0.1(0.1)}{0.865} = 0.011 . \qquad (5\text{–}64)$$

The expected cost, exclusive of the cost of using the checkout equipment, when action a_1 is taken is $0.011(20,000) = \$220$ and is $\$500$ if action a_2 is taken. The expected costs, including the cost of making the checkout, are then $\$270$ and $\$550$, respectively. The appropriate action to take when the reading is f_1 is then a_1, that is, install without alignment and the expected cost is $\$270$, so that $\bar{K}_{11} = \$270$ when we denote by $\mathfrak{R}_1$ the experiment in which the checkout equipment is used.

Consider next the case where the checkout equipment gives the reading f_2. Then the posterior probabilities of the states of nature are

$$p(e_1|f_2) = \frac{p(f_2|e_1)p_1}{p(f_2)} \; ; \quad p(e_2|f_2) = \frac{p(f_2|e_2)p_2}{p(f_2)} \; , \tag{5–65}$$

and

$$p(f_2) = p(f_2|e_1)p_1 + p(f_2|e_2)p_2 = 0.05(0.9) + 0.9(0.1) = 0.135 \; .$$

Thus

$$p(e_1|f_2) = \frac{0.045}{0.135} = 0.333; \quad p(e_2|f_2) = \frac{0.090}{0.135} = 0.667 \; .$$

The expected cost if action a_1 is taken (exclusive of the cost of the checkout) is $0.667(20,000) = \$13,340$ or $\$13,390$ including the cost of the checkout. If action a_2 is taken, the cost including the checkout cost is $\$550$. The smallest of these is $\$550$, so that $\bar{K}_{12} = \$550$, and the optimal action to take is a_2. The expected cost before the checkout is performed, given that the optimal action is taken for either reading after the checkout is performed, is

$$\bar{K}_1 = p(f_1)\bar{K}_{11} + p(f_2)\bar{K}_{12} = 0.865(270) + 0.135(550) \; ,$$

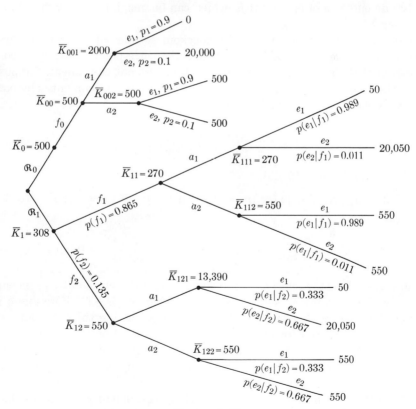

Figure 5–14

so that $\overline{K}_1 = \$308$. Now $\overline{K}_0 = \$500$; hence the expected savings per system installed resulting from use of the checkout equipment is \$192. The decision tree, which illustrates the steps in performing the computations, is shown in Figure 5–14.

5–17 COMPARISON OF BAYESIAN AND CLASSICAL APPROACHES

In closing our brief discussion of modern decision theory it will be instructive to compare the theory developed here with the classical approach studied in Chapter 4. The first thing to note is that the Bayesian approach we have discussed can be applied to a much wider class of problems than is treated by the classical approach. Bayesian techniques can be applied to single-stage or sequential decision problems and both to situations where a random experiment is performed to gain information about the state of nature and those where no experiment is to be performed. The classical approach is concerned only with single-stage decision problems in which an experiment is performed to gain information about the state of nature.

Let us then compare the two approaches to treating single-stage decision problems in which an experiment is performed to gain information about the state of nature. Even here the Bayesian approach is considerably more general than the classical approach, since in the Bayesian approach any number of actions can be considered and it is possible to select among various experiments, including the possibility of not performing any experiment. The classical approach considers only situations in which one of two alternative actions is to be selected (or in certain cases three actions are considered). Furthermore, the only experiment selection possible is the sample size. The type of experiment must be specified, and it is not possible to determine whether or not an experiment should be performed at all. Additional restrictions in the classical approach arise in the way the states of nature are described and in the experiment to be performed, since it is assumed that the state of nature can be characterized numerically. The random experiment performed is one in which the value of some random variable is observed, the random variable being such that the state of nature is some parameter in the distribution of this random variable, typically the mean.

Consider now a comparison of the two methods for a problem to which either can be applied. It will be assumed that the decision maker wishes to select precisely one of two actions a_1 or a_2. The appropriate action to take depends upon the expected value μ_x of some random variable x. An experiment is to be performed which will yield a value of $\bar{x}$. It is desired to determine the sample size to be used in this experiment, and the decision rule to be used in determining the action to be taken. Let us suppose that large values of μ_x favor action a_1 while small values favor a_2. The classical ap-

proach to solving this problem was discussed in Section 4–10. First two values μ_0 and μ_1 of μ_x are chosen such that if $\mu_x = \mu_0$, action a_1 should be taken and if $\mu_x = \mu_1 < \mu_0$, action a_2 should be taken. The problem is then looked at as testing the hypothesis H_0 that $\mu_x = \mu_0$ against the alternative hypothesis H_1 that $\mu_x = \mu_1$. The probabilities α and β of rejecting H_0 and H_1 when they are true are specified, and this then determines the sample size n and a number δ such that if $\bar{x} \leq \delta$, action a_2 is taken and if $\bar{x} > \delta$, action a_1 is taken. This covers the classical approach. In effect, the classical approach proceeds as if there can exist only two states of nature, which are $\mu_x = \mu_0$ and $\mu_x = \mu_1$. The classical approach does not make any use of prior probabilities or of costs or profits.

Let us now give the Bayesian approach for treating the problem. First the states of nature would be specified. Conceivably it would be considered to be adequate to suppose that μ_x can only take on two different values, perhaps μ_0 and μ_1, but normally more than two possible states of nature would be allowed for. Next the prior probability of each of these states of nature would be estimated. Then a payoff table would be constructed which gives the profit or cost, exclusive of the cost of the experiment, for each of the two actions and each possible state of nature. After this the cost of performing the experiment for each sample size would be determined, and then the theory of Section 5–15 would be applied to determine the best experiment, that is, the best sample size to use. Simultaneously, the optimal action to take for each possible outcome of the experiment would be determined. This decision rule for the action to take would have the form: Take action a_2 if $\bar{x} \leq \delta$ and a_1 otherwise. Thus both methods lead to precisely the same type of decision rule and both determine a sample size n. This is done in quite different ways, however, and it will not in general be true that the values of n and δ obtained in the two different ways will be the same. The Bayesian approach requires considerably more numerical computation than the classical approach, but it also includes much more detail about the nature of the situation.

REFERENCES

1. Chernoff, H., and L. E. Moses, *Elementary Decision Theory*. Wiley, New York, 1959.

This book, written at an elementary level, discusses exclusively cases where experiments are performed to gain information about the state of nature. Considerable attention is given to methods, which do not make use of prior probabilities.

2. Hadley, G., *Introduction to Probability and Statistical Decision Theory*. Holden-Day, Inc., San Francisco, 1967.

3. Schlaifer, R., *Probability and Statistics for Business Decisions*. McGraw-Hill, New York, 1959.

The first book which attempted to show in some detail how decision theory could be used in solving practical problems. Very little mathematical background is assumed, and mathematics is avoided as much as possible.

PROBLEMS

Section 5–2

1. Why is it convenient to define the actions in such a way that the decision maker will be selecting precisely one action? Suppose that the problem is one which involves the stocking of two different types of bread. If this is to be treated as a single problem, how would the actions have to be defined?

2. An oil company is planning to purchase two or less new tankers for its fleet. There are three different sizes of tanker that can be purchased. Make a list of all the actions implied by this situation.

3. Re-solve Problem 2 if each of two sizes of tanker can be purchased from two different ship construction firms, while the remaining size can be obtained from just a single supplier.

4. An oil company is planning to bid on a piece of property which may contain oil. What are the actions here? What are the states of nature? How might the number of states of nature be reduced to a reasonable level?

5. A firm is studying the possibility of marketing a new product. How might the states of nature be defined here?

6. A firm is studying the possibility of investing a considerable sum of money for an offshore tin-mining venture in Thailand. How might the states of nature be defined in this case?

7. Draw a decision tree for the bread-stocking example.

8. Describe an actual situation where the probabilities of the states of nature depend on the action selected.

9. The medical profession is trying to decide whether to use a new type of radiation treatment for cancer patients. It is difficult to say what sorts of harmful side effects might be caused. How might the states of nature be described in this case?

Section 5–3

1. Are there circumstances under which a decision maker might prefer a contract with one possible outcome being a rather large profit, even though the expected profit from the contract is less than that for another contract whose best outcome does not yield such a large profit? Describe any such circumstances.

2. What factors, other than monetary factors, might influence a firm's decision as to where to build a plant?

3. What factors, other than profit maximization, might enter into the price a firm charges for a given product?

4. How can a firm justify producing a product on which it loses money?

5. Give a specific example, different from that presented in the text, illustrating a case where maximization of expected profit might not lead to the selection of the action most preferred by the decision maker.

6. Give an example of a problem where maximization of expected profits cannot be used because monetary considerations do not appear in any way in the problem.

7. Show that if $U\{i\}$ is a random variable representing the profit to be obtained from some situation if action a_i is taken, the same action will be taken if the expected profit is maximized or if the expected value of $-U\{i\}$ is minimized. Show that $-U\{i\}$ can be interpreted as a cost. With this interpretation, show that any problem can be solved either by maximizing the expected profit or by minimizing the expected cost.

8. A drug firm is attempting to decide whether or not to release a new drug. It is never easy to determine what harmful side effects might ultimately be caused by the drug. What are the difficulties in determining the profits or costs if something turns out to be seriously wrong after the drug is in widespread use?

Sections 5–4 and 5–5

1. Solve the decision problem whose payoff table is shown in Table 5–14.

2. A businessman is planning to ship a used machine to his plant in South America, where he would like to use it for the next five years. He is trying to decide whether to overhaul the machine before sending it. The cost of overhaul is $2000. If the machine fails when in operation in South America, it will cost him a total of $5000 in lost production and repair costs. He estimates that the probability that it will fail is 0.5 if he does not overhaul it and is 0.1 if he does overhaul it. Neglecting the possibility that the machine might fail more than once in South America, should he overhaul the machine before sending it?

TABLE 5–14

	e_1	e_2	e_3	e_4
a_1	0.2 / −6	0.3 / 3	0.4 / 2	0.1 / −1
a_2	0.4 / 5	0.3 / 0	0.2 / −2	0.1 / 4
a_3	0.3 / 6	0.1 / −5	0.2 / 1	0.4 / 4

3. What difficulties would be encountered in dealing with Problem 2 if one had to allow for the possibility that the machine might fail more than once?

4. A candy store must decide how many of a large, expensive chocolate rabbit to stock for the Easter season. The rabbit costs $4.00 and sells for $7.50. Any rabbits not sold by Easter can be sold afterward at 30 percent below cost. The owner does not feel that he suffers any loss if a customer demands one of these rabbits when he is out of stock. Over the past five years, the store sold 4, 3, 5, 2 and 4 of these rabbits in the respective Easter seasons. In the three years with the highest sales all rabbits stocked were sold. No records were kept, however, of demands which could not be filled. On the basis of these meager data, the owner feels that the following is his best estimate of the probability function for the demand:

$$p(0) = 0.05, \quad p(1) = 0.10, \quad p(2) = 0.15, \quad p(3) = 0.20,$$
$$p(4) = 0.30, \quad p(5) = 0.15, \quad p(6) = 0.05.$$

Under the assumption that it is impossible to order additional rabbits, how many should be stocked?

5. Re-solve Problem 4 under the assumption that there is a cost to the owner of $2.00 for each rabbit demanded which he cannot supply.

6. An individual is offered the opportunity to invest $10,000 in a wildcat drilling operation in Texas. If the well turns out to produce an average of between 3000 and 5000 barrels per day of crude petroleum for a period of a year, he will be paid $100,000 in return for his investment. If it produces an average of between 1500 and 3000 barrels per day, he will be paid $50,000. If the average is less than 1500 barrels per day, he will be paid nothing and will simply lose his investment. There is no chance that the well will produce an average of more than 5000 barrels per day. He estimates that the probability of an average production of less than 1500 barrels per day is 0.6, of between 1500 and 3000 barrels per day, 0.3, and greater than 3000, 0.1. Should the individual invest or not?

7. A greeting card store is trying to decide how many boxes of a special expensive Christmas card to stock. The cards are not sold individually but as a box of 20 cards. These cards cost $8.00 per box and retail for $13.50 per box. Each box not sold during the season can be sold after the holidays at $6.00 per box. There will be no problems due to customers demanding these cards when they are sold out because customers, rather than request a special design, simply look over the selection and choose from what there is. However, the shop has limited capital and does not like to invest in things that will not sell, because it can earn more by not making such mistakes. The owner feels that for each $8.00 invested in a box of these cards which is not sold during the season, he could have earned $2.00 profit by investing it in other cards. The owner estimates the probability function for the number of boxes of these cards that will be demanded as:

$$p(0) = 0.03, \quad p(1) = 0.04, \quad p(2) = 0.08, \quad p(3) = 0.15,$$
$$p(4) = 0.20, \quad p(5) = 0.20, \quad p(6) = 0.15, \quad p(7) = 0.05,$$
$$p(8) = 0.05, \quad p(9) = 0.03, \quad p(10) = 0.02.$$

How many boxes should he stock at the beginning of the season if it is impossible to place re-orders?

8. For many decision problems encountered in the real world, each of the possible states of nature is very complicated and is revealed over a considerable period of time. In such cases the decision maker cannot know what each or even one of the states of nature really is. Give an example of such a situation and discuss the relevance of the models developed here in such cases. What are some of the procedures often used to try to avoid explaining in detail to the decision maker what each of the states of nature is?

9. What difficulties would be introduced into Example 2 of Section 5–5 if a customer's annoyance carried over to future weeks?

10. A department store buyer is trying to decide whether to stock a blue or orange high-fashion silk dress. The manufacturer only accepts orders for these dresses in multiples of 50. The buyer is not considering the purchase of more than 50 dresses, so that his problem reduces to one of deciding whether to obtain 50 blue or 50 orange dresses. For the purposes of making this decision, the buyer feels that it is adequate to assume that sales of either type of dress will only have one of the four possible values: 10, 20, 30 or 50. Based on past experience and his feelings about the market for the present year, he estimates the probabilities of each of these events to be 0.20, 0.20, 0.50 and 0.10 respectively for the blue dress and 0.10, 0.30, 0.40 and 0.20 for the orange dress. The cost of 50 blue silk dresses is $2500 and of 50 orange ones is $2900. A blue silk dress sells for $125 and an orange one for $130. Any dresses left over at the end of the season can be sold for $40. Which dress should the buyer stock?

Section 5–6

1. Apply each of the five rules discussed in this section to the first example of Section 5–5, and determine the action to be taken in each case. For the optimism index case, determine the action for each possible value of α.

2. Apply each of the five rules discussed in this section to the second example of Section 5–5, and determine the action to be taken in each case. For the optimism index case, determine the action for each possible value of α. Compare the results obtained here with that in Section 5–5.

3. Consider the single stage decision problem for which the payoff table is Table 5–15. Apply each of the five rules discussed in the text and determine the action to be taken in each case. For the optimism index case, determine the action for each possible value of α.

TABLE 5–15

	e_1	e_2	e_3	e_4	e_5	e_6
a_1	15	4	8	-3	0	12
a_2	6	11	-4	2	9	1
a_3	7	6	18	4	-18	7
a_4	-12	9	6	18	10	-2

4. Consider the following payoff table:

TABLE 5–16

	e_1	e_2
a_1	0	100,000
a_2	0.01	0.01

Show that the max-min criterion says to select action a_2. Does this make sense?

5. If one knows nothing about the probabilities of the states of nature, might this sometimes cause difficulties in deciding what the states of nature are, that is, in deciding what events are impossible? Can you provide an example of this?

6. Develop an example which shows the same sort of behavior for the min-max regret criterion as that exhibited in Problem 4 for the max-min criterion.

7. Construct an example where there are four actions and the min-max regret criterion selects action a_3, and where, in addition, if action a_1 is made unavailable, so that only actions a_2, a_3, a_4 remain, then the min-max regret criterion selects action a_2. Explain why this could be considered an undesirable characteristic of a decision rule.

8. Consider the payoff table for a single stage decision problem given in Table 5–17.

TABLE 5–17

	e_1	e_2	e_3	e_4	e_5	e_6
a_1	0	1	1	1	1	1
a_2	1	0	0	0	0	0

What does the optimism index criterion say about the action to be selected? Does this seem reasonable?

9. What criticisms can be made of the Laplace criterion?

Section 5–7

1. Consider the single-stage decision problem whose payoff table is given in Table 5–18.

TABLE 5–18

	e_1	e_2	e_3	e_4
a_1	4	-2	1	6
a_2	5	7	0	-7
a_3	2	1	8	0

Suppose that the decision maker uses a mixed strategy which selects action a_1 with probability 0.3, action a_2 with probability 0.5, and action a_3 with probability 0.2. What is the decision maker's expected profit if the state of nature turns out to be e_1? What is it if the state of nature turns out to be e_4?

2. For the example of Problem 1, set up, but do not try to solve, the linear programming problem whose solution will yield a max-min mixed strategy for the problem.

3. Show how the problem of generalizing the max-max criterion to max-max mixed strategies can be handled, and formulate a problem whose solution yields a max-max mixed strategy. This is a linear programming problem also.

4. Show how the min-max regret criterion can be generalized to min-max regret mixed strategies, and formulate a problem whose solution will yield a min-max regret mixed strategy. This is another linear programming problem.

5. Show how the optimism index criterion can be generalized to optimism index mixed strategies, and formulate a problem whose solution will yield an optimism index mixed strategy. This is also a linear programming problem.

6. Use the results of Problem 3 to formulate the linear programming problem whose solution yields a max-max mixed strategy for the example of Problem 1.

7. Use the results of Problem 4 to formulate the linear programming problem whose solution yields a min-max regret mixed strategy for the example of Problem 1.

8. Use the results of Problem 5 to formulate the linear programming problem whose solution yields an optimism index mixed strategy for a given α for the example of Problem 1.

Section 5–8

1. Consider the single-stage decision problem whose payoff matrix is shown in Table 5–19. Illustrate geometrically the set X of all points $(x(q), y(q))$ generated

TABLE 5–19		
	e_1	e_2
a_1	7	1
a_2	3	5

TABLE 5–20		
	e_1	e_2
a_1	1	9
a_2	5	7

by all mixed strategies. In particular, determine the points corresponding to $q = 0.2$ and $q = 0.7$. Determine the max-min mixed strategy and illustrate the procedure geometrically.

2. Consider the single-stage decision problem whose payoff matrix is shown in Table 5–20. Illustrate geometrically the set of all points $(x(q), y(q))$ generated by all mixed strategies. In particular, determine the points corresponding to $q = 0.3$ and $q = 0.5$. Determine the max-min mixed strategy and illustrate the procedure geometrically.

3. Consider the single-stage decision problem whose payoff table is that shown in Table 5–21 below. Illustrate the set of points (x, y) generated by all mixed strategies having only q_2 and q_4 or both different from 0.

TABLE 5–21	e_1	e_2
a_1	1	8
a_2	6	1
a_3	4	2
a_4	3	5
a_5	2	6

TABLE 5–22	e_1	e_2
a_1	3	2
a_2	8	−4
a_3	6	0
a_4	5	1
a_5	4	1.5

TABLE 5–23	e_1	e_2
a_1	2	8
a_2	3	7
a_3	3	5
a_4	2.5	3
a_5	1	6

4. Illustrate geometrically the set X of points (x, y) generated by all mixed strategies for the problem whose payoff table is given in Table 5–21. Determine the max-min mixed strategy for this problem, and illustrate geometrically.

5. Illustrate geometrically the set X of points (x, y) generated by all mixed strategies for the problem whose payoff table is given in Table 5–22. Determine the max-min mixed strategy for this problem, and illustrate geometrically.

6. Illustrate geometrically the set X of points (x, y) generated by all mixed strategies for the problem whose payoff table is given in Table 5–23. In determining X, do not eliminate any pure strategies which are dominated by others. Determine the set of all max-min mixed strategies for this problem and illustrate geometrically.

7. Determine the set of admissible mixed strategies for the problems whose payoff tables are given in Tables 5–21, 5–22, and 5–23, and illustrate geometrically the subsets of X which these represent. Also, show that a max-min mixed strategy is admissible.

8. Prove that a max-min mixed strategy must always be an admissible mixed strategy.

9. Show how the problem of determining a min-max regret mixed strategy can be handled geometrically in the case where there are only two states of nature in precisely the same manner as the determination of max-min mixed strategies. Illustrate by finding the min-max regret mixed strategy for the example of Table 5–8.

10. Determine the min-max regret mixed strategy for the example of Table 5–20 using geometrical analysis.

11. Determine the min-max regret mixed strategy for the problem whose payoff table is given in Table 5–21. Use geometrical analysis to do this.

12. Determine the min-max regret mixed strategy for the problem whose payoff table is given in Table 5–22. Use geometrical analysis to do this.

13. Determine the min-max regret mixed strategy for the problem whose payoff table is given in Table 5–23. Use geometrical analysis to do this.

14. Show how the problem of determining a max-max strategy can be handled geometrically in the case where there are only two states of nature. Illustrate by finding the max-max mixed strategy for the example of Table 5–19.

15. Determine by geometrical analysis the max-max strategy for the problem whose payoff table is given in Table 5–20.

16. How should the notion of admissible mixed strategies be defined when using a regret table?

Section 5–9

1. Use geometrical analysis to determine the Bayes' mixed strategy for the example of Table 5–19 for Section 5–8.

2. Use geometrical analysis to determine the Bayes' mixed strategy for the example of Table 5–20 for Section 5–8.

3. Use geometrical analysis to determine the Bayes' mixed strategy for the problem whose payoff matrix is given by Table 5–21 in the problems for Section 5–8.

4. Solve Problem 3 using Table 5–22 instead of Table 5–8.

5. Solve Problem 3 using Table 5–23 instead of Table 5–8.

6. Prove that a Bayes' mixed strategy must be an admissible mixed strategy.

7. Prove that at least one pure strategy will always be a max-max mixed strategy.

8. Show that the generalization of the Laplace criterion to mixed strategies leads to the determination of a Bayes' strategy for the case where if there are m states of nature, each state has the probability $1/m$ of occurring.

Sections 5–10 through 5–13

1. A decision maker is considering what action to take in a decision situation which is repeated frequently and for which there are three states of nature e_1, e_2 and e_3. From historical data he has observed that e_1 occurred 15 percent of the time; e_2, 30 percent of the time; and e_3, 55 percent of the time. The state of nature is determined before he makes a decision, and an experiment $\Re$, which has two outcomes f_1 and f_2, is performed to gain additional information. On performing $\Re$, the outcome f_2 is obtained. If $p(f_2|e_1) = 0.30$, $p(f_2|e_2) = 0.50$ and $p(f_2|e_3) = 0.10$, what probabilities should be used for the states of nature in deciding what action to select? Give an intuitive explanation as to why these are the appropriate probabilities. Does the result depend in any way on how many outcomes $\Re$ has?

2. A decision maker is considering a decision problem in which there are four states of nature to which he has assigned the following prior probabilities: $p_1 = 0.20$, $p_2 = 0.25$, $p_3 = 0.35$, $p_4 = 0.20$. To obtain more information about the state of nature, he performs an experiment $\Re$ which yields the outcome f_1, the conditional probabilities for which are

$$p(f_1|e_1) = 0.10, \quad p(f_1|e_2) = 0.80, \quad p(f_1|e_3) = 0.10, \quad p(f_1|e_4) = 0.05.$$

What are the probabilities of the states of nature which the decision maker should use in selecting the action to take?

3. A decision maker is considering a decision problem with three states of nature to which he has assigned the prior probabilities $p_1 = 0.30$, $p_2 = 0.50$ and $p_3 = 0.20$. He performs an experiment $\Re$ to obtain more information about the state of nature. $\Re$ yields the outcome f_1 for which the conditional probabilities are $p(f_1|e_1) = 0.4$, $p(f_1|e_2) = 0.4$, $p(f_1|e_3) = 0.4$. What are the posterior probabilities for the states of nature? What is the intuitive explanation for this? Can it be the case that some outcomes of $\Re$ will not give any information about the state of nature while others will? Construct a numerical example to illustrate this.

4. Consider a decision problem in which the prior probabilities of the states of nature are denoted by p_j. Suppose now that a random experiment $\Re_1$ is performed to gain information on the state of nature and it yields the outcome f_{1k}. Assume that Bayes' law is used to compute posterior probabilities $p(e_i|f_{1k})$. Imagine that another random experiment $\Re_2$ is then performed to gain still more information about the state of nature and yields the outcome f_{2h} (the probability of which may depend on the outcome of $\Re_1$ as well as on the state of nature). Bayes' law is then applied to compute posterior probabilities $p(e_i|f_{1k},f_{2h})$, using the $p(e_i|f_{1k})$ as prior probabilities. Alternatively, one could consider $\Re_1$ and $\Re_2$ to represent a single random experiment $\Re$. Then Bayes' law could be applied to compute posterior probabilities $p_1(e_i|f_{1k},f_{2h})$, using the p_j as prior probabilities. Prove that it is always true that

$$p(e_i|f_{1k},f_{2h}) = p_1(e_i|f_{1k},f_{2h}) \ .$$

In other words, prove that if $\Re$ can be decomposed into two or more random experiments, the results of these experiments can all be used at once or can be digested sequentially one after the other and the results obtained will be the same in either case. Hint: The experiment which determines the state of nature, $\Re_1$ and $\Re_2$, can be looked at collectively as a three-stage experiment.

5. For a single-stage decision problem having four states of nature, an experiment $\Re$ is performed to obtain more information and yields the outcome f_1, the conditional probabilities for which are

$$p(f_1|e_1) = 0.1, \quad p(f_1|e_2) = 0.8, \quad p(f_1|e_3) = 0.1, \quad p(f_1|e_4) = 0.05 \ .$$

Determine the posterior probabilities for the states of nature when the prior probabilities are

(a) $p_1 = 0.3$, $p_2 = 0.1$, $p_3 = 0.4$, $p_4 = 0.2$;
(b) $p_1 = 0.2$, $p_2 = 0.6$, $p_3 = 0.1$, $p_4 = 0.1$.

6. For a single-stage decision problem having four states of nature, an experiment $\Re$ is performed to obtain more information and yields the outcome f_1, the conditional probabilities for which are

$$p(f_1|e_1) = 0.3, \quad p(f_1|e_2) = 0.6, \quad p(f_1|e_3) = 0.4, \quad p(f_1|e_4) = 0.2 \ .$$

Determine the posterior probabilities for the states of nature when the prior probabilities are

(a) $p_1 = 0.2$, $p_2 = 0.4$, $p_3 = 0.3$, $p_4 = 0.1$;
(b) $p_1 = 0.5$, $p_2 = 0.2$, $p_3 = 0.1$, $p_4 = 0.2$.

7. What action should the independent operator of Section 5–12 take if the seismic experiment yields the outcome f_2? What is his expected profit in this case?

8. What action should the independent operator take if the seismic experiment yields the outcome f_1? What is his expected profit in this case?

9. What action should the independent operator take if the seismic experiment yields the outcome f_3 and the conditional probabilities are not those shown in Table 5–9 but are instead: $p(f_3|e_1) = 0.10$, $p(f_3|e_2) = 0.30$ and $p(f_3|e_3) = 0.20$?

10. Suppose that the independent oil operator decides that, in addition to performing a seismic study, he will do a magnetic study of the region. This employs a magnetometer, which gives an indication of the earth's magnetic field. For simplicity, suppose that the magnetometer gives only three readings, g_1 (high), g_2 (medium) and g_3 (low), and that the conditional probabilities of these readings, given the state of nature, are those in Table 5–24. The outcome of the magnetic study does not depend in any way on the outcome of this seismic experiment. What should the operator do when the seismic experiment yields the outcome f_2 and the magnetometer gives the reading g_3? What is his expected profit in this case?

11. For the situation described in Problem 10, what action should the operator take if the seismic experiment yields the outcome f_3 and the magnetometer gives the reading g_1? What is the expected profit in this case?

TABLE 5–24

	e_1	e_2	e_3
g_1	0.60	0.30	0.20
g_2	0.30	0.60	0.50
g_3	0.10	0.10	0.30

12. A pharmaceutical manufacturer is trying to decide whether to undertake a widespread marketing and sales campaign for a new drug. The decision hinges on the question of how many people may be adversely affected by the drug. The state of nature can be imagined to be described by p, the probability that the drug will adversely affect any particular individual, that is, the long-run fraction of individuals who will be adversely affected. It is felt that for purposes of making a decision, it can be imagined that p can take on three values, 0.0005, 0.002 or 0.01. Let us refer to the three states of nature as e_1, e_2 and e_3, respectively. If the drug is marketed, the profits are \$4,000,000, \$1,000,000 and −\$100,000 when the state of nature turns out to be e_1, e_2 or e_3, respectively. If the drug is not marketed, the loss is \$600,000. Based on the extensive animal experiments, the firm estimates the prior probabilities of the states of nature to be 0.60, 0.35 and 0.05, respectively. To gain more information about the reaction on humans, clinical tests on 500 human patients were carried

out, and it was found that 3 showed adverse reaction to the drug. Should the firm market the drug?

13. A publisher is trying to decide whether to publish a manuscript he has received. In making a decision he feels that it is sufficient to imagine that there are just four states of nature, which are: e_1, the book will sell an average of 500 copies per year; e_2, the average annual sales will be 1500 copies; e_3, the average annual sales will be 3000 copies; and e_4, the average annual sales will be 10,000 copies. The prior probabilities which he assigns to these states of nature are respectively 0.30, 0.40, 0.25 and 0.05. The discounted profits if he publishes the book are, for each of the states of nature, $-\$30,000$, $\$4000$, $\$12,000$ and $\$100,000$. The discounted profit is 0 if he does not publish it. To gain additional information, he sends the book to a reviewer who says that he likes the manuscript. From previous experience with the reviewer, the publisher feels that the conditional probabilities of this response from the reviewer for the different states of nature are respectively 0.10, 0.30, 0.50, 0.10· Should he publish the book?

14. What action should the radio producer of Section 5–13 take if he finds ten defectives in the sample rather than 25? What is the expected cost? What is the expected cost of uncertainty if the cost of performing the experiment is \$15?

15. Suppose that the radio producer of Section 5–13 takes a sample of 2000 resistors instead of 1000. What action should he take if 35 defectives are found in sample?

16. Suppose that on reviewing his records, the radio producer of Section 5–13 finds that he made an error in his original assessment and decides that the prior probabilities should be $p_1 = 0.70$, $p_2 = 0.2995$ and $p_3 = 0.0005$. Under these conditions, what action should be taken when he finds 25 defectives in the sample?

17. Suppose that the radio producer of Section 5–13 decided that a lot fraction defective of 0.02 was also a possible state of nature. He now assigns prior probabilities of 0.70, 0.25, 0.025 and 0.025 to the lot fraction defectives 0.001, 0.005, 0.02 and 0.03, respectively. What action should he take in this case if 25 defectives are found in the sample?

18. Consider a decision problem in which there are m states of nature, the prior probability for the jth being p_j. An experiment $\mathfrak{R}$ is now performed to gain information about the state of nature. Suppose that it is convenient to treat the outcome of $\mathfrak{R}$ as a continuous random variable y, and let $f(y|e_j)$ be the density function for y when the state of nature is e_j. If the observed value of y on performing $\mathfrak{R}$ is ζ, use an intuitive argument to show that the posterior probability $p(e_j|\zeta)$ is given by

$$p(e_j|\zeta) = \frac{f(\zeta|e_j)p_j}{\sum_{u=1}^{m} f(\zeta|e_u)p_u},$$

which is the appropriate form of Bayes' law to be used in this case.

19. Consider the farmer discussed in Example 3 for Section 4–8. Suppose that, to describe the states of nature, the amount by which the expected yield in bushels per

acre when the new fertilizer is used exceeds that for the standard fertilizer is specified. For the problem at hand the farmer believes it is sufficient to suppose that there
are just four states of nature: -25, 0, 40 and 100. The negative number indicates
that the new fertilizer is not as good as the current one. The farmer believes that the
prior probabilities for these states of nature are 0.10, 0.30, 0.50 and 0.10,
respectively. Corn sells for $0.50 per bushel. The use of the new fertilizer costs $10
per acre more than the standard one. On performing the experiment referred to in
Section 4–8, the farmer finds that $v = 25$. Should he use the new fertilizer or
continue using the standard fertilizer?

20. Consider the firm discussed in Section 4–9 which is contemplating the purchase
of some Diamond Head property. Suppose that the decision maker involved feels it
adequate to assume that the fraction of the voters who will favor a park will have
only one of the values 0.40, 0.45, 0.55 or 0.60. The prior probabilities which he assigns to these states of nature are 0.10, 0.30, 0.50 and 0.10. If the land is purchased
now and there is no park, a profit of five million dollars will be obtained, whereas, if
there is a park, there will be a loss of $500,000. If an option is taken on the land and
there is no park, there will be a profit of three million dollars, whereas if there is a
part there will be a loss of $50,000. If the land is not purchased, the profit is 0 in
either case. Suppose that the poll referred to in Section 4–9 yields the result that 0.56
of those interviewed will vote for a park. What action should the firm take?

Section 5–14

1. Consider a single-stage decision problem for which the payoff table is that shown
in Table 5–25. The state of nature is determined before an action is selected, and a
random experiment $\Re$ is performed to gain more information about the state of nature. The conditional probabilities for the outcomes of $\Re$ are given in Table 5–26. If
the prior probabilities for the states of nature are $p_1 = 0.8$ and $p_2 = 0.2$, determine
an optimal strategy S^* for the decision maker to use.

TABLE 5–25

	e_1	e_2
a_1	-5	20
a_2	-3	15
a_3	0	10

TABLE 5–26

	e_1	e_2
f_1	0.6	0.3
f_2	0.3	0.4
f_3	0.1	0.3

2. For the situation outlined in Problem 1, construct a strategy payoff table and
determine the max-min, max-max, and Laplace strategies; the min-max regret
strategy; and the optimism index strategy for every α, $0 \le \alpha \le 1$.

3. Use a graphical procedure to determine the max-min mixed strategy for the
situation outlined in Problem 1, using the strategy payoff table constructed in
Problem 2.

4. Use a graphical procedure to determine the mix-max regret mixed strategy for
the situation outlined in Problem 1.

5. Determine the optimal strategy for the situation described in Problem 1, using the strategy payoff table constructed in Problem 2. Also, show that the same strategy is obtained as in Problem 1.

6. Use a graphical method and the strategy payoff table constructed in Problem 2 to determine the Bayes' mixed strategy. Show that this strategy is the same as that obtained in Problem 1.

7. Utilize the results of Problem 4 to determine a Bayes' mixed stragegy using regrets. Call this a Bayes' regret mixed strategy. Show that this strategy is the same as that obtained in Problem 1.

8. In Table 5–12 determine the Laplace strategy and the optimism index strategy for each α, $0 \leq \alpha \leq 1$.

9. When the prior probabilities are given by (5–29), show that S_{10} is the optimal strategy in Table 5–12.

Sections 5–15 and 5–16

1. Consider the independent oil producer example studied in Section 5–12. If a seismic experiment costs $3000 to perform, is it worthwhile to perform it in the case under consideration?

2. Consider the independent oil producer example studied in Section 5–12 and the magnetometer experiment described in Problem 10 for Sections 5–10 through 5–13. If the seismic experiment costs $3000 and the magnetic study $4000, which experiment is preferable if one and only one of them is to be performed?

3. For the situation outlined in Problem 2, is it worthwhile to perform both the seismic and magnetometer experiments?

4. If it cost $30,000 to perform the clinical test involving human beings referred to in Problem 12 for Sections 5–10 through 5–13, would you have recommended that the test be made?

CHAPTER 6

Analysis of Variance

6-1 COMPARISON OF SEVERAL PRODUCTS OR TREATMENTS

In Section 4–7 and in some later sections we studied situations where it was of interest to compare two products, the purpose being to try to decide whether one seemed to be superior to the other. Sometimes situations arise where it is of interest to compare more than two products or treatments. For example, the Army may be interested in evaluating several competing brands of batteries, or a farmer may be interested in comparing several different brands of fertilizer, or a plastics manufacturer may be interested in studying several different antistatic coatings to be used on a certain type of sheet plastic. We wish to study problems of this general nature in the present chapter.

Let us begin with an example. Suppose that the Army wishes to test four competing brands of transistor battery. This type of battery is widely used in the field in various pieces of communications equipment. It is of interest to decide if there is any significant difference in the expected operating life between the various brands. To do this we shall suppose that six batteries of each brand are tested and the operating life is determined. In practice we might expect many more than six batteries to be tested, but to keep the computations in the example simple, it will be assumed that only six of each type is tested. The results of testing 24 batteries are shown in Table 6–1. The six numbers in any particular column give the operating lifetimes of the six batteries of that particular brand which were tested.

In order to decide if there is any difference in the expected operating life between the various brands, it is first necessary to estimate the expected operating life for each brand from the data. To do this we merely add up the six numbers in the column corresponding to that brand and divide by six.

TABLE 6–1

OPERATING LIFETIMES OF
BATTERIES TESTED IN THOU-
SANDS OF HOURS

Brand

A	B	C	D
1.40	1.20	0.70	0.90
1.00	0.80	1.30	1.10
0.90	1.00	1.20	0.80
1.10	1.30	0.90	1.20
0.70	0.80	1.00	0.60
1.30	1.10	0.90	1.00

TABLE 6–2

SUMMARY OF EXPERIMENTAL
RESULTS

Brand

	A	B	C	D
$\bar{\xi}$	1.067	1.033	1.000	0.933
s	0.26	0.21	0.22	0.22

In Table 6–2 we have given the resulting estimate $\bar{\xi}$ of the expected life of each brand (in the units of thousands of hours). We have also given the estimate s of the standard deviation of the life of each brand (in the units of thousands of hours). There is a difference of about 134 hours in the average life of brand A, the one with the greatest average lifetime, and brand D, the one with the smallest average lifetime. This is a fairly significant difference, being more than ten percent of the average life, and thus it might seem reasonable to conclude that there is indeed a difference in the expected life. However, there still remains the question as to whether these differences are real or whether they could have merely occurred by chance when, in fact, the mean life is essentially the same for each brand. Let us now examine this in a little more detail.

The standard deviation of $\bar{x}$, the average life of the six batteries tested of any particular brand is roughly $s/\sqrt{6}$, where s is the estimate of the standard deviation of the life given in Table 6–2. The estimate of $\sigma_{\bar{x}}$ for brand A is then $0.26/\sqrt{6} = 0.106$ and is $0.22/\sqrt{6} = 0.090$ for brand D. These values are fairly close together and we see that $\bar{\xi}_A - \bar{\xi}_D = 0.134$ is roughly $1.4\sigma_{\bar{x}}$, so that the largest and smallest average lifetimes are separated by about 1.4 standard deviations of $\bar{x}$. This is not an exceptionally large difference when viewed in this way, and suggests that perhaps we cannot be too sure the differences are not merely due to chance. If we com-

pute a 95 percent confidence interval for μ_A, the expected life of brand A, using the t distribution with $\nu = 5$, we find that the end points of this interval are

$$1.07 - 2.57(0.106) = 0.80; \qquad 1.07 + 2.57(0.106) = 1.34$$

since the value of t is 2.57. Similarly the end points of the confidence interval for μ_D, the expected life of D, are

$$0.93 - 2.57(0.090) = 0.70; \qquad 0.93 + 2.57(0.090) = 1.16 .$$

There is a considerable amount of overlap in these confidence intervals, as can be seen in Figure 6–1, and this again suggests that we cannot be too certain that there is any real difference in the means.

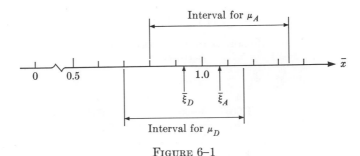

FIGURE 6–1

To examine the matter further, we can apply the theory developed in Section 4–7 and test the hypothesis that brand A and brand D have the same expected life. If this hypothesis is rejected, then we must conclude that there does seem to be a difference in the expected lifetimes of the various brands. If this test is not significant then a test involving any other pair of brands will not be significant either because the difference of the average lifetimes will be less than $\bar{\xi}_A - \bar{\xi}_D$. Hence, if the test is not significant we will be unable to claim, at whatever level of significance is used, that there is any substantial difference between the mean lifetimes of the various brands.

Let us then use the test of Section 4–7 to help us decide if there appears to be any significant difference between the expected life of a battery of brand A and one of brand D. The value $n = 6$ is rather small to use the normal distribution when σ must be replaced by s. However, we shall do this. There is some justification in feeling that the s values are reasonably accurate since roughly the same value was obtained for each brand, which suggests that they may all have the same standard deviation. We shall use the actual values obtained, however. Let $\bar{x}$ and $\bar{y}$ represent the average lifetimes of the six batteries tested of brand A and D, respectively. Then

write $v = \bar{x} - \bar{y}$. We wish to test the null hypothesis that $\mu_v = 0$. From Section 4–7, v is essentially normally distributed with

$$\sigma_v = \sqrt{\frac{\sigma_A^2 + \sigma_D^2}{6}}$$

where σ_A^2 and σ_D^2 are the variances of the lifetimes of A and D, respectively. In place of σ_A^2 we shall use s_A^2 and in place of σ_B^2 we shall use s_B^2. Thus our estimate of σ_v is

$$\frac{1}{2.45} \sqrt{0.067 + 0.046} = 0.14 .$$

Let us use a significance level of 0.05. The number λ such that $1 - \Phi(\lambda) = 0.025$ is $\lambda = 1.96$, and the critical values of v are

$$\delta_1 = -1.96(0.14) = -0.28; \qquad \delta_2 = 1.96(0.14) = 0.28 .$$

Now the observed value of v is 0.14 which is not greater than or equal to δ_2. Thus the test is not significant at the five percent level and we cannot reject the null hypothesis that brands A and D and hence all four brands have the same expected life.

The difficulty in attempting to decide if there is any significant difference in the expected life of the various brands is the result of testing so few batteries. If a large number of each type of battery were tested, it would not be difficult to distinguish if there is a significant difference in the expected useful life of the different brands. While there is no reason that a large number of batteries could not be tested, there are many practical problems where it would not be possible, for one reason or another, to make n large. Thus in certain areas, one is frequently confronted with situations similar to the one just analyzed. The procedure of Section 4–7 could be employed in the manner described above to try to decide if there exists a significant difference between the expected value of some random variable for several different brands of a product or several different treatments. However, there is another procedure, called analysis of variance, which can be used in this situation. It has the advantage that it considers all brands simultaneously and also that it can be readily generalized to more complicated situations. The analysis of variance procedure is the one that would normally be used if it were desired to test a hypothesis concerning the equality of the expected lifetimes of the various brands of batteries. We shall introduce the analysis of variance technique in the next section.

6–2 ANALYSIS OF VARIANCE—INTRODUCTION

We noted in the previous section that whether or not the differences in the observed average lifetimes of the different brands could be considered to be

significant depended on how these differences compared with the estimates of the standard deviations of the lifetimes. Let us now investigate this in a little more detailed way. Suppose that the null hypothesis is true and each brand has the same expected life. Suppose, in addition, that the standard deviation of the operating life is the same for each brand. Call this value σ. Finally, assume that the distribution of the life of each brand is the same and is, in fact, a normal distribution. Given these assumptions, it then follows that when the null hypothesis is true, the random variables $x\{A\}$, $x\{B\}$, $x\{C\}$ and $x\{D\}$ representing the life of a battery of type A, B, C and D respectively all have the same normal distribution. In other words, there is just a single random variable x representing the life of any battery. The brand of the battery has no influence on the distribution of x. Thus all 24 observed lifetimes recorded in Table 6–1 can be thought of as 24 values of x obtained from 24 independent trials of the experiment which measures the lifetime of a battery.

Let us now examine what the implications of the null hypothesis being true are when the above assumptions are made. In this case, any one of the numbers s_A^2, s_B^2, s_C^2 or s_D^2 given in Table 6–2 will be an estimate of the variance of σ^2 of x based on six observations. The average of these will then be an estimate of σ^2 also, that is,

$$s^2 = \frac{1}{4}(s_A^2 + s_B^2 + s_C^2 + s_D^2) \tag{6–1}$$

is an estimate (unbiased) of σ^2, since the expected value of the random variable whose value is s^2 is $\frac{1}{4}$ the sum of the expected values of the random variables whose values are s_A^2, ..., s_D^2 and each of these has an expected value σ^2. Hence, the expected value of the random variable whose value is s^2 is $4\sigma^2/4 = \sigma^2$ and this random variable yields an unbiased statistic for estimating σ^2.

Consider next the random variable $\bar{x}$ which represents the average of six observed values of x. In Table 6–2 are given four observed values of $\bar{x}$ if the null hypothesis and above assumptions are true. These are $\bar{\xi}_A$, $\bar{\xi}_B$, $\bar{\xi}_C$ and $\bar{\xi}_D$, since each of these is the average of the lifetimes of six batteries. The fact that they are averages of lifetimes of different brands is irrelevant if the null hypothesis is true, since all brands have the same distribution of battery life. Now we know from our previous work that the variance of $\bar{x}$ is σ^2/n, that is, $\sigma^2/6$. Furthermore, we can estimate the variance of $\bar{x}$ directly using the four observed values of $\bar{x}$ given in Table 6–2. All we do is use (3–37). Let $\bar{s}^2$ be the estimate of $\sigma_{\bar{x}}^2$ obtained in this way. Since $\sigma^2 = n\sigma_{\bar{x}}^2 = 6\sigma_{\bar{x}}^2$, we see that $6\bar{s}^2$ should provide us with another estimate of σ^2. If the null hypothesis is true, as we have been assuming, then $6\bar{s}^2$ and s^2 should be roughly the same value, since both are estimates of σ^2.

Consider now the case where the means are not all equal, but it is still true that the standard deviation of the battery life is the same for each brand and that the lifetime is normally distributed in each case. The only change we are allowing to take place is that the means are no longer required to be the same. Now s^2 given by (6–1) is still an estimate of the common variance σ^2, provided all brands have the same variance, even if the means are different, since it is the average of the estimates of the variance for each brand. It is important to keep in mind that s^2 will continue to provide an unbiased estimate of the variance even if the means are different. However, $6\bar{s}^2$ is no longer an estimate of σ^2 since $\bar{\xi}_A, \bar{\xi}_B, \bar{\xi}_C$ and $\bar{\xi}_D$ are no longer observed values of a single random variable $\bar{x}$ but are observed values of as many as four different random variables. In this case $\bar{s}^2$ will reflect the differences in the means as well as the random variations, and thus $6\bar{s}^2$ will tend to be greater than σ^2 and thus greater than s^2. We can now see how it may be possible to introduce a statistic for testing the null hypothesis that the means are equal.

Let Q_1 be the random variable whose value is $6\bar{s}^2$ and Q_2 the random variable whose value is s^2. Consider now the random variable

$$F = \frac{Q_1}{Q_2}. \tag{6–2}$$

The observed value of F in the above situation is then $6\bar{s}^2/s^2$. From what we have noted above, if the null hypothesis is true, then the observed value of F will be roughly 1, whereas, if the hypothesis is not true, then F will tend to be greater than 1. Now in the event that the null hypothesis is true, so that the means are all equal, and in addition the lifetimes are normally distributed, all having the same variance, it is possible to determine the distribution of F. This distribution is called the F distribution. The F distribution depends on two parameters denoted by ν_1 and ν_2; ν_1 is called the degrees of freedom for Q_1 in (6–2), and ν_2 is called the degrees of freedom for Q_2. The degrees of freedom for Q_1 is one less than the number of $\bar{\xi}$'s used in obtaining $\bar{s}^2$, that is, $\nu_1 = 4 - 1 = 3$. The number of degrees of freedom of Q_2 is the number of brands times one less than the number of batteries tested, that is, $\nu_2 = 4(6-1) = 20$. Given the distribution of F we can then test the null hypothesis that all the means are equal just like any other hypothesis. If $F \geq \delta$ we reject the hypothesis and otherwise we do not. The value of δ is determined from the significance level α. In Table I at the end of the text are given the critical values δ for two different significance levels $\alpha = 0.05$ and 0.01 and for various values of ν_1 and ν_2.

Let us now apply the test just developed to the data of Table 6–1. From Table 6–2 and (6–1),

$$s^2 = \tfrac{1}{4}(0.068 + 0.044 + 0.048 + 0.048) = 0.052$$

and (note that many significant figures are needed here to obtain any accuracy in $\bar{s}^2$)

$$\bar{s}^2 = \tfrac{1}{12}[4(1.1385 + 1.0671 + 1.0000 + 0.8705) - (4.033)^2]$$
$$= \tfrac{1}{12}[16.3044 - 16.2651] = \tfrac{1}{12}(0.0393) = 0.0033 ,$$

so

$$6\bar{s}^2 = 0.020 .$$

Hence

$$F = \frac{0.020}{0.052} = 0.38 ,$$

which is less than 1. Thus even without a table we are quite certain that the test will not be significant at either the five or one percent level. In fact, the critical value δ for $\alpha = 0.05$ with $\nu_1 = 3$ and $\nu_2 = 20$ is from Table I, $\delta = 3.10$. Thus we reach the same conclusion here that we reached in the previous section. We cannot reject the null hypothesis that there is no difference between the means. The procedure which we introduced here for testing the hypothesis that the means are all equal by comparing two different estimates of the variance σ^2 is referred to as *analysis of variance*.

6–3 ONE-VARIABLE ANALYSIS OF VARIANCE

In this section we shall generalize the analysis of variance method introduced in the previous section to the case where k different products are being compared and it is of interest to know if there is any difference in the expected value of some physical characteristic associated with each of the products. To investigate this we perform a random experiment in which the value of the characteristic is measured for n units of each of the k products, giving a total of kn observed values of the characteristic. The experimental results can be arranged to form a table like that shown in Table 6–3. Each column refers to a particular product. We have used ξ_{ij} to denote the observed value of the characteristic on the ith unit of product j. The subscript j thus refers to the product and the subscript i to the particular unit of the product under consideration. In the last row we have given the $\bar{\xi}$'s, $\bar{\xi}_j$ being the estimate of the expected value of the physical property for product j. Note that

$$\bar{\xi}_j = \frac{1}{n} \sum_{i=1}^{n} \xi_{ij} = \frac{1}{n} [\xi_{1j} + \cdots + \xi_{nj}] . \tag{6–3}$$

Let $x\{j\}$ be the random variable representing the value of the physical characteristic for any unit of product j. Then for each j, n values of $x\{j\}$

TABLE 6–3

FORMAT FOR PRESENTING EXPERIMENTAL
RESULTS—ONE-VARIABLE CASE

Product

1	2	3	$\cdots$	j	$\cdots$	k
ξ_{11}	ξ_{12}	ξ_{13}		ξ_{1j}		ξ_{1k}
ξ_{21}	ξ_{22}	ξ_{23}		ξ_{2j}		ξ_{2k}
$\vdots$	$\vdots$	$\vdots$		$\vdots$		$\vdots$
ξ_{i1}	ξ_{i2}	ξ_{i3}		ξ_{ij}		ξ_{ik}
$\vdots$	$\vdots$	$\vdots$		$\vdots$		$\vdots$
ξ_{n1}	ξ_{n2}	ξ_{n3}	$\cdots$	ξ_{nj}	$\cdots$	ξ_{nk}
$\bar{\xi}_1$	$\bar{\xi}_2$	$\bar{\xi}_3$	$\cdots$	$\bar{\xi}_j$	$\cdots$	$\bar{\xi}_k$

are observed since n units of each product are tested. We shall denote by $x\{i, j\}$ the random variable representing $x\{j\}$ for the ith unit of j tested; $x\{i, j\}$ is simply $x\{j\}$ for the ith unit. Then $\bar{\xi}_j$ is the observed value of the random variable

$$\bar{x}\{j\} = \frac{1}{n} \sum_{i=1}^{n} x\{i, j\} = \frac{1}{n} [x\{1, j\} + \cdots + x\{n, j\}] . \qquad (6\text{--}4)$$

Let us now consider the analysis of variance procedure for testing the hypothesis that each product has the same expected value of the physical property under consideration, that is, each $x\{j\}$ has the same expected value. Two important assumptions must be made concerning the $x\{j\}$ in order to use the analysis of variance method. These are:

ASSUMPTION 1. Each $x\{j\}$ is normally distributed.

ASSUMPTION 2. All of the $x\{j\}$ have the same variance, call it σ^2.

The analysis of variance procedure cannot be employed under all circumstances. The above two assumptions must be reasonably close to being satisfied if it is to be used. For the example considered in the last two sections, it seems reasonable that these assumptions are satisfied to a fair approximation. In general, however, there is no reason why the variances of the $x\{j\}$ for different products should all be the same.

We shall now suppose the two assumptions are satisfied. Then

$$s_j^2 = \frac{1}{n - 1} \sum_{i=1}^{n} (\xi_{ij} - \bar{\xi}_j)^2 \qquad (6\text{--}5)$$

is, for each j, an estimate of the variance σ^2 of the property under consideration. The argument given in the previous section then shows that

$$s^2 = \frac{1}{k} \sum_{j=1}^{k} s_j^2 = \frac{1}{k} (s_1^2 + \cdots + s_k^2) \qquad (6\text{-}6)$$

is also an estimate of σ^2 (which, on the average, should be better than any one of the s_j^2 since more observations are used). On combining (6-6) and (6-5) we can write

$$s^2 = \frac{1}{k(n-1)} \sum_{j=1}^{k} \left[\sum_{i=1}^{n} (\xi_{ij} - \bar{\xi}_j)^2 \right]. \qquad (6\text{-}7)$$

Now s^2 is the observed value of a random variable which we called Q_2 in the previous section; Q_2 can be expressed in terms of the random variables $x\{i, j\}$ as

$$Q_2 = \frac{1}{k(n-1)} \sum_{j=1}^{k} \left[\sum_{i=1}^{n} (x\{i, j\} - \bar{x}\{j\})^2 \right]$$

$$= \frac{1}{k(n-1)} [(x\{1, 1\} - \bar{x}\{1\})^2 + \cdots + (x\{n, 1\} - \bar{x}\{1\})^2$$

$$+ \cdots + (x\{1, k\} - \bar{x}\{k\})^2 + \cdots + (x\{n, k\} - \bar{x}\{k\})^2], \qquad (6\text{-}8)$$

and Q_2 is an unbiased statistic for estimating σ^2 regardless of whether or not the means are all equal.

When the null hypothesis is true, another estimate of σ^2 can be obtained from the $\bar{\xi}_j$. Let

$$\bar{\bar{\xi}} = \frac{1}{k} \sum_{j=1}^{k} \bar{\xi}_j = \frac{1}{k} (\bar{\xi}_1 + \cdots + \bar{\xi}_k). \qquad (6\text{-}9)$$

Then $\bar{\bar{\xi}}$ is the average of all nk observed values of the physical property, and $\bar{\bar{\xi}}$ is the observed value of the random variable

$$\bar{\bar{x}} = \frac{1}{k} \sum_{j=1}^{k} \bar{x}\{j\} = \frac{1}{kn} \sum_{j=1}^{k} \sum_{i=1}^{n} x\{i, j\}. \qquad (6\text{-}10)$$

An estimate of σ^2/n is then

$$\bar{s}^2 = \frac{1}{k-1} \sum_{j=1}^{k} (\bar{\xi}_j - \bar{\bar{\xi}})^2, \qquad (6\text{-}11)$$

and $n\bar{s}^2$ is thus another estimate of σ^2. Note that $n\bar{s}^2$ is the observed value of the random variable

$$Q_1 = \frac{n}{k-1} \sum_{j=1}^{k} (\bar{x}\{j\} - \bar{\bar{x}})^2. \qquad (6\text{-}12)$$

If the null hypothesis is true, then $n\bar{s}^2$ and s^2 should have roughly the same value whereas, if the means are not all the same, $n\bar{s}^2$ will tend to be larger than σ^2. However, s^2 will still be an estimate of σ^2 even if the means are different, provided that all brands have the same variance, that is, provided assumption 2 is satisfied. Thus, we should accept the null hypothesis if $n\bar{s}^2/s^2$ is approximately 1 or less and should reject the null hypothesis if $n\bar{s}^2/s^2$ becomes too large. The basic question is how large this ratio must be before the null hypothesis is to be rejected. To answer this question we must know the distribution of the random variable of which $n\bar{s}^2/s^2$ is the observed value, given that the null hypothesis is true. The random variable

$$F = \frac{Q_1}{Q_2} \tag{6–13}$$

is the random variable whose value is $n\bar{s}^2/s^2$, Q_1 and Q_2 being given by (6–12) and (6–8) respectively. It is possible to find the distribution of this random variable when assumptions 1 and 2 hold and the null hypothesis is satisfied.

This distribution is, as we have indicated previously, called the *F distribution*, and Table I at the end of the text contains the information needed to test the null hypothesis either at the five or one percent levels. The *F* distribution involves two parameters ν_1 and ν_2; ν_1 is called the degrees of freedom of Q_1 and

$$\nu_1 = k - 1 , \tag{6–14}$$

which is the denominator of Q_1 in (6–12). Similarly, ν_2 is called the degrees of freedom of Q_2 and

$$\nu_2 = k(n - 1) , \tag{6–15}$$

which is the denominator of Q_2 in (6–8).

To test the null hypothesis that all the means are equal in a given case, we first determine ν_1 and ν_2 from (6–14) and (6–15). Then for the given significance level α we determine the critical value δ of F. Next $\bar{s}^2$ and s^2 are determined from the data using (6–7) and (6–11) (a more convenient format for computation is that of (3–37)). Finally, $n\bar{s}^2/s^2$ is determined. If $n\bar{s}^2/s^2 \geq \delta$, the null hypothesis is rejected. Otherwise, the null hypothesis is not rejected. The numerical computations needed to perform this analysis of variance are rather tedious to carry out by hand. They can, however, be done very easily on a computer.

The reader may have wondered why, in the above analysis, instead of using s^2 for the good estimate of σ^2, we did not simply treat all the observed values of the physical property as being observed values of a single random variable x and then estimate the variance using

$$s_*^2 = \frac{1}{kn - 1} \sum_{j=1}^{k} \sum_{i=1}^{n} (\xi_{ij} - \bar{\bar{\xi}})^2 . \tag{6–16}$$

This is merely the normal way of estimating σ^2 if kn values of x are available. The reason this was not done lies in the fact that if the means are not all equal s_*^2 will not be an estimate of σ^2. We needed one estimate that was independent of whether or not the means were equal, and s^2 provided such an estimate. Observe that s_*^2 is the observed value of the random variable

$$Q = \frac{1}{kn-1} \sum_{j=1}^{k} \sum_{i=1}^{n} (x\{i,j\} - \bar{\bar{x}})^2 . \tag{6-17}$$

There is an interesting connection between the random variables Q, Q_1 and Q_2 which we shall now obtain.

Note that

$$x\{i,j\} - \bar{\bar{x}} = x\{i,j\} - \bar{x}_j + \bar{x}_j - \bar{\bar{x}} , \tag{6-18}$$

since all we have done is add and subtract $\bar{x}_j$. Then

$$(x\{i,j\} - \bar{\bar{x}})^2 = (x\{i,j\} - \bar{x}_j + \bar{x}_j - \bar{\bar{x}})^2$$
$$= (x\{i,j\} - \bar{x}_j)^2 + 2(x\{i,j\} - \bar{x}_j)(\bar{x}_j - \bar{\bar{x}}) + (\bar{x}_j - \bar{\bar{x}})^2 . \tag{6-19}$$

Hence

$$\sum_{j=1}^{k} \sum_{i=1}^{n} (x\{i,j\} - \bar{\bar{x}})^2 = \sum_{j=1}^{k} \sum_{i=1}^{n} (x\{i,j\} - \bar{x}_j)^2$$
$$+ 2\sum_{j=1}^{k} \sum_{i=1}^{n} (x\{i,j\} - \bar{x}_j)(\bar{x}_j - \bar{\bar{x}}) + \sum_{j=1}^{k} \sum_{i=1}^{n} (\bar{x}_j - \bar{\bar{x}})^2 . \tag{6-20}$$

Now

$$\sum_{i=1}^{n} (x\{i,j\} - \bar{x}_j)(\bar{x}_j - \bar{\bar{x}}) = (\bar{x}_j - \bar{\bar{x}}) \sum_{i=1}^{n} (x\{i,j\} - \bar{x}_j)$$
$$= 0 ,$$

since

$$\sum_{i=1}^{n} (x\{i,j\} - \bar{x}_j) = \sum_{i=1}^{n} x\{i,j\} - n\bar{x}_j ,$$

which on using the definition of $\bar{x}_j$ yields 0. Hence

$$2 \sum_{j=1}^{k} \sum_{i=1}^{n} (x\{i,j\} - \bar{x}_j)(\bar{x}_j - \bar{\bar{x}}) = 0 .$$

Next

$$\sum_{i=1}^{n} (\bar{x}_j - \bar{\bar{x}})^2 = (\bar{x}_j - \bar{\bar{x}})^2 \sum_{i=1}^{n} 1 = n(\bar{x}_j - \bar{\bar{x}})^2 .$$

Consequently, (6-20) can be written

$$\sum_{j=1}^{k} \sum_{i=1}^{n} (x\{i,j\} - \bar{\bar{x}})^2 = \sum_{j=1}^{k} \sum_{i=1}^{n} (x\{i,j\} - \bar{x}_j)^2 + n \sum_{j=1}^{k} (\bar{x}_j - \bar{\bar{x}})^2 . \tag{6-21}$$

Equation (6–21) is an interesting and important one. It says that the total variation in the observed values of the physical property with respect to the grand mean $\bar{\bar{x}}$ is the sum of two terms, one representing the variations within brands, and the other representing the variations of the brand means from the grand mean. This notion of breaking up the total variation in the observed values into a sum of terms, each of which represents a particular form of variation, is crucial in more complex applications of the analysis of variance method. If we denote the random variable on the left in (6–21) by W and the ones on the right by W_2 and W_1 respectively, then $W = W_2 + W_1$ and

$$Q = \frac{W}{(kn-1)} ; \qquad Q_2 = \frac{W_2}{k(n-1)} ; \qquad Q_1 = \frac{W_1}{(k-1)} . \qquad (6\text{–}22)$$

Thus

$$Q = \frac{k(n-1)}{kn-1} Q_2 + \frac{k-1}{kn-1} Q_1 . \qquad (6\text{–}23)$$

The random variable F can also be expressed in terms of W_2 and W_1 as

$$F = \frac{W_1}{W_2} \frac{k(n-1)}{k-1} . \qquad (6\text{–}24)$$

The notion of degrees of freedom is associated with the variables W_1 and W_2 just as with Q_1 and Q_2. The degrees of freedom of W_1 is $k-1$, and the degrees of freedom of W_2 is $k(n-1)$. We also associate degrees of freedom with Q and W. The degrees of freedom of Q and W is $kn-1$, the denominator in (6–17). Note that just as $W = W_1 + W_2$, the degrees of freedom of W is equal to the degrees of freedom of W_1 plus the degrees of freedom of W_2.

In carrying out the computations to test the null hypothesis that all the means are equal, it is often convenient to construct a little table called an analysis of variance table, the format for which is shown in Table 6–4. The term one-variable is used in the title for the table, since the only variation

TABLE 6–4

ONE-VARIABLE ANALYSIS OF VARIANCE

Source of Variation	Degrees of Freedom	Value of Sum of Squares Random Variable	Estimate of σ^2	F
Between Brands	$k-1$	W_1	$W_1/(k-1)$	$\dfrac{W_1}{W_2}\dfrac{k(n-1)}{k-1}$
Within Brands	$k(n-1)$	W_2	$W_2/k(n-1)$	
Total	$kn-1$	$W = W_1 + W_2$		

TABLE 6–5

ANALYSIS OF VARIANCE TABLE FOR EXAMPLE

Source of Variation	Degrees of Freedom	Value of Sum of Squares Random Variable	Estimate of σ^2	F
Between Brands	3	0.060	0.020	0.38
Within Brands	20	1.040	0.052	
Total	23	1.100		

being studied is the use of different brands, and the variable refers to the brand used. For the example we have been studying the formal of Table 6–4 reduces to what is shown in Table 6–5.

6–4 EXPERIMENTAL DESIGN

The design of the experiment to compare the lifetimes of several different brands of batteries was fairly simple. We merely determined the operating life of n batteries of each brand and then used the analysis of variance approach. We did not consider what value of n should be used. We shall not attempt to examine this problem. Instead we shall examine some problems which arise if we wish to compare several products in cases where the product can be thought of as a treatment on something else. Thus we may be interested in comparing several different brands of fertilizers. To do so we must treat some suitable land and grow an appropriate crop. Equally well, we might be interested in comparing several different brands of toothpaste by having a number of different children use them, or the adsorption of several different brands of dye on a particular type of plastic.

The new difficulty encountered in treatment problems is that not only can there exist variations between brands and variations in quality within brands (that is, changes in quality from one unit of a brand to another), but in addition, there can be variations in the material being treated. One must be very careful to be sure that the variations in the material treated do not in one way or another bias the experiment and lead to erroneous conclusions. Let us provide some examples.

Ideally, the farmer or farmer's cooperative would like to test the various fertilizers on the same piece of land. However, even if the effects of one fertilizer did not carry over from one year to the next, it would still be necessary to test the different fertilizers in different growing seasons. If this were done, however, the question would then arise as to whether any differences observed might be attributable to differences in weather conditions rather than to differences in the fertilizers. Such an experiment would also require an excessively long time to perform. Thus, the experiment would normally be performed in a given year using different pieces of land

for the different fertilizers. When different pieces of land are used, care must be taken that differences in land quality or other factors do not bias the results. For example, it would not be wise to try one fertilizer on one farm and the other fertilizers on other farms because then the question would arise as to whether any observed differences would be due to differences in the land (or in the methods of cultivation used, or perhaps in the seed used) rather than reflecting differences in the fertilizers themselves.

How, then, can one design the experiment in this case? It is not possible to eliminate variations in land or possibly in seed, but an attempt can be made to average out such variations to minimize their influence. The precise way in which this is done varies somewhat depending on the situation. Generally, however, the notion of *randomization* is important. Let us now explain this by use of some examples.

Consider a farmer who is interested in comparing several fertilizers. Suppose that the farm under consideration is a large one located in the Midwest region of the U.S. where the land is rather flat and uniform. It will be imagined that the farm is large enough that all fertilizers can be tested simultaneously in a single year. It will also be assumed that there are no known significant differences in land quality from one part of the farm to another. What the farmer might then do is divide up the land into a number of plots of equal size, this number being perhaps ten or twenty times the number of fertilizers to be tested, depending on what a convenient size of plot might be. To be specific suppose that there are four fertilizers and the farmer subdivides the land into 40 plots. Next the plots would be numbered from 1 to 40, perhaps as shown in Figure 6–2.

To decide what fertilizers to use on what plots, ten random numbers whose values came from the numbers 1 to 40 would be selected, and the first fertilizer, call it A, would then be used on these plots. Next, ten random numbers from the remaining 30 numbers would be selected, and the second fertilizer B would be used on these plots. Finally, ten random numbers would be selected from the remaining 20 numbers, and fertilizer C would be used on these. Fertilizer D would be used on the remaining ten plots. The results might then be as shown in Figure 6–2. The fertilizers are assigned to the plots in a random way, and this is what we mean by use of the

1 A	2 D	3 B	4 C	5 A	6 D	7 B	8 C	9 A	10 B
11 C	12 A	13 C	14 D	15 B	16 B	17 B	18 D	19 C	20 C
21 A	22 A	23 D	24 C	25 D	26 A	27 D	28 C	29 B	30 A
31 B	32 D	33 A	34 C	35 B	36 D	37 B	38 A	39 D	40 C

FIGURE 6–2

term randomization. It is very desirable to assign the fertilizers to the plots in a random way to avoid the possibility of biasing the results in some way. For example, if the first ten plots were used for A, the next ten for B, and so on, a bias might be introduced because of a streak of poor land somewhere. Of course, it is conceivable that the results will still be biased just by chance on using random selection, but nothing can be done about this.

If there are known significant differences in quality of the land on the farm, then the above procedure for assigning fertilizer to plots might not be too good because all plots for one type of fertilizer might end upon one type of land by chance. When there are known land differences, it would be advisable first to divide the land up into sections of more or less uniform quality. Then these sections would be divided up into plots and fertilizers would be assigned to the plots in the manner discussed above. In this way, each fertilizer would be tested on at least one plot of each type of land, and the sort of difficulty referred to above cannot occur. In a similar vein, if several different farmers are involved in testing the fertilizers, each farmer should test all the fertilizers on at least one plot to avoid bias due to land, cultivation methods or seed used.

Problems similar to those described above are encountered elsewhere—in industry, for example, when treatments are being studied. To illustrate, consider the adsorption of different brands of dye on a plastic. The plastic is produced as a continuous sheet. However, its properties are not absolutely uniform from one point on the sheet to another. Generally speaking, the plastic tends to be more uniform from one side of the sheet to the other than it does along its length. This suggests that the following type of experimental design be used. Several full width strips are cut from the plastic as shown in Figure 6–3. Then from each such strip several pieces are cut, these being numbered as shown in Figure 6–3. To be specific, suppose that

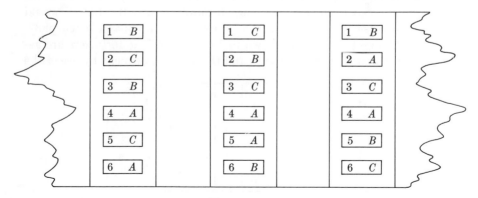

FIGURE 6–3

three brands of dye are to be tested. Then for each set of six pieces cut from a given strip a random assignment of the dyes to the pieces is used, each dye being used on precisely two pieces. This might result in the sort of assignment shown in Figure 6–3. Once again it is desirable to randomize the assignment of the dyes to each set of pieces, because even though the plastic is relatively uniform from one edge to the other there still can be slight systematic variations which could bias the experiment if for example, die A was always used on the edge pieces. The procedure suggested here is really just the same as that suggested for the fertilizers in the case where not all land is of the same quality.

In this section we have discussed some of the simplest notions involved in designing experiments involving comparison of treatments. To design a good experiment it is necessary to have clearly in mind what results are desired and what other factors enter in which could bias the results. It is then necessary to design the experiment so as to minimize these biases. It is often not easy to avoid all biases, because they are sometimes subtle to detect. Thus even in the simple experiment of testing batteries, there could be biases due to the fact that the various pieces of equipment in which the batteries were tested are not identical in using battery power.

6–5 THE BAYESIAN APPROACH

We shall now consider the Bayesian approach to the problem of comparing the various brands of batteries or, more generally, to the problem of comparing several products or treatments. Recall that it was pointed out in Chapter 4 that one must be careful in testing an hypothesis that an attempt is being made to answer the question of interest and not some other related question. Let us now ask what is the real question that the Army is trying to answer in testing the batteries. What it really wants to know, presumably, is which brand should be used in its equipment. Does testing an hypothesis concerning the equality of the mean operating life of the various brands answer this question? The answer generally is no, since no consideration has been given to the costs of the various brands. If they all had the same cost, then the test of the null hypothesis would probably suffice, since if the null hypothesis is rejected, this would indicate that the brand with the greatest average life should be used, while if the null hypothesis is not rejected, this would be interpreted as meaning that it doesn't much matter which brand is used.

The situation becomes more complicated, however, when the costs are different. If the null hypothesis is rejected, it does not necessarily follow that the battery with the longest average life should be adopted, since it may be more expensive than the others. Equally well, if the null hypothesis is not rejected, it is not entirely clear that the cheapest battery should be

used. The Bayesian approach is capable of treating the case where the costs
are different, as well as providing an alternative and more complete treatment of the case where the costs are the same. Let us then illustrate the
Bayesian approach to the problem.

It is very easy to describe in general terms the Bayesian approach to the
battery testing problem. First, for each battery type, the decision maker
determines the prior distribution for the expected life. This prior distribution can vary from one brand to another. Next the experiment is performed
to determine the average life for each brand. Precisely the same experiment
described in Section 6–1 can be used in the Bayesian approach. The experimental results are then combined with the prior distributions by means of
Bayes' law to obtain the posterior distributions of the expected life for the
brands. Finally, the expected cost of using each brand is determined. The
brand to use is the one yielding the minimum expected cost.

The Bayesian procedure will next be illustrated in detail by extending the
numerical example which we have been considering. Suppose that, based
on previous experience, the decision maker concerned believes that the expected life of each brand can have only two values $\mu_1 = 1.05$ or $\mu_2 = 0.95$
(in thousands of hours), that is, it will either be 1050 hours or 950 hours.
The prior probabilities which he assigns to each of these states of nature is
shown in Table 6–6. He does not, it will be noted, feel that the prior probabilities are the same for each brand.

TABLE 6–6

PRIOR PROBABILITIES FOR EXAMPLE

State of Nature

Brand	μ_1	μ_2
A	0.70	0.30
B	0.50	0.50
C	0.40	0.60
D	0.30	0.70

Next Bayes' law is used to determine the posterior probabilities. Let us
now examine how this is done, since it introduces something new that we
have not considered before. If p_v is the prior probability of the state of
nature μ_v and f is the simple event that occurred on performing the experiment for a given brand, then the standard form of Bayes' law becomes

$$p(\mu_v|f) = \frac{p(f|\mu_v)p_v}{p(f|\mu_1)p_1 + p(f|\mu_2)p_2}. \quad (6-25)$$

Consider now the conditional probability $p(f|\mu_v)$ which is the probability
that the experiment yields the outcome f given that μ_2 is the state of nature.
For the case under consideration f is the average life of the six batteries of

the brand under consideration that were tested, so that f is the observed value of what we previously called $\bar{x}\{j\}$, if brand j is the one under consideration; $\bar{x}\{j\}$ is given by (6–4). If the state of nature is μ_v, then $\bar{x}\{j\}$ is essentially normally distributed with mean μ_j and standard deviation $\sigma/\sqrt{6}$. The decision maker believes that each brand will have the same standard deviation σ, and as an estimate of σ he will use $\sqrt{s^2}$, where s^2 is given by (6–6). Now $s^2 = 0.052$, so that his estimate of the standard deviation of $\bar{x}\{j\}$ is $(0.052/6)^{1/2} = 0.093$. Thus each $\bar{x}\{j\}$ will be taken to be normally distributed with mean μ_v and standard deviation 0.093.

At this point we encounter a difficulty since $\bar{x}\{j\}$ is now being treated as a continuous random variable whose density function is $n(\bar{x}\{j\}; \mu_v, 0.093)$. In (6–25), however, a probability $p(\bar{\xi}_j|\mu_v)$ appears (since f is the value of $\bar{x}\{j\}$, that is, $\bar{\xi}_j$). The question then arises as to what should be done. To see what to do, we must recall that the notion of a continuous random variable is an approximation and that the density function represents the outline of the histogram. The probability that $\bar{x}\{j\}$ has any particular value $\bar{\xi}_j$ is then the area of a rectangle on the histogram, and this area is essentially $n(\bar{\xi}_j; \mu_v, 0.093)h$ where h is the spacing between the different values of $\bar{x}\{j\}$. Thus

$$p(\bar{\xi}_j|\mu_v) \doteq n(\bar{\xi}_j; \mu_v, 0.093)h ,$$

and hence Bayes' law becomes

$$p(\mu_v|\bar{\xi}_j) = \frac{n(\bar{\xi}_j; \mu_v, 0.093)p_v}{n(\bar{\xi}_j; \mu_1, 0.093)p_1 + n(\bar{\xi}_j; \mu_2, 0.093)p_2} \tag{6–26}$$

since h cancels out. We have thus seen how to convert Bayes' law to a form which can be used when the outcome of the experiment is described by a normally distributed random variable.

Now

$$n(\bar{\xi}_j; \mu_v, 0.093) = \frac{1}{0.093} \varphi\left(\frac{\bar{\xi}_j - \mu_v}{0.093}\right) \tag{6–27}$$

and thus Bayes' law becomes

$$p(\mu_v|\bar{\xi}_v) = \frac{p_v\varphi\left(\dfrac{\bar{\xi}_j - \mu_v}{0.093}\right)}{p_1\varphi\left(\dfrac{\bar{\xi}_j - \mu_1}{0.093}\right) + p_2\varphi\left(\dfrac{\bar{\xi}_j - \mu_2}{0.093}\right)} \tag{6–28}$$

since the factor $1/0.093$ cancels out. It is (6–28) that will be used to compute the posterior probabilities. Note that it is the standardized normal *density* function, not the cumulative function, which appears in (6–28). Let us illustrate the computation for brand A, where $\mu_1 = 0.70$ and $\mu_2 =$

TABLE 6–7

POSTERIOR PROBABILITIES FOR EXAMPLE

State of Nature

Brand	μ_1	μ_2
A	0.837	0.163
B	0.594	0.406
C	0.400	0.600
D	0.164	0.836

0.30. From Table 6–2, the average life of the six batteries tested was $\bar{\xi}_A = 1.07$ thousands of hours. Now

$$\frac{\bar{\xi}_A - \mu_1}{0.093} = \frac{1.067 - 1.050}{0.093} = 0.18; \qquad \frac{\bar{\xi}_A - \mu_2}{0.093} = \frac{1.067 - 0.950}{0.093} = 1.26 .$$

Table F gives values of $\varphi(t)$. Thus

$$\varphi(0.18) = 0.392; \quad \varphi(1.26) = 0.180 .$$

Consequently, for brand A

$$p(\mu_1|1.07) = \frac{0.70(0.392)}{0.70(0.392) + 0.30(0.180)} = \frac{0.274}{0.274 + 0.054} = 0.837$$

and

$$p(\mu_2|1.07) = 0.163 .$$

The posterior probabilities for the other brands are presented in Table 6–7.

Given the posterior probabilities, all that remains is to compute the expected cost of using each battery. In Table 6–8 are given the cost per battery of each brand. Let us now consider the determination of the expected cost. Suppose that a total of 50,000 thousands of hours, that is, 50,000,000 hours, of operation per year will be required of the batteries in various pieces of equipment. If the state of nature is μ_v, then the expected number of batteries needed should be $50,000/\mu_v$, since μ_v is the average life of a battery and the expected number of batteries that will be needed taking into account the two different possibilities for the expected life is

$$50,000 \left[\frac{1}{\mu_1} p(\mu_1|\bar{\xi}) + \frac{1}{\mu_2} p(\mu_2|\bar{\xi}) \right], \qquad (6\text{--}29)$$

TABLE 6–8

COST PER BATTERY OF VARIOUS BRANDS
IN DOLLARS

Brand	A	B	C	D
Cost	4.25	4.00	3.85	3.85

where $p(\mu_v|\xi)$ is the posterior probability that the state of nature is μ_v. The expected cost per year is then the cost per battery times (6–29).* Thus the expected annual cost if brand A is used is

$$4.25(50,000)\left[\frac{1}{1.05}(0.837) + \frac{1}{0.95}(0.163)\right]$$
$$= 202,500[0.797 + 0.172] = \$206,000.$$

Similarly, the expected costs for brands B, C and D are respectively \$198,-500, \$195,000 and \$199,500. Thus it would appear that brand C should be used. The differences in expected cost between the various brands are rather small in percentage terms but fairly large in terms of actual dollar values.

In the above example, it was assumed that the decision maker specified initially the prior distribution for the expected life of each brand. For certain types of problems it is not possible to give a prior distribution for the expected value of the physical property of interest for each brand. Thus the farmer could not give for each brand of fertilizer the prior distribution for the expected yield, since this varies from season to season depending on the growing conditions, and presumably it would not be easy for the farmer to decide precisely how any given set of growing conditions would influence the yield. What would normally be done when the prior distribution for the mean of each brand cannot be specified is to specify instead a prior distribution for the expected differences between brands. Thus the farmer might select one fertilizer as a standard, and then for each of the other fertilizers he would compare this fertilizer with the selected standard and would specify a prior distribution for the expected difference in yields between the given fertilizer and the standard. The fertilizer to select would then be the one giving the greatest expected increase in profit over the standard fertilizer (the standard, of course, could be selected if the expected increase in profit was negative for all of the other fertilizers).

6–6 TWO-VARIABLE ANALYSIS OF VARIANCE

We shall conclude this chapter by illustrating in one special case how the analysis of variance procedure can be generalized to treat more complicated situations. Sometimes it is of interest to construct experiments in which the effects of two different variables are considered simultaneously. For example, a farmer may be interested in comparing several different fertilizers for

---

* The careful reader will note that the number of batteries required per year is a random variable even if the state of nature is known. We have in (6–29) merely used the expected number of batteries for each state of nature. The proof that this procedure is valid is given in reference [4] on pp. 445–447.

use on a particular vegetable crop and also several different insect sprays. He could attempt to test the fertilizers and insect sprays separately, but such a procedure might require several years to carry out. A more efficient procedure would be to test both simultaneously. Suppose that there are k different fertilizers and n different insect sprays. Imagine that the farmer's land is quite uniform. Then he might subdivide a large plot of land into nk plots of equal size. On each plot he would then use one of the fertilizers and one of the insect sprays. A different combination would be used on each. Since there are nk different ways to combine the fertilizers and sprays, each possible combination would be tried on one plot.

Let ξ_{ij} be the yield from the plot on which spray i and fertilizer j is used. The results of the experiment can then be arranged as shown in Table 6–9. This table looks identical to Table 6–3, but the interpretation is now differ-

TABLE 6–9

FORMAT FOR PRESENTING EXPERIMENTAL RESULTS—
TWO-VARIABLE CASE

		Product 1						
		1	2	3	$\ldots$	j	$\ldots$	k	
	1	ξ_{11}	ξ_{12}	ξ_{13}	$\ldots$	ξ_{1j}	$\ldots$	ξ_{1k}	$\bar{\xi}_{1\cdot}$
	2	ξ_{21}	ξ_{22}	ξ_{23}		ξ_{2j}		ξ_{2k}	$\bar{\xi}_{2\cdot}$
Product 2	$\vdots$	$\vdots$	$\vdots$	$\vdots$		$\vdots$		$\vdots$	
	i	ξ_{i1}	ξ_{i2}	ξ_{i3}		ξ_{ij}		ξ_{ik}	$\bar{\xi}_{i\cdot}$
	$\vdots$	$\vdots$	$\vdots$	$\vdots$		$\vdots$		$\vdots$	
	n	ξ_{n1}	ξ_{n2}	ξ_{n3}	$\ldots$	ξ_{nj}	$\ldots$	ξ_{nk}	$\bar{\xi}_{n\cdot}$
		$\bar{\xi}_{\cdot1}$	$\bar{\xi}_{\cdot2}$	$\bar{\xi}_{\cdot3}$	$\ldots$	$\bar{\xi}_{\cdot j}$	$\ldots$	$\bar{\xi}_{\cdot k}$	$\bar{\bar{\xi}}$

ent. The average of the values in column j will be denoted by $\bar{\xi}_{\cdot j}$ and the average of the values in row i by $\bar{\xi}_{i\cdot}$. The location of the dot is used to distinguish between row and column averages. If it was not used, we would not know, for example, whether $\bar{\xi}_4$ referred to the average of the values in column 4 or in row 4.

Let us suppose that the farmer is interested in trying to decide if there is any significant differences, so far as yield goes, between the various brands of fertilizer, and also whether there is any significant difference in yield between the various brands of insect sprays. For the sort of situation being considered, there should be no interaction between the effects of the fertilizers and that of the insect sprays, that is, if a given insect spray i gives an expected increase in yield of α_i, it should do so regardless of which fertilizer is used. Similarly, if a given fertilizer j gives an expected increase in yield

of β_j, it should do so regardless of which spray is used. The net resulting expected increase in yield on using spray i and fertilizer j should be the sum of the effects of each one, that is, $\alpha_i + \beta_j$.

If it is assumed that the effects are additive in the manner just described, and if, in addition, it is assumed that each ξ_{ij} is the observed value of a normally distributed random variable $x\{i, j\}$ with variance σ^2 (note carefully that all these random variables are assumed to have the same variance), then the analysis of variance method can be generalized in the following manner to yield tests of significance for differences in the fertilizers and in the sprays. Note that $\bar{\xi}_{\cdot j}$ and $\bar{\xi}_{i\cdot}$ are observed values of random variables $\bar{x}\{\cdot j\}$ and $\bar{x}\{i\cdot\}$ and $\bar{\bar{\xi}}$, the average of all nk of the ξ_{ij}, is the observed value of a random variable $\bar{\bar{x}}$. Then it can be shown in a manner similar to that which led to (6–21) that

$$\sum_{j=1}^{k} \sum_{i=1}^{n} (x\{i,j\} - \bar{\bar{x}})^2 = n \sum_{j=1}^{k} (\bar{x}\{\cdot j\} - \bar{\bar{x}})^2 + k \sum_{i=1}^{n} (\bar{x}\{i\cdot\} - \bar{\bar{x}})^2 + W_4 \,.$$

$$(6\text{--}30)$$

This can be written $W = W_1 + W_3 + W_4$. Here W_1 is the same as in Section 6–3 and W_2 of that section is given by $W_2 = W_3 + W_4$. Now the total variation W has been broken up into a variation W_1 between fertilizers, a variation W_3 between sprays, and the remaining random variation W_4, as we have called it.

Let

$$Q_1 = \frac{W_1}{k-1} \,; \qquad Q_3 = \frac{W_3}{n-1} \,; \qquad Q_4 = \frac{W_4}{(n-1)(k-1)} \,. \qquad (6\text{--}31)$$

It then happens to be true (although we shall not try to prove it) that the observed value of Q_4 is an unbiased estimate of σ^2 regardless of whether there exist differences between the fertilizers and/or the sprays. Furthermore, the observed value of Q_1 provides an unbiased estimate of σ^2 if there are no differences between the fertilizers. This is true regardless of whether or not there exist differences between the sprays. However, if there are differences between the fertilizers, the observed value of Q_1 will not provide an unbiased estimate of σ^2 but will instead tend to be larger than σ^2. Similarly, the observed value of Q_3 provides an unbiased estimate of σ^2 if there are no differences between the sprays (this is true regardless of whether or not there exist differences between the fertilizers), but does not provide an unbiased estimate if there are differences between the sprays. In the latter case, the observed value of Q_3 will tend to be larger than σ^2.

Consider the random variables

$$F_1 = \frac{Q_1}{Q_4} \,; \qquad F_2 = \frac{Q_3}{Q_4} \,. \qquad (6\text{--}32)$$

From the discussion just given, we see that we would want to reject the null hypothesis that there are no significant differences between the fertilizers if F_1 is too large, and similarly we would want to reject the null hypothesis that there is no difference between the sprays if F_2 is too large. Now once again F_1 and F_2 have an F distribution. The numerator of F_1 has $k - 1$ degrees of freedom while the numerator of F_2 has $n - 1$ degrees of freedom. The denominator of both F_1 and F_2 has $(n - 1)(k - 1)$ degrees of freedom. Given this fact we can then test either one of the null hypotheses that there exist no differences between the fertilizers or no differences between the sprays at a given level of significance α. For example, to test the null hypothesis that there is no difference between the fertilizers at the significance level $\alpha = 0.05$, the critical value δ of F_1 would be found from Table I corresponding to a numerator with $k - 1$ and a denominator with $(n - 1) \times (k - 1)$ degrees of freedom. Then if the observed value of F_1 satisfies $F_1 \geq \delta$, the null hypothesis is rejected. Otherwise, the null hypothesis is not rejected. The details of the computations can be summarized in an analysis of variance table, the format for which is shown in Table 6–10.

TABLE 6–10

TWO-VARIABLE ANALYSIS OF VARIANCE

Source of Variation	Degrees of Freedom	Value of Sum of Squares Random Variable	Estimate of σ^2	F Variable
Between Fertilizers	$k - 1$	W_1	$\dfrac{W_1}{k - 1}$	$\dfrac{W_1}{W_4}(n - 1)$
Between Sprays	$n - 1$	W_3	$\dfrac{W_3}{n - 1}$	$\dfrac{W_3}{W_4}(k - 1)$
Random Component	$(n - 1)(k - 1)$	W_4	$\dfrac{W_4}{(n - 1)(k - 1)}$	
Total	$nk - 1$	W		

We can reinterpret the example dealing with different brands of batteries to yield an example to which the above theory can be applied. Let us now suppose that not only did the experiment whose results are given in Table 6–1 test different brands of batteries, but in addition, tested them in different types of equipment. We shall suppose that each row of Table 6–1 refers to a different type of equipment, so that the batteries were tested in six different types of equipment. Now we are interested in testing the null hypothesis that there is no difference between the brands and also the null hypothesis that there is no difference in the expected battery life in the different types of equipment.

It is easily checked that the row means in Table 6–1 are 1.05, 1.05, 0.975, 1.125, 0.775 and 1.075 respectively. Also $W_3 = 0.308$ and $W_4 = 0.732$. Hence

$$F_1 = (n - 1)\frac{W_1}{W_4} = 5\left(\frac{0.060}{0.732}\right) = 0.410;$$

$$F_2 = (k - 1)\frac{W_3}{W_4} = 3\left(\frac{0.308}{0.732}\right) = 1.263 \ .$$

The results can be presented in the analysis of variance table shown in Table 6–11. It is clear even without using a table of the F distribution that the null hypothesis concerning differences in the brands will not be rejected. For $\nu_1 = 5$, $\nu_2 = 15$ and $\alpha = 0.05$ the critical value of F is 2.90. Thus the hypothesis that there is no difference in expected battery life in the different types of equipment cannot be rejected either. In other words, the experiment does not yield any evidence at the five percent level of significance that there is any significant difference between the expected lifetimes of the various brands or that the expected lifetime of a battery differs from one piece of equipment to another.

TABLE 6–11

TWO-VARIABLE ANALYSIS OF VARIANCE TABLE FOR EXAMPLE

Source of Variation	Degrees of Freedom	Value of Sum of Squares Random Variable	Estimate of σ^2	F Variable
Between Brands	3	0.060	0.020	0.410
Between Equipment	5	0.308	0.0615	1.263
Random Component	15	0.732	0.0487	
Total	23	1.100		

In the above example only one battery of each brand was tested in each type of equipment. It no doubt seems reasonable to the reader that a number of batteries of a given brand should be tested in each type of equipment. This would clearly improve the experiment and is referred to as *replication*. With replication more than a single value of each random variable $x\{i, j\}$ is obtained. Replication would also be desirable in testing the fertilizers and sprays to average out land differences if these were significant. The method of analysis of variance can be generalized without difficulty to handle experiments in which there is replication. Replication is essential if the model assumed above is modified to one where, instead of the effects of the two

variables being additive, there can be interactions between the two variables. Thus, for example, if a manufacturer was studying how several different types of catalysts and several different temperatures influenced the yield of a chemical reaction, it is not necessarily reasonable to think of the effects of the catalyst and temperature as being additive, because these two variables could have an influence on each other, and one particular combination of catalyst and temperature might be much better than what would be expected under the assumption of additive effects. Once again the analysis of variance method can be generalized to treat cases where interactions are allowed for. However, we shall not consider this generalization or any of the many other generalizations which can be made.

REFERENCES

1. Davies, O. L. ed., *Design and Analysis of Industrial Experiments*. Hafner, New York, 1956.

Discusses in some detail the design of experiments in the chemical industry and the manner in which analysis of variance techniques can be applied.

2. Freund, J. E., *Modern Elementary Statistics*, 3rd ed. Prentice-Hall, Englewood Cliffs, N.J., 1967.

Gives an elementary discussion of analysis of variance.

3. Freund, J. E., *Mathematical Statistics*. Prentice-Hall, Englewood Cliffs, N.J., 1962.

Gives a more advanced treatment of analysis of variance than is given in reference [2].

4. Hadley, G., *Introduction to Probability and Statistical Decision Theory*. Holden-Day, San Francisco, 1967.

5. Hoel, P. G., *Elementary Statistics* 2nd ed. Wiley, New York, 1966.

PROBLEMS

Sections 6-1 through 6-3

1. A large machine shop has tested three different brands of cutting tools on its large high-speed lathes. The useful life of each tool was determined. Because of the time involved, it was not possible to test a large number of tools of each brand; only five of each were tested. The results are given in Table 6–12. Use the method sug-

TABLE 6–12

USEFUL LIFE IN HOURS OF CUTTING TOOLS

	Brand	
A	B	C
120	250	200
160	140	300
98	200	140
200	175	190
110	130	235

gested in Section 6–1 with $\alpha = 0.05$ to determine if there seems to be any significant difference in the expected life of the brands.

2. Re-solve Problem 1 using the analysis of variance method and $\alpha = 0.05$. Does it seem reasonable here that the variance of the tool life is the same for each brand?

3. A farmer tests each of four different brands of fertilizer on six two-acre plots of tomatoes. The yield of tomatoes on each of these plots is given in Table 6–13. Use the analysis of variance method with $\alpha = 0.05$ to test the null hypothesis that there is no significant difference between the brands of fertilizer.

TABLE 6–13

YIELD IN TONS OF TOMATOES

Fertilizer

A	B	C	D
58	47	58	51
64	59	60	47
40	50	45	42
51	61	50	61
60	42	52	50
55	49	48	57

4. A trucking firm is interested in evaluating three different brands of fuel pumps. It equips 21 trucks with these pumps, seven trucks being outfitted with brand A, seven with brand B and seven with brand C. The life of each fuel pump is determined and the results are presented in Table 6–14. Use analysis of variance and $\alpha = 0.05$ to test the null hypothesis that there is no difference between the brands. What assumptions would need to be satisfied if the experiment is to yield meaningful results?

TABLE 6–14

AVERAGE LIFE IN MONTHS FOR FUEL PUMPS

Brand

A	B	C
15.1	18.3	13.2
12.6	15.5	15.9
17.8	17.6	14.1
14.3	19.1	12.8
10.2	12.7	10.2
19.1	14.6	9.1
16.4	17.8	13.2

5. The mathematics department in a university is interested in comparing four different calculus texts. To do so it selects at random 120 students and then assigns them at random to four sections of 30 students each. Each of the sections uses a different calculus text and at the end of the year the average grade on the final

TABLE 6–15

AVERAGE GRADES ON CALCULUS FINAL
EXAMINATION

Text			
A	B	C	D
72	80	74	68
80	90	81	70
64	70	72	60

examination is determined. This procedure is repeated for three years. The results are given in Table 6–15. Use analysis of variance and $\alpha = 0.05$ to determine if there appears to be any significant difference in the tests.

6. The model on which the one variable analysis of variance is based is one where $x\{i,j\}$ is a normally distributed random variable with mean μ_j and variance σ^2 (σ^2 being independent of i and j). Now write $\mu = (\Sigma_{j=1}^{k} \mu_j)/k$ so that μ is the average of the means for the different brands. Then write $\mu_j = \mu + \delta_j$. Show that it must be true that

$$\sum_{j=1}^{k} \delta_j = 0 .$$

The δ_j are sometimes referred to as the differential effects. Consider now $\bar{x}\{j\}$. Show that the expected value of this random variable is $\mu + \delta_j$. Similarly, show that the expected value of $\bar{\bar{x}}$ is μ. Thus show that $y\{j\} = \bar{x}\{j\} - \bar{\bar{x}}$ is an unbiased statistic for estimating δ_j and $\bar{x}\{u\} - \bar{x}\{v\}$ is an unbiased statistic for estimating $\mu_u - \mu_v$.

7. Consider the situation described in Problem 6. Show that Q_2 given by (6–8) is an unbiased statistic for estimating σ^2 regardless of what values the δ_j have.

8. Consider the situation described in Problem 6. Show that the expected value of Q_1 is

$$\sigma^2 + \frac{n}{k-1} \sum_{j=1}^{k} \delta_j^2 .$$

Thus conclude that Q_1 is an unbiased statistic for estimating σ^2 when all $\delta_j = 0$ and the expected value of Q_1 is greater than σ^2 when the means are not equal. Hint: Write $(\bar{x}\{j\} - \bar{\bar{x}})^2 = [(\bar{x}\{j\} - \mu - \delta_j) - (\bar{\bar{x}} - \mu) + \delta_j]^2$.

Section 6–4

1. A fertilizer manufacturer is interested in testing four new blends of fertilizer for use in drier climates. There is a considerable variety in the crops, the soils and the irrigation techniques used in the areas where these fertilizers might be applied. Discuss the design of an experiment which could be used to test the null hypothesis that there is no significant difference in the fertilizers.

2. A toothpaste manufacturer is interested in testing out several different formulations of a fluoride toothpaste to determine if there appears to be any significant differences among them. Suggest an experiment for doing this.

3. A manufacturer wishes to try out four alternative procedures for assembling a new product. The entire product is assembled by a single individual. Of course, many individuals in the plant will be at work assembling the product. Suggest a suitable experiment for determining whether or not there appears to be any significant difference in the expected times required for assembly using the various procedures.

Section 6–5

1. For Problem 1 of Sections 6–1 through 6–3, which tool should be used if their cost is \$35 for A, \$30 for B and \$40 for C? The decision maker believes that the expected life of a tool will have only three values: $\mu_1 = 130$, $\mu_2 = 150$ and $\mu_3 = 200$ hours. The prior probabilities which he assigns to each state of nature for each brand are shown in Table 6–16.

TABLE 6–16

PRIOR PROBABILITIES

State of Nature

		μ_1	μ_2	μ_3
	A	0.30	0.50	0.20
Brand	B	0.40	0.50	0.10
	C	0.20	0.50	0.30

2. For Problem 3 of Sections 6–1 through 6–3, determine which fertilizer should be used if the following data apply. The costs per acre of applying fertilizers A, B, C and D are respectively \$20, \$18, \$19 and \$20. Tomatoes sell for \$75 per ton. To express the prior distributions, the farmer gives his opinion as to how each of the other fertilizers compare with B. He feels that if μ is the expected increase (or decrease) in yield per acre as compared to B, μ will only have the values -5, 0, 2 or 5 tons. Denote these states of nature by μ_1, μ_2, μ_3 and μ_4. The prior probabilities which he assigns to these states of nature for the other fertilizers are given in Table 6–17.

3. For Problem 4 of Sections 6–1 through 6–3, determine which fuel pump should be used if the following data apply. The costs of brands A, B and C are respectively \$25, \$30 and \$23. The decision maker feels that it is adequate to assume that there will be only three states of nature (the same ones for each brand) corresponding to

TABLE 6–17

PRIOR PROBABILITIES

State of Nature

		μ_1	μ_2	μ_3	μ_4
	A	0.20	0.50	0.40	0.10
Fertilizer	C	0.30	0.60	0.08	0.02
	D	0.20	0.60	0.10	0.10

TABLE 6–18

PRIOR PROBABILITIES

State of Nature

		μ_1	μ_2	μ_3
	A	0.30	0.40	0.30
Brand	B	0.40	0.40	0.20
	C	0.50	0.40	0.10

three different values for the expected life of a pump. These are $\mu_1 = 14.0$, $\mu_2 = 16.0$ and $\mu_3 = 18$ months. The prior probabilities of the state of nature for each brand are given in Table 6–18.

Section 6–6

1. Consider the situation described in Problem 1 for Sections 6–1 through 6–3. Suppose now that each row in Table 6–11 refers to a different type of machine, so that the tools were tested on five different types of machines. Given this new interpretation, test at the five percent level the null hypotheses that there is no difference between the tools and that there is no difference between the expected useful life on the various machines. Also construct an analysis of variance table.

2. Consider the situation described in Problem 3 for Sections 6–1 through 6–3. Suppose now that each row in Table 6–12 refers to a different type of insect spray, so that the experiment really tested four fertilizers and six insect sprays. Given this new interpretation, test at the five percent level the null hypotheses that there is no difference between the fertilizers and no difference between the sprays. Also construct an analysis of variance table.

3. Consider the situation described in Problem 4 for Sections 6–1 through 6–3. Suppose now that each row of Table 6–13 refers to a truck used on a different route, so that the fuel pumps were tried on seven different routes. Given this new interpretation, test at the five percent level the null hypotheses that there is no difference between the brands and that the expected pump life is the same on all routes. Also construct an analysis of variance table.

4. Consider the situation described in Problem 5 for Sections 6–1 through 6–3. Suppose now that a different instructor was used in each of the three years. However, in any given year, the same instructor taught all four sections. Given this additional information, test at the five percent level the null hypotheses that there is no difference between the texts and also that there is no difference between the instructors. Construct an analysis of variance table for the experiment.

5. For the basic model discussed in the text we assumed that the effects on the expected yield of the fertilizer and spray were additive. In other words, the expected value μ_{ij} of $x\{i, j\}$ could be written $\alpha_i + \beta_j$. A more convenient way to express this is to write $\mu_{ij} = \mu + \epsilon_i + \delta_j$, where

$$\mu = \frac{1}{kn} \sum_{i=1}^{n} \sum_{j=1}^{k} \mu_{ij}; \qquad \sum_{i=1}^{n} \epsilon_i = 0; \qquad \sum_{j=1}^{k} \delta_j = 0.$$

Show then that the expected value of $\bar{x}\{\cdot j\}$ is $\mu + \delta_j$ and the expected value of $\bar{\bar{x}}$ is μ, so that $\bar{x}\{\cdot j\} - \bar{\bar{x}}$ is an unbiased statistic for estimating δ_j. Similarly show that $\bar{x}\{i\cdot\} - \bar{\bar{x}}$ is an unbiased statistic for estimating ϵ_i.

6. Show that the random variable W_4 of (6–30) is given by

$$W_4 = \sum_{j=1}^{k} \sum_{i=1}^{n} (x\{i,j\} - \bar{x}\{i\cdot\} - \bar{x}\{\cdot j\} + \bar{\bar{x}})^2 .$$

7. Review Problems 5 and 6 and show that the expected value of Q_1 is $\sigma^2 + n(k-1)^{-1}\Sigma_j{}_{=1}^{k} \delta_j^2$ the expected value of Q_3 is $\sigma^2 + k(n-1)^{-1}\Sigma_i{}_{=1}^{n} \epsilon_i^2$ and the expected value of Q_4 is σ^2, where Q_1, Q_3 and Q_4 are given by (6–31). Note that the expected value of Q_4 is independent of what values the ϵ_i and δ_j have. Hint: Use the same sort of procedure introduced in Problem 8 for Sections 6–1 through 6–3.

CHAPTER 7

Regression and Correlation Analysis

7–1 FUNCTIONAL RELATIONSHIPS

One of the principal objectives of the physical sciences is to discover and characterize the laws of nature. Frequently the laws of nature can be conveniently characterized by specifying how one variable is related to one or more other variables. We have made frequent use of functions in our previous studies and have noted how useful the notion of a function is. The relationship between variables implied by some law of nature is often representable as a function. Thus, for example, one law of nature is that the distance d a body released from a point above the surface of the earth will fall in a time period of length t is

$$d = \frac{1}{2} g t^2 , \qquad (7\text{–}1)$$

where g, called the gravitational constant, is a number whose value varies slightly from one point on the globe to another, but has the approximate value 32.0 when t is measured in seconds and d in feet. If this value is used in (7–1), we obtain

$$d = 16 t^2 . \qquad (7\text{–}2)$$

This simple formula expresses d as a function of t, and given any value of t, it is possible to compute the distance d that the body will fall in that time. Here we have a very simple example of how a law of nature can be expressed in a *quantitatively* useful form by means of a function.

It is easy to give many other examples which show how laws of nature can be expressed by functional relationships. Let us provide two other sim-

ple examples. If we have a mass m_0 of a radioactive material at time 0, the mass m which has not undergone radioactive decay by a time t is

$$m = m_0 e^{-\lambda t}, \tag{7-3}$$

where λ is a constant which depends on the type of material and e is the same number which appeared in the Poisson and normal distributions.

A metallic rod will increase in length when the temperature of the rod is increased. The length y of the rod is related to the temperature x by a function of the form

$$y = ax + b, \tag{7-4}$$

where a and b depend on the nature of the rod.

The above examples illustrate the fact that a variety of functions arise in representing the laws of nature. It is often convenient to represent these functions graphically, just as we represented density functions. This can be done in precisely the same way. Let us illustrate by determining the curve which represents (7–2). To do this we select several values of t, as shown in Table 7–1. For each t we compute the corresponding value of d,

Table 7–1

$$d = 16t^2$$

t	0	1	2	3	4	5
d	0	16	64	144	256	400

using (7–2). Thus when $t = 3$, $d = 144$. Next, for each t, the "point" representing the pair of numbers (t, d) is located as shown in Figure 7–1. Finally, a smooth curve is drawn through these points as shown in Figure 7–1. The form of the curves representing (7–3) and (7–4) are shown in Figures 7–2 and 7–3, respectively. A curve like that shown in Figure 7–1 is called a *parabola*, that shown in Figure 7–2 is called an *exponential curve* and that in Figure 7–3 is called a *straight line*.

The great advantage in expressing the laws of nature mathematically

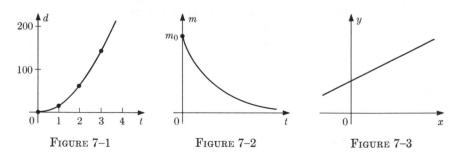

FIGURE 7–1   FIGURE 7–2   FIGURE 7–3

through the use of quantitative functional relationships is that it then becomes possible to use these laws to make calculations concerning the real world. These calculations are really predictions as to what will happen if certain conditions hold, for example, predictions about how far a body will fall in a given time or how much of a radioactive material will decay in a given time or what the length of a rod will be when its temperature is specified. It is this ability to predict which gives man some control over his environment and makes a technological society possible.

Just as there are laws of nature in the physical world, there are also "laws of nature" in the social sciences, business and economics. In this chapter we wish to examine how these laws in the social sciences can be ex-expressed as functional relationships between variables in a quantitative form useful for making predictions.

7–2 LAWS IN THE SOCIAL SCIENCES

Let us now try to explain in a little bit more detail what we mean by laws in the social sciences, business or economics—laws which can be expressed in a quantitatively useful form. We shall do this by giving some examples. Tomatoes are grown in large quantities in the San Joaquin valley of California. In a given region, the total yield of tomatoes can be predicted fairly accurately if the total amount of water received either from rainfall or irrigation is specified. Equally well, in Iowa the weight of corn-fed pigs can be estimated fairly well if their age is specified. Similarly, the demand for certain types of aircraft spare parts can be related to the number of hours that aircraft will be flying. Also, the number of road casualties is directly related to the number of licensed vehicles. Each of these relationships can be quite useful if it can be expressed in a quantitative form.

As a final example, let us give one which illustrates in an especially clear manner the usefulness of such quantitative relationships. Consider a producer of various chemical sprays and powders for use on farms and orchards. The demand for a given type of fungicide in a given month in a given region can be predicted with surprising accuracy if the number of inches of rainfall in that area for the previous month is known. The reason for this is that fungus growth depends strongly on the rainfall. A precise knowledge of the functional relationship for each marketing area is extremely useful in production planning, since the product is manufactured no more than one month in advance of sales because of its perishable nature. Given the rainfall in each region, the demand for the fungicide can be estimated for each region, and finally these can be aggregated to determine production schedules.

There is one important difference between the laws of nature in the social

sciences, economics, and business and those in the natural sciences. This
difference lies in the relative imprecision of the laws in the social science
areas. It is normal to think of physical laws as being deterministic, so that,
for example, a body will always fall exactly $16t^2$ feet in t seconds. Each time
the experiment is performed, the body will fall exactly $16t^2$ feet in time t. On
the other hand, the laws in the social sciences, economics and business are
not normally thought of as being strictly deterministic. Chance always plays
a role. Thus, knowing the rainfall in its various marketing regions this
month, the producer of fungicides cannot predict exactly what the demand
will be in the coming month. Even if the rainfall pattern were identical in a
given month for two different years, the demands in the following months
would not necessarily be identical. It would then appear that the laws dealt
with in the social science areas are of a different kind than those encountered in the physical sciences, the laws in the physical sciences being deterministic and those in the social sciences, economics and business being
probabilistic in nature.

This is not really true, however. The laws in the physical sciences are also
probabilistic. If the distance that a body falls in a time t is measured with
great accuracy, it will not be exactly the same each time the experiment is
performed. However, the differences are extremely small by ordinary standards of measurement. The difference between the laws in the natural sciences
and in the social sciences, then, is more one of degree than of kind. Both are
probabilistic, but the laws in the natural sciences can often be treated as
deterministic to a very good degree of approximation, while this need not
be true at all in the social science areas.

It is important to recognize this imprecision of the laws encountered in
the social sciences, economics and business resulting from the appearance of
chance effects. Nothing can be done about this normally. It is simply part of
the nature of the situations dealt with. These laws or relationships between
variables are frequently very useful, in spite of the fact that they are not
deterministic. Of course, the less deterministic they are the less valuable
they are.

We shall be interested in trying to determine functions which represent
in quantitative terms some business law. A function such as $y = 3x + 2$ is
deterministic in the sense that, given a value of x, the value of y is determined exactly. How can functions then be used to represent laws which are
probabilistic? This is done in the same way as in the natural sciences. If the
law relating pounds of fungicide y demanded in a given region in a given
month to the rainfall x in the previous month has the form

$$y = 200,000x + 40,000 , \qquad (7\text{--}5)$$

then for the given x we compute y. The value ζ' of y so obtained is our estimate of the demand, just as $d = 16t^2$ is our estimate of the distance a body

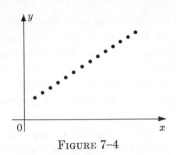

FIGURE 7-4

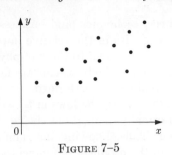

FIGURE 7-5

falls in time t. The actual demand ζ will not usually be precisely equal to ζ'. We can think of y as a random variable whose value is influenced by the value of x. The functional relationship between y and x then assists us in predicting what value y will take on. We shall be more specific later about the precise interpretation of the prediction.

The general procedure for determining the functional form of some law is the same in either the natural sciences or social sciences. One first constructs a model, which leads to a description of the general nature of the function, and then, by experiment, the parameters are determined. The imprecision in the laws in the social science areas shows up very clearly when we gather experimental data to use in determining the parameters in the function. If the length y of a rod is measured at a number of different temperatures x and the resulting points (x, y) were represented geometrically, something like Figure 7-4 would be obtained, while if the demand y for fungicide in a given region was determined for different values of the rainfall x, the result would look something like that shown in Figure 7-5. Diagrams like Figure 7-4 and 7-5 on which experimental data are shown are sometimes referred to as *scatter diagrams*. Typically, in the physical sciences there is relatively little "scatter" in such a diagram, as indicated by Figure 7-4, while in the social sciences, business and economics there is frequently quite a bit of scatter, as indicated by Figure 7-5.

One of the things we wish to do in this chapter is show how to obtain numerical values for the parameters appearing in some function $y = f(x)$ relating two variables, this function being the quantitative representation of some relevant law. This will be done by using experimental data. The experimental data will be of a form which provides values of y for different values of x. If represented geometrically, these data might look like that shown in Figure 7-6. For any given values of the parameters, the graphical representation of the function $y = f(x)$ will then be a curve such as curve 1 shown in Figure 7-6. If the parameters are changed, a different curve, perhaps the curve labeled 2 in Figure 7-6, will be obtained. What we desire to do is find those values of the parameters such that the resulting curve, and hence the resulting function, best fits the experimental data.

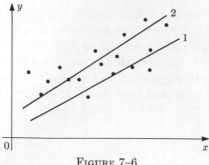

FIGURE 7-6

7-3 LINEAR FUNCTIONS

We shall confine our attention almost exclusively to situations where it is believed that the law of interest can be represented quantitatively by a linear function. Suppose that we are studying a law which can be expressed by specifying how a variable y is related to a variable x, that is, by specifying a function $y = f(x)$. A function $y = f(x)$ is said to be a *linear function* if it has the form

$$y = ax + b \qquad\qquad (7\text{-}6)$$

for all values of x that are of interest, and a and b are specified numbers. No matter what values are assigned to a and b in (7-6), the resulting function is called a linear function. Thus

$$y = 300{,}000x - 27 \quad \text{and} \quad y = -6x + 40$$

are linear functions. The numbers a and b in a linear function (7-6) are what we referred to as parameters in the previous section. The function is not completely determined until numerical values of a and b are specified. Any function which does not have the form (7-6) is called a *nonlinear function*. Thus $y = 3x^2$, $y = \sqrt{x}$ and $y = 1/x$ are nonlinear functions.

Let us now study briefly the graphical representation of linear functions. *The important characteristic of linear functions is that they are always represented graphically by a straight line.* To illustrate this in a specific case, consider the linear function (also frequently called a linear equation) $y = 2x - 3$. In Table 7-2 we have selected several different values of x and

TABLE 7-2

$$y = 2x - 3$$

x	-3	-1	1	3	5	7
y	-9	-5	-1	3	7	11

computed the corresponding values of y. In Figure 7–7 we have shown the points (x, y). To obtain the curve representing $y = 2x - 3$ we draw a smooth curve through these points. Now if a ruler is placed along these points, it will be found that they all lie along the edge of the ruler. Thus to draw a smooth curve through the points it is only necessary to run a pencil along the edge of the ruler, and thus the resulting curve is a straight line, as shown in Figure 7–7.

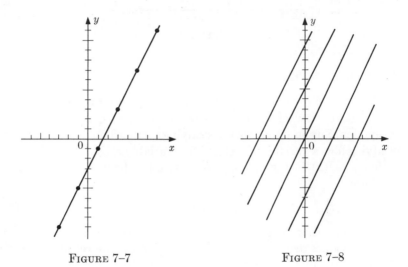

FIGURE 7–7   FIGURE 7–8

For any values of a and b, the curve representing $y = ax + b$ will be a straight line. It is now instructive to note how the line is moved around when the parameters a and b are changed. In Figure 7–8, we have shown the lines representing $y = 2x + b$ for different values of b. These lines are what we call parallel straight lines. The value of b is the value of y for $x = 0$, and hence the line crosses the y-axis at $y = b$. This is illustrated in Figure 7–8. Thus when b is changed in $y = 2x + b$, this merely translates the corresponding line to yield a new one parallel to the original line.

Consider next the lines representing $y = ax + 1$ for various values of a. In Figure 7–9 we have shown these lines for several different values of a. The value of a determines the *slope* of the line, that is, how steep the line is. If $a > 0$, then the line slopes upward to the right, the slope or steepness increasing as a increases. If $a = 0$, the line is horizontal, and if $a < 0$, the line slopes downward to the right. The number a is referred to as the *slope* of the line representing $y = ax + b$ and is also referred to as the slope of the function or equation $y = ax + b$. The number a tells how much y changes for a one-unit increase in x. That is, if x is replaced by $x + 1$, then y is replaced by $y + a$. Thus if $a = 2$, y increases 2 units for a one-unit increase

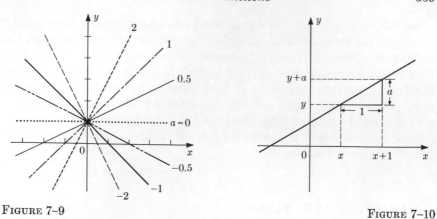

FIGURE 7–9   FIGURE 7–10

in x, and if $a = -1.5$, y decreases 1.5 units for a one-unit increase in x. This is illustrated in Figure 7–10.

Our interest will focus mainly on linear functions, simply because they are the most useful type of function for representing laws in the social sciences. It is seldom true that a law can be represented accurately by a linear function over the entire range of values that the variable x may take on. The important observation, however, is that we are seldom interested in the entire range of values of x, but instead only with a relatively limited range of values of x. Over this limited range of values, the assumption that the law can be represented by a linear function is often quite adequate and, in fact, our understanding of the real world is often insufficient to allow us to develop any better model even if we wanted to. To illustrate how it may be possible to represent a function which is nonlinear when all values of x are considered by a linear function over a relatively limited range of values of x, consider Figure 7–11. Between x_1 and x_2 the function can be adequately represented by the straight line 1, and between x_3 and x_4 it can be represented by the straight line 2. It is important to note that different straight lines, that is, different linear functions, may be needed for different ranges of x. This is clearly indicated in Figure 7–11.

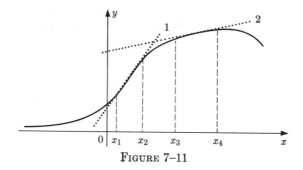

FIGURE 7–11

Sometimes the linearity of a functional relationship can be increased by not using the most obvious variables, but instead by using perhaps the logarithms of these variables. The reader familiar with logarithms will note that both nonlinear functions (7–1) and (7–3) can be converted to linear functions by taking logarithms of both sides of the equations. Equation (7–1) can also be converted to a linear equation by introducing the variable $u = t^2$, since the function then becomes $d = gu/2$. In most cases, however, a linear function can be used with the natural variables if the range of variation of x is limited sufficiently.

7–4 LEAST-SQUARES LINES

We would now like to consider a situation where we have constructed a model to represent some law and we are ready to determine the parameters in the resulting function. Let us suppose that this law can be expressed quantitatively by relating the value of a variable y to a variable x through a function $y = f(x)$. Assume, furthermore, that as our model we have decided to use a linear function $y = ax + b$ over the range of values of x which are of interest. To complete the specification of the function, it is necessary to give numerical values to a and b. Let us now turn our attention to the determination of a and b. As we have noted earlier, the values of these parameters are determined by making use of experimental data.

Suppose that in the real world we have observed the value that the variable y took on when x had the values $\xi_1, \xi_2, \ldots, \xi_n$. Denote by ζ_j the value of y when $x = \xi_j$. We do not require that all the ξ_j be different. It may be true that $\xi_1 = \xi_{17} = 6.4$, for example. In such cases, it is not necessarily true that if $\xi_i = \xi_k$ then $\zeta_i = \zeta_k$, because as we have noted previously, the laws of interest are not deterministic. What we are saying then is that we have observed n pairs of values $(\xi_1, \zeta_1), (\xi_2, \zeta_2), \ldots, (\xi_n, \zeta_n)$. In the fungicide example that we considered previously, the ξ_j may refer to the rainfall in n different months in a given marketing region. Then ζ_j will be the demand for the fungicide in the month following that when the rainfall was ξ_j. We shall refer to the n pairs of numbers (ξ_j, ζ_j) as *data points*. Each can be represented graphically by a point, and when all of them are plotted we obtain a scatter diagram which may look something like Figure 7–5.

In assigning numerical values to a and b we would like to determine the best values of these parameters. What do we mean by best? We shall now investigate this notion in a little more detail. For any specific values of a and b, the predicted value of y when $x = \xi_j$ is $a\xi_j + b$. Let us write $\zeta'_j = a\xi_j + b$, so that ζ'_j is the predicted value of y. The value of y actually observed in the real world when $x = \xi_j$ was ζ_j. The number $\epsilon_j = \zeta_j - \zeta'_j$ is then the error that we would have made if we had predicted the value of

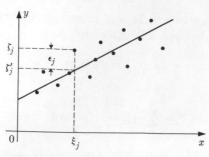

FIGURE 7–12

y for $x = \xi_j$ using $y = ax + b$. The situation can be illustrated graphically as shown in Figure 7–12, where we have shown the points representing the data points (ξ_j, η_j) and the line representing the function $y = ax + b$ under consideration. The error ϵ_j is then a measure of the vertical distance of the point (ξ_j, η_j) from the line. If $\epsilon_j > 0$ the point lies above the line, and if $\epsilon_j < 0$ the point lies below the line. If $\epsilon_j = 0$, the point lies on the line.

It now seems reasonable that a and b should be chosen so as to make the vertical distances of the data points (ξ_j, ζ_j) from the line be as small as possible. More precisely, it seems reasonable to minimize the sum of these distances. The sum of these distances is not $\Sigma_{j=1}^{n} \epsilon_j$ because the ϵ_j are positive if the points lie above the line and negative when they lie below. Thus $\Sigma_{j=1}^{n} \epsilon_j$ could be 0 even though all the points were far away from the line. What we really want to minimize is $\Sigma_{j=1}^{n} |\epsilon_j|$, where $|\epsilon_j| = \epsilon_j$ if $\epsilon_j \geq 0$ and $|\epsilon_j| = -\epsilon_j$ if $\epsilon_j < 0$; $|\epsilon_j|$ is called the magnitude of ϵ_j. It is $|\epsilon_j|$ that gives the vertical distance of (ξ_j, η_j) from the line.

Now it turns out to be clumsy mathematically to try to minimize $\Sigma_{j=1}^{n} |\epsilon_j|$. For this reason, it is convenient to introduce the same sort of measure used when we defined the variance. Instead of minimizing $\Sigma_{j=1}^{n} |\epsilon_j|$ we shall minimize

$$\theta = \sum_{j=1}^{n} \epsilon_j^2, \qquad (7\text{–}7)$$

that is, the sum of the squares of the errors will be minimized. Note that $\epsilon_j^2 > 0$ whenever $\epsilon_j \neq 0$, so that whenever the point does not lie on the line $\epsilon_j^2 > 0$. Thus $\theta > 0$ whenever at least one point does not lie on the line representing $y = ax + b$.

What we wish to do is to determine the values of a and b which yield the smallest value of θ. Clearly it will not be possible, in general, to select a and b so that each data point will be on the line, and therefore, in general, it is not possible to make $\theta = 0$. However, some values of a and b yield much

smaller values of θ than others. It is of interest to find that pair of values which yields the smallest possible value of θ. Now

$$\epsilon_j = \zeta_j - \zeta_j' = \zeta_j - a\xi_j - b$$

and hence

$$\theta = \sum_{j=1}^{n} (\zeta_j - a\xi_j - b)^2 . \tag{7-8}$$

By use of (7–8), we can determine θ directly for any given values of a and b, since the data points (ξ_j, η_j) have been specified.

There exist a variety of ways to determine the values of a and b which minimize θ. We shall not go through the details of the mathematics involved, but will instead simply give the formulas for computing the values of a and b which minimize θ. These are the following:

$$a = \frac{n \sum_{j=1}^{n} \xi_j \zeta_j - \left(\sum_{j=1}^{n} \xi_j \right) \left(\sum_{j=1}^{n} \zeta_j \right)}{n \sum_{j=1}^{n} \xi_j^2 - \left(\sum_{j=1}^{n} \xi_j \right)^2} ; \tag{7-9}$$

$$b = \frac{\left(\sum_{j=1}^{n} \zeta_j \right) \left(\sum_{j=1}^{n} \xi_j^2 \right) - \left(\sum_{j=1}^{n} \xi_j \right) \left(\sum_{j=1}^{n} \xi_j \zeta_j \right)}{n \sum_{j=1}^{n} \xi_j^2 - \left(\sum_{j=1}^{n} \xi_j \right)^2} . \tag{7-10}$$

The equations (7–9) and (7–10) are rather fearsome looking at first glance. The thing which makes them look so complicated is the appearance of so many summation signs. In actuality, there is nothing difficult about them. If we write

$$u_1 = \sum_{j=1}^{n} \xi_j; \quad u_2 = \sum_{j=1}^{n} \zeta_j; \quad u_{12} = \sum_{j=1}^{n} \xi_j \zeta_j; \quad u_{11} = \sum_{j=1}^{n} \xi_j^2;$$

$$\Delta = n u_{11} - u_1^2 , \tag{7-11}$$

then (7–9) and (7–10) can be written

$$a = \frac{1}{\Delta} (n u_{12} - u_1 u_2); \quad b = \frac{1}{\Delta} (u_2 u_{11} - u_1 u_{12}) . \tag{7-12}$$

The reader should not attempt to memorize (7–9) and (7–10) or even (7–11) and (7–12). It is much simpler merely to refer to them when they are

TABLE 7–3

ξ_i	ξ_j^2	ζ_i	$\xi_i\zeta_i$
ξ_1	ξ_1^2	ζ_1	$\xi_1\zeta_1$
.	.	.	
.	.	.	
.	.	.	
ξ_n	ξ_n^2	ζ_n	$\xi_n\zeta_n$
u_1	u_{11}	u_2	u_{12}

needed. There is a convenient tabular format which can be used in carrying out the computation needed to determine a and b. This is shown in Table 7–3. The numbers in the last row are the sums of the numbers in the corresponding columns. We shall illustrate the use of this format in the next section.

The function $y = ax + b$, where a and b have been determined from n data points (ξ_i, η_i) by minimizing θ (the sum of the squares of the errors), is called a *least-squares linear function* or *regression function*, and the line which represents this function geometrically is referred to as a *least-squares line* or *regression line*. The reader may well be wondering at this point why we bother going through some rather tedious computations to determine a and b when it is possible simply by sight, using a transparent ruler, to do a rather good job of passing a line through the data points. As a matter of fact it is possible to do this, and if one were only interested in doing this once, it would probably be quite adequate.

The real usefulness of the method we have been developing, however, comes in situations where it is not convenient to construct a scatter diagram and then draw a line by eye through the points. One situation where this would be difficult would be that where hundreds or thousands of functions $y = ax + b$ were needed. The need to do this actually arises fairly frequently in certain applications. The other situation is that where more than two variables are involved. In this latter case, geometric methods cannot be applied; however, the least-squares method can. It is rather tedious to determine least-squares functions when the computations are to be made by hand. For this reason, computers are frequently used in making the computations, especially in the types of situations, such as those we have just referred to, where many functions are needed or more than two variables are involved.

There is another reason that it is of interest to study the least-squares procedure. Although we are restricting our attention to linear functions, the least-squares method can be applied equally well to determining the param-

eters in a nonlinear function. Thus, for example, if the gravitational constant g in (7–1) were not known, it could be estimated by measuring the distance d_j that a body falls in a time t_j for a number of different t_j and then determining the value of g which minimizes

$$\theta = \sum_{j=1}^{n} \left(d_j - \frac{1}{2} g t_j^2 \right)^2 . \qquad (7\text{–}13)$$

The parabola $d = \frac{1}{2} g t^2$ for the g so obtained would be called a least-squares parabola.

In determining a and b using (7–12), one must be careful to carry enough significant figures to avoid a loss in accuracy sufficiently great that serious errors are made in determining a and b. Often many more significant figures must be carried than are needed in the final answer. The reason for this is as follows. Note that both a and b are the quotients of two terms, these terms in turn being the difference of two other terms. Thus, for example, Δ is the difference of nu_{11} and u_1^2. Now it will often turn out that nu_{11} and u_1^2 are both rather large numbers, but the difference is not large. Thus it might be true that $nu_{11} = 47{,}285$ and $u_1^2 = 47{,}211$, so that $\Delta = 74$. If only three significant figures had been used, then $nu_{11} = 47{,}300$ and $u_1^2 = 47{,}200$, and according to this $\Delta = 100$. Thus a very sizable percentage error has been made in Δ, even though the percentage errors in nu_{11} and u_1^2 are very small. To make clear how serious the errors can be if an insufficient number of significant figures is carried, suppose that $nu_{12} = 267{,}411$; $u_1 u_2 = 266{,}000$; $nu_{11} = 27{,}249$, $u_1^2 = 26{,}416$. Then

$$nu_{12} - u_1 u_2 = 267{,}411 - 266{,}000 = 1411;$$
$$\Delta = nu_{11} - u_1^2 = 27{,}249 - 26{,}416 = 833 ,$$

so by (7–12), $a = 1411/833 = 1.695$. However, if all computations had been made to only three significant figures, then $nu_{12} - u_1 u_2 = 267{,}000 - 266{,}000 = 1000$; $\Delta = 27{,}000 - 26{,}000 = 1000$; and we would obtain $a = 1000/1000 = 1.000$, which is greatly in error.

7–5 EXAMPLES

In this section we shall give two simple examples illustrating how a least-squares linear function or least-squares line can be determined.

1. Suppose that we are given three points (1, 2), (4, 6) and (7, 7). Let us determine the least-squares line through these points (that is, determined by these points). By using the format of Table 7–3, we obtain Table 7–4. The last row gives the sums needed.

TABLE 7–4

COMPUTATIONS FOR EXAMPLE

ξ_i	ξ_i^2	ζ_i	$\xi_i\zeta_i$
1	1	2	2
4	16	6	24
7	49	7	49
Sums 12	66	15	75

Then by (7–11), since $n = 3$ here, $\Delta = 3(66) - (12)^2 = 198 - 144 = 54$.
Hence by (7–12)

$$a = \frac{3(75) - 12(15)}{54} = \frac{45}{54} = 0.834;$$

$$b = \frac{15(66) - (12)(75)}{54} = \frac{90}{54} = 1.665,$$

and the least-squares function is

$$y = 0.834x + 1.665 . \tag{7–14}$$

The least-squares line, along with the given points, is shown in Figure 7–13.

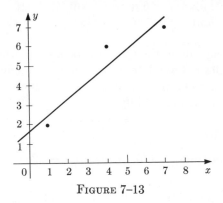

FIGURE 7–13

Let us now determine the errors and the sum of the squares of the errors.
Since $\epsilon_i = \zeta_i - a\xi_i - b$, we see that

$$\epsilon_1 = 2 - 0.834(1) - 1.665 = -0.499 ;$$
$$\epsilon_2 = 6 - 0.834(4) - 1.665 = 0.999 ;$$
$$\epsilon_3 = 7 - 0.834(7) - 1.665 = -0.503 .$$

Then the minimum value of θ is

$$\theta = (-0.499)^2 + (0.999)^2 + (-0.503)^2 = 1.501 .$$

Note that here none of the data points actually lies on the least-squares line.

2. Consider once again the firm which produces fungicides. Suppose that in a given marketing region the rainfall in inches in a number of different months, not necessarily consecutive, has been measured and recorded in the first column of Table 7–5. In the third column of Table 7–5 is recorded the

TABLE 7–5

FUNGICIDE DEMAND EXAMPLE

	ξ_i	ξ_i^2	ζ_i	$\xi_i \zeta_i$
	2.5	6.25	120	300
	1.0	1.00	70	70
	1.5	2.25	70	105
	2.1	4.41	90	189
	0.7	0.49	30	21
	1.8	3.24	100	180
	2.3	5.29	130	299
	1.2	1.44	40	48
	1.6	2.56	90	144
	0.4	0.16	40	16
Sums	15.1	27.09	780	1372

demand for the fungicide in thousands of pounds for the following month. Suppose that it is desired to relate the demand for fungicide in the region to the rainfall in the previous month using a linear function. Let us determine the least-squares line determined by these data. Table 7–5 can be used conveniently to determine the various sums needed.

By (7–11)

$$\Delta = 10(27.09) - (15.1)^2 = 270.9 - 228.0 = 42.9 ,$$

so

$$a = \frac{10(1372) - 15.1(780)}{42.9} = \frac{1942}{42.9} = 45.3; \qquad (7\text{–}15)$$

$$b = \frac{780(27.09) - 15.1(1372)}{42.9} = \frac{413}{42.9} = 9.64 . \qquad (7\text{–}16)$$

Thus the least-squares function is

$$y = 45.3x + 9.64 . \qquad (7\text{–}17)$$

Hence if the rainfall in some month is 2.8 inches, the estimate of the demand in the following month is

$$y = 45.3(2.8) + 9.64 = 136 ,$$

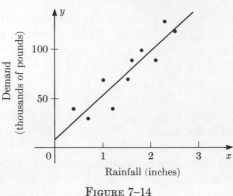

FIGURE 7–14

or 138,000 pounds. The data points and the least squares line are shown in Figure 7–14.

7–6 THE NATURE OF THE MODEL

The method of least squares can be used to estimate the parameters a and b in the function $y = ax + b$ from the set of data points, regardless of the precise interpretation of the model in which $y = ax + b$ is being used. Let us now examine in a little bit more detail the sort of model that one may have in mind when using $y = ax + b$. Generally speaking, y will always be interpreted as a random variable. The basic assumption of the model is that the distribution of y changes with the value of x, and hence a knowledge of the value of x gives us information about the value of y. There are many ways in which x could influence the distribution of y. The usual assumption is that x influences the value of μ_y, the expected value of y. Often, although not always, it is assumed that σ_y^2 is independent of x. In such cases it is also usually assumed that x does not change the shape of the density function for y; all x does is influence μ_y. For many cases of interest, it is reasonable to assume that y is normally distributed, and the model is often one which assumes y is normally distributed with a fixed variance, but with a mean that depends on the value of x.

How does the mean of y vary with x? The assumption typically made in a model is that μ_y is related to x by a linear function, that is,

$$\mu_y = ax + b. \tag{7–18}$$

In other words, the model that one often has in mind is that where y is a normally-distributed random variable with a fixed standard deviation σ and whose mean is related to the value of x by (7–18). Usually the model will not

give the values of a and b. These must be determined from experimental data. For a model of this type, what we are really doing when determining a least-squares function $y = ax + b$ is estimating the parameters a and b in (7–18). What this means is that the resulting function $y = ax + b$ is not really being used to estimate y (which is a random variable), but instead to estimate μ_y from the given value of x. Thus it would be more accurate to write the least-squares function as $\mu_y = ax + b$, but this is seldom done. We can represent graphically the sort of model we have been considering, as shown in Figure 7–15. It is supposed that $\mu_y = ax + b$; this line is shown in the figure. The variable y is a normally-distributed random variable with mean $\mu_y = ax + b$ and standard deviation σ, which is independent of x. This has been indicated in Figure 7–15 by showing the density function for y for several different values of x.

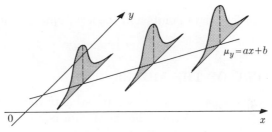

FIGURE 7–15

The model we have just discussed is not the only possible model that we might have in mind when determining a regression line. We have noted this above. It is not necessarily a complete model either. Sometimes, additional assumptions, which are needed if certain types of analyses are to be performed, are introduced. It is worth noting that it is possible to use a regression function for prediction purposes, and people often do, without having any specific model in mind whatever, or at best only the vaguest notion of what the model might be. Sometimes this is adequate; all that is needed is the fact that the value of y can be estimated better using a regression function than not using such a function.

Frequently, however, we are concerned about how much the observed value of y will deviate from $ax + b$. Thus, for example, in the scheduling of the fungicide production, it is not necessarily true that the precise amount estimated from the regression lines will be scheduled. The demand may turn out to be for more than this amount, and hence it may be desirable to schedule somewhat more in order to have a safety stock. To decide how much should be scheduled for safety stock, it is necessary to know something about the errors that can be made, that is, the distribution of forecast errors. To study this latter problem it is usually necessary to have a more

precise model in mind. The forecast error is by definition the random variable ϵ,

$$\epsilon = y - ax - b \,, \tag{7–19}$$

that is, the difference between the value of y and the predicted value $ax + b$. In many practical problems, it is of considerable interest to know as much as possible about the distribution of ϵ. We shall not attempt to study ϵ in any more detail now, although we shall return to it again later in this chapter.

7–7 MULTIVARIABLE PROBLEMS

The types of laws we have been studying have the characteristic that they can be expressed quantitatively by specifying a function which relates a variable y to another variable x. It is by no means true that all laws can be expressed in such a simple manner. Let us consider an example. The demand for steel is strongly dependent on what is taking place in several key industries, such a automobile manufacture, construction (building and highway) and heavy equipment. If we let u be the current operating level of the auto industry (in some convenient units), v the current level of building construction, w the current level of highway construction and x the current level of heavy-equipment production, each of these requires a certain rate of input of steel and the total rate of usage of steel in these industries will be the sum of the rate of usage in each of these industries, so that if y is the total rate of steel usage in these industries, then it should be true that

$$y = au + bv + cw + dx \,. \tag{7–20}$$

If the parameters a, b, c and d are known, then it is possible to estimate the total rate of demand for steel coming from these industries (which will be a very large fraction of the total demand) at some future point in time if the levels of activity in the four industries can be estimated. Thus a knowledge of the function (7–20) could be quite useful to planners in the steel industry.

In the example just given, it was necessary to introduce several variables, not merely two, since the demand for steel depends on what is taking place in several different industries. We say that when y is given by (7–20), then y is a function of four variables u, v, w and x. The function (7–20) is again referred to as a linear function (and also a linear equation). To represent laws quantitatively it is often necessary to introduce more than two variables and to make one of these a function of the remaining variables. It is easy to think of a variety of other examples in addition to the one just given. For example, the yield in a chemical reaction is a function of the temperature and pressure; the size of a crop is a function of the amount of fertilizer used, the amount of rain and the amount of sunshine; the sales of a product

depend on its price, on the prices of competing products and on the amount of advertising. Very frequently it is reasonable to adopt as a model one in which the law is represented quantitatively by a linear function such as

$$y = a_1 x\{1\} + \cdots + a_n x\{n\} + b \qquad (7\text{-}21)$$

where we have shown the general case in which y is a function of n variables $x\{1\}, x\{2\}, \ldots, x\{n\}$.

To estimate the parameters appearing in (7-21), it is again necessary to make use of experimental data, and once again the method of least squares provides a convenient way to estimate these parameters. A function (7-21) whose parameters have been determined in this way is referred to as a *least-squares linear function* or *a regression function* (regression equation). The formulas for determining the parameters are quite complicated to write out explicitly and we shall not give them. The task of making the numerical computations by hand is also very cumbersome, and for this reason a computer would almost always be used. While it was possible to determine by eye a good line which passes through a set of points when only two variables are involved, this cannot be done when more than two variables appear. Hence some mathematical technique must be used, and it is for such problems that the real usefulness of the method of least squares becomes apparent.

7-8 CORRELATION

Let us return again to a study of laws which can be represented quantitatively by a linear function $y = ax + b$. Once we have this function it seems reasonable to ask how useful it is. In other words, of what help is it to know the value of x when we are trying to predict μ_y? We now wish to study one way of answering this question.

Recall that we are thinking of y as a random variable. Suppose for the moment that we ignore the dependence of μ_y on x, and instead merely decide to build a model for y which does not involve x. To do this the values $\zeta_1, \ldots, \zeta_n$ of y which are available from experimental data would be used to construct a histogram for y. It is very likely that we could approximate this with a normal curve. As the mean μ_y^* of this normal distribution we would use

$$\mu_y^* = \bar{\zeta} = \frac{1}{n} \sum_{j=1}^{n} \zeta_j, \qquad (7\text{-}22)$$

and as the standard deviation σ_y^*, we would use

$$s_y^* = \left[\frac{1}{n-1} \sum_{j=1}^{n} (\zeta_j - \bar{\zeta})^2 \right]^{1/2} \qquad (7\text{-}23)$$

In other words, we would use the estimates of these quantities obtained from the experimental data. This is the model that would be used for making any probability computations concerning y. If we were asked to estimate y, the only thing we could do would be to use $\mu_y^* = \bar{\zeta}$ as the estimate. In general, of course, this estimate would not be correct and the standard deviation of the actual values of y about this estimate would be essentially (7–23) if the model is reasonably accurate.

Consider next the situation where the influence of x is taken into account. We shall use a model of the type described in Section 7–6, where it is imagined that $\mu_y = ax + b$ and $\sigma_y = \sigma$, a number which is independent of x. We can also imagine that y is normally distributed although we shall not need to use this directly. For this model let us imagine that a and b have been estimated by least-squares analysis. Consider next the problem of estimating σ. If we had k different observations of y for a given value of x, say ξ, then since our estimate of μ_y for this x is $ax + b$, we would estimate σ using

$$\left[\frac{1}{k-1} \sum_{j=1}^{k} (\zeta_j - a\xi - b)^2\right]^{1/2} \tag{7–24}$$

However, in general the values of x will all be different, or perhaps a very small number of them will have the same value. However, if σ_y is independent of x, then it occurs to us that in this case the fact that the values of x are different should not matter particularly, if in each case we use the estimate of the mean appropriate to that x. Thus it would appear that, for σ in our model, we can use

$$s = \left[\frac{1}{n-1} \sum_{j=1}^{n} (\zeta_j - a\xi_j - b)^2\right]^{1/2} = \left[\frac{1}{n-1} \sum_{j=1}^{n} \epsilon_j^2\right]^{1/2}, \tag{7–25}$$

since $\epsilon_j = \zeta_j - a\xi_j - b$. In other words, we can estimate σ from the errors ϵ_j, as indicated in (7–25). This is indeed a valid procedure.*

With our current model, where the variation of μ_y with x is taken into account, what we are really doing is constructing a conditional probability model. This conditional model is one in which $\mu_y = ax + b$ and $\sigma_y = \sigma$. For this model, when asked to predict the value of y, we would use $\mu_y = ax + b$, when x is known. Once again, y will not, in general, turn out to be the value predicted. However, the standard deviation of the actual values about the predicted values will now be σ which will, we hope, be less than σ_y^*, the value in the model where the influence of x is not taken into account.

---

* To obtain an unbiased estimate of σ, it turns out that we should divide by $n-2$ in (7–25) rather than $n-1$. However, for our purposes here, it is convenient to use $n-1$ When n is reasonably large, the difference is negligible.

We can now see one way of characterizing the usefulness of the law which relates y to x. This can be done by computing

$$\rho = \frac{s}{s_y^*}$$

or

$$\rho^2 = \left(\frac{s}{s_y^*}\right)^2 = \frac{\sum_{j=1}^{n} \epsilon_j^2}{\sum_{j=1}^{n} (\zeta_j - \bar{\zeta})^2}. \tag{7-26}$$

That is, the ratio of our estimates of the standard deviations, or the variances, in the two models can be used. If the use of the law relating y to x reduces the variance or standard deviation significantly so that s is small with respect to s^*, then the law should be a useful one. On the other hand, if s is almost as large as s_y^*, then the law is of relatively little value. Now both ρ and ρ^2 lie between 0 and 1, so what we are saying is that if ρ or ρ^2 is close to 0 the law is useful, whereas if ρ or ρ^2 is close to 1, the law is of little use.

We can think of ρ^2 as being the fraction of the total variance which is not explained by the law relating y and x, since σ^2 represents the variance not explained and $(\sigma_y^*)^2$ is the total variance. Hence

$$r^2 = 1 - \rho^2 = 1 - \left(\frac{s}{s_y^*}\right)^2 \tag{7-27}$$

is the fraction of the total variance in y which can be explained by the law, that is, which can be explained by taking into account the variation of μ_y with x. Now r^2 can be used as a measure of the goodness of the model. The value of r^2 lies between 0 and 1 also, that is, $0 \leq r^2 \leq 1$. The use of r^2 has an advantage over ρ or ρ^2 in that the larger value of r^2 the better the law is, whereas a value of ρ or ρ^2 close to 0 indicated a useful law. It is typical to use r^2 in practice rather than ρ or ρ^2. In addition to using r^2, another measure, which is essentially the square root of r, is also used. The precise measure used is

$$r = \begin{cases} \sqrt{r^2} & \text{if } a \geq 0 \\ -\sqrt{r^2} & \text{if } a < 0, \end{cases} \tag{7-28}$$

where a is the parameter appearing in $y = ax + b$. Thus r is a number whose magnitude is the square root of r^2 and whose sign is plus if the slope of the line representing $y = ax + b$ is positive and minus if the slope of the line representing $y = ax + b$ is negative. The reason for attaching a sign to r will appear below. When r is defined as in (7-28), then $-1 \leq r \leq 1$, and

values of r close to -1 or 1 indicate a law which is useful. The number r is referred to as the *correlation coefficient*.

We have just suggested that r^2, given by (7–27) as

$$r^2 = 1 - \frac{\displaystyle\sum_{j=1}^{n} \epsilon_j^2}{\displaystyle\sum_{j=1}^{n} (\zeta_j - \bar{\zeta})^2}, \qquad (7\text{–}29)$$

or r, given by (7–28), can be used to provide a numerical measure of the usefulness of some law. Recall that r^2 gives the fraction of the total variance of y (based on the experimental data available) which can be removed or explained by taking into account the variation of μ_y with x; r is in magnitude the square root of r^2. Now by (7–27), $r^2 = 1$ if and only if $\rho^2 = 0$, and $\rho^2 = 0$ if and only if $s^2 = 0$. But by (7–25), $s^2 = 0$ if and only if each error $\epsilon_j = 0$, that is, each data point lies on the line representing $y = ax + b$, and in this case our model would say that y could be determined exactly from a knowledge of x. If this is not the case, then $0 \le r^2 < 1$ and $-1 < r < 1$.

It is possible to compute r^2 and r for any given set of data points and any linear function $y = ax + b$ (or any function) using (7–29). Normally the computation is made only for $y = ax + b$ in the case where a and b are determined by a least-squares analysis. In this case it is possible to use the equations (7–9) and (7–10) to obtain a formula for computing r directly which is much simpler to use than (7–29). This formula, which we shall not derive in detail from (7–29), is

$$r = \frac{n \displaystyle\sum_{j=1}^{n} \xi_j \zeta_j - \left(\displaystyle\sum_{j=1}^{n} \xi_j\right)\left(\displaystyle\sum_{j=1}^{n} \zeta_j\right)}{\sqrt{n \displaystyle\sum_{j=1}^{n} \xi_j^2 - \left(\displaystyle\sum_{j=1}^{n} \xi_j\right)^2} \sqrt{n \displaystyle\sum_{j=1}^{n} \zeta_j^2 - \left(\displaystyle\sum_{j=1}^{n} \zeta_j\right)^2}}, \qquad (7\text{–}30)$$

or on using (7–11)

$$r = \frac{n u_{12} - u_1 u_2}{\sqrt{\Delta}\,\sqrt{\Delta^*}}, \qquad (7\text{–}31)$$

where

$$\Delta^* = n u_{22} - u_2^2; \quad u_{22} = \sum_{j=1}^{n} \zeta_j^2. \qquad (7\text{–}32)$$

In (7–30) or (7–31) the same sums appear that are needed in the least-squares determination of a and b. One additional sum is needed in (7–30) or (7–31). This is the sum of the squares of the values of y.

We can now make two observations concerning (7–30) and (7–31). The first is that the sign of r is automatically generated in these formulas. We need not worry about what it is. In fact, this is the real reason that a sign was originally attached to r. Secondly, it should be observed that r can be computed from (7–30) or (7–31) without a knowledge of the least-squares linear function. Only the data points are needed. A knowledge of the function is not needed because the values of a and b were used to obtain (7–30). It should be kept in mind that (7–29) is general, while (7–30) holds only for the case of a linear function with a and b determined by the least-squares method. In Figures 7–16, 7–17 and 7–18 we have illustrated cases where

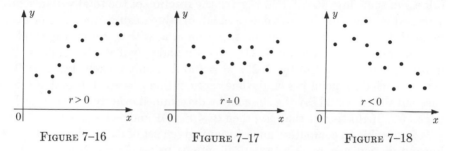

FIGURE 7–16   FIGURE 7–17   FIGURE 7–18

$r > 0$, $r \doteq 0$ and $r < 0$ respectively. The number r or r^2 is simply a numerical measure of how useful a law is, and the reader should not try to attach any greater significance to it than this. It should be noted that r only tells us how well the regression function does at explaining data that have been collected. There is no guarantee that r will remain unchanged as more data are generated.

It is important to note that r has meaning only if there is a sound basis for believing that there is a relationship between two variables. By chance the correlation coefficient for an arbitrarily selected pair of variables might suggest a useful law, when it is clear from other considerations that no such law exists. Thus over a certain period of time the correlation coefficient for the number of visitors to Yellowstone National Park on one day and the Dow Jones industrial average on the following day might, by chance, be significantly different from 0, but this does not imply the existence of any causal relationship. Misleading results can also be obtained sometimes because both variables y and x are influenced by some other variable and there is no direct relationship between y and x. Thus over a period of years the correlation coefficient for the salary of the president of United Airlines and the sales of Oldsmobile 98's may be very high, but this does not mean that the president's salary depends on the sales of Oldsmobile 98's. This correlation occurs because both are increasing with time as a result of an expanding economy. Let us conclude by computing the correlation coefficient for the examples of Section 7–5.

EXAMPLES. *1.* Let us determine r for Example 1 of Section 7–5. From Table 7–4, $u_1 = 12$, $u_{11} = 66$, $u_2 = 15$ and $u_{12} = 75$; also $\Delta = 54$. Now

$$u_{22} = \sum_{j=1}^{n} \zeta_j^2 = 4 + 36 + 49 = 89 ,$$

and

$$\Delta^* = 3(89) - 225 = 42 .$$

Next

$$\sqrt{\Delta} = \sqrt{54} = 7.34; \quad \sqrt{\Delta^*} = \sqrt{42} = 6.48 .$$

Hence, from (7–31),

$$r = \frac{225 - (12)(15)}{7.34(6.48)} = \frac{45}{47.5} = 0.947 ,$$

and $r^2 = 0.896$.

2. Now we shall determine r for the second example of Section 7–5. From Table 7–5, $u_1 = 15.1$, $u_{11} = 27.09$, $u_2 = 780$ and $u_{12} = 1372$; also $\Delta = 42.9$. The reader should check that $u_{22} = 71{,}400$ and hence

$$\Delta^* = 10(71{,}400) - (780)^2 = 105{,}600 .$$

Next

$$\sqrt{\Delta} = \sqrt{42.9} = 6.55; \quad \sqrt{\Delta^*} = \sqrt{105{,}600} = 324.8 .$$

Thus by (7–31)

$$r = \frac{10(1372) - 15.1(780)}{6.55(324.8)} = \frac{1942}{2120} = 0.917 ,$$

and $r^2 = 0.84$. Thus 84 percent of the variability of the demand for fungicide in the months recorded can be explained by the rainfall in the previous month. Thus the law relating fungicide demand to rainfall in the previous month seems to be a useful one and should be helpful in making plans for production.

REFERENCES

1. Freund, J. E., *Modern Elementary Statistics*, 3rd ed. Prentice-Hall, Englewood Cliffs, N.J., 1967.

2. Hoel, P. G., *Elementary Statistics, 2nd ed.* Wiley, New York, 1966.

3. Johnston, J., *Econometric Methods.* McGraw-Hill, New York, 1963.

A good discussion of the use of regression models in economics. A knowledge of calculus is needed.

PROBLEMS

Section 7–3

In Problems 1 through 10 illustrate the line which represents the given linear function. Determine the slope of the line.

1. $y = x$;	**2.** $y = 2x$;	**3.** $y = -x$;
4. $y = x + 1$;	**5.** $y = -2x + 1$;	**6.** $y = 3x - 2$;
7. $y = -0.5x - 1$;	**8.** $y = 7 - 4x$;	**9.** $y = 0.4x + 17$;
10. $y = -1.5x - 6$.		

Sections 7–4 and 7–5

1. Determine the change in the values of a and b from those obtained in Example 1 if the following two data points are added to the three given in the text. The additional points are $(3, 3)$ and $(8, 9)$.

2. Determine the change in the values of a and b from those obtained in Example 2 if the data points $(3, 200)$, $(1, 50)$ and $(2.8, 180)$ are added to those given in Table 7–5.

3. Determine the ϵ_i and θ for the least-squares line of Example 2.

4. Determine the least-squares line which passes through the points $(0, 1)$, $(-2, -1)$, $(3, 4)$ and $(10, 15)$. Compute the ϵ_i and θ for this line. Illustrate the line and data points graphically.

5. Determine the least-squares line which passes through the points $(1, 6)$, $(2, 4)$, $(5, -1)$ and $(15, -20)$. Compute the ϵ_i and θ for this line. Illustrate the line and the data points graphically.

6. A plumbing firm has noted that roughly five months after construction is begun on large buildings plumbing fixtures are required. It is interested in knowing if this observation could be used as an aid in planning production. To investigate this, it determined for the Manhattan marketing area the dollar volume of construction x started in a number of different months, and the sales of plumbing fixtures y five months later. The results, expressed in millions of dollars, are given in Table 7–6.

TABLE 7–6

PLUMBING FIXTURE SALES

| x | 0.7 | 3.4 | 2.5 | 1.4 | 0.8 | 6.1 | 4.0 | 0.9 | 1.8 |
| y | 0.1 | 0.6 | 0.4 | 0.3 | 0.2 | 1.4 | 1.0 | 0.4 | 0.5 |

Determine the least-squares linear function implied by the data. Illustrate graphically the data points and the least-squares line. Does it appear that there is a useful relationship here which could be of assistance in planning production? If new construction commitments in January were 3 million dollars, what is your estimate of the May demand for the company's plumbing fixtures in this region?

7. Explain in detail how the plumbing firm of Problem 6 might proceed to set production schedules in any one of its plants.

8. Determine the ϵ_i and θ for Problem 6.

9. An agricultural research center is studying the yield of tomatoes as a function of the amount of a given type of fertilizer used. The tests have been conducted under various growing conditions and on different soils. The yields y in tons of tomatoes per acre as a function of the tons per acre x of fertilizer used are given in Table 7–7. Construct a scatter diagram for these data. Over what range of values

TABLE 7–7

FERTILIZER DATA

x	y	x	y	x	y
0.25	18.0	2.85	26.5	0.90	23.0
0.50	21.0	2.50	26.0	1.10	24.5
1.00	25.0	2.20	25.0	1.60	26.0
1.50	27.0	2.00	25.5	2.00	29.0
2.00	28.0	1.50	24.0	2.60	28.0
2.50	29.0	1.00	21.0	3.10	29.5
3.00	29.5	0.80	21.5	3.50	30.0
4.00	30.0	0.50	19.5	4.00	29.0
4.50	29.0	0.25	17.0	4.50	28.0
3.10	27.0	0.60	20.8	2.50	27.0

of x does it appear that the behavior can be represented by a linear function? What happens when x is too large? Determine the set of data points having x lying in the interval where a linear function might be used to relate y to x, then determine the least-squares function implied by these data. Illustrate the least-squares line on the scatter diagram obtained above. Does the least-squares function appear useful for relating yield to fertilizer use?

10. A book publisher is interested in the relationship between the demand y for calculus books in any given year and the number of high-school graduates x in the previous year. He collects data which gives the sales y of his calculus books for a given year in thousands of copies and the number of high-school graduates x in the previous year in millions of students. These data are presented in Table 7–8. Determine the least-squares linear function implied by these data. Illustrate graphically the data points and the least-squares line. Does this seem to be a sound law to use for predicting book sales? What factors are omitted? What influence might these factors have?

TABLE 7–8

CALCULUS TEXT DATA

x	y	x	y
1.8	4	2.9	20
2.0	6	3.2	28
2.4	10	3.4	32
2.7	15	3.2	36

11. A manufacturer of transistor batteries has noted that there appears to be a relationship between the increase y in sales (in millions of batteries) during one three-month period over the previous period and the sales x of transistor radios (in units of millions of radios) in the previous three-month period. Data obtained over the past several years are given in Table 7–9. Determine the least-squares linear

TABLE 7–9

TRANSISTOR BATTERY DATA

x	y	x	y
0.35	0.25	0.50	0.30
0.20	0.15	0.30	0.20
0.45	0.30	0.55	0.35
1.20	0.60	1.65	0.90
0.43	0.20	0.40	0.30
0.18	0.18	0.25	0.25
0.60	0.40	0.75	0.40
1.45	0.80	1.70	0.95

function implied by these data. Illustrate graphically the data points and the least-squares line. Explain why it should be the change in sales volume and not the actual volume of sales that is related to sales of transistor radios in the previous period. If quarterly sales of the batteries were 8 million in the period just prior to that corresponding to the first value given for y, determine actual sales in the last period for which y is given. Can you use the regression function to predict actual sales in the last period? Is the prediction very accurate?

12. A manufacturer of electrical appliances has noted that changes in the demand rate at the retail level are not noted in factory orders for two or more months, on account of the fact that retailers order from local distributors, who in turn order from regional distributors. It is these regional distributors who order from the factory, and a significant period of time is required for the changes taking place at the retail level to pass through the system and reach the factory. The manufacturer has noted that it is the change in the orders received in any month that is related to the change in sales in a previous month. The delay seems to be roughly two months. However, the change in factory orders will not be precisely equal to

the change in sales two months earlier, because the lag is not exactly two months and, more importantly, because distributors and retailers will also be adjusting their inventories upward or downward as a result of retail sales. The manufacturer has collected sales data for refrigerators over a number of months, giving the change in factory orders y in thousands of units from the previous month and the change in retail sales x in thousands of units two months earlier. The data are given in Table 7–10. Determine the least-squares linear function implied by these data.

TABLE 7–10

REFRIGERATOR SALES

x	y	x	y	x	y
3.0	3.5	−1.2	−1.6	−0.5	−0.1
1.2	1.0	−2.5	−3.2	1.2	1.8
0.5	0.4	−2.0	−3.0	2.0	2.0
0	−0.1	−1.5	−1.8	1.7	1.9

Illustrate graphically the data points and the least-squares line so obtained. What interpretation can you give to the slope of the line here?

13. For the situation considered in Problem 12, describe the sort of data-gathering system that would be needed to enable the company to make use of the law referred to there. Note that if it takes longer to get the data on retail sales than it does for changes to be transmitted back to the plant, the results of Problem 12 cannot be used. Explain how the law could be used as an aid in production scheduling. Assume that production is scheduled one month in advance. Should precisely the amount predicted from the least-squares function be scheduled? Explain the sort of reasoning that retailers and distributors would use in adjusting their inventories. What factors might limit the usefulness of the law?

14. The final step in the manufacture of graphite electrodes, to be used, for example, in furnaces in steel mills, is graphitizing in an electric furnace. This converts the coke in the electrode into graphite. The principal reason for graphitizing is to reduce the resistance of the electrode so that it will not burn up in use. The graphitizing process is a slow one and the electrodes must remain in the electric furnace for several weeks before the graphitizing is completed. A large producer of these electrodes has been following the practice of keeping all electrodes in the graphitizing furnace for four weeks. Now the producer knows that some customers need electrodes with a lower resistance than other customers. The plant is currently operating at full capacity and is having trouble meeting all demands. It occurs to the production manager that it might be possible to increase production significantly by not keeping all electrodes in the graphitizing furnaces for four weeks, but instead producing some batches with higher resistance which have remained in the furnace for a shorter length of time, but which would be quite satisfactory for some customers. A study is then initiated to determine how the resistance of an electrode varies with the length of time it spends in the graphitizing furnace. The results are given in Table 7–11, where x is the time in days spent in the graphitizing furnace

TABLE 7–11

ELECTRODE RESISTANCE

x	y	x	y	x	y
10	4.0	5	4.8	24	1.7
15	3.0	11	3.7	21	2.5
20	2.3	18	2.2	19	1.8
25	1.4	22	1.6	15	3.4
30	0.9	28	1.2	9	4.5

and y is the resistance in the special units used by the company for characterizing the resistance of these electrodes. Determine the least-squares linear function determined by these data. Illustrate graphically the data points and the least-squares line obtained. Does there appear to be a useful law here?

15. What is the slope of the least-squares line obtained in Problem 14? What practical interpretation can you give to the slope? Could it be possible that y is related to x by a linear function for arbitrarily large values of x?

16. The amount that a machine which fills boxes of detergent actually puts into the box is determined by setting a dial. The readings on the dial indicate the number of ounces which should go into the box. However, it has been found that the dial is not completely accurate, and it is desired to calibrate the dial by making a series of experimental tests. The results are given in Table 7–12, where x is the

TABLE 7–12

FILLING-MACHINE CALIBRATION

x	y	x	y	x	y
20.0	21.1	32.0	33.1	23.0	23.8
22.0	22.8	35.0	36.2	25.0	25.7
25.0	25.9	37.0	38.0	30.0	31.1
28.0	29.0	40.0	41.0	35.0	36.0
30.0	30.8	45.0	46.1	40.0	40.8

dial setting (in ounces) and y is the actual quantity (in ounces) placed in the box. Determine the least-squares linear function implied by these data. Illustrate graphically the data points and the least-squares line.

17. The manufacturer of graphite electrodes referred to in Problem 14 uses a ball mill to grind the coke it receives from refineries down to the desired particle size. The life of the steel balls in the mill depends on the carbon content of the steel. A steel mill is interested in studying how the lifetime of balls in ball mills varies with the carbon content. The lifetime will, of course, depend on what is being processed in the mills and in precisely how they are used. The steel mill collects data from a number of its customers and the results are presented in Table 7–13,

TABLE 7–13

BALL LIFE DATA

x	y	x	y	x	y
0.01	4	0.01	6	0.02	2
0.015	8	0.03	4	0.04	6
0.02	5	0.04	9	0.05	5
0.03	7	0.05	10	0.01	3
0.035	13	0.035	7	0.025	5.5

where x is the percent by weight of carbon in the balls and y is the life in months. Determine the least-squares line implied by these data. Illustrate graphically the data points and the least-squares line.

18. In a steel mill, after steel coming out of the electric furnace is poured into ingots, the ingots are taken to the forge shop, where the ingots are shaped on the forge. Before going to the forge, the ingots must be preheated to bring them up to a temperature that is suitable for forging. The time that an ingot must spend in preheat depends on its temperature when it is placed into the furnace. If ingots are brought to the forge shop immediately after being poured, the preheat time will be low. However, if for one reason or another they are allowed to stand and cool after being poured, the preheat time will be greater. The preheat ovens in a steel mill's forge shop are currently being used at about full capacity, and the foreman is studying the question of how much capacity could be increased if ingots were always brought to the forge shop promptly after being poured. To answer this he must know how the preheat time needed changes with the temperature of the ingots. This time will depend somewhat on other factors, such as precisely how the furnace is loaded, but it will mainly be a function of the temperature of the incoming ingots. To gain information about the relationship some experiments are performed which generate the data given in Table 7–14. Here x is the temperature in degrees Fahren-

TABLE 7–14

INGOT PREHEAT TIMES

x	y	x	y	x	y
100	4.3	600	3.0	100	4.0
200	3.8	550	3.2	200	3.6
300	3.5	450	3.3	300	3.3
400	3.3	250	3.6	400	3.2
500	3.2	150	3.8	450	3.1

heit, and y is the preheat time in hours. Determine the least-squares linear function implied by these data. Illustrate graphically the data points and the least-squares line.

19. A furniture manufacturer feels that it can improve its production scheduling by using the fact that sales of furniture in a given quarter are related to the number of new homes on which construction was started two quarters earlier. To investigate this it decides to study the Los Angeles area. It determines furniture sales y for this area in each of a number of quarters and also determines the number of new houses that were started in construction two quarters earlier. The data are given in Table 7–15, where y refers to sales in thousands of dollars and x refers to new

TABLE 7–15

FURNITURE SALES DATA

x	y	x	y	x	y
2.0	120	5.0	250	3.0	250
1.5	240	4.0	230	4.5	280
3.0	200	3.5	200	6.0	300
1.1	180	2.0	180	1.2	120

housing starts in thousands. Determine the least-squares linear function implied by these data. Illustrate graphically the data points and the least-squares line. What can you conclude from this figure?

20. A government bureau is interested in the relation between hog production in one year and the price of pork in the previous year. It collects the data shown in Table 7–16, where x is the change in dollars in the average price per pound in a given

TABLE 7–16

HOG PRODUCTION DATA

x	y	x	y	x	y
-0.10	0	0.10	1.2	0.02	0.5
0.20	2.0	0.15	3.0	-0.03	0.2
0	1.2	0.20	2.5	-0.08	-1.3
-0.20	-3.1	0.12	2.0	-0.12	-2.5
-0.05	-1.0	0.05	-0.5	-0.18	-3.2

year from the previous year and y is the change in hog production in the following year from the given year (in millions of pounds). Note that x and y are the changes in prices and production, not the levels. Determine the least-squares linear function implied by these data. Illustrate graphically the data points and the least-squares line.

21. There exists a theory in meteorology that there is a relation between volcanic eruptions in any given year and the amount of sunlight that reaches the surface of the earth in the following year. The reasoning is that the fine dust particles thrown into the upper levels of the atmosphere by the eruptions limit the quantity of solar radiation reaching the earth. This theory has been used, in fact, to explain the ap-

TABLE 7–17

VOLCANIC ERUPTIONS

x	y	x	y	x	y
10	87	5	94	9	84
3	98	2	100	6	89
2	99	1	101	2	101
3	102	15	88	1	99
0	101	3	96	3	98
1	99	2	99	0	103

pearance of the ice ages. To test this theory over some years, the data shown in Table 7–17 were gathered. In this table, x is the magnitude of volcanic eruptions (on a suitable scale) in a given year and y gives the average pyrheliometer reading (an instrument which measures solar radiation) for the United States as a percentage of the normal reading for the following year. Determine the least-squares linear function implied by these data. Illustrate graphically the data and the least-squares line.

22. Table 7–18 shows the increase in road casualties (in thousands) with the num-

TABLE 7–18

ROAD CASUALTY DATA

Year	1947	1948	1949	1950	1951	1952	1953	1954	1955	1956
Vehicles	3.52	3.73	4.11	4.41	4.62	4.90	5.29	5.77	6.41	6.92
Casualties	166	153	177	201	216	208	227	238	268	268

ber of licensed vehicles (in millions) for the United Kingdom. Determine the least-squares linear function implied by these data. What is the intuitive interpretation of the slope of this line? What would be your prediction for road casualties if the number of licensed vehicles was 8.40 million?

23. An insurance company is interested in studying the relationship between accident rates and age for its male auto insurance policy holders. Table 7–19 gives the age of a number of drivers and the number of accidents they had in the past three

TABLE 7–19

ACCIDENT RATE DATA

Age	Accidents	Age	Accidents	Age	Accidents
40	1	50	4	18	4
42	2	47	1	20	3
25	3	42	1	22	8
20	1	23	2	55	2
24	6	37	1	60	1
22	2	65	2	39	2

years. The data include only drivers who had at least one accident in the last three years. Determine the least-squares linear function implied by these data, taking the variable x to refer to the age of the drivers. Illustrate graphically the least-squares line and the data points. Does there appear to be a useful law here?

24. Show that (7–9) and (7–10) can be used to determine a and b if a nonlinear law of the form $y = at^2 + b$ is used. Determine a and b when the observed values of y and t are those given in Table 7–20. Hint: Introduce a new variable $x = t^2$.

TABLE 7–20

DATA FOR NONLINEAR LAW

t	0	1	2	3	4	5
y	0	16	64	144	256	400

25. Determine θ, the sum of the squares of the errors for the nonlinear regression function determined in Problem 24. Illustrate the resulting function graphically.

26. Apply the method developed in Problem 24 to determine the least-squares non-linear regression function $y = at^2 + b$ as determined by the data given in Table 7–21.

TABLE 7–21

DATA FOR NONLINEAR LAW

t	y	t	y	t	y
−3	0.8	1	0.1	5	2.6
−2	0.3	2	0.5	6	4.0
−1	0.2	3	0.9	7	4.5
0	0.1	4	1.8	8	6.6

27. Re-solve Problem 25 for the situation described in Problem 26.

Section 7–6

1. Discuss what the general model considered in this section means in terms of the concrete situation referred to in Problem 9 for Sections 7–4 and 7–5. Does it seem reasonable that σ_y should be considered to be constant here?

2. Discuss what the general model considered in this section means in terms of the concrete situation referred to in Problem 11 for Sections 7–4 and 7–5. Does it seem reasonable that σ_y should be considered to be constant here?

3. Discuss what the general model considered in this section means in terms of the concrete situation referred to in Problem 14 for Sections 7–4 and 7–5. Does it seem reasonable that σ_y should be considered to be constant here?

Section 7–7

1. An agricultural research center is studying how the yield of strawberries in the Pacific Northwest varies with the amounts of two different fertilizers that are used

in various combinations. If a linear function is used to relate the yield to the amounts of each fertilizer used, what form will this function have? What sort of experimental arrangement would you suggest here for gaining the data that are of interest?

2. Multivariable problems frequently arise not only because several different variables are needed, but also because a variable y may depend not merely on the value of x at a particular time, but instead on the values of x at several different points in time. The values of x at the different points in time are then treated as different variables in the least squares analysis. For example, the sales of the plumbing firm referred to in Problem 6 for Sections 7–4 and 7–5 might be more accurately predicted by taking into account the volume of construction begun seven, six and five months ago. Construct a linear model (that is, a model using a linear function) which would provide for this.

3. Discuss how the observation in Problem 2 might be used to improve the estimates of transistor battery sales in Problem 11 for Sections 7–4 and 7–5.

4. Discuss how the observation in Problem 2 might be used to improve the estimation of factory orders in Problem 12 for Sections 7–4 and 7–5.

Section 7–8

In Problems 1 through 12 determine r and r^2, using (7–30), for the situation referred to in the problem for Sections 7–4 and 7–5 whose number is given.

1. 6;   **2.** 9;   **3.** 10;   **4.** 11;   **5.** 12;   **6.** 14;
7. 16;   **8.** 17;   **9.** 18;   **10.** 19;   **11.** 20;   **12.** 21.

13. If it is assumed that the model discussed in Section 7–6 is valid, show that an estimate s of σ obtained from n data points can be computed using

$$s = \left[\frac{(1 - r^2)}{n - 1} \sum_{j=1}^{n} (\zeta_j - \bar{\zeta})^2 \right]^{1/2}.$$

Hint: Use (7–27).

In Problems 14 through 25 estimate the σ introduced in Section 7–6 for the situation referred to in the problem for Sections 7–4 and 7–5 whose number is given.

14. 6;   **15.** 9;   **16.** 10;   **17.** 11;   **18.** 12;   **19.** 14;
20. 16;   **21.** 17;   **22.** 18;   **23.** 19;   **24.** 20;   **25.** 21.

26. Show that $r = a\sqrt{\Delta/\Delta^*}$, where a is given by (7–9). Thus conclude that if two different models having the same set of ξ_j also have the same a in the linear recursion function, it must be true that if $r_1 > r_2$, r_1 and r_2 being the correlation coefficients for the models, that $s_1^* < s_2^*$, s_1^* and s_2^* referring to (7–23) for the respective models.

CHAPTER 8

Time Series

8–1 **INTRODUCTION**

Most of the variables of interest to businessmen, economists and the general public have the characteristic that the values which they take on change with time. The time behavior of a variety of variables is of great interest. Thus a financial vice-president may be interested in the time behavior of interest rates, the marketing manager may be concerned with the change through time of sales, the production manager with the time dependence of raw materials prices and wage rates, the investor with the change in the Dow Jones average through time and the housewife with the variation over time of the price of beef. A function which gives the value of some variable of interest as a function time over some interval is referred to as a *time series*. It is called a time series because it is typical to visualize the function graphically as a curve in which time is represented on the horizontal axis and the value of the variable on the vertical axis, as shown in Figure 8–1. Intui-

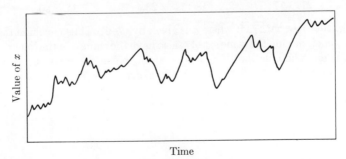

Time

FIGURE 8–1

396

tively, we think of the time series as the curve, although from a mathematical point of view it is the function which the curve represents. The graphic representation is often the most vivid and useful way to represent a time series, and is the method we shall normally employ. An alternative means of representation would be to use a table, where in the first column are listed values of time and in the second column the corresponding values of the variables.

In this chapter we wish to study certain aspects of time series. In particular, we wish to show how probability theory can be applied in the analysis of time series and in the construction of models for time series.

8–2 NATURE OF TIME SERIES

To start off, it is interesting to analyze the various types of behavior which may be exhibited by time series. The simplest sort of behavior is that illustrated in Figure 8–2, which shows the population of the United States as a

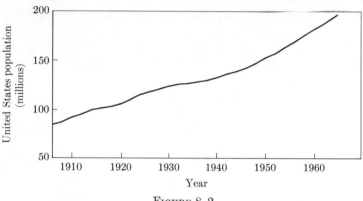

FIGURE 8–2

function of time. Here we have an example of a time series which shows a very smooth upward trend. A somewhat different sort of behavior is shown in Figure 8–3, which gives the number of employed persons as a function of time as determined by the Department of Commerce. Once again there is an upward trend, but superimposed on this trend is a seasonal pattern. A time series is said to exhibit a seasonal pattern if the value of the variable increases and decreases according to the season of the year. Still another sort of behavior is illustrated in Figure 8–4, which shows the metal mineral production in the United States as a function of time. The output scale gives the fraction of the 1957 output. In this case there is no very distinct upward trend with time. Neither is there an obvious seasonal pattern. However, there are very pronounced cyclical fluctuations, which depend on

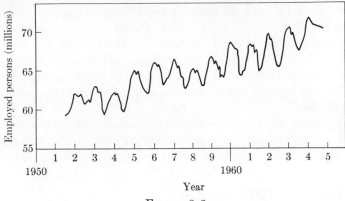

Year

FIGURE 8–3

the state of the economy and the requirements for military purposes. Here, then, is an example where the time series exhibits cyclical fluctuations which are not seasonal but depend on other factors, such as the state of the economy. One final possibility is that illustrated in Figure 8–5, which shows the daily demand for a particular type of battery at one manufacturer's warehouse. In this case there is no discernible pattern whatever to the time series. We might say that it wanders about some average value in an aimless or random way.

An arbitrary time series may contain elements of each of the four features illustrated in Figures 8–2 to 8–5. That is, there may be contained in the time series a long-term trend, which will be imagined to be represented by either a smooth increase or decrease over time. Superimposed on this may be a cyclical behavior caused by business cycle or other cycles causing phenomena. In addition to this, there may be a seasonal pattern. Finally,

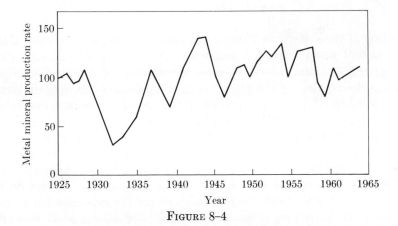

Year

FIGURE 8–4

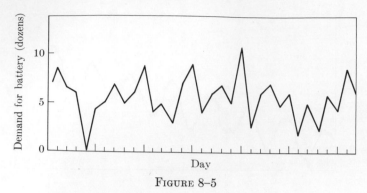

FIGURE 8–5

superimposed on all the above may be an aimless or random pattern which we shall refer to as noise.* What we are suggesting is that it is sometimes convenient to imagine that a time series consists of four separate parts:

(1) long-term trend;
(2) cyclical oscillations;
(3) seasonal pattern;
(4) random noise.

Stated in still a different way, what we are saying is that if x is the variable of interest and $x(t)$ is used to denote its value at time t, then $x(t)$ can, in general, be imagined to be the sum of four other variables which are: (1) $\tau(t)$, the long-term trend; (2) $c(t)$, the cyclical part; (3) $s(t)$, the seasonal part; and (4) $\epsilon(t)$, the noise. Then

$$x(t) = \tau(t) + c(t) + s(t) + \epsilon(t) . \qquad (8\text{–}1)$$

In any given time series, one or more of the variables in (8–1) may always be 0. Thus if there is no seasonal pattern, then $s(t) = 0$ for all t. For example, the time series shown in Figure 8–6 can be thought of as the sum of the three time series shown in Figure 8–7, that is, $x(t)$ is composed of a trend plus a seasonal pattern plus noise, and for each t, $x(t) = \tau(t) + s(t) + \epsilon(t)$. In general, the seasonal pattern is imagined to take place about the trend curve, and this is the reason $s(t)$ is negative in the low part of the season in Figure 8–7. Similarly, the noise is imagined to modify the normal seasonal pattern, and this is the reason $\epsilon(t)$ is negative at times in Figure 8–7. This is merely a convention, but it is often a convenient one.

The concept of breaking down a time series into the four components listed above is mainly of intuitive value. Its usefulness in any particular

---

* The term noise arose in electrical engineering, where random disturbances superimposed on electromagnetic waves result in noise such as that heard on a radio as the result of outside disturbances. Lightning is particularly noticeable in AM sets, for example. In television sets noise results in a deterioration of the picture quality in one way or another.

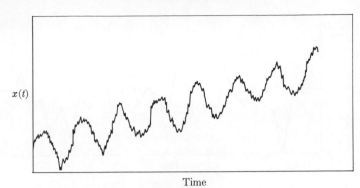

Time

FIGURE 8–6

case depends on how clearly defined the various components are. The mathematical problems encountered if one wishes to develop a rigorous mathematical procedure for breaking down a time series into its components are considerable, because in general it is not clear, for any given t, how much of $x(t)$ should be allocated to each of the four parts $\tau(t)$, $c(t)$, $s(t)$ and $\epsilon(t)$. For example, in some time series the cyclical behavior and the seasonal patterns are so intertwined that it is not a simple matter to separate them. If the various components are clearly defined, as in Figure 8–6, then it is fairly simple to break down the time series, at least approximately, just using simple graphical procedures. First a trend line is drawn in. Then this is subtracted from $x(t)$. Next the seasonal pattern is sketched over the resulting curve. This is then subtracted out, and what is left is the noise. The results are shown in Figure 8–7. If there is also a cyclical pattern, then one more step would be needed.

We noted above that, in any given time series, one or more of the four components listed above, such as a seasonal pattern, may be absent. Although this may seem unlikely to the reader at the moment, the one component which is never completely absent is the noise component. The

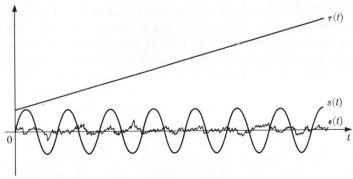

FIGURE 8–7

reader may feel that Figure 8–2 provides a contradiction to this statement. However, if we considerably expanded the horizontal and vertical scales and looked at the population on a day-to-day basis, the time series would then look something like Figure 8–5, and would appear to exhibit a considerable amount of randomness when viewed on this level. The random fluctuations would be small in magnitude compared to the actual size of the population, and in this case we say that the noise level is small with respect to the signal level. The first three components of a time series are often referred to as the *signal* in contradistinction to the fourth one, the noise. The term signal comes from electrical engineering, just as does the term noise. The signal refers to that part of the time series which is of interest and which carries the information, while the noise is the additional random component of the time series which is introduced by the equipment or the surroundings. Although the precise meaning of these terms as used in electrical engineering does not carry over in every detail to the time series of interest to the businessman, the economist or the general public, the analogy is sufficiently good that the terms signal and noise are often used for economic time series, especially by those working in the operations research or management science areas.

Generally speaking, the signal portion of a time series is considered to be essentially deterministic and more or less predictable, whereas the noise is random and can only be described probabilistically. As we shall see in more detail later, one of the basic problems in working with time series is to try to predict the future behavior of the time series. If the noise magnitude is relatively small compared with the signal strength, then it is to be expected that relatively accurate predictions can be made. However, as the noise level increases, the ability to make precise predictions decreases. Unfortunately, for many time series of interest in business and economics, the noise level is very high and this makes prediction difficult. It also follows, then, that for such time series probability theory and statistics would be expected to play a major role in any analysis of the series. This is indeed true and, in fact, the entire concept of prediction will have to be fitted into a probabilistic framework.

Example. In Figure 8–8 we have shown a time series which exhibits a seasonal pattern and in which there is relatively little noise. In Figure 8–9 is shown a time series having the same seasonal pattern but with a much higher noise level. It is fairly clear intuitively that it is more difficult to predict anything about the value of the time series in Figure 8–9 at any future point in time than it is for that in Figure 8–8.

Before going on it may be helpful to make one observation about the way time series are represented graphically. For the time series of Figures 8–2 and 8–3, it is perfectly legitimate to think of a value of the variable x

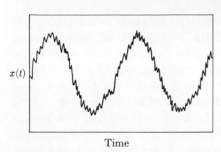

Time

FIGURE 8-8

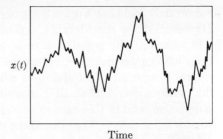

Time

FIGURE 8-9

being associated with each value of the time t, that is, for each t there is a population of the United States and for each t there is some number of employed persons. However, the situation is somewhat different for Figure 8-5. Here the variable x is the demand for the battery over some interval of time—a day. In reality then, x is defined only for values of t which are 1, 2, 3, 4, . . . , corresponding to day 1, day 2, ... The way we would, from our previous studies, represent $x(t)$ as a function of t would be to use a bar diagram, as shown in Figure 8-10. This is indeed a legitimate and the most accurate way to represent the situation. Alternatively, to show that $x(t)$ has the same value for an entire day, one could use a representation like that shown in Figure 8-11, where a horizontal line is drawn from $t - \frac{1}{2}$ to $t + \frac{1}{2}$. Very often, however, the graphical representation used would be that shown in Figure 8-12. Here the points representing the demand on the various days are plotted and are then joined together by straight line segments. This is the representation used in Figure 8-5.

Figures 8-10, 8-11 and 8-12 illustrate different ways of representing the same time series. Although any one may be encountered, the representation in Figure 8-12 is the form used most frequently. It is necessary to be careful in using the representation of Figure 8-12, however, since only the points on the curve corresponding to $t = 1, 2, 3, . . .$ have any meaning. The value of $x(t)$ read from the graph at $t = 2.5$, for example, has no meaning. It is not the demand on day 2, and neither is it the demand on day 3. The

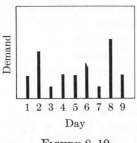

1 2 3 4 5 6 7 8 9
Day

FIGURE 8-10

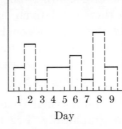

1 2 3 4 5 6 7 8 9
Day

FIGURE 8-11

1 2 3 4 5 6 7 8 9
Day

FIGURE 8-12

reader should have no trouble interpreting the graphical representation of time series once the point made in this paragraph is understood.

8–3 PROBABILISTIC MODELS OF TIME SERIES

In this section we would like to investigate briefly the problems involved in constructing a model of the process which generates a time series. We noted in Section 8–2 that randomness is always present in any time series and is usually rather important. Thus the model must, in general, be a probabilistic model. The characteristic of a time series is that $x(t)$, the value of the variable under consideration at any future time t, is not known and cannot be predicted exactly. It is convenient to imagine that the value this variable will take on at time t will be determined by the outcome of a random experiment. In other words, for each future time t, it is convenient to imagine that $x(t)$ is a *random variable*. For each future time we can then introduce a random variable $x(t)$ which represents the value of x at time t. In this way we obtain a whole collection of random variables. By associating with each future t a random variable $x(t)$ we have automatically defined a function. The domain of the function is the set of values of t and the range is a set of random variables. The image of a particular t is the random variable $x(t)$. This function is often referred to as a *random function*. A random function is nothing but a whole collection of random variables, one for each time. A random function is a generalization of the notion of a random variable. Random functions play essentially the same role in the analysis of time series that random variables did in our earlier studies.

We have previously referred to the physical processes which serve to generate a value of a random variable. We shall refer to the process which operates through time to generate values of each of the random variables $x(t)$ as a *stochastic process*. The notion of a stochastic process is a generalization of the notion of a random experiment. Often a stochastic process can be conveniently imagined to consist of the performance of a succession of random experiments, each experiment determining the value of one random variable $x(t)$. With this interpretation a stochastic process becomes a generalization of the notion of a two-stage random experiment to a multistage random experiment with an arbitrarily large number of stages.

The basic problem in formulating a model of a stochastic process is to provide a means for describing probabilistically the set of random variables $x(t)$ and the relations between them. A convenient way to characterize a stochastic process is to imagine that if the behavior of the process in the past is known, then the value for the random variable $x(t)$ for some future t is determined by the outcome of a random experiment $\Re$. It is by specifying the nature of $\Re$ that the stochastic process is characterized. In general, the model to be used for $\Re$ will depend on all past values of x, as well as on

t and the current time, call it t_1. The model for $\mathfrak{R}$ can thus be very complicated. We shall not attempt to consider in detail how these models of $\mathfrak{R}$ can be obtained. We shall, however, give one simple example.

Let us consider a stochastic process or time series in which the random variables $x(t)$ are defined only for $t = 1, 2, 3, \ldots$, that is, we shall be considering a case like that illustrated in Figure 8–5 or Figures 8–10 to 8–12. To be specific, and to make the problem interesting, let us construct a model for the stock market. Let the random variable $x(t)$ be the value of one of the market averages, such as the Dow Jones average or the New York Times average, at the end of day t. We use an average rather than an individual stock in order to obtain a representation of the behavior of the market as a whole. Now if $y(t + 1)$ is the change in the average during day $t + 1$ from its previous close, then $x(t + 1)$, the closing value at the end of day $t + 1$, is related to $x(t)$ and $y(t + 1)$ by

$$x(t + 1) = x(t) + y(t + 1) . \tag{8–2}$$

If now we specify the nature of the random experiment which determines $y(t + 1)$ for each t, then we shall have completely specified the nature of the stochastic process. We imagine that on each day a random experiment is performed which determines the change in the average from its closing value on the previous day. If we know the closing average had the value ξ_0 at the end of day 0 (any day can be selected as day 0), and if the changes on days $1, 2, \ldots, n$ were $\zeta_1, \ldots, \zeta_n$, then the value of the average at the end of day n is

$$x(n) = \xi_0 + \zeta_1 + \cdots + \zeta_n ,$$

so that by knowing the initial value and the changes, the value of the average at the end of any later day can be determined.

What now is the nature of $\mathfrak{R}$? In general, the model for $\mathfrak{R}$ might depend on the entire past history of the exchange and on t, and hence a different model for $\mathfrak{R}$ might be needed for each different day. Let us instead ask: What is the simplest imaginable model that could be used for $\mathfrak{R}$? The simplest case is that where $\mathfrak{R}$ is completely independent of past history and of t. In other words, the *same* random experiment $\mathfrak{R}$ is performed every day, and each day represents an independent trial of $\mathfrak{R}$. Thus the changes $\zeta_1, \ldots, \zeta_n$ on n successive days can be looked upon as the n values taken on by a random variable y associated with $\mathfrak{R}$ in n independent trials of $\mathfrak{R}$. With this model, $y(t + 1)$ in (8–2) does not depend on t and can simply be denoted by y, so that

$$x(t + 1) = x(t) + y . \tag{8–3}$$

It will be instructive to study this model in a little bit more detail. What shall we use as the distribution for y when $\mathfrak{R}$ is performed? Clearly y can

take on a large number of different values, both positive and negative. The simplest imaginable model for y is to assume that y is normally distributed with mean μ and standard deviation σ. We now have formulated a complete model for the stochastic process which represents the closing value of one of the stock market averages. Before evaluating the usefulness of this model, we shall illustrate the sort of behavior it will exhibit by performing a small-scale simulation.

EXAMPLE. Let us suppose that for the model just developed, $\mu = 0.25$ and $\sigma = 2$. This means that the expected increase per day in the closing average is 0.25 points and the standard deviation of the change is 2 points. Assume that the Dow Jones average is the one being used and that its value at the end of day 0 is 900. Denote by μ_n the expected value of $x(n)$, the closing Dow Jones average at the end of day n. Then by (8–3),

$$\mu_1 = 900 + 0.25; \quad \mu_2 = \mu_1 + 0.25 = 900 + 0.50;$$
$$\mu_3 = \mu_2 + 0.25 = 900 + 0.75 \ .$$

Thus, in general,

$$\mu_n = 900 + 0.25n \ , \tag{8–4}$$

and we have determined the expected value of the average at the end of day n.

Let us next simulate the operation of the market over a period of 50 days. To do this, all we need to do is generate 50 random numbers from the normal distribution $n(y; 0.25, 2)$. These will represent the change in the Dow Jones average on each of the 50 days. Recall from Section 2–18 that to generate these random numbers we first generate 50 random numbers from the uniform distribution over the unit interval. These are then converted to random numbers from the normal distribution, using the cumulative normal distribution $N(x; 0.25, 2)$. If θ is one of the random numbers from the uniform distribution and ζ is the corresponding random number from the normal distribution, then ζ is the number such that

$$\theta = N(\zeta; 0.25, 2) \ .$$

How do we find ζ, given θ? To do this it is convenient to recall that

$$N(\zeta; 0.25, 2) = \Phi\left(\frac{\zeta - 0.25}{2}\right) \ .$$

Now it is easy to determine, from Table E, the number λ such that $\theta = \Phi(\lambda)$. However, given λ, we see that

$$\frac{\zeta - 0.25}{2} = \lambda \quad \text{or} \quad \zeta = 2\lambda + 0.25 \ .$$

We have now shown how the 50 random numbers from the normal distribution can be generated. If these are denoted by $\zeta_1, \ldots, \zeta_{50}$, then

$$x(n) = 900 + \zeta_1 + \cdots + \zeta_n = x(n - 1) + \zeta_n.$$

In this way we generate the closing Dow Jones average at the end of each day.

The details of the simulation are given in Table 8–1. To obtain the random numbers from the uniform distribution over the unit interval we used Table J. The first three digits of each number in the last section of 50 numbers were used. To compute λ when $\theta < 0.500$, observe that in this case λ is negative and $\Phi(\lambda) = 1 - \Phi(-\lambda)$. Thus $\Phi(-\lambda) = 1 - \theta$, and $1 - \theta > 0.500$. Then $-\lambda$ is determined directly from Table E. For example, when $n = 2$, $\theta = 0.176$. Then $1 - \theta = 0.824$, and from Table E, $-\lambda = 0.93$, so $\lambda = -0.93$, which is the value tabulated.

TABLE 8–1

STOCK MARKET SIMULATION

n	θ	λ	ζ	$x(n)$	n	θ	λ	ζ	$x(n)$
1	.654	.40	1.05	901.05	26	.002	−2.88	−5.51	901.10
2	.176	−.93	−1.61	899.44	27	.455	−.11	.03	901.13
3	.174	−.94	−1.63	897.81	28	.642	.36	.97	902.10
4	.509	.02	.29	898.10	29	.962	1.77	3.79	905.89
5	.580	.20	.65	898.75	30	.026	−1.94	−3.63	902.26
6	.769	.74	1.73	900.48	31	.699	.52	1.29	903.55
7	.730	.61	1.47	901.95	32	.268	−.62	−.99	902.56
8	.571	.18	.61	902.56	33	.662	.42	1.09	903.65
9	.402	−.25	−.25	902.31	34	.291	−.55	−.85	902.80
10	.165	−.97	−1.69	900.62	35	.369	−.33	−.41	902.39
11	.801	.85	1.95	902.57	36	.872	1.14	2.53	904.92
12	.356	−.37	−.49	902.08	37	.766	.73	1.71	906.63
13	.177	−.93	−1.61	900.47	38	.139	−1.08	−1.91	904.72
14	.080	−1.41	−2.57	897.90	39	.944	1.59	3.43	908.15
15	.453	−.12	.01	897.91	40	.564	.16	.57	908.72
16	.223	−.76	−1.27	896.64	41	.098	−1.29	−2.33	906.39
17	.211	−.80	−1.35	895.29	42	.205	−.82	−1.39	905.00
18	.782	.78	1.81	897.10	43	.142	−1.07	−1.89	903.11
19	.143	−1.07	−1.89	895.21	44	.685	.48	1.21	904.32
20	.537	.09	.43	895.64	45	.464	−.09	.07	904.39
21	.743	.65	1.55	897.19	46	.567	.17	.59	904.98
22	.998	2.88	6.01	903.20	47	.962	1.77	3.79	908.77
23	.774	.75	1.75	904.95	48	.788	.80	1.85	910.62
24	.772	.75	1.75	906.70	49	.543	.11	.47	911.09
25	.432	−.17	−.09	906.61	50	.145	−1.06	−1.87	909.22

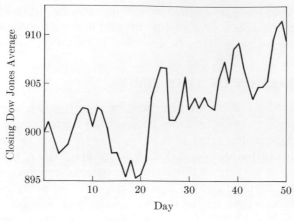

FIGURE 8–13

The closing Dow Jones averages are shown graphically in Figure 8–13. It will be observed that this figure looks remarkably similar to the graphs to be found in the New York Times or Wall Street Journal showing the behavior of the Dow Jones average. Note that after some trading sessions there is a decline in prices over a number of days. This is then followed by a brisk rally followed by some additional days of so-called trading sessions. Each day market analysts attempt to provide profound explanations for the behavior of the averages. Interestingly enough, however, most of the behavior exhibited by the stock market is characteristic of a system in which each day an independent trial of the same random experiment is performed. It is indeed true that after a given day is over some explanation can be found for certain events that have taken place. These are normally of no value, however, in indicating what will occur tomorrow, and hence the behavior tomorrow can *a priori* be imagined to be the result of a random experiment, even though after tomorrow is over an explanation can be found for some things which have occurred.

The above example has illustrated, by means of simulation, the sort of behavior our stock market model will exhibit. There has been considerable attention in recent years devoted to constructing models of the stock market. Interestingly enough, the results of all this research suggest that the simple model which we have constructed seems, with very minor modifications, to be the most accurate one. In other words, a knowledge of the past behavior of the market is of essentially no value in determining what will take place tomorrow. To a considerable extent the market behaves as if each day an independent trial of a random experiment is performed whose outcome determines the change in the averages.

The model we developed in this section, which considers the process that

generates a time series as a succession of independent trials of some fixed random experiment, is a very useful one, and many situations which arise in practice can be imagined to be described in this way.

8–4 AGGREGATION OF TIME SERIES

Often so many time series are generated by the operation of some firm or other organization that it becomes necessary to aggregate groups of them together, forming a new set of time series, in order to reduce the number to a level which can be comprehended by a decision maker. For example, a large corporation may make thousands of products, and each of these may be marketed in many regions. There exists a time series representing sales of each product in each marketing region. The total number of such time series might exceed 100,000. Clearly it would be out of the question for any one individual to examine each of these in any detail. The marketing managers in the various regions may be familiar with the time series for the products under their responsibility in their region, but no one can be familiar with the details of all of them for each product and every region. Nonetheless, there are corporation officials who are vitally concerned with overall sales of various product classes and in various regions. To give them some helpful information, it is typical to aggregate a number of series together to form a new series. Thus total sales of a given product may be provided by summing the sales of this product over all marketing regions. Alternatively, a number of products may be aggregated together and sales for this class of products given. Generally speaking, the higher the level of an executive, the more aggregated are the time series he studies. The reason for this is that, the higher the level, the broader will be the range of questions that the executive must be concerned with—not just sales, but inventories, costs, capital investment and so on. To study the details of all these areas would involve millions of time series, and this is the reason that there must be aggregation.

It is not just the businessman who must use aggregated time series. Almost everyone does so. The investor cannot follow the behavior of each and every one of the more than 1000 stocks on the New York Exchange. Instead, he relies on one of the market averages, such as the Dow Jones, to give him a feel for what the market is doing. Labor unions with cost-of-living clauses in their contracts must rely on an aggregate series, such as the cost-of-living index. Perhaps the grandest aggregate of all is the gross national product series, which gives an indication of the total output of the economy.

The important thing to realize in using time series which are aggregates of several other time series is that they must be used with great caution. Aggregate series can be extremely misleading and can thus cause an indi-

vidual using them as a basis for making decisions to commit serious errors. It is easy to provide many examples illustrating this. For example, an executive studying a series which gives total investment in inventories may correctly feel that the investment has been about what it should be. This may lead him to conclude that inventories are being well managed, when in fact they may be very poorly managed. The inventories of some items may be far too high while those of others are insufficient. As another example, the total sales of some product may be showing a significant growth over time, thus leading an executive to believe that things are going well with this product. In reality, however, sales may be declining significantly in certain marketing areas where a new product made by a competitor has been introduced, and this sales decline will spread to other regions as the new product is introduced there. Unfortunately, there does not, in general, exist any means by which a number of time series can be aggregated together to form a new time series which may not be very misleading. Information is always lost on aggregating, and this information may on occasions be useful.

The simplest way to form an aggregate time series is to add together a number of other time series. Thus to get total sales of a firm for all products, the sales for each of the products are added together. A more complicated way of aggregating time series is to perform some sort of averaging of a number of other time series. Thus a firm may be interested in the average cost per ton of some raw material which it purchases at different prices at a number of different locations. How is the time series representing the average price generated? There are a large number of different ways in which this could be done. Suppose that there are n prices $p_1(t), \ldots, p_n(t)$ which it is desired to average to form a new time series $r(t)$. The simplest sort of average would be a direct arithmetic average, given by

$$r(t) = \frac{1}{n} \left[p_1(t) + \cdots + p_n(t) \right] . \qquad (8\text{–}5)$$

An alternative way to compute the average price would be to compute the total dollar amount spent on the raw material divided by the total amount in tons of the raw material purchased. Suppose that $w_j(t)$ is the tons purchased at price $p_j(t)$. Then the total number of tons purchased is

$$W(t) = \sum_{j=1}^{n} w_j(t) , \qquad (8\text{–}6)$$

and the total cost is

$$\sum_{j=1}^{n} w_j(t) p_j(t) ,$$

so that the average in this case becomes

$$r(t) = \frac{1}{W(t)} \sum_{j=1}^{n} w_i(t)p_i(t) \ . \tag{8-7}$$

These two methods of averaging can lead to quite different time series $r(t)$. Which method should be used? It cannot be claimed that one is always superior to the other. However, (8-7) does have the desirable feature, not possessed by (8-5), that if $r(t)$ is multiplied by the total amount purchased, there results the total cost of the purchases; (8-7) is defined so that this is true.

Almost every method of averaging a set of time series $p_1(t), \ldots, p_n(t)$ to yield a new time series $r(t)$ can be represented by the equation

$$r(t) = \alpha_1(t)p_1(t) + \cdots + \alpha_n(t)p_n(t) = \sum_{j=1}^{n} \alpha_j(t)p_i(t) \ , \tag{8-8}$$

where the $\alpha_j(t)$ are non-negative numbers which sum to 1, that is, $\alpha_j(t) \geq 0$ and

$$\alpha_1(t) + \cdots + \alpha_r(t) = \sum_{j=1}^{n} \alpha_j(t) = 1 \ . \tag{8-9}$$

The $\alpha_j(t)$ are referred to as weights. Different sorts of averages can be obtained by using different weights $\alpha_j(t)$. Both (8-5) and (8-7) are special cases of (8-8). In (8-5) $\alpha_j(t) = 1/n$, and in (8-7) $\alpha_j(t) = w_j(t)/W(t)$. Note that in each case the α_j are non-negative and sum to 1. Sometimes, as in the computation of stock market averages, not all of the relevant time series are used in computing the average. For example, in the case of the stock market, only a relatively small number of stocks are used in determining the average. This case is included in (8-8) also, since some of the $\alpha_j(t)$ in (8-8) may be 0. If $\alpha_k(t) = 0$ for all t, then the time series $p_k(t)$ is not used in computing the average.

The various stock market averages provide some interesting illustrations of the way averages can be computed and some of the problems that can be encountered. The New York Times and Dow Jones averages both started out as arithmetic averages of the prices of certain selected stocks. For example, the Dow Jones industrial average uses 30 selected industrial issues. However, troubles were soon encountered due to things like stock splits and stock dividends. To illustrate the problem caused by stock splits, suppose that only two stocks are used and that at the close on a given day they are selling for 50 and 36, so that the average is 43. Imagine now that on the next day the stock selling for 50 splits. Immediately after the split the price is 25. Suppose now that by the end of the day it is selling for 26, while the other is selling for 38. If an arithmetic average of the two prices is taken now, 32 is

obtained and the average has dropped drastically while in reality both stocks have increased in value.

To get around this problem the New York Times introduces multipliers, and computes

$$\frac{1}{2}[38 + 2(26)] = 45 .$$

In other words, a multiplier is introduced to convert the split stock to the value it would have had based on the original number of shares. The Dow Jones average uses a different method for handling this problem. The arithmetic average just before the split was 43. Suppose now that instead of adding 50 and 36, we add 25 and 36 (25 being the value of the split stock immediately after the split). This yields 61. We now find that number which when divided into 61 yields 43 (the average before the split). This is $61/43 = 1.418$. In future days then, the average is determined by adding the closing prices and dividing by 1.418. Thus on the day when the stocks closed at 26 and 38, the Dow Jones method would yield the average

$$\frac{1}{1.418}[38 + 26] = 45.1 ,$$

which is now slightly different from the average the New York Times uses. After taking account of stock splits, the average is still given by (8–8). However, the $\alpha_i(t)$ no longer sum to 1. As more and more splits occur, the averages corrected in the manner just described become less and less averages, and in reality do not in any sense represent averages of the current stock prices. Some of the multipliers in the New York Times average are now more than 30, and the divisor in the Dow Jones average is less than 5 (it started out at 30). This illustration has shown how averages may quickly become meaningless even though they may have had meaning when originally constructed. In the above we did not even consider all sorts of other problems, such as stock dividends and mergers, which arise in the computation of stock market averages and which complicate things even more.

The Standard and Poor's index is close to being an application of (8–7). It multiplies the closing price of each stock by the number of shares outstanding and then adds the resulting numbers to obtain the total value of all shares of the companies used in the index. However, instead of dividing by the total number of shares outstanding to determine an average price per share, it divides by the value in the period 1941–1943 of all shares of these companies, thus giving the value today relative to the value in the 1941–1943 period. Finally, this number is multiplied by 10. The result is the value of all the shares today, based on their having a value of 10 in the period 1941–1943. Since each price is multiplied by the number of shares outstanding, stock splits are automatically accounted for.

8-5 INDEX SERIES

In studying a time series representing the time behavior of some variable x, it is often of interest to know how $x(t)$, the value of x at time t, compares with $x(t_0)$, the value of x at some other time t_0. A convenient way to do this is to compute the ratio $x(t)/x(t_0) = f(t)$. The number $f(t)$ expresses $x(t)$ as a fraction of $x(t_0)$. Instead of using $f(t)$, it is often convenient to use $g(t) = 100\, f(t)$, which expresses $x(t)$ as a percentage of $x(t_0)$. Now observe that for every t for which $x(t)$ is known it is possible to compute

$$g(t) = 100\, \frac{x(t)}{x(t_0)} . \tag{8-10}$$

In this way a new time series is generated which gives the value $g(t)$. This time series is called the *index time series* for $x(t)$ when t_0 is used as a *base*. Note from (8-10) that $g(t_0) = 100$.

To obtain a clear indication of how the values of x compare with the value at t_0, it is often convenient to determine the time series showing how g changes with t. Similarly, the series for g is also useful when interest centers on percentage changes in $x(t)$. The change in x from t_0 to t is $x(t) - x(t_0)$, and the percentage change with respect to the value at t_0 is

$$100 \left[\frac{x(t) - x(t_0)}{x(t_0)} \right] = 100 \left[\frac{x(t)}{x(t_0)} - 1 \right] = g(t) - 100 . \tag{8-11}$$

Thus if $g(t) = 135$, then x at time t has increased by 35 percent over its value at t_0. Since percentage changes in a variable are often of great interest, it is frequently useful to construct the index time series (8-10). The variable x may be anything—a price of a product, a quantity used or an employment level, for example.

EXAMPLE. In Figure 8-14 we have shown the price of filet steak in Alice Springs, Australia, from 1945 to 1963. Let us construct a price index for this type of steak. Any point in time can be used as a base. Suppose we select January, 1950. From Figure 8-14 we see that the price in January 1950 was 3 shillings per pound. Thus the index $g(t)$ for any other time t is

$$g(t) = 100\, \frac{p(t)}{3} = 33.3\, p(t) ,$$

where $p(t)$ is the price read from Figure 8-14. The resulting price index is shown graphically in Figure 8-15. It is easy to determine the percentage changes in the price of steak from their 1950 values using Figure 8-15. Note, however, that all information about the actual price of steak is lost in Figure 8-15, since only percentage values are shown. If we were interested in the actual price variation, then Figure 8-14 would have to be used.

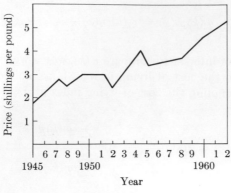

FIGURE 8–14

Indices are used a great deal in practice, especially by governments in providing economic information. The three best known indices in the United States are the consumer price index, the wholesale price index and the index of industrial production. The indices are all computed using (8–10). However, it is normally true that the time series representing x is really an aggregate of a number of other time series. Let us now study such cases. We shall use as an example the computation of a cost-of-living index for a particular family.

Consider a given family which lives in a specified city. Each month the family pays out a certain sum for various goods and services, including food, clothes, rent, utilities, medical care and entertainment. Suppose that there are n different goods and services which the family uses. Let the prices per unit for these in month k be $p_1(k)$, . . . , $p_n(k)$. Also let $q_1(k)$, . . . , $q_n(k)$ be the number of units of each of the goods and services purchased in month k. Then total expenditures $w(k)$ in month k are

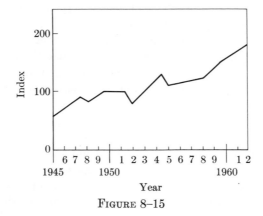

FIGURE 8–15

$$w(k) = \sum_{j=1}^{n} q_i(k)p_i(k) = q_1(k)p_1(k) + \cdots + q_n(k)p_n(k) . \quad (8\text{--}12)$$

Suppose it is now of interest to compare $w(k)$, which we can refer to as the cost of living, with the cost of living in month k_0. This can be done conveniently by determining the cost-of-living index $g(k)$,

$$g(k) = 100 \frac{w(k)}{w(k_0)} = 100 \frac{\sum_{j=1}^{n} q_j(k)p_j(k)}{\sum_{j=1}^{n} q_j(k_0)p_j(k_0)} . \quad (8\text{--}13)$$

Thus the cost-of-living index is simply computed using (8–10); however, $w(k)$ is an aggregate of a number of other series. Note that $w(k)$ can differ from $w(k_0)$ because some of the prices p_i have changed or the quantities used q_i have changed or both.

The above gives the general idea of how the Bureau of Labor Statistics might compute the consumer price index. However, a great many complications arise in practice. Let us now investigate some of these briefly. In the above we considered just a single family living in a specified city. The actual consumer price index, however, attempts to represent a general average for families all over the country. Thus $w(k)$ is not the actual expenditure for a given family but is a weighted average of expenditures of a large number of families in different cities. There are many ways in which this weighting could be done. We shall not attempt to consider the procedure in detail. However, one important simplification used is that only changes in prices are considered. Changes in the quantities of the various goods and services are not taken into account. In terms of (8–13), this means that $q_i(k_0)$ is used for $q_i(k)$, that is, the weights associated with the prices are not changed. For this reason, the index does not directly measure cost-of-living changes but only price changes. This is why it is referred to as a price index. Nonetheless, it is often used as a cost-of-living index by labor unions which have cost-of-living clauses in their contracts, meaning that when the consumer price index increases by a certain percentage there is an automatic wage adjustment. The price index tends to overstate cost-of-living increases, since when some prices p_i increase, consumers will tend to lower the q_i for these items and substitute other goods or goods of lower quality.

The price of any given good will vary from city to city at a given time and even from one part of a city to another or from store to store within a given part of a city. Thus the $p_i(k)$ used in computing the price index must themselves be averages. It is, of course, impossible to determine all the prices for a given good and hence the $p_i(k)$ will be averages of some sample of prices. Other difficulties in constructing a meaningful index arise because

of the introduction of new products and because of changes in quality or other characteristics of existing products. Thus the task of computing a cost-of-living index based on (8–13), which is conceptually very simple when just a single family is involved, becomes quite complicated when an attempt is made to obtain an index representative of the country as a whole. Just as time series which involve aggregates of other time series must be used with great caution, indices for aggregated time series must also be used with great caution. This applies, of course, to the well-known indices such as the consumer price index. The index gives one a rough general idea of what has been happening to consumer prices, but not much more than this.

The behavior of the consumer price index over time is shown in Figure 8–16. The base for this series is not a particular month but an average

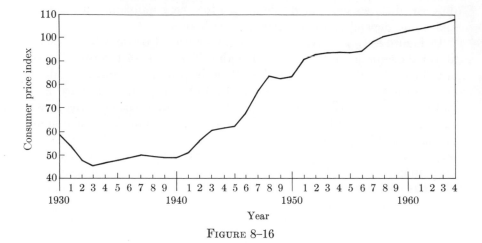

Year

FIGURE 8–16

value over the period 1957–1959 (this is often written 1957–1959 = 100). The consumer price index provides a measure of the inflation (or deflation) taking place over time, since when the index increases, this means that a given collection of goods costs more and inflation has occurred. The justification for this interpretation is the fact that the weights q_j are not changed over time and thus the consumer is always buying the same "basket of goods and services."

The fact that the price index is a measure of inflation introduces one important use for this index. Note that if the index has the value 125 at some point in time t, this means roughly that \$125 is needed to buy the same collection of goods and services that could be obtained for \$100 in 1957–1959. Thus one dollar at time t is worth only $100/125 = 0.80$ of what it was during 1957–1959, or \$0.80 in 1957–1959 would buy what \$1.00 does at

time t. This fact is often used to *deflate* or convert to *real terms* time series which are expressed in monetary units.

Let $x(t)$ be the value of some time series expressed in monetary terms (x might be wages or profits, for example). Then if $r(t)$ is the consumer price index at time t,

$$y(t) = 100 \frac{x(t)}{r(t)} \qquad (8\text{--}14)$$

is roughly the value of $x(t)$ in terms of 1957–1959 dollars. Hence the time series which gives y as a function of t expresses x in terms of the purchasing power of 1957–1959 dollars, and hence $y(t)$ is the deflated value of $x(t)$, or $y(t)$ is the value of $x(t)$ expressed in real terms (inflation having been removed). It should be kept in mind that this procedure only approximately deflates the series for x. To do so exactly would be quite difficult because a different deflated series might be needed for every individual or company and would depend on their precise buying habits.

In Figure 8–17 we have shown the gross national product (GNP) series and on the same figure the real GNP series. The value of the real GNP at any time is found by dividing the GNP by the price index read from Figure 8–16 at the same time and multiplying by 100. If the wholesale price index had been used to deflate the GNP, rather than the consumer price index, the resulting real GNP series would not be quite the same as the one shown in Figure 8–17. This only serves to emphasize again that the deflation is only approximate. The consumer price index is especially useful for determining real wages. In Table 8–2 we have shown the gross average weekly earnings for workers in manufacturing industries for several years and the corresponding real wages. The number in the real wage column for any row is found by dividing the wages in that column by the price index and multi-

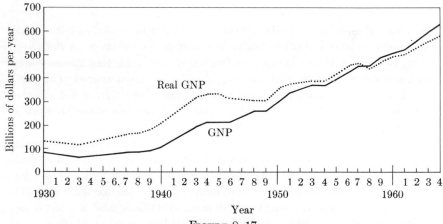

FIGURE 8–17

TABLE 8–2

MONEY AND REAL WAGES

Year	Wages	Price Index	Real Wages
1940	25.00	48.8	51.30
1945	44.20	62.7	70.50
1950	58.30	83.8	69.50
1955	75.70	93.3	81.20
1960	89.70	103.1	87.00
1962	96.60	105.4	91.70
1964	103.00	108.1	95.40

plying the result by 100. In Table 8–2, the base for the real wages is 1957–1959. It is seen that real wages did not quite double between 1940 and 1964, whereas money wages went up by more than a factor of four.

8–6 FORECASTING

A problem which arises frequently in business, economics or everyday life is that of predicting or forecasting the value that some variable will take on at a specified future time. Thus a manufacturer may be interested in forecasting the demand for one of his products three months in the future, or he may be interested in predicting the production capacity he will need two years from now, or he may wish to estimate the firm's cash balance position at the end of the quarter. Equally well an economist may be interested in predicting the GNP for the coming year or the President may be interested in predicting military costs for some future period of time. If the current time is t, it is desired to estimate the value that a variable x will take on at some future time $t + \tau$. Now $x(t + \tau)$, the symbolism for the variable x at $t + \tau$ must, in general, be imagined to be a random variable. The value that x takes on at time $t + \tau$ cannot be known with certainty. Thus there exists no way to predict the value of $x(t + \tau)$ and it would only be a fortuitous coincidence if any prediction turned out to be exactly correct. What we are really interested in predicting, then, is $\mu_x(t + \tau)$, the expected value of $x(t + \tau)$.

Since nothing is known about the future, the only information that can be used in making forecasts is information about what has taken place in the past. Any sort of information available out of the past for which there is sound reason to believe that it might influence $\mu_x(t + \tau)$ can be used in making the forecast. Although a great many factors may influence $\mu_x(t + \tau)$, it is rarely possible in any practical situation ever to take into account more than a few of these. Nonetheless, there exists a fair variety of ways in which a forecast can be made.

The simplest methods are those which base the forecast on the values of x at previous times and do not use any other information. In other words, only historical data concerning the previous values of x are used in forecasting the future. Many techniques exist for forecasting $\mu_x(t + \tau)$ using only historical data concerning x. The most elementary approach is merely to plot graphically the values of x at previous times, draw a smooth curve through these points and then by eye extrapolate this curve to the desired future point in time. Examples illustrating such an approach are shown in Figures 8–18, 8–19 and 8–20. Such an approach is still used very frequently.

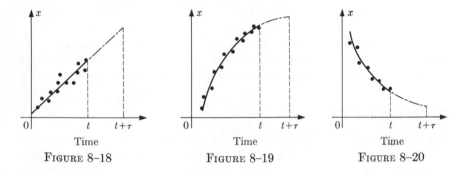

FIGURE 8–18   FIGURE 8–19   FIGURE 8–20

Alternatively, various mathematical techniques can be used to *extrapolate* historical data into the future. We shall investigate one of these techniques in more detail in the next two sections.

In addition to using the historical behavior of x in making the forecast, it is sometimes found convenient to use information regarding the historical behavior of one or more other variables as well. We have already discussed in some detail how this can be done when we studied least-squares or regression analysis. An attempt is made to determine one or more variables whose current or past values have a very significant influence on $\mu_x(t + \tau)$. Thus, for example, we studied a case where the rainfall in one month influenced very significantly the expected demand for fungicide in the following month. The variables that have an influence on $\mu_x(t + \tau)$ may, of course, depend on τ. A knowledge of past rainfall data, for example, would be of essentially no value in estimating the demand for fungicide in the spring quarter two years from the current time. What this points out is that different forecasting methods may be needed, depending on how far into the future it is desired to carry the forecast. It is typical when forecasting $\mu_x(t + \tau)$ from the past history of one or more other variables to assume that $\mu_x(t + \tau)$ can be expressed in terms of these variables through a linear function. Thus one important application of the material developed in Chapter 7 is in the area of forecasting. The basic problem when using such methods is to determine one or more variables which have a direct, physi-

cally explainable influence on $\mu_x(t + \tau)$. It is very dangerous, as has already been noted in Chapter 7, to try to forecast $\mu_x(t + \tau)$ using variables for which the regression function appears to be useful for this purpose but for which it is not possible to provide a sound explanation as to why these variables should influence $\mu_x(t + \tau)$.

Certain types of forecasts require still more complicated procedures than those we have yet discussed. We shall refer to these as multistage forecasts. To illustrate the sort of situation we have in mind here, consider a large integrated chemical company. A particular basic chemical is used in a variety of the firm's plants in the manufacture of a number of other products; it is also sold to other customers. In order to estimate future expected demands for this chemical, it is first necessary to estimate the demands for the various products made by the firm in which this chemical is a raw material. These demands can then be converted into requirements for the chemical under consideration. Then demands from outside customers are forecast, and all these things are aggregated together to obtain the demand forecast for the chemical under consideration.

In certain cases, it is necessary to carry this procedure of estimating the demand for certain products and then obtaining from this the demand for others through several stages, and that is why we have referred to the process as multistage forecasting. The task of predicting requirements of a product which has many end uses is a rather complicated one. Indeed, large firms have recently shown considerable interest in using models of the entire economy, such as Leontief input-output models (the reader need not be concerned if he is not familiar with such models) to aid them in making forecasts. These models take into account the transfer of products between industries, the output of one industry becoming raw materials for other industries. In this way the requirements for any basic raw material can ultimately be traced through to requirements for final consumer-type goods. By predicting demands for these, it is possible using a model of the economy to determine the requirements of the basic raw material that must be met to satisfy the ultimate consumer demands.

The subject of forecasting is often subdivided into *long-range* and *short-range* forecasting. There is really no clear-cut distinction between the two, and what is short range and long range depends on the variable being forecast. For many cases, but certainly not for all, a year can be used as the dividing line between short- and long-range forecasts, so that those which project less than a year into the future are termed short-range or short-term, while those which project more than a year into the future are called long-range forecasts. The various techniques that we have referred to in the above paragraphs can be used either for short- or for long-range forecasts.

However, the further into the future the projections are made, the more

important factors other than mere extrapolation of historical data become. These are factors to which, at best, the theory of probability and statistics is only partially applicable. The events which have the greatest influence on the distant future are often dramatic or revolutionary changes in the structure of things, perhaps in technology, government or the structure of markets. The very nature of these changes makes them unpredictable in the sense we have been using the term. Indeed, one major problem lies in trying to guess what sorts of changes of this type might conceivably occur. This is a very difficult but important problem, because the greatest failures of long-term forecasting have often arisen because of the occurrence of some completely unexpected event. We shall not attempt to discuss these aspects of long-range forecasting, since there is really not too much that can be said about them in general terms. The remainder of our discussion of forecasting will center on routine short-range forecasting. This is in itself a very important area.

8–7 MOVING AVERAGES

There are a variety of models that are in general use for making routine forecasts. The type of model that must be used depends on the nature of the time series being forecast. We shall not attempt to study routine forecasting in detail. Instead we shall study only the simplest of all routine forecasting methods, which is usually referred to as a *moving average*. We shall study this method in some detail and show how it can be used to make forecasts on a routine basis. In practice, it is often necessary to make forecasts for thousands of different items. If this is to be carried out frequently, it is necessary that the computations be carried out automatically on high-speed computers. The moving average method we shall consider is one which can easily be mechanized for use on a computer. Typically it is the demand for something that is being forecast, and thus in our discussions we shall treat the problem as one of forecasting demands. In actuality, however, the method could be used to forecast any variable, not merely demands.

First, we shall introduce some notation that will be used throughout the remainder of our discussion of forecasting. Let us suppose that time is divided up into periods. The length of a period might be a month, a day or a week, for example. During each period there will occur a definite demand for each item under consideration. The historical data for any particular item will then consist of a set of numbers $d_1, d_2, \ldots, d_j$ which are the demands in periods $1, 2, \ldots, j$. The period called 1 may be chosen arbitrarily; we can imagine that it is the earliest period for which we have any data on the item. With this convention increasing time is represented by increasing values of j. The last period for which the demand for the item is known is then what we have called j in the above enumeration. The basic problem is,

given these data, to estimate the expected demand in period $j + k$. This is often referred to as forecasting the demand in period $j + k$. It should be kept in mind, however, that it is the expected demand which is being forecast. One other point is worth emphasizing now. We shall usually refer to the demand for an item rather than sales of the item. Sales are equivalent to demand if there is always sufficient stock available and orders are filled very promptly.

The situation we are considering is represented diagrammatically for the case where $k = 3$ in Figure 8–21. In the above discussion we have not indi-

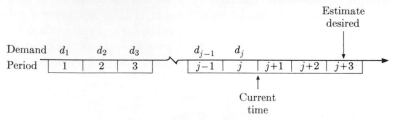

FIGURE 8–21

cated where on the time scale the current time. The current time may, for example, be the end of period j. If this is true, it is desired to estimate the demand three periods ahead. This is the situation shown in Figure 8–21. It is not necessarily true that the current time is at the end of period j; it could be at any time between the end of period j and the beginning of period $j + k$. The current time could not be earlier than the end of period j since the demand in period j is assumed to be known. In order for the current time to be at the end of period j, there cannot be any lag in reporting demands. If there were a one-period lag in tabulating and reporting demand data, then the current time could not be earlier than the end of period $j + 1$. Our analysis will not need to be concerned with the precise location of the current time in Figure 8–21, although clearly the location of the current time influences the value of k, and therefore it is very important to know in practice. The basic problem we shall consider is that of estimating the expected demand in period $j + k$, given the demands in periods 1 through j. Only the demands in periods 1 through j will be used in making the forecast.

The type of situation we are envisaging is one where each period an estimate of the demand is made for a period lying a specified number of periods in the future. Thus the process is a repetitive one which is carried out every day or week or month, depending on the length of the period. Often, in the past, forecasts would only be made annually and updated quarterly. However, use of computers makes it possible to make new forecasts at the beginning of each period for those items where this is helpful, and this is the

sort of situation we shall be studying. Note that each time a forecast for a given item is made, one additional piece of information is available over and above that used in the previous forecast. This additional information is the demand in the period following the most recent period used in the previous forecast. Thus if this time $d_1, \ldots, d_j$ are used in making the forecast, the next time $d_1, \ldots, d_{j+1}$ can be used.

We are now ready to discuss forecasting using moving averages. Denote by d'_{j+k} the estimate of the demand in period $j + k$. We are really estimating μ_{j+k}, the expected demand in period $j + k$, and it might seem more appropriate to denote the estimate by μ_{j+k}, using a notation to indicate that it is the expected value being estimated. This is not typically done, however, and thus we shall use the more common type of symbolism just introduced. A moving average uses for d'_{j+k} the average of the most recent M periods' demand, where M can be chosen arbitrarily. In other words,

$$d'_{j+k} = \frac{d_j + d_{j-1} + \cdots + d_{j-M+1}}{M} = \frac{1}{M} \sum_{i=0}^{M-1} d_{j-i} . \qquad (8\text{-}15)$$

Note that the estimate of d'_{j+k} is independent of k, that is, the estimate of the demand for any future period is the same and is given by the right-hand side of (8-15).

We can imagine that the process generating the demands for any item is a stochastic process, and it is of interest to ask for what sort of stochastic process it would be reasonable to make forecasts using a moving average (8-15). What (8-15) says is that our estimate of the future expected demand is nothing but the average of past demands. This suggests a process in which the expected demand is the same in every period. More specifically, we can imagine that each period a random experiment $\mathfrak{R}$ is performed which determines the demand in that period, so that the sequence of demands is generated by a sequence of independent trials of $\mathfrak{R}$. This is the same sort of situation hypothesized in the construction of the stock market model (the demands here corresponding to changes in the Dow Jones average). The demand in any period is then a random variable. However, the distribution of this random variable does not change from one period to the next and, in particular, its expected value μ does not change. The random variable $d\{n\}$ representing the demand in period n can then be written

$$d\{n\} = \mu + \epsilon\{n\} , \qquad (8\text{-}16)$$

where $\epsilon\{n\}$ is the random variable which represents the deviation of the demand from its expected value. If this sort of process does indeed generate the demands, then what we are attempting to do in using a forecasting system is to estimate μ.

Recall now that in Chapter 3 we suggested that to estimate the expected value of some random variable x, a logical procedure is to perform a number of independent trials of the associated random experiment and average the values of x so obtained. Now the observed values of the demand in different periods correspond to observed values of x, and (8–15) is then precisely the equation that should be used to estimate μ if the most recent M periods of demand history are to be used in the estimate.

If our model of the stochastic process is one where μ does not change from period to period, why do we bother making a new estimate of μ each period? There might be a justification for doing this if we merely incorporated the new piece of data and always averaged the demand for all the previous periods, since then, presumably, we would be improving our estimate as time went by. This is not what is done, however. The estimate is always made using the most recent M periods' demand. The answer is that we really do not believe that μ will remain absolutely constant throughout time. Instead, it will change somewhat as time goes on and hence a forecasting system is needed which will detect these changes. This is why a new estimate is made each period. It also explains why the older data are dropped from the estimates as time goes on. They are dropped because they no longer are representative of the random experiment (which changes with time) currently generating the demands. We can now begin to see the logic involved in using moving averages for forecasting.

The use of a moving average in making forecasts is most useful when the expected demand is changing very slowly, so that the change in the expected demand from one period to the next is quite small. It is possible, of course, that over a considerable period of time the expected demand will change rather drastically. This need not be of concern, provided the change in μ over the time intervals involved in making any one forecast is small. The method of moving averages can, of course, be used to make forecasts even if the above assumptions are not satisfied, but it might be expected that its performance would then have some undesirable features. In the next section we shall study how the moving average method behaves under different circumstances and, in addition, we shall illustrate in more detail precisely how to use (8–15).

8–8 PROPERTIES OF MOVING AVERAGES

Recall that we make a new forecast each period, and we always wish to forecast the demand in a period which lies a given number of periods in the future. Thus if this time we are estimating the expected demand in period $j + k$, the previous forecast was for the expected demand in period $j + k - 1$. Now according to (8–15),

$$d'_{j+k} = \frac{d_j + d_{j-1} + \cdots + d_{j-M+1}}{M} \; ; \qquad (8\text{--}17)$$

$$d'_{j+k-1} = \frac{d_{j-1} + d_{j-2} + \cdots + d_{j-M}}{M} \, . \qquad (8\text{--}18)$$

We can now make the very useful observation that it is not really necessary to add M numbers to obtain d'_{j+k}. Indeed d'_{j+k} may be conveniently computed by modifying d'_{j+k-1}. Suppose that in the numerator of (8–17) we add and subtract d_{j-M} to yield

$$\begin{aligned} d'_{j+k} &= \frac{d_j + d_{j-1} + \cdots + d_{j-M+1} + d_{j-M} - d_{j-M}}{M} \\ &= \frac{d_j}{M} + \frac{d_{j-1} + d_{j-2} + \cdots + d_{j-M}}{M} - \frac{d_{j-M}}{M} \, . \qquad (8\text{--}19) \end{aligned}$$

Now the middle term in (8–19) is precisely d'_{j+k-1}. Hence

$$d'_{j+k} = d'_{j+k-1} + \frac{d_j - d_{j-M}}{M} , \qquad (8\text{--}20)$$

and here we have a formula which allows us to compute the estimate for period $j + k$ from the estimate for period $j + k - 1$, d_j and d_{j-M}. It is normally much easier to use (8–20) than to use (8–18).

Let us now illustrate how (8–20) can be used to make routine sales forecasts on a period-by-period basis. We shall give an example in which sales are forecast using a moving average. Then the forecasts will be compared with the actual sales, and in this way we can evaluate the performance of the forecasting technique. In the second column of Table 8–3 are given the sales, in hundreds of gallons, of a particular type of industrial latex wall paint on a month-by-month basis for a total of 40 months. These sales data are represented graphically in Figure 8–22 by the solid curve. Suppose now that at the end of each month j it is desired to predict the sales

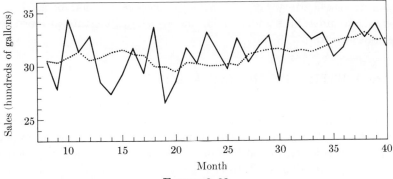

Month

Figure 8–22

(we shall refer to sales rather than demand here) for this paint in month $j + 3$, that is, the sales are being predicted three months in advance. Assume that at the end of month j, the sales in month j are known. To forecast the sales in month $j + 3$, a moving average will be used with $M = 5$. Thus the last five months' sales are averaged to estimate the sales three months from now, so

$$d'_{j+3} = \frac{1}{5}(d_j + d_{j-1} + d_{j-2} + d_{j-3} + d_{j-4})$$

$$= \frac{1}{5}(d_j - d_{j-5}) + d'_{j+2} \tag{8–21}$$

where d'_{j+2} is the estimate of the sales in month $j + 2$.

Given the sales data in column 2 of Table 8–3, it is possible to use (8–21) to generate a sequence of forecasts of sales. The first month for which a sales forecast is possible is month 8. The sales forecast for month 8 is simply the average demand in months 1 through 5, that is,

$$d'_8 = \frac{1}{5}(30.5 + 29.4 + 28.6 + 31.0 + 33.2) = 30.54 \,.$$

It is not possible to use (8–20) to obtain the first forecast. Equation (8–17) must be used to determine the first one. However, (8–20) can be used for all forecasts after the first one. It is now possible to forecast the demand in each of the following months. The forecast of the demand in month 9 is the average demand in months 2 through 6. However, from (8–21), we know that the forecast for month 9 can be computed from the forecast for month 8 and the actual sales in months 1 and 6 by

$$d'_9 = d'_8 + \frac{1}{5}d_6 - \frac{1}{5}d_1 = 30.54 + 5.90 - 6.10 = 30.34 \,.$$

This procedure can be continued to generate the estimates of the demands in each of the following months.

The computations can be conveniently carried out, as shown in Table 8–3. Let us first explain how the headings in Table 8–3 are to be interpreted. The heading for any column gives the number that appears in row j of that column, j being determined by the numbers which appear in the first column. Thus if the heading of a column is d_j, this means that the demand for period j appears in row j of the column. If $d_{j-5}/5$ appears as the heading, this means that the demand in period $j - 5$ divided by 5 appears in row j (not row $j - 5$) of the column. In more detail, the columns in Table 8–3 not yet described are as follows. In column 3 we list $d_j/5$. For any row, the entry in this column is found by dividing the entry in column 2 by 5. In column 4 we have tabulated $d_{j-5}/5$. This column is obtained merely by sliding column 3 down by five rows. In column 5 the forecast sales have been listed.

Table 8–3

Moving average forecast example

j	d_j	$d_j/5$	$d_{j-5}/5$	d'_{j+3}	d'_j	θ_j
1	30.5	6.10				
2	29.4	5.88				
3	28.6	5.72				
4	31.0	6.20				
5	33.2	6.64		30.5		
6	29.5	5.90	6.10	30.3		
7	31.7	6.34	5.88	30.8		
8	30.6	6.12	5.72	31.2	30.5	0.1
9	27.8	5.56	6.20	30.6	30.3	−2.5
10	34.5	6.90	6.64	30.8	30.8	3.7
11	31.3	6.26	5.90	31.2	31.2	0.1
12	32.8	6.56	6.34	31.4	30.6	2.2
13	28.6	5.72	6.12	31.0	30.8	−2.2
14	27.4	5.48	5.56	30.9	31.2	−3.8
15	29.2	5.84	6.90	29.9	31.4	−2.2
16	31.7	6.34	6.26	30.0	31.0	0.7
17	29.4	5.88	6.56	29.3	30.9	−1.5
18	33.6	6.72	5.72	30.3	29.9	3.7
19	26.5	5.30	5.48	30.1	30.0	−3.5
20	28.3	5.66	5.84	29.9	29.3	−1.0
21	31.7	6.34	6.34	29.9	30.3	1.4
22	30.0	6.00	5.88	30.0	30.1	−0.1
23	33.2	6.64	6.72	29.9	29.9	3.3
24	31.4	6.28	5.30	30.9	29.9	1.5
25	29.8	5.96	5.66	31.2	30.0	−0.2
26	32.5	6.50	6.34	31.4	29.9	2.6
27	30.4	6.08	6.00	31.5	30.9	−0.5
28	31.6	6.32	6.64	31.1	31.2	0.4
29	32.8	6.56	6.28	31.4	31.4	1.4
30	28.5	5.70	5.96	31.2	31.5	−3.0
31	34.9	6.98	6.50	31.6	31.1	3.8
32	33.5	6.70	6.08	32.3	31.4	2.1
33	32.4	6.48	6.32	32.4	31.2	1.2
34	33.0	6.60	6.56	32.5	31.6	1.4
35	30.8	6.16	5.70	32.9	32.3	−1.5
36	31.6	6.32	6.98	32.3	32.4	−0.8
37	34.0	6.80	6.70	32.4	32.5	1.5
38	32.7				32.9	−0.2
39	33.9				32.3	1.6
40	31.7				32.4	−0.7

The forecast is given in the row corresponding the end of the month at which it was determined. Thus the number in row j of column 5 is the forecast of sales in month $j + 3$. To compute the number in row j of column 5, the number in row $j - 1$ of column 5 is added to the number in column 3 of row j and then the number in column 4 of row j is subtracted from this. The numbers in column 5 were computed correct to two places after the decimal, but were rounded to one place after the decimal for entry in the table. In this way they are expressed to the same number of significant figures as the given demands.

The method of computation just outlined is merely a verbal description of the use of (8–21). Column 6 is obtained by moving column 5 down three rows. In column 6, then, the forecast demand is listed in the row corresponding to the month for which it is the forecast. Thus by comparing the numbers in columns 2 and 6 of any row, we can compare actual sales with the forecast sales for that month. The dotted curve in Figure 8–22 gives the forecast sales obtained from the moving average method. We start with month 8 in Figure 8–22 because this is the first month for which a forecast is available. The stochastic process generating the sales illustrated in Figure 8–22 is of a type where a moving average might reasonably be used in forecasting. The expected sales volume does not seem to be absolutely constant, but appears to be increasing fairly slowly with time.

Not only is it of interest to know the forecast of the paint sales, it is also of considerable interest for production planning and inventory control purposes to know something about the errors that can occur in the forecast. The error made in the forecast for any particular month can be looked upon as a random variable, call it θ. What we are interested in is determining the distribution of θ. In order to be able to make any statements about θ, it is necessary to introduce an important assumption. This is the assumption that the distribution of θ does not change from month to month. To be specific let us define θ as

$$\theta = d\{j\} - d'\{j\} , \qquad (8\text{–}22)$$

so that θ is equal to the actual demand minus the predicted demand for period j. Both $d\{j\}$ and $d'\{j\}$ are random variables, and the assumption that the distribution of θ does not change is equivalent to saying that the distribution of $d\{j\} - d'\{j\}$ is independent of j. If the model originally hypothesized in Section 8–7 is valid, then θ will be independent of j (why?), and our above assumption is exactly correct. However, if the process is changing through time, we cannot be sure that θ will be independent of j. Nonetheless, when the mean is changing very slowly, θ will be independent of j to a reasonable approximation.

Denote by θ_j the error in the forecast for period j. Thus

$$\theta_j = d_j - d'_j . \qquad (8\text{–}23)$$

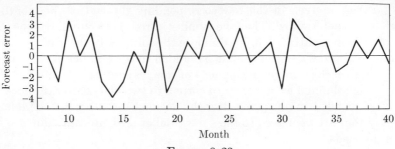

Month

FIGURE 8–23

In column 7 of Table 8–3 we have given the forecast errors. To compute the forecast error for period j it is only necessary to subtract the number in column 6 from the number in column 2. Therefore, for month 8

$$\theta_8 = 30.6 - 30.5 = 0.1 .$$

The forecast errors are shown graphically in Figure 8–23. If the distribution of θ does not change with j, then each θ_j is an observed value of θ. Hence the θ_j can be used to construct a histogram for θ. This histogram is shown in Figure 8–24. The average of all of the θ_j then gives an estimate of μ_θ, the expected value of θ. If all the numbers in column 7 of Table 8–3 are added together and then divided by 33, the result is 0.184, and this is our estimate of μ_θ. It thus appears that μ_θ is close to 0, and hence the moving average in this case yields an essentially unbiased estimate of the expected demand. If the model of Section 8–7 were correct, then we know, of course, that the moving average is exactly an unbiased estimate of the expected demand. As we shall see below, we would expect for the situation illustrated in Figure 8–22 that μ_θ would be positive, as our estimate of it did indeed turn out to be.

We have noted that a moving average seems especially appropriate for

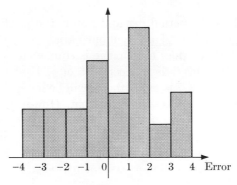

FIGURE 8–24

forecasting in the case where the expected demand does not change from one period to the next. It is interesting to study in more detail how the moving average method will behave when the expected demand does not remain constant over time. To do so, it is convenient to study very simple situations. There are two of these which display with considerable clarity the behavior of moving averages. Consider first a situation where for all past time the sales have remained constant at 40 units per month. Sud⸱ denly and discontinuously, they change to 60 units per month and ⸱emain at 60 units per month for all future time. Let us imagine that the change from 40 to 60 takes place during the period we call 1. Then monthly sales for the product are represented by the solid curve in Figure 8-25. Suppose

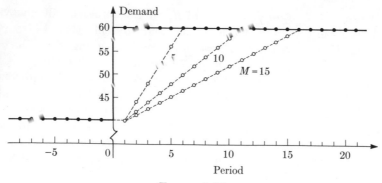

FIGURE 8-25

that a moving average is being used to forecast sales, and to be specific, sales for the coming month are being forecast, that is, at the end of month $j - 1$ the sales for month j are forecast. Imagine also that sales in month $j - 1$ are known at the end of month $j - 1$, so that the forecast of demand in month j is

$$d'_j = \frac{d_{j-1} + d_{j-2} + \cdots + d_{j-M}}{M}. \qquad (8\text{-}24)$$

Let us now illustrate the behavior of this forecasting rule for several different values of M. We shall work things out in detail for the case of $M = 5$. Up to $j = 0$, we shall have $d'_j = 40$, since all previous d_j's are 40. Hence up to $j = 0$, the forecast error will always be 0. However, a change takes place at $j = 1$, since $d_1 = 60$ and

$$d'_1 = d'_0 + \frac{1}{5}(d_0 - d_{-5}) = 40 + \frac{1}{5}(40 - 40) = 40.$$

Thus although $d_1 = 60$, $d'_1 = 40$. Next

$$d_2' = d_1' + \frac{1}{5}(d_1 - d_{-4}) = 40 + 4 = 44 \; ;$$

$$d_3' = d_2' + \frac{1}{5}(d_2 - d_{-3}) = 44 + 4 = 48 \; ;$$

$$d_4' = d_3' + \frac{1}{5}(d_3 - d_{-2}) = 48 + 4 = 52 \; ;$$

$$d_5' = d_4' + \frac{1}{5}(d_4 - d_{-1}) = 52 + 4 = 56 \; ;$$

$$d_6' = d_5' + \frac{1}{5}(d_5 - d_0) = 56 + 4 = 60 \; ;$$

$$d_7' = d_6' + \frac{1}{5}(d_6 - d_1) = 60 + \frac{1}{5}(0) = 60 \, .$$

Thus when $j \geq 7$, $d_7' = d_7 = 60$, and again the forecast will always be correct. We see, then, that when the demand suddenly changes from one value to another, there is a transition of five periods during which the moving average forecast is in the process of catching up with the change. The forecasts are represented by the dashed curve for $M = 5$ in Figure 8-25. In Figure 8-25 we have also shown the forecast curves for $M = 10$ and 15. It always takes the moving average M periods to catch up with the change, although ultimately it does so. Thus the larger M is the slower will be the response of the forecast system to changes of the type illustrated in Figure 8-25. Changes of this type are, for obvious reasons, often referred to as *step function changes*.

Let us next study the *response* of moving average forecasts to a different type of change in sales. Suppose that for all past months the sales have been 40 units per month. Then suddenly sales begin to increase at the rate of 5 units per month. This begins in month 1, so that $d_1 = 45$, $d_2 = 50$, $d_3 = 55$ and so on. This sales pattern is shown in Figure 8-26. Such a sales pattern is often referred to as a *ramp pattern*. We shall now study how a moving average forecast behaves here. As in the discussion of Figure 8-25, we shall assume that it is of interest to forecast the sales for the coming month, and that the sales for a given month are known at the end of the month. Then the forecasting equation is again (8-24). To illustrate how (8-24) behaves here, the case of $M = 5$ will be worked out in detail. Up to $j = 0$, $d_j' = d_j = 40$ and there is no forecast error. However, a change takes place in period 1 because the demand increases to 45, whereas $d_1' = 40$. Next

$$d_2' = d_1' + \frac{1}{5}(d_1 - d_{-4}) = 40 + 1 = 41$$

$$d_3' = d_2' + \frac{1}{5}(d_2 - d_{-3}) = 41 + 2 = 43$$

$$d_4' = d_3' + \frac{1}{5}(d_3 - d_{-2}) = 43 + 3 = 46$$

$$d_5' = d_4' + \frac{1}{5}(d_4 - d_{-1}) = 46 + 4 = 50$$

$$d_6' = d_5' + \frac{1}{5}(d_5 - d_0) = 50 + 5 = 55$$

$$d_7' = d_6' + \frac{1}{5}(d_6 - d_1) = 55 + 5 = 60 \ .$$

For all $j \geq 7, d_j' = d_{j+1}' + 5$, so that the forecast increases by 5 units per month just as does the demand. However, the forecast is not correct. In fact, the forecast will always be 15 units less than the actual demand when $j \geq 7$. The curve representing the forecast demand in this case is the dashed curve for $M = 5$ in Figure 8–26. We have also shown in Figure 8–26 the forecasts obtained using $M = 10$ and 15. It will be noted that the

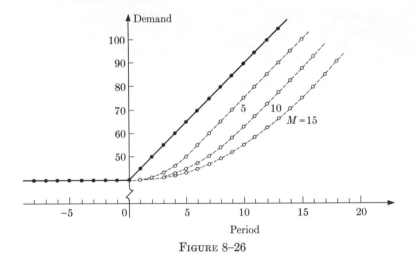

FIGURE 8–26

larger the value of M, the greater will be the forecast error. Thus we see that a moving average forecast never catches up with a ramp-type demand function.

The discussion presented above shows that a moving average forecast cannot adjust to or *track* precisely situations where the expected demand is changing with time. Thus if the expected demand is increasing with time, the moving average forecast will always lag behind and the forecast errors will tend to be positive, that is, the expected forecast error μ_θ will be positive. This explains why we would expect μ_θ to be positive in Figure 8–22, since Figure 8–22 illustrates a case where the expected demand seems to be

increasing very slowly. If the expected demand were decreasing over time then it would be expected that $\mu_\theta < 0$.

We have not explained as yet how the value of M is to be selected if a moving average is to be used for forecasting the demand for a given item. Let us now examine briefly this problem. Suppose first that the demands are generated by a stochastic process of the type hypothesized in Section 8–8, this model being one where the demands are generated by independent trials of a random experiment $\mathfrak{R}$, the same random experiment being performed each month. Then for every j, the random variable $d\{j\}$ representing the demand in period j can be written

$$d\{j\} = \mu + \epsilon\{j\} . \tag{8–25}$$

In particular, $d\{j + k\} = \mu + \epsilon\{j + k\}$. Suppose now that the demand in period $j + k$ is being forecast using an average of the demands in periods $j, j - 1, \ldots, j - M + 1$. The forecast of the demand can also be looked upon as a random variable, call it $d'\{j + k\}$, which is

$$d'\{j + k\} = \frac{1}{M}(d\{j\} + \cdots + d\{j - M + 1\}) . \tag{8–26}$$

On using (8–25) for each $d\{j\}$ in (8–26) we obtain

$$d'\{j + k\} = \mu + \frac{1}{M}(\epsilon\{j\} + \cdots + \epsilon\{j - M + 1\}) . \tag{8–27}$$

The random variable θ representing the forecast error can then be written

$$\theta = d\{n + k\} - d'\{n + k\}$$

$$= \epsilon\{n + k\} - \frac{1}{M}(\epsilon\{n\} + \cdots + \epsilon\{n - M + 1\}) . \tag{-288}$$

Now according to our model the $\epsilon\{j\}$ are independent random variables and all have the same variance, call it σ_ϵ^2. Then according to (2–53), σ_θ^2, the variance of θ, is

$$\sigma_\theta^2 = \sigma_\epsilon^2 + \frac{1}{M^2}(\sigma_\epsilon^2 + \cdots + \sigma_\epsilon^2) = \sigma_\epsilon^2 + \frac{M}{M^2}\sigma_\epsilon^2 ,$$

so

$$\sigma_\theta^2 = \left(1 + \frac{1}{M}\right)\sigma_\epsilon^2 . \tag{8–29}$$

The idea now suggests itself that M should be chosen so as to make the variance of the forecast error as small as possible. According to (8–29), the M which does this is the largest M that can be conveniently used. It is interesting to note from (8–29) that the variance of the forecast error will never be less than σ_ϵ^2, the variance of the demand on performing $\mathfrak{R}$.

In other words, the variance of the forecast error cannot be reduced below the variance of the noise. We have noted earlier that in practice we do not really expect the model used to obtain (8–29) to hold exactly. The expected demand μ will usually change with time. We have also noted that when μ changes the moving average will not track precisely, but will instead lag behind. The amount of this lag increases with M. Now when the moving average lags, this introduces a bias into the estimate of the expected demand, or equivalently into the forecast error. A bias in the forecast is clearly very undesirable, because it means that, on the average, the plans based on it will be wrong. We are faced with a dilemma. The variance of the forecast error will tend to be reduced as M is increased, but the bias in the forecasts will tend to increase with M. What must be done, then, is to choose an M which yields a suitable compromise. The proper M depends on the precise application, and we shall not attempt to consider in any greater detail how to select this value of M.

There is another factor which often influences the selection of M. Note that to use the moving average forecast, the M most recent periods' demand must be stored in the computer. If forecasts are being made for N different items, then at least NM numbers must be kept in the computer. When N is large, the computer processing time can become rather significant if M is made too large. Thus computer requirements often limit the size of M. In practice, a reasonable value of M is frequently selected and no attempt is made to perform a detailed study to find the best M for each item. Instead, the same M will be used for a whole group of items.

This will conclude our discussion of moving averages. We have noted that moving averages are not too suitable when there is a trend in the demand. Equally well, moving averages are not suitable when there is a strong seasonal pattern to the demand and this pattern is what is actually being forecast. The reader is asked to study the way moving average forecasts behave with a seasonal demand pattern in the problems. These limitations just referred to severely limit the applicability of the method of moving averages. It is not difficult to develop more general models which can be used when trends and/or seasonal patterns are present. We shall not attempt to consider such generalizations, however. Details concerning them can be found in the references below.

REFERENCES

References 5, 6 and 9 give elementary discussions of time series and index numbers. References 1, 6, 7 and 8 discuss routine forecasting, while 2, 3, 4 and 10 are concerned with models of the stock market.

1. Brown, R. G., *Smoothing, Forecasting, and Prediction of Discrete Time Series.* Prentice-Hall, Englewood Cliffs, N.J., 1963.

2. Cootner, P. H. (ed.), *The Random Character of Stock Market Prices*. M.I.T. Press, Cambridge, Mass., 1964.

3. Fama, E. F., "The Behavior of Stock Market Prices," *Journal of Business*, Vol. XXXVIII, Jan. 1965, pp. 34–105.

4. Fama, E. F., "Random Walks in Stock Market Prices," *Financial Analysts Journal*, Sept.–Oct. 1965, Vol. 21, pp. 55–59.

5. Freund, J. E., and F. J. Williams, *Elementary Business Statistics: The Modern Approach*. Prentice-Hall, Englewood Cliffs, N.J., 1964.

6. Hadley, G., *Introduction to Business Statistics*. Holden-Day, San Francisco, 1968.

7. Hadley, G. and T. M. Whitin, *Analysis of Inventory Systems*. Prentice-Hall, Englewood Cliffs, N.J., 1963.

8. Holt, Modigliani, Muth, and Simon, *Planning Production, Inventories, and Work Force*. Prentice-Hall, Englewood Cliffs, N.J., 1960.

9. Neter, J., and W. Wasserman, *Fundamental Statistics for Business and Economics*. Allyn and Bacon, Boston, 1966.

10. Osborne, M. F. M., "Brownian Motion in the Stock Market," *Operations Research*, Vol. 7, Mar.–Apr. 1959, pp. 145–173.

PROBLEMS

Section 8–2

1. What would the "noise" represent in the time series for employed persons? What might be the explanation for the seasonal pattern? Can you detect any evidence of a longer-term cyclical pattern in the series?

2. The nature of the long-term trend in a time series sometimes changes rather drastically in a short period of time. Can you give any actual examples where this has occurred? What sorts of factors can cause such changes in the trend?

3. From the appropriate government statistics determine the monthly rainfall in your city over the past five years and illustrate the time series graphically. Which of the four elements of a time series seem to be present in the rainfall data?

4. Re-solve Problem 3 using hours of sunshine instead of rainfall. You may measure the hours of sunshine in any manner which seems convenient.

5. Determine the price behavior of U.S. Steel stock over the past ten years. Illustrate this time series graphically. Which of the four basic elements of a time series seem to be present?

6. Look up the number of passenger miles flown by U.S. airlines in the continental U.S. on a monthly basis over the past five years. Illustrate this time series graphically. Which of the four basic elements of a time series seem to be present?

7. Explain some of the reasons why it may not be possible to break down a time series rigorously into the form (8–1). Hint: Are the effects always additive? Can there be interactions between the various components? Is there always a logical basis for separating out various elements?

8. Suppose that once per day an individual plays a game in which a roulette wheel with the numbers -5, -4, -3, -2, -1, 0, 1, 2, 3, 4, 5 painted on it is spun. The amount the player wins (he loses if the number is negative) is the number at which the wheel stops. The sequence of his daily winnings then generates a time series. The claim is made that this time series consists of noise only and none of the other elements is present. Is this true? What does the noise refer to here?

9. Attempt to break down the time series shown in Figure 8–3 for employed persons into its various components, that is, separate out the trend, seasonal pattern and noise. It may be helpful to look up the actual data in doing this.

Section 8–3

1. Continue the simulation of the stock market for another fifty days by selecting some new random numbers from Table J. Present the results graphically, along with those obtained in Table 8–1.

2. Use a different set of fifty random numbers to re-do the simulation in Table 8–1. Represent the results graphically and compare with Figure 8–13.

3. Look up the daily change in the Dow Jones average for the past three months. Construct a histogram for these changes. Determine the mean change and the standard deviation. Draw on the same figure the curve representing the normal density function with the same mean and standard deviation. Does it appear that the changes may be normally distributed?

4. For the stock market model constructed in the text, determine the expected value of the Dow Jones average after n days have passed and also the standard deviation of the Dow Jones average after n days. On a single figure show the curves representing μ_n, $\mu_n + \sigma_n$ and $\mu_n - \sigma_n$, where μ_n and σ_n are the mean and standard deviation, respectively, after n days.

5. Explain why a model of the stock market which involved each day the performance of an *independent* trial of some random experiment which determined the Dow Jones average, rather than the change in the Dow Jones average, would not be realistic, even if the mean and variance of the random variable were allowed to change with time.

6. Is it possible to view the model of the stock market constructed in the text in an alternative form, where the experiment performed each day determined not the change in but the actual value of the Dow Jones average? If this can be done, describe the experiment $\mathcal{R}_n$ to be performed on day n. Hint: Review Problem 5 and note that $\mathcal{R}_1, \ldots, \mathcal{R}_n$ must be considered as an n-stage experiment rather than as independent trials. What information from $\mathcal{R}_1, \ldots, \mathcal{R}_{n-1}$ is needed in characterizing $\mathcal{R}_n$? Is anything other than the outcome of $\mathcal{R}_{n-1}$ needed?

7. Consider a stochastic process in which the random function $x(t)$ is defined only at discrete times $t = 1,2,3, \ldots$, so that the stochastic process is defined by the sequence of random variables $x\{1\}$, $x\{2\}$, $x\{3\}$, $\ldots$. The stochastic process is called a *Markov process*, if the conditional density function for $x\{n\}$, given that $x\{1\} = \xi_1, \ldots, x\{n - 1\} = \xi_{n-1}$, is the same as the conditional density function for $x\{n\}$, given that $x\{n - 1\} = \xi_{n-1}$, the values of $x\{1\}$ to $x\{n - 2\}$ not being

specified. In other words, in constructing a conditional probability model for $x\{n\}$, a knowledge of $x\{n-1\}$ contains just as much information as a knowledge of the values of all previous variables, so that only the value of the immediately preceding variable is of any use in determining a model for $x\{n\}$. If we let the $x\{j\}$ here be the closing Dow Jones averages, show that, for the model developed in the text, the behavior of the Dow Jones average is a Markov process.

8. A stochastic process of the type described in Problem 7 is called a *martingale process* if $\mu_n = \xi_{n-1}$; that is, the expected value of $x\{n\}$ is simply the observed value of the immediately preceeding variable $x\{n-1\}$. Under what conditions does the model introduced in the text lead to representing the Dow Jones average as a martingale process?

9. Suppose that we modified the model of the stock market introduced in the text in such a way that the experiment which determines the change in the Dow Jones average is not the same one every day. Assume for the new model that the mean change for the experiment performed on day n is equal to the actual change on the previous day. Take the standard deviation always to be a constant. Simulate the behavior of the Dow Jones average for fifty days using this model and present the results graphically. Assume that the change in the Dow Jones average on the day prior to starting the simulation was 1.0. Use the same standard deviation used in the text. Does this seem like a reasonable model? What objections can be raised?

Section 8-4

1. In discussing the output over time of a large company which makes many products, it is typical to express the aggregate output as a dollar value rather than in units. Why is this? What sorts of distortions can arise in using the dollar value rather than physical output?

2. A chemical firm purchases a particular organic chemical at five different locations for use in five different plants. The current prices per ton at these five locations are: $p_1 = \$87.30$, $p_2 = \$91.40$, $p_3 = \$76.50$, $p_4 = \$80.00$ and $p_5 = \$94.70$. The annual tonnage requirements at the five plants are $w_1 = 500$, $w_2 = 1000$, $w_3 = 10,000$, $w_4 = 900$ and $w_5 = 1500$. Compute the average price per ton using (8-5) and (8-7).

3. For the situation described in Problem 2, suppose that the price at location 3 jumps to $82.70 per ton. Determine the new averages according to (8-5) and (8-7). Also determine the change in the average price in each case from that determined in Problem 2.

4. Time series are available which give labor productivity (perhaps dollars of output per man hour). The labor productivity has been increasing through the years. Does this mean that workers are working harder now than they used to? How is this time series to be interpreted?

5. Suppose that a given company used in one of market averages purchases another company also used in computing the average. What sort of difficulties could this cause, depending on the way the average is computed? Illustrate in the case where there are three stocks in the average, A, B and C, which yesterday sold for 20, 30 and 100, with the number of shares outstanding being one million, twenty million

and one million. Suppose that today company C is merged into company B by supplying the stockholders of C with $3\frac{1}{3}$ shares of B for every share of C they held. How would the division of the Dow Jones average have to be changed? What should be done with the New York Times average and the Standard and Poor Index?

6. Suggest some ways, different from those discussed in the text, in which aggregate time series may be misleading to decision makers.

7. Why is it that the α_j in (8–8) should essentially always sum to 1 when an average is being computed?

Section 8–5

1. Suppose that we have an index series $g(t)$ for something (this might be the price index for rice in Hong Kong, for example). Let t_0 be the base time used in computing the index. Suppose now that it is desired to change to a new base t_1. Explain how the new index series $\hat{g}(t)$ may be obtained from $g(t)$. Look up the consumer price index and generate a price index which uses January, 1965, as a base.

2. Suppose that an index series $g(t)$ is available for some variable $x(t)$, but that the time series for x is not available. Let t_0 be the base used in determining $g(t)$. Assume that it is desired to compute the percentage change in x from t_1 to t_2 (relative to t_1). Show how to do this using the series $g(t)$.

3. Suppose that the consumer price index for Los Angeles is higher than for Portland. Does this tell us anything about the relative cost of living in Los Angeles and in Portland? Could it be true that the cost of living is greater in Portland than in Los Angeles? What does a comparison of the two price indices tell us?

4. In some reference work, such as the *U.S. Book of Facts, Statistics and Information*, look up the consumer price indices for Los Angeles and Portland and plot both of these on the same graph.

5. Assume that $p_1(t)$ and $p_2(t)$ are the prices of two commodities and

$$p(t) = \alpha_1 p_1(t) + \alpha_2 p_2(t)$$

is some sort of an average price. It is desired to determine an index series $g(t)$ for $p(t)$ using t_0 as a base. Can this be done if only index series $g_1(t)$ for $p_1(t)$ and $g_2(t)$ for $p_2(t)$ are known for arbitrary bases? Can it be done if $p_1(t_0)/p_2(t_0)$ is known? If your answer is yes in either case, show in detail how to determine $g(t)$ from $g_1(t)$ and $g_2(t)$. If your answer is no, explain why it cannot be done.

6. Look up the price indices for food, rent, and apparel and upkeep, then illustrate these three graphically on the same figure. Has one of these consistently shown a higher rate of increase than the others?

7. On the same figure illustrate the consumer price index and the wholesale price index. Has one of these been rising more rapidly than the other? Can one conclude anything about the behavior of wholesaler's profits from this? Explain in detail whatever statement you make.

8. From the *U.S. Book of Facts, Statistics and Information* or some other suitable reference work, determine the price index for cornflakes, round steak, milk and bananas, using 1940 as a base. Illustrate all these graphically on the same figure.

9. From a suitable reference, determine the index of farm output per man hour and illustrate this graphically.

10. Construct an index for the total volume of construction business, using 1940 as a base.

11. Use both the index for rent and the consumer price index for San Francisco to determine an index for rents in real terms, that is, deflate the rent index for San Francisco.

12. Let $p_1(t)$ and $p_2(t)$ be the prices of two commodities. Also let $p(t) = 0.5[p_1(t) + p_2(t)]$, and let $g(t)$ be the index series for $p(t)$ with base t_0. Next define the new series $w(t) = 50p_1(t) + 13p_2(t)$ and let $g^*(t)$ be the index series for $w(t)$ with base t_0. Assume $p_1(t_0) = 10$ and $p_2(t_0) = 30$. Both $g(t)$ and $g^*(t)$ can be thought of as composite price indices for $p_1(t)$ and $p_2(t)$. Show by example that there exist values of $p_1(t)$ and $p_2(t)$ such that the index $g(t) > 1$ while $g^*(t) < 1$, that is, one index has increased while the other has decreased.

13. The prices of milk (per quart), meat (per pound) and vegetables (per pound) over nine years are as shown in Table 8–4. A single individual uses per year 150

TABLE 8–4

COMMODITY PRICES

Year	1955	1956	1957	1958	1959	1960	1961	1962	1963
Milk	0.17	0.18	0.19	0.21	0.21	0.23	0.24	0.24	0.25
Meat	0.65	0.70	0.73	0.78	0.80	0.82	0.90	0.92	0.94
Vegetables	0.09	0.09	0.10	0.09	0.10	0.11	0.11	0.10	0.12

quarts of milk, 200 pounds of meat and 350 pounds of vegetables. Construct a cost-of-living index based on these three commodities for this individual, using 1955 as a base. Assume the quantities used per year do not change.

14. Explain how you would determine a cost-of-production index for a certain product which requires a number of raw materials, labor of various types, work on several different machines and final crating. What difficulties are encountered due to the fact that the quantities of raw materials, labor and so on used for any given unit of the product are not always the same but are instead random variables?

Section 8–6

1. How would you attempt to forecast the number of planes to be needed by a domestic airline in 1975? Discuss in detail the procedure you would use.

2. How would you attempt to forecast the demand for sulfuric acid in 1972? Discuss in detail the procedure you would use.

3. How would you attempt to forecast the demand for auto gasoline in the U.S. in 1975? Discuss in detail the procedure you would use.

4. Look up data which give the number of union members in each year over the past twenty years. By extrapolating this curve, what would be the number of union

members you predict for the year 1975? Does this seem like a reasonable way to make the forecast?

5. Look up data which give electrical energy consumption in your city (or in the U.S.) over the past twenty years. If this curve is extrapolated to 1975, what is your estimate of the energy consumption in 1975? Does this seem like a sound way to make the forecast?

Sections 8–7 and 8–8

1. Re-solve the example of Table 8–3, using a moving average forecast with $M = 10$. Illustrate the results graphically. Also illustrate graphically the forecast errors.

2. From the results of Problem 1 determine a histogram of the forecast errors. Compute the mean and standard deviation of the forecast error. Also compute the mean and standard deviation of the forecast error obtained in Table 8–3 and compare the results. On the figure with the histogram, draw the normal density function with the same mean and standard deviation as the histogram.

3. Consider the demand data given in Table 8–5. Use a moving average with $M = 5$ to obtain d'_{j+3}. Illustrate the results graphically, as in Figure 8–22. Also illustrate graphically the forecast errors. Does the moving average work well here?

TABLE 8–5

MONTHLY SALES OF LARGE ELECTRODE USED IN STEEL INDUSTRY

j	d_j	j	d_j	j	d_j	j	d_j
1	34	11	36	21	41	31	42
2	32	12	38	22	44	32	50
3	32	13	36	23	40	33	52
4	35	14	34	24	42	34	48
5	34	15	40	25	40	35	47
6	31	16	38	26	38	36	48
7	36	17	39	27	40	37	52
8	34	18	36	28	44	38	50
9	32	19	38	29	41	39	50
10	35	20	36	30	44	40	54

4. From the results of Problem 3 determine a histogram of the forecast errors. Also draw on this figure the curve representing the normal density function with the same mean and standard deviation.

5. Re-solve Problem 3 using $M = 10$.

6. Re-solve Problem 4 for $M = 10$.

7. Re-solve Problem 3 using the data of Table 8–6.

8. Re-solve Problem 4, using the results of Problem 7.

9. Re-solve Problem 3, using $M = 12$ and the demand data in Table 8–6. Is there something special about using a moving average with $M = 12$ on data which exhibit a seasonal pattern?

Table 8–6

Monthly Sales in Thousands of Cases
of a Desert Tan Attache Case

j	d_j	j	d_j	j	d_j	j	d_j
1	57.7	19	50.6	37	70.3	55	55.8
2	49.7	20	62.2	38	56.5	56	69.3
3	46.7	21	59.3	39	55.5	57	70.7
4	45.0	22	68.0	40	48.6	58	80.0
5	47.0	23	73.2	41	45.7	59	85.0
6	45.1	24	64.5	42	48.2	60	72.0
7	49.7	25	69.1	43	59.9	61	66.8
8	59.6	26	53.2	44	70.9	62	58.9
9	56.6	27	45.6	45	67.7	63	56.7
10	65.6	28	44.1	46	78.0	64	47.9
11	69.8	29	50.6	47	77.6	65	55.7
12	68.1	30	45.2	48	72.6	66	56.3
13	67.0	31	54.2	49	61.0	67	57.7
14	55.0	32	63.4	50	59.0	68	72.0
15	52.3	33	57.9	51	52.8	69	70.0
16	43.7	34	75.0	52	48.7	70	85.0
17	47.1	35	75.1	53	46.8	71	90.0
18	47.4	36	77.3	54	55.4	72	76.0

10. Re-solve Problem 4, using the results of Problem 9.

11. Derive in detail the curves for $M = 10$ and 15 shown in Figure 8–25.

12. Derive in detail the curves for $M = 10$ and 15 shown in Figure 8–26.

13. For the data of Table 8–3, use a moving average with $M = 5$ to forecast the expected demand in the next period rather than three periods ahead. Illustrate the results graphically as in Figure 8–22. Also illustrate graphically the forecast errors.

14. Re-solve Problem 13 using $M = 10$.

15. Suppose that for all past time and up through period 0, the demand for some item has been 40 units per period. Then in period 1 and all later periods the demand is given by $d_j = 40 + 0.2j^2$. Imagine that at the end of period j the demand in the next period is forecast, using a moving average with $M = 5$. Determine how the forecast behaves in this case and illustrate the results graphically as in Figure 8–26.

16. Suppose that the demand for some item has been increasing by α units per period for a long time. If a moving average is used at the end of period j, show that the value of this moving average is

$$d_j - \frac{\alpha}{2}(M - 1),$$

so that the moving average lags the most recent demand used in the average by $\alpha(M - 1)/2$.

Tables

TABLE A

NUMBER OF COMBINATIONS C_r^n OF n THINGS TAKEN r AT A TIME

n \ r	2	3	4	5	6	7	8	9	10	11	12	13
2	1											
3	3	1										
4	6	4	1									
5	10	10	5	1								
6	15	20	15	6	1							
7	21	35	35	21	7	1						
8	28	56	70	56	28	8	1					
9	36	84	126	126	84	36	9	1				
10	45	120	210	252	210	120	45	10	1			
11	55	165	330	462	462	330	165	55	11	1		
12	66	220	495	792	924	792	495	220	66	12	1	
13	78	286	715	1,287	1,716	1,716	1,287	715	286	78	13	1
14	91	364	1,001	2,002	3,003	3,432	3,003	2,002	1,001	364	91	14
15	105	455	1,365	3,003	5,005	6,435	6,435	5,005	3,003	1,365	455	105
16	120	560	1,820	4,368	8,008	11,440	12,870	11,440	8,008	4,368	1,820	560
17	136	680	2,380	6,188	12,376	19,448	24,310	24,310	19,448	12,376	6,188	2,380
18	153	816	3,060	8,568	18,564	31,824	43,758	48,620	43,758	31,824	18,564	8,568
19	171	969	3,876	11,628	27,132	50,388	75,582	92,378	92,378	75,582	50,388	27,132
20	190	1,140	4,845	15,504	38,760	77,520	125,970	167,960	184,756	167,960	125,970	77,520
21	210	1,330	5,985	20,349	54,264	116,280	203,490	293,930	352,716	352,716	293,930	203,490
22	231	1,540	7,315	26,334	74,613	170,544	319,770	497,420	646,646	705,432	646,646	497,420
23	253	1,771	8,855	33,649	100,947	245,157	490,314	817,190	1,144,066	1,352,078	1,352,078	1,144,066
24	276	2,024	10,626	42,504	134,596	346,104	735,471	1,307,504	1,961,256	2,496,144	2,704,156	2,496,144
25	300	2,300	12,650	53,130	177,100	480,700	1,081,575	2,042,975	3,268,760	4,457,400	5,200,300	5,200,300
26	325	2,600	14,950	65,780	230,230	657,800	1,562,275	3,124,550	5,311,735	7,726,160	9,657,700	10,400,600

TABLE B

VALUES OF THE BINOMIAL PROBABILITY FUNCTION $b(x; n, p)$

n	x	$p = .05$	$p = .1$	$p = .2$	$p = .3$	$p = .4$	n	x	$p = .05$	$p = .1$	$p = .2$	$p = .3$	$p = .4$
2	0	.9025	.8100	.6400	.4900	.3600	8	0	.6634	.4305	.1678	.0576	.0168
	1	.0950	.1800	.3200	.4200	.4800		1	.2793	.3826	.3355	.1977	.0896
	2	.0025	.0100	.0400	.0900	.1600		2	.0515	.1488	.2936	.2965	.2090
3	0	.8574	.7290	.5120	.3430	.2160		3	.0054	.0331	.1468	.2541	.2787
	1	.1354	.2430	.3840	.4410	.4320		4	.0004	.0046	.0459	.1361	.2322
	2	.0071	.0270	.0960	.1890	.2880		5		.0004	.0092	.0467	.1239
	3	.0001	.0010	.0080	.0270	.0640		6			.0011	.0100	.0413
4	0	.8145	.6561	.4096	.2401	.1296		7			.0001	.0012	.0079
	1	.1715	.2916	.4096	.4116	.3456		8				.0001	.0007
	2	.0135	.0486	.1536	.2646	.3456	9	0	.6302	.3874	.1342	.0404	.0101
	3	.0005	.0036	.0256	.0756	.1536		1	.2985	.3874	.3020	.1556	.0605
	4		.0001	.0016	.0081	.0256		2	.0629	.1722	.3020	.2668	.1612
5	0	.7738	.5905	.3277	.1681	.0778		3	.0077	.0446	.1762	.2668	.2508
	1	.2036	.3280	.4096	.3602	.2592		4	.0006	.0074	.0661	.1715	.2508
	2	.0214	.0729	.2048	.3087	.3456		5		.0008	.0165	.0735	.1672
	3	.0011	.0081	.0512	.1323	.2304		6		.0001	.0028	.0210	.0743
	4		.0005	.0064	.0284	.0768		7			.0003	.0039	.0212
	5			.0003	.0024	.0102		8				.0004	.0035
6	0	.7351	.5314	.2621	.1176	.0467		9					.0003
	1	.2321	.3543	.3932	.3025	.1866	10	0	.5987	.3487	.1074	.0282	.0060
	2	.0305	.0984	.2458	.3241	.3110		1	.3151	.3874	.2684	.1211	.0403
	3	.0021	.0146	.0819	.1852	.2765		2	.0746	.1937	.3020	.2335	.1209
	4	.0001	.0012	.0154	.0595	.1382		3	.0105	.0574	.2013	.2668	.2150
	5		.0001	.0015	.0102	.0369		4	.0010	.0112	.0881	.2001	.2508
	6			.0001	.0007	.0041		5	.0001	.0015	.0264	.1029	.2007
7	0	.6983	.4783	.2097	.0824	.0280		6		.0001	.0055	.0368	.1115
	1	.2573	.3720	.3670	.2471	.1306		7			.0008	.0090	.0425
	2	.0406	.1240	.2753	.3176	.2613		8			.0001	.0014	.0106
	3	.0036	.0230	.1147	.2269	.2903		9				.0001	.0016
	4	.0002	.0026	.0287	.0972	.1935		10					.0001
	5		.0002	.0043	.0250	.0774							
	6			.0004	.0036	.0172							
	7				.0002	.0016							

Table C

Values of the binomial probability function $b(x; n, p)$ for $p = 0.5$

n	x	p = .5	n	x	p = .5	n	x	p = .5	n	x	p = .5	n	x	p = .5
2	0	.2500	13	0	.0001	18	0	.0000	23	2	.0000	27	3	.0000
	1	.5000		1	.0016		1	.0001		3	.0002		4	.0001
3	0	.1250		2	.0095		2	.0006		4	.0011		5	.0006
	1	.3750		3	.0349		3	.0031		5	.0040		6	.0022
4	0	.0625		4	.0873		4	.0117		6	.0120		7	.0066
	1	.2500		5	.1571		5	.0327		7	.0292		8	.0165
	2	.3750		6	.2095		6	.0708		8	.0584		9	.0349
5	0	.0312	14	0	.0001		7	.1214		9	.0974		10	.0629
	1	.1562		1	.0009		8	.1669		10	.1364		11	.0971
	2	.3125		2	.0056		9	.1855		11	.1612		12	.1295
6	0	.0156		3	.0222	19	1	.0000	24	2	.0000		13	.1494
	1	.0938		4	.0611		2	.0003		3	.0001	28	3	.0000
	2	.2344		5	.1222		3	.0018		4	.0006		4	.0001
	3	.3125		6	.1833		4	.0074		5	.0025		5	.0004
7	0	.0078		7	.2095		5	.0222		6	.0080		6	.0014
	1	.0547	15	0	.0000		6	.0518		7	.0206		7	.0041
	2	.1641		1	.0005		7	.0961		8	.0438		8	.0116
	3	.2734		2	.0032		8	.1442		9	.0779		9	.0257
8	0	.0039		3	.0139		9	.1762		10	.1169		10	.0489
	1	.0312		4	.0417	20	1	.0000		11	.1488		11	.0800
	2	.1094		5	.0916		2	.0002		12	.1612		12	.1133
	3	.2188		6	.1527		3	.0011	25	2	.0000		13	.1395
	4	.2734		7	.1964		4	.0046		3	.0001		14	.1494
9	0	.0020	16	0	.0000		5	.0148		4	.0004	29	4	.0000
	1	.0176		1	.0002		6	.0370		5	.0016		5	.0002
	2	.0703		2	.0018		7	.0739		6	.0053		6	.0009
	3	.1641		3	.0085		8	.1201		7	.0143		7	.0029
	4	.2461		4	.0278		9	.1602		8	.0322		8	.0080
10	0	.0010		5	.0667		10	.1762		9	.0609		9	.0187
	1	.0098		6	.1222	21	1	.0000		10	.0974		10	.0373
	2	.0439		7	.1746		2	.0001		11	.1328		11	.0644
	3	.1172		8	.1964		3	.0006		12	.1550		12	.0967
	4	.2051	17	0	.0000		4	.0029	26	3	.0000		13	.1264
	5	.2461		1	.0001		5	.0097		4	.0002		14	.1445
11	0	.0005		2	.0010		6	.0259		5	.0010	30	4	.0000
	1	.0054		3	.0052		7	.0554		6	.0034		5	.0001
	2	.0269		4	.0182		8	.0970		7	.0098		6	.0006
	3	.0806		5	.0472		9	.1402		8	.0233		7	.0019
	4	.1611		6	.0944		10	.1682		9	.0466		8	.0055
	5	.2256		7	.1484	22	1	.0000		10	.0792		9	.0133
12	0	.0002		8	.1855		2	.0001		11	.1151		10	.0280
	1	.0029					3	.0004		12	.1439		11	.0509
	2	.0161					4	.0017		13	.1550		12	.0806
	3	.0537					5	.0063					13	.1115
	4	.1208					6	.0178					14	.1354
	5	.1934					7	.0407					15	.1445
	6	.2256					8	.0762						
							9	.1186						
							10	.1542						
							11	.1682						

TABLE D

VALUES OF THE POISSON PROBABILITY FUNCTION $p(x; \beta)$

x	.1	.2	.3	.4	.5	.6	.7	.8	.9	1.0
					β					
0	.9048	.8187	.7408	.6703	.6065	.5488	.4966	.4493	.4066	.3679
1	.0905	.1637	.2222	.2681	.3033	.3293	.3476	.3595	.3659	.3679
2	.0045	.0164	.0333	.0536	.0758	.0988	.1217	.1438	.1647	.1839
3	.0002	.0011	.0033	.0072	.0126	.0198	.0284	.0383	.0494	.0613
4		.0001	.0003	.0007	.0016	.0030	.0050	.0077	.0111	.0153
5				.0001	.0002	.0004	.0007	.0012	.0020	.0031
6							.0001	.0002	.0003	.0005
7										.0001

x	1	2	3	4	5	6	7	8	9	10
					β					
0	.3679	.1353	.0498	.0183	.0067	.0025	.0009	.0003	.0001	.0000
1	.3679	.2707	.1494	.0733	.0337	.0149	.0064	.0027	.0011	.0005
2	.1839	.2707	.2240	.1465	.0842	.0446	.0223	.0107	.0050	.0023
3	.0613	.1804	.2240	.1954	.1404	.0892	.0521	.0286	.0150	.0076
4	.0153	.0902	.1680	.1954	.1755	.1339	.0912	.0572	.0337	.0189
5	.0031	.0361	.1008	.1563	.1755	.1606	.1277	.0916	.0607	.0378
6	.0005	.0120	.0504	.1042	.1462	.1606	.1490	.1221	.0911	.0631
7	.0001	.0034	.0216	.0595	.1044	.1377	.1490	.1396	.1171	.0901
8		.0009	.0081	.0298	.0653	.1033	.1304	.1396	.1318	.1126
9		.0002	.0027	.0132	.0363	.0688	.1014	.1241	.1318	.1251
10			.0008	.0053	.0181	.0413	.0710	.0993	.1186	.1251
11			.0002	.0019	.0082	.0225	.0452	.0722	.0970	.1137
12			.0001	.0006	.0034	.0113	.0264	.0481	.0728	.0948
13				.0002	.0013	.0052	.0142	.0296	.0504	.0729
14				.0001	.0005	.0022	.0071	.0169	.0324	.0521
15					.0002	.0009	.0033	.0090	.0194	.0347
16						.0003	.0014	.0045	.0109	.0217
17						.0001	.0006	.0021	.0058	.0128
18							.0002	.0009	.0029	.0071
19							.0001	.0004	.0014	.0037
20								.0002	.0006	.0019
21								.0001	.0003	.0009
22									.0001	.0004
23										.0002
24										.0001

TABLE E

CUMULATIVE STANDARDIZED NORMAL DISTRIBUTION $\Phi(t)$

This table gives $\Phi(t)$
for various values of t.

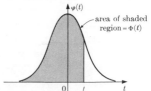

area of shaded region $= \Phi(t)$

t	.00	.01	.02	.03	.04	.05	.06	.07	.08	.09
.0	.5000	.5040	.5080	.5120	.5160	.5199	.5239	.5279	.5319	.5359
.1	.5398	.5438	.5478	.5517	.5557	.5596	.5636	.5675	.5714	.5753
.2	.5793	.5832	.5871	.5910	.5948	.5987	.6026	.6064	.6103	.6141
.3	.6179	.6217	.6255	.6293	.6331	.6368	.6406	.6443	.6480	.6517
.4	.6554	.6591	.6628	.6664	.6700	.6736	.6772	.6808	.6844	.6879
.5	.6915	.6950	.6985	.7019	.7054	.7088	.7123	.7157	.7190	.7224
.6	.7257	.7291	.7324	.7357	.7389	.7422	.7454	.7486	.7517	.7549
.7	.7580	.7611	.7642	.7673	.7704	.7734	.7764	.7794	.7823	.7852
.8	.7881	.7910	.7939	.7967	.7995	.8023	.8051	.8078	.8106	.8133
.9	.8159	.8186	.8212	.8238	.8264	.8289	.8315	.8340	.8365	.8389
1.0	.8413	.8438	.8461	.8485	.8508	.8531	.8554	.8577	.8599	.8621
1.1	.8643	.8665	.8686	.8708	.8729	.8749	.8770	.8790	.8810	.8830
1.2	.8849	.8869	.8888	.8907	.8925	.8944	.8962	.8980	.8997	.9015
1.3	.9032	.9049	.9066	.9082	.9099	.9115	.9131	.9147	.9162	.9177
1.4	.9192	.9207	.9222	.9236	.9251	.9265	.9279	.9292	.9306	.9319
1.5	.9332	.9345	.9357	.9370	.9382	.9394	.9406	.9418	.9429	.9441
1.6	.9452	.9463	.9474	.9484	.9495	.9505	.9515	.9525	.9535	.9545
1.7	.9554	.9564	.9573	.9582	.9591	.9599	.9608	.9616	.9625	.9633
1.8	.9641	.9649	.9656	.9664	.9671	.9678	.9686	.9693	.9699	.9706
1.9	.9713	.9719	.9726	.9732	.9738	.9744	.9750	.9756	.9761	.9767
2.0	.9772	.9778	.9783	.9788	.9793	.9798	.9803	.9808	.9812	.9817
2.1	.9821	.9826	.9830	.9834	.9838	.9842	.9846	.9850	.9854	.9857
2.2	.9861	.9864	.9868	.9871	.9875	.9878	.9881	.9884	.9887	.9890
2.3	.9893	.9896	.9898	.9901	.9904	.9906	.9909	.9911	.9913	.9916
2.4	.9918	.9920	.9922	.9925	.9927	.9929	.9931	.9932	.9934	.9936
2.5	.9938	.9940	.9941	.9943	.9945	.9946	.9948	.9949	.9951	.9952
2.6	.9953	.9955	.9956	.9957	.9959	.9960	.9961	.9962	.9963	.9964
2.7	.9965	.9966	.9967	.9968	.9969	.9970	.9971	.9972	.9973	.9974
2.8	.9974	.9975	.9976	.9977	.9977	.9978	.9979	.9979	.9980	.9981
2.9	.9981	.9982	.9982	.9983	.9984	.9984	.9985	.9985	.9986	.9986
3.0	.9987	.9987	.9987	.9988	.9988	.9989	.9989	.9989	.9990	.9990
3.1	.9990	.9991	.9991	.9991	.9992	.9992	.9992	.9992	.9993	.9993
3.2	.9993	.9993	.9994	.9994	.9994	.9994	.9994	.9995	.9995	.9995
3.3	.9995	.9995	.9995	.9996	.9996	.9996	.9996	.9996	.9996	.9997
3.4	.9997	.9997	.9997	.9997	.9997	.9997	.9997	.9997	.9997	.9998

The entries from 3.49 to 3.61 all equal .9998.
The entries from 3.62 to 3.89 all equal .9999.
All entries from 3.90 and up equal 1.0000.

TABLE F

STANDARDIZED NORMAL DENSITY FUNCTION $\varphi(t) = \dfrac{1}{\sqrt{2\pi}}\, e^{-t^2/2}$

This table gives values of $\varphi(t)$ for various values of t.

t	.00	.01	.02	.03	.04	.05	.06	.07	.08	.09
.0	.3989	.3989	.3989	.3988	.3986	.3984	.3982	.3980	.3977	.3973
.1	.3970	.3965	.3961	.3956	.3951	.3945	.3939	.3932	.3925	.3918
.2	.3910	.3902	.3894	.3885	.3876	.3867	.3857	.3847	.3836	.3825
.3	.3814	.3802	.3790	.3778	.3765	.3752	.3739	.3725	.3712	.3697
.4	.3683	.3668	.3653	.3637	.3621	.3605	.3589	.3572	.3555	.3538
.5	.3521	.3503	.3485	.3467	.3448	.3429	.3410	.3391	.3372	.3352
.6	.3332	.3312	.3292	.3271	.3251	.3230	.3209	.3187	.3166	.3144
.7	.3123	.3101	.3079	.3056	.3034	.3011	.2989	.2966	.2943	.2920
.8	.2897	.2874	.2850	.2827	.2803	.2780	.2756	.2732	.2709	.2685
.9	.2661	.2637	.2613	.2589	.2565	.2541	.2516	.2492	.2468	.2444
1.0	.2420	.2396	.2371	.2347	.2323	.2299	.2275	.2251	.2227	.2203
1.1	.2179	.2155	.2131	.2107	.2083	.2059	.2036	.2012	.1989	.1965
1.2	.1942	.1919	.1895	.1872	.1849	.1826	.1804	.1781	.1758	.1736
1.3	.1711	.1691	.1669	.1647	.1626	.1604	.1582	.1561	.1539	.1518
1.4	.1497	.1476	.1456	.1435	.1415	.1394	.1374	.1354	.1334	.1315
1.5	.1295	.1276	.1257	.1238	.1219	.1200	.1182	.1163	.1145	.1127
1.6	.1109	.1092	.1074	.1057	.1040	.1023	.1006	.0989	.0973	.0957
1.7	.0940	.0925	.0909	.0893	.0878	.0863	.0848	.0833	.0818	.0804
1.8	.0790	.0775	.0761	.0748	.0734	.0721	.0707	.0694	.0681	.0669
1.9	.0656	.0644	.0632	.0620	.0608	.0596	.0584	.0573	.0562	.0551
2.0	.0540	.0529	.0519	.0508	.0498	.0488	.0478	.0468	.0459	.0449
2.1	.0440	.0431	.0422	.0413	.0404	.0396	.0387	.0379	.0371	.0363
2.2	.0355	.0347	.0339	.0332	.0325	.0317	.0310	.0303	.0297	.0290
2.3	.0283	.0277	.0270	.0264	.0258	.0252	.0246	.0241	.0235	.0229
2.4	.0224	.0219	.0213	.0208	.0203	.0198	.0194	.0189	.0184	.0180
2.5	.0175	.0171	.0167	.0163	.0158	.0154	.0151	.0147	.0143	.0139
2.6	.0136	.0132	.0129	.0126	.0122	.0119	.0116	.0113	.0110	.0107
2.7	.0104	.0101	.0099	.0096	.0093	.0091	.0088	.0086	.0084	.0081
2.8	.0079	.0077	.0075	.0073	.0071	.0069	.0067	.0065	.0063	.0061
2.9	.0060	.0058	.0056	.0055	.0053	.0051	.0050	.0048	.0047	.0046
3.0	.0044	.0043	.0042	.0040	.0039	.0038	.0037	.0036	.0035	.0034
3.1	.0033	.0032	.0031	.0030	.0029	.0028	.0027	.0026	.0025	.0025
3.2	.0024	.0023	.0022	.0022	.0021	.0020	.0020	.0019	.0018	.0018
3.3	.0017	.0017	.0016	.0016	.0015	.0015	.0014	.0014	.0013	.0013
3.4	.0012	.0012	.0012	.0011	.0011	.0010	.0010	.0010	.0009	.0009

STUDENT t DISTRIBUTION

This table gives the value λ of t
for various values of the degrees of
freedom ν such that the area to
the right of λ is equal to α.

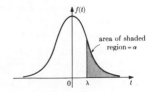

ν \ α	0.25	0.125	0.05	0.025	0.0125	0.005	0.0025
1	1.00000	2.4142	6.3138	12.706	25.452	63.657	127.32
2	0.81650	1.6036	2.9200	4.3027	6.2053	9.9248	14.089
3	0.76489	1.4226	2.3534	3.1825	4.1765	5.8409	7.4533
4	0.74070	1.3444	2.1318	2.7764	3.4954	4.6041	5.5976
5	0.72669	1.3009	2.0150	2.5706	3.1634	4.0321	4.7733
6	0.71756	1.2733	1.9432	2.4469	2.9687	3.7074	4.3168
7	0.71114	1.2543	1.8946	2.3646	2.8412	3.4995	4.0293
8	0.70639	1.2403	1.8595	2.3060	2.7515	3.3554	3.8325
9	0.70272	1.2297	1.8331	2.2622	2.6850	3.2498	3.6897
10	0.69981	1.2213	1.8125	2.2281	2.6338	3.1693	3.5814
11	0.69745	1.2145	1.7959	2.2010	2.5931	3.1058	3.4966
12	0.69548	1.2089	1.7823	2.1788	2.5600	3.0545	3.4284
13	0.69384	1.2041	1.7709	2.1604	2.5326	3.0123	3.3725
14	0.69242	1.2001	1.7613	2.1448	2.5096	2.9768	3.3257
15	0.69120	1.1967	1.7530	2.1315	2.4899	2.9467	3.2860
16	0.69013	1.1937	1.7459	2.1199	2.4729	2.9208	3.2520
17	0.68919	1.1910	1.7396	2.1098	2.4581	2.8982	3.2225
18	0.68837	1.1887	1.7341	2.1009	2.4450	2.8784	3.1966
19	0.68763	1.1866	1.7291	2.0930	2.4334	2.8609	3.1737
20	0.68696	1.1848	1.7247	2.0860	2.4231	2.8453	3.1534
21	0.68635	1.1831	1.7207	2.0796	2.4138	2.8314	3.1352
22	0.68580	1.1816	1.7171	2.0739	2.4055	2.8188	3.1188
23	0.68531	1.1802	1.7139	2.0687	2.3979	2.8073	3.1040
24	0.68485	1.1789	1.7109	2.0639	2.3910	2.7969	3.0905
25	0.68443	1.1777	1.7081	2.0595	2.3846	2.7874	3.0782
26	0.68405	1.1766	1.7056	2.0555	2.3788	2.7787	3.0669
27	0.68370	1.1757	1.7033	2.0518	2.3734	2.7707	3.0565
28	0.68335	1.1748	1.7011	2.0484	2.3685	2.7633	3.0469
29	0.68304	1.1739	1.6991	2.0452	2.3638	2.7564	3.0380
30	0.68276	1.1731	1.6973	2.0423	2.3596	2.7500	3.0298
40	0.68066	1.1673	1.6839	2.0211	2.3289	2.7045	2.9712
60	0.67862	1.1616	1.6707	2.0003	2.2991	2.6603	2.9146
120	0.67656	1.1559	1.6577	1.9799	2.2699	2.6174	2.8599
∞	0.67449	1.1503	1.6449	1.9600	2.2414	2.5758	2.8070

Taken with permission from "Tables of the Percentage Points of the Incomplete Beta
Function," C. M. Thompson, *Biometrika*, Vol. 32, 1941.

TABLE H

χ² DISTRIBUTION

This table gives the value δ of χ^2 for various values of the degrees of freedom ν such that the area to the right of δ is equal to α.

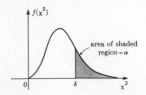

ν \ α	0.995	0.975	0.050	0.025	0.010	0.005
1	0.0⁴3927	0.0³9821	3.84146	5.02389	6.63490	7.87944
2	0.010025	0.050636	5.99147	7.37776	9.21034	10.5966
3	0.071721	0.215795	7.81473	9.34840	11.3449	12.8381
4	0.206990	0.484419	9.48773	11.1433	13.2767	14.8602
5	0.411740	0.831211	11.0705	12.8325	15.0863	16.7496
6	0.675727	1.237347	12.5916	14.4494	16.8119	18.5476
7	0.989265	1.68987	14.0671	16.0128	18.4753	20.2777
8	1.344419	2.17973	15.5073	17.5346	20.0902	21.9550
9	1.734926	2.70039	16.9190	19.0228	21.6660	23.5893
10	2.15585	3.24697	18.3070	20.4831	23.2093	25.1882
11	2.60321	3.81575	19.6751	21.9200	24.7250	26.7569
12	3.07382	4.40379	21.0261	23.3367	26.2170	28.2995
13	3.56503	5.00874	22.3621	24.7356	27.6883	29.8194
14	4.07468	5.62872	23.6848	26.1190	29.1413	31.3193
15	4.60094	6.26214	24.9958	27.4884	30.5779	32.8013
16	5.14224	6.90766	26.2962	28.8454	31.9999	34.2672
17	5.69724	7.56418	27.5871	30.1910	33.4087	35.7185
18	6.26481	8.23075	28.8693	31.5264	34.8053	37.1564
19	6.84398	8.90655	30.1435	32.8523	36.1908	38.5822
20	7.43386	9.59083	31.4104	34.1696	37.5662	39.9968
21	8.03366	10.28293	32.6705	35.4789	38.9321	41.4010
22	8.64272	10.9823	33.9244	36.7807	40.2894	42.7956
23	9.26042	11.6885	35.1725	38.0757	41.6384	44.1813
24	9.88623	12.4001	36.4151	39.3641	42.9798	45.5585
25	10.5197	13.1197	37.6525	40.6465	44.3141	46.9278
26	11.1603	13.8439	38.8852	41.9232	45.6417	48.2899
27	11.8076	14.5733	40.1133	43.1944	46.9630	49.6449
28	12.4613	15.3079	41.3372	44.4607	48.2782	50.9933
29	13.1211	16.0471	42.5569	45.7222	49.5879	52.3356
30	13.7867	16.7908	43.7729	46.9792	50.8922	53.6720
40	20.7065	24.4331	55.7585	59.3417	63.6907	66.7659
50	27.9907	32.3574	67.5048	71.4202	76.1539	79.4900
60	35.5346	40.4817	79.0819	83.2976	88.3794	91.9517
70	43.2752	48.7576	90.5312	95.0231	100.425	104.215
80	51.1720	57.1532	101.879	106.629	112.329	116.321
90	59.1963	65.6466	113.145	118.136	124.116	128.299
100	67.3276	74.2219	124.342	129.561	135.807	140.169

Taken with permission from "Tables of the Percentage Points of the χ^2 Distribution," C. M. Thompson, *Biometrika*, Vol. 32, 1941.

TABLE I

Values of the F distribution for $\alpha = 0.01$

This table gives values δ of F such that the probability that $F \geqq \delta$ is 0.01.

Degrees of freedom of numerator

ν_2 \ ν_1	1	2	3	4	5	6	7	8	9	10	12	15	20	24	30	40	60	120	∞
1	4,052	5,000	5,403	5,625	5,764	5,859	5,928	5,982	6,023	6,056	6,106	6,157	6,209	6,235	6,261	6,287	6,313	6,339	6,366
2	98.5	99.0	99.2	99.2	99.3	99.3	99.4	99.4	99.4	99.4	99.4	99.4	99.4	99.5	99.5	99.5	99.5	99.5	99.5
3	34.1	30.8	29.5	28.7	28.2	27.9	27.7	27.5	27.3	27.2	27.1	26.9	26.7	26.6	26.5	26.4	26.3	26.2	26.1
4	21.2	18.0	16.7	16.0	15.5	15.2	15.0	14.8	14.7	14.5	14.4	14.2	14.0	13.9	13.8	13.7	13.7	13.6	13.5
5	16.3	13.3	12.1	11.4	11.0	10.7	10.5	10.3	10.2	10.1	9.89	9.72	9.55	9.47	9.38	9.29	9.20	9.11	9.02
6	13.7	10.9	9.78	9.15	8.75	8.47	8.26	8.10	7.98	7.87	7.72	7.56	7.40	7.31	7.23	7.14	7.06	6.97	6.88
7	12.2	9.55	8.45	7.85	7.46	7.19	6.99	6.84	6.72	6.62	6.47	6.31	6.16	6.07	5.99	5.91	5.82	5.74	5.65
8	11.3	8.65	7.59	7.01	6.63	6.37	6.18	6.03	5.91	5.81	5.67	5.52	5.36	5.28	5.20	5.12	5.03	4.95	4.86
9	10.6	8.02	6.99	6.42	6.06	5.80	5.61	5.47	5.35	5.26	5.11	4.96	4.81	4.73	4.65	4.57	4.48	4.40	4.31
10	10.0	7.56	6.55	5.99	5.64	5.39	5.20	5.06	4.94	4.85	4.71	4.56	4.41	4.33	4.25	4.17	4.08	4.00	3.91
11	9.65	7.21	6.22	5.67	5.32	5.07	4.89	4.74	4.63	4.54	4.40	4.25	4.10	4.02	3.94	3.86	3.78	3.69	3.60
12	9.33	6.93	5.95	5.41	5.06	4.82	4.64	4.50	4.39	4.30	4.16	4.01	3.86	3.78	3.70	3.62	3.54	3.45	3.36
13	9.07	6.70	5.74	5.21	4.86	4.62	4.44	4.30	4.19	4.10	3.96	3.82	3.66	3.59	3.51	3.43	3.34	3.25	3.17
14	8.86	6.51	5.56	5.04	4.70	4.46	4.28	4.14	4.03	3.94	3.80	3.66	3.51	3.43	3.35	3.27	3.18	3.09	3.00
15	8.68	6.36	5.42	4.89	4.56	4.32	4.14	4.00	3.89	3.80	3.67	3.52	3.37	3.29	3.21	3.13	3.05	2.96	2.87
16	8.53	6.23	5.29	4.77	4.44	4.20	4.03	3.89	3.78	3.69	3.55	3.41	3.26	3.18	3.10	3.02	2.93	2.84	2.75
17	8.40	6.11	5.19	4.67	4.34	4.10	3.93	3.79	3.68	3.59	3.46	3.31	3.16	3.08	3.00	2.92	2.83	2.75	2.65
18	8.29	6.01	5.09	4.58	4.25	4.01	3.84	3.71	3.60	3.51	3.37	3.23	3.08	3.00	2.92	2.84	2.75	2.66	2.57
19	8.19	5.93	5.01	4.50	4.17	3.94	3.77	3.63	3.52	3.43	3.30	3.15	3.00	2.92	2.84	2.76	2.67	2.58	2.49
20	8.10	5.85	4.94	4.43	4.10	3.87	3.70	3.56	3.46	3.37	3.23	3.09	2.94	2.86	2.78	2.69	2.61	2.52	2.42
21	8.02	5.78	4.87	4.37	4.04	3.81	3.64	3.51	3.40	3.31	3.17	3.03	2.88	2.80	2.72	2.64	2.55	2.46	2.36
22	7.95	5.72	4.82	4.31	3.99	3.76	3.59	3.45	3.35	3.26	3.12	2.98	2.83	2.75	2.67	2.58	2.50	2.40	2.31
23	7.88	5.66	4.76	4.26	3.94	3.71	3.54	3.41	3.30	3.21	3.07	2.93	2.78	2.70	2.62	2.54	2.45	2.35	2.26
24	7.82	5.61	4.72	4.22	3.90	3.67	3.50	3.36	3.26	3.17	3.03	2.89	2.74	2.66	2.58	2.49	2.40	2.31	2.21
25	7.77	5.57	4.68	4.18	3.86	3.63	3.46	3.32	3.22	3.13	2.99	2.85	2.70	2.62	2.53	2.45	2.36	2.27	2.17
30	7.56	5.39	4.51	4.02	3.70	3.47	3.30	3.17	3.07	2.98	2.84	2.70	2.55	2.47	2.39	2.30	2.21	2.11	2.01
40	7.31	5.18	4.31	3.83	3.51	3.29	3.12	2.99	2.89	2.80	2.66	2.52	2.37	2.29	2.20	2.11	2.02	1.92	1.80
60	7.08	4.98	4.13	3.65	3.34	3.12	2.95	2.82	2.72	2.63	2.50	2.35	2.20	2.12	2.03	1.94	1.84	1.73	1.60
120	6.85	4.79	3.95	3.48	3.17	2.96	2.79	2.66	2.56	2.47	2.34	2.19	2.03	1.95	1.86	1.76	1.66	1.53	1.38
∞	6.63	4.61	3.78	3.32	3.02	2.80	2.64	2.51	2.41	2.32	2.18	2.04	1.88	1.79	1.70	1.59	1.47	1.32	1.00

Degrees of freedom of denominator

* Reproduced with permission from M. Merrington and C. M. Thompson, "Tables of percentage points of the inverted beta (F) distribution," *Biometrika*, vol. 33 (1943).

TABLE I'

VALUES OF THE F DISTRIBUTION FOR $\alpha = 0.05$

This table gives values δ of F such that the probability that $F \geq \delta$ is 0.05

Degrees of freedom of numerator

v_2 \ v_1	1	2	3	4	5	6	7	8	9	10	12	15	20	24	30	40	60	120	∞
1	161	200	216	225	230	234	237	239	241	242	244	246	248	249	250	251	252	253	254
2	18.5	19.0	19.2	19.2	19.3	19.4	19.4	19.4	19.4	19.4	19.4	19.4	19.4	19.5	19.5	19.5	19.5	19.5	19.5
3	10.1	9.55	9.28	9.12	9.01	8.94	8.89	8.85	8.81	8.79	8.74	8.70	8.66	8.64	8.62	8.59	8.57	8.55	8.53
4	7.71	6.94	6.59	6.39	6.26	6.16	6.09	6.04	6.00	5.96	5.91	5.86	5.80	5.77	5.75	5.72	5.69	5.66	5.63
5	6.61	5.79	5.41	5.19	5.05	4.95	4.88	4.82	4.77	4.74	4.68	4.62	4.56	4.53	4.50	4.46	4.43	4.40	4.37
6	5.99	5.14	4.76	4.53	4.39	4.28	4.21	4.15	4.10	4.06	4.00	3.94	3.87	3.84	3.81	3.77	3.74	3.70	3.67
7	5.59	4.74	4.35	4.12	3.97	3.87	3.79	3.73	3.68	3.64	3.57	3.51	3.44	3.41	3.38	3.34	3.30	3.27	3.23
8	5.32	4.46	4.07	3.84	3.69	3.58	3.50	3.44	3.39	3.35	3.28	3.22	3.15	3.12	3.08	3.04	3.01	2.97	2.93
9	5.12	4.26	3.86	3.63	3.48	3.37	3.29	3.23	3.18	3.14	3.07	3.01	2.94	2.90	2.86	2.83	2.79	2.75	2.71
10	4.96	4.10	3.71	3.48	3.33	3.22	3.14	3.07	3.02	2.98	2.91	2.85	2.77	2.74	2.70	2.66	2.62	2.58	2.54
11	4.84	3.98	3.59	3.36	3.20	3.09	3.01	2.95	2.90	2.85	2.79	2.72	2.65	2.61	2.57	2.53	2.49	2.45	2.40
12	4.75	3.89	3.49	3.26	3.11	3.00	2.91	2.85	2.80	2.75	2.69	2.62	2.54	2.51	2.47	2.43	2.38	2.34	2.30
13	4.67	3.81	3.41	3.18	3.03	2.92	2.83	2.77	2.71	2.67	2.60	2.53	2.46	2.42	2.38	2.34	2.30	2.25	2.21
14	4.60	3.74	3.34	3.11	2.96	2.85	2.76	2.70	2.65	2.60	2.53	2.46	2.39	2.35	2.31	2.27	2.22	2.18	2.13
15	4.54	3.68	3.29	3.06	2.90	2.79	2.71	2.64	2.59	2.54	2.48	2.40	2.33	2.29	2.25	2.20	2.16	2.11	2.07
16	4.49	3.63	3.24	3.01	2.85	2.74	2.66	2.59	2.54	2.49	2.42	2.35	2.28	2.24	2.19	2.15	2.11	2.06	2.01
17	4.45	3.59	3.20	2.96	2.81	2.70	2.61	2.55	2.49	2.45	2.38	2.31	2.23	2.19	2.15	2.10	2.06	2.01	1.96
18	4.41	3.55	3.16	2.93	2.77	2.66	2.58	2.51	2.46	2.41	2.34	2.27	2.19	2.15	2.11	2.06	2.02	1.97	1.92
19	4.38	3.52	3.13	2.90	2.74	2.63	2.54	2.48	2.42	2.38	2.31	2.23	2.16	2.11	2.07	2.03	1.98	1.93	1.88
20	4.35	3.49	3.10	2.87	2.71	2.60	2.51	2.45	2.39	2.35	2.28	2.20	2.12	2.08	2.04	1.99	1.95	1.90	1.84
21	4.32	3.47	3.07	2.84	2.68	2.57	2.49	2.42	2.37	2.32	2.25	2.18	2.10	2.05	2.01	1.96	1.92	1.87	1.81
22	4.30	3.44	3.05	2.82	2.66	2.55	2.46	2.40	2.34	2.30	2.23	2.15	2.07	2.03	1.98	1.94	1.89	1.84	1.78
23	4.28	3.42	3.03	2.80	2.64	2.53	2.44	2.37	2.32	2.27	2.20	2.13	2.05	2.01	1.96	1.91	1.86	1.81	1.76
24	4.26	3.40	3.01	2.78	2.62	2.51	2.42	2.36	2.30	2.25	2.18	2.11	2.03	1.98	1.94	1.89	1.84	1.79	1.73
25	4.24	3.39	2.99	2.76	2.60	2.49	2.40	2.34	2.28	2.24	2.16	2.09	2.01	1.96	1.92	1.87	1.82	1.77	1.71
30	4.17	3.32	2.92	2.69	2.53	2.42	2.33	2.27	2.21	2.16	2.09	2.01	1.93	1.89	1.84	1.79	1.74	1.68	1.62
40	4.08	3.23	2.84	2.61	2.45	2.34	2.25	2.18	2.12	2.08	2.00	1.92	1.84	1.79	1.74	1.69	1.64	1.58	1.51
60	4.00	3.15	2.76	2.53	2.37	2.25	2.17	2.10	2.04	1.99	1.92	1.84	1.75	1.70	1.65	1.59	1.53	1.47	1.39
120	3.92	3.07	2.68	2.45	2.29	2.18	2.09	2.02	1.96	1.91	1.83	1.75	1.66	1.61	1.55	1.50	1.43	1.35	1.25
∞	3.84	3.00	2.60	2.37	2.21	2.10	2.01	1.94	1.88	1.83	1.75	1.67	1.57	1.52	1.46	1.39	1.32	1.22	1.00

Degrees of freedom of denominator

* This table is reproduced with permission from M. Merrington and C. M. Thompson, "Tables of percentage points of the inverted beta (F) distribution," *Biometrika*, vol. 33 (1943), by permission of the *Biometrika* trustees.

Table J

Random Digits

10097	32533	76520	13586	34673	54876	80959	09117	39292	74945
37542	04805	64894	74296	24805	24037	20636	10402	00822	91665
08422	68953	19645	09303	23209	02560	15953	34764	35080	33606
99019	02529	09376	70715	38311	31165	88676	74397	04436	27659
12807	99970	80157	36147	64032	36653	98951	16877	12171	76833
66065	74717	34072	76850	36697	36170	65813	39885	11199	29170
31060	10805	45571	82406	35303	42614	86799	07439	23403	09732
85269	77602	02051	65692	68665	74818	73053	85247	18623	88579
63573	32135	05325	47048	90553	57548	28468	28709	83491	25624
73796	45753	03529	64778	35808	34282	60935	20344	35273	88435
98520	17767	14905	68607	22109	40558	60970	93433	50500	73998
11805	05431	39808	27732	50725	68248	29405	24201	52775	67851
83452	99634	06288	98083	13746	70078	18475	40610	68711	77817
88685	40200	86507	58401	36766	67951	90364	76493	29609	11062
99594	67348	87517	64969	91826	08928	93785	61368	23478	34113
65481	17674	17468	50950	58047	76974	73039	57186	40218	16544
80124	35635	17727	08015	45318	22374	21115	78253	14385	53763
74350	99817	77402	77214	43236	00210	45521	64237	96286	02655
69916	26803	66252	29148	36936	87203	76621	13990	94400	56418
09893	20505	14225	68514	46427	56788	96297	78822	54382	14598

Reproduced with permission from "A Million Random Digits with 100,000 Normal Deviates," by the RAND Corp., The Free Press, New York (1955).

TABLE K

SQUARE ROOTS

n	$\sqrt{n}$	$\sqrt{10n}$	n	$\sqrt{n}$	$\sqrt{10n}$	n	$\sqrt{n}$	$\sqrt{10n}$
1	1.0000	3.1623	34	5.8310	18.439	67	8.1854	25.884
2	1.4142	4.4721	35	5.9161	18.708	68	8.2462	26.077
3	1.7321	5.4772	36	6.0000	18.974	69	8.3066	26.268
4	2.0000	6.3246	37	6.0828	19.235	70	8.3666	26.458
5	2.2361	7.0711	38	6.1644	19.494	71	8.4261	26.646
6	2.4495	7.7460	39	6.2450	19.748	72	8.4853	26.833
7	2.6458	8.3666	40	6.3246	20.000	73	8.5440	27.019
8	2.8284	8.9443	41	6.4031	20.248	74	8.6023	27.203
9	3.0000	9.4868	42	6.4807	20.494	75	8.6603	27.386
10	3.1623	10.000	43	6.5574	20.736	76	8.7178	27.568
11	3.3166	10.488	44	6.6332	20.976	77	8.7750	27.749
12	3.4641	10.954	45	6.7082	21.213	78	8.8318	27.928
13	3.6056	11.402	46	6.7823	21.448	79	8.8882	28.107
14	3.7417	11.832	47	6.8557	21.679	80	8.9443	28.284
15	3.8730	12.247	48	6.9282	21.909	81	9.0000	28.460
16	4.0000	12.649	49	7.0000	22.136	82	9.0554	28.636
17	4.1231	13.038	50	7.0711	22.361	83	9.1104	28.810
18	4.2426	13.416	51	7.1414	22.583	84	9.1652	28.983
19	4.3589	13.784	52	7.2111	22.804	85	9.2195	29.155
20	4.4721	14.142	53	7.2801	23.022	86	9.2736	29.326
21	4.5826	14.491	54	7.3485	23.238	87	9.3274	29.496
22	4.6904	14.832	55	7.4162	23.452	88	9.3808	29.665
23	4.7958	15.166	56	7.4833	23.664	89	9.4340	29.833
24	4.8990	15.492	57	7.5498	23.875	90	9.4868	30.000
25	5.0000	15.811	58	7.6158	24.083	91	9.5394	30.166
26	5.0990	16.125	59	7.6811	24.290	92	9.5917	30.332
27	5.1962	16.432	60	7.7460	24.495	93	9.6437	30.496
28	5.2915	16.733	61	7.8102	24.698	94	9.6954	30.659
29	5.3852	17.029	62	7.8740	24.900	95	9.7468	30.822
30	5.4772	17.321	63	7.9373	25.100	96	9.7980	30.984
31	5.5678	17.607	64	8.0000	25.298	97	9.8489	31.145
32	5.6569	17.889	65	8.0623	25.495	98	9.8995	31.305
33	5.7446	18.166	66	8.1240	25.690	99	9.9499	31.464

TABLE L

GREEK ALPHABET

Name	Lowercase	Capital	Name	Lowercase	Capital
alpha	α	A	nu	ν	N
beta	β	B	xi	ξ	Ξ
gamma	γ	Γ	omicron	o	O
delta	δ	Δ	pi	π	Π
epsilon	ϵ	E	rho	ρ	P
zeta	ζ	Z	sigma	σ	Σ
eta	η	H	tau	τ	T
theta	θ or ϑ	Θ	upsilon	υ	Υ
iota	ι	I	phi	φ or ϕ	Φ
kappa	κ	K	chi	χ	X
lambda	λ	Λ	psi	ψ	Ψ
mu	μ	M	omega	ω	Ω

Index

)CIAL SCIENCE LIBRARY

Manor Road Building
Manor Road

302562635W